Genotype x Environment Interactions and its Implications for Plant Breeding

Reza Mohammadi
Mohammad Reza Jalal Kamali
Ahmed Amri • Zakaria Kehel
Behzad Sadeghzadeh
Editors

Genotype x Environment Interactions and its Implications for Plant Breeding

Springer

Editors
Reza Mohammadi
Cereal Research Department
Dryland Agricultural Research Institute
(DARI), AREEO
Kermanshah, Iran

Mohammad Reza Jalal Kamali
Global Wheat Program
International Maize and Wheat Improvement
Center (CIMMYT)
Karaj, Iran

Ahmed Amri
Genetic Resources Department
International Centre for Agricultural
Research in the Dry Areas (ICARDA)
Rabat, Morocco

Zakaria Kehel
Biodiversity and Integrated Gene Management
International Centre for Agricultural Research
in the Dry Areas (ICARDA)
Rabat, Morocco

Behzad Sadeghzadeh
School of Agriculture, Food and
Ecosystem Sciences (SAFES)
The University of Melbourne
Melbourne, VIC, Australia

ISBN 978-981-95-5663-2 ISBN 978-981-95-5664-9 (eBook)
https://doi.org/10.1007/978-981-95-5664-9

This Springer imprint is published by the registered company Springer Nature Singapore Pte Ltd.
The registered company address is: 152 Beach Road, #21-01/04 Gateway East, Singapore 189721, Singapore

In Memory of Dr. Leonardo Crespo Herrera

While preparing this book, we were deeply saddened by the loss of one of our most dedicated and principled contributors, Dr. Leonardo Crespo Herrera, who passed away due to serious medical issues. Just a week before his passing away, he informed us that his health had deteriorated to the point that he could no longer complete the assigned chapter. His absence is profoundly felt.

As a Senior Scientist in Bread Wheat Improvement Program with CIMMYT's Global Wheat Program, Dr. L. Crespo Herrera made exceptional contributions to enhancing wheat productivity and resilience for farmers around the world. Since joining CIMMYT in 2015, he devoted his career to advancing food security through innovative wheat research. His work transformed breeding strategies, resulting in the development of wheat varieties that are more productive, disease-resistant, and adaptable to changing environmental conditions. Guided by a deep commitment to improving crop performance, he addressed complex challenges in yield stability and climate resilience, significantly strengthening the global wheat breeding landscape.

Dr. L. Crespo Herrera was instrumental in driving innovation through extensive collaborations with national and international partners. His impact reached far beyond the laboratory—shaping agricultural policy, supporting farming communities, and building more resilient wheat production systems worldwide. A passionate mentor, he nurtured young scientists and inspired a new generation to pursue agricultural innovation with integrity and purpose. His legacy will continue to influence the field for years to come.

We express our sincere condolences to his family, friends, colleagues, and all who had the honor of knowing and working with him.

R. Mohammadi
M. R. Jalal Kamali
A. Amri
Z. Kehel
B. Sadeghzadeh

Foreword

Crops' grain yield is the lifeblood of global food security. As populations grow and climates shift, the challenge before agriculture is clear: produce more food, more consistently, under increasingly unpredictable conditions. At the heart of this challenge lies the genotype × environment interaction (GEI)—the ever-changing relationship between a crop's genetic potential and the environment in which it grows.

For plant breeders, GEI is both obstacle and opportunity. It can confound selection, turning a top-performing variety in one environment into a disappointment in another. Yet it also offers the chance to tailor crops to specific regions, or to develop varieties that combine high yield with resilience under diverse and stressful conditions. In an era marked by intensifying droughts, heat waves, shifting pest and disease pressures, and evolving soil/crop management practices, understanding and managing GEI has never been more urgent.

Multi-environment trials (METs), the backbone of modern breeding programs, reveal the full complexity of GEI. Strong interaction effects can reduce the heritability of grain yield and slow genetic gain. Too often, environmental responses shaped by crop management are overlooked, even though management is deeply woven into the environmental fabric. To truly exploit GEI, breeders must see the whole picture—integrating genetic insights with agronomic, ecological, and environmental knowledge.

This book, edited by Dr. Reza Mohammadi, Dr. Mohammad Reza Jalal Kamali, Dr. Ahmed Amri, Dr. Zakaria Kehel, and Dr. Behzad Sadeghzadeh, brings together thought-provoking alternative concepts, cutting-edge research, practical breeding strategies, and advanced analytical approaches to tackle GEI head-on. It shows how negative interactions can be mitigated and positive ones harnessed to accelerate progress.

More than a technical manual, this book also offers a roadmap for building climate-resilient agriculture. It is an excellent resource for plant breeders, agronomists, researchers, and students dedicated to securing a stable, sustainable food supply for a world in transition.

Emeritus Principal Scientist & Former Director of Global Wheat Program, CIMMYT Int., Texcoco, Mexico — Hans Joachim Braun

Wheat Expert and Senior Advisor, Distinguished Scientist (Emeritus) & Former Head of Global Wheat Improvement, CIMMYT Int., Texcoco, Mexico — Ravi Prakash Singh

Preface

The phenomenon of genotype-by-environment interaction (GEI) has long been recognized as a challenge and cornerstone in plant breeding programs, with its significance first emerged in the early twentieth century. Plant breeders soon observed that the performance of genotypes was not consistent across diverse environments, challenging the notion of absolute genetic superiority. This realization underscored the need to develop cultivars that not only produce high yield but also perform reliably across variable conditions. As the plant breeding objectives evolved, so too did the understanding that adaptability and stability are conditional traits—shaped and performed profoundly by environmental context.

Over the decades, advances in statistical methodologies—particularly the development of mixed models and factor-analytic techniques—have enabled plant breeders to dissect and model GEI with increasing precision. These tools have enhanced the efficiency of selection, allowing for more informed decisions in cultivar development. Yet, despite this progress, our analysis of current research practices reveals a notable gap: while GEIs are widely acknowledged, there remains limited effort to identify their underlying causes. Understanding the drivers of these interactions—especially their repeatability within target populations of environments (TPEs) assessed through multi-environment trials (METs)—is crucial for designing more effective and predictive breeding strategies.

Given that traits such as grain yield are formed continuously throughout the growing season by a dynamic interplay of biotic and abiotic stresses, it becomes essential to examine how these factors contribute to differential genotype performance. Are certain genotypes consistently superior due to inherent resilience, or do they merely respond favorably to specific environmental cues? Addressing such questions requires moving beyond mere detection of GEI effects toward a deeper, biologically and ecologically informed interpretation.

Stability, a key criterion in METs, is commonly used to evaluate the consistency of genotype performance across environments, particularly for economically important traits like yield and disease resistance. While stability metrics are well-established for easily measured traits, their application to complex, resource-intensive phenotypes remains challenging. Moreover, a central debate persists among plant breeders: should the goal be the development of broadly adapted "mega-cultivars," or should breeding programs focus on tailored genotypes optimized for specific TPEs? The biological basis of stability—whether rooted in GEI or governed by specific genetic mechanisms—remains an exciting area for exploration.

A comprehensive understanding of the genetic and environmental components influencing GEI is vital for designing plant breeding programs that deliver resilient, high-performing cultivars. This demands not only advanced statistical approaches beyond traditional analysis tools but also integration with knowledge from disciplines such as physiology, ecology, agronomy, and molecular biology. Univariate and multivariate analysis methods have long served as the backbone of GEI analysis, yet the integration of genotypic and environmental covariates through analytical (biological and ecological) modeling offers a more insightful pathway to uncovering the causes of interaction effects. Incorporating climate typing, envirotyping, and high-throughput phenotyping can further enhance our ability to predict genotype performance under current and future climate scenarios.

Equally important is unraveling the genetic architecture underlying GEI. Achieving predictive power in plant breeding requires identifying genes or quantitative trait loci (QTLs) whose effects vary across environments—i.e., Gene × E or QTL × E interactions. While existing methods have successfully detected environment-dependent QTL expression, they often fall short in explaining which specific environmental factors drive these changes. When environmental data are available, approaches such as factorial regression offer valuable insights into how climatic variables modulate genetic effects.

This book aims to provide a concise yet comprehensive overview of GEI concepts, emphasizing both empirical and analytical (biologically and ecologically meaningful) interpretations, with a strong focus on performance stability analysis across a range of crop species and environments. We present the most up-to-date univariate and multivariate statistical methods used in the analysis of GEI, discussing their applications, strengths, and limitations in modern plant breeding programs. Additionally, the book highlights available software tools and statistical platforms that facilitate robust GEI assessment and genetic association studies.

By bridging theory and practice, this book seeks to equip plant breeders, agronomists, researchers, and students with the knowledge and tools necessary to better understand, interpret, and exploit GEI in plant breeding programs. It is our hope that the insights offered here will inspire more targeted, efficient, and forward-looking approaches in crop improvement programs—ultimately contributing to the development of adaptable, resilient and high stable yield cultivars for meeting the challenges of a changing climate and world.

Kermanshah, Iran — Reza Mohammadi
Karaj, Iran — Mohammad Reza Jalal Kamali
Rabat, Morocco — Ahmed Amri
Rabat, Morocco — Zakaria Kehel
Melbourne, VIC, Australia — Behzad Sadeghzadeh

Contents

Editors and Contributors

About the Editors

Reza Mohammadi holds a Ph.D. in genetics and plant breeding and serves as an associate professor at the Dryland Agricultural Research Institute (DARI). With 25 years of experience in wheat breeding, particularly for dryland conditions in Iran, his research primarily focuses on genetic resources in durum and bread wheat. He has led over 100 research projects and has been instrumental in the development of new wheat cultivars. He has authored or co-authored more than 120 publications in peer-reviewed journals and also has experience writing book chapters and editing books.

Mohammad Reza Jalal Kamali earned his Ph.D. from the University of Western Australia and has 45+ years of experience in wheat breeding. He played a pivotal role in Iran's National Wheat Breeding Program and later contributed to the Global Wheat Program at CIMMYT. His research has led to the release of seven improved wheat cultivars. He has co-authored 3 books, 4 book chapters, and 70+ peer-reviewed articles. He has been honored with multiple national and international awards and has held significant positions, including president of the Crop Science Society of Iran. Currently, he is the Editor-in-Chief of two scientific journals, *Seed and Plant* (in Persian) and *Crop Breeding* (in English), as well as editorial board member of three scientific journals.

Ahmed Amri holds a Ph.D. in genetics and plant breeding from Kansas State University. He has contributed to the release of numerous barley, wheat, and triticale varieties in his tenure at INRA-Morocco. At ICARDA, he has led projects focused on dryland agrobiodiversity, established gene banks, and strengthened pre-breeding efforts in various crops. He has published 97+ peer-reviewed journal articles and has provided extensive training in genetic resources conservation.

Zakaria Kehel is a senior biometrician and research team leader at ICARDA, with a Ph.D. from the University of Córdoba, Spain. His expertise lies in germplasm enhancement, predictive breeding, and data science. He has over 15 years of experience in genetic resources conservation and breeding data management and has held key roles at Bayer Crop Science and CIMMYT.

Behzad Sadeghzadeh earned his Ph.D. from The University of Western Australia and is a research fellow at The University of Melbourne, specializing in crop molecular breeding. With a strong background in breeding for drought resilience and biofortification, he has published over 100 peer-reviewed papers. He has collaborated extensively with international research institutes like ICARDA and CIMMYT and has significant experience in teaching and supervising graduate students.

Contributors

Admas Alemu Department of Plant Breeding, Swedish University of Agricultural Sciences, Alnarp, Sweden

Amaresh ICAR-Sugarcane Breeding Institute, Coimbatore, Tamil Nadu, India

Alberto J. Andrade Instituto de Biología de la Altura, UNJu, SS de Jujuy, Argentina

A. Anna Durai ICAR-Sugarcane Breeding Institute, Coimbatore, Tamil Nadu, India

M. Arif ICAR-Indian Institute of Farming Systems Research, Meerut, Uttar Pradesh, India

Twahira Begum Agrotechnology and Rural Development Division, CSIR-North East Institute of Science and Technology, Jorhat, Assam, India

Harun Bektas Department of Agricultural Biotechnology, Faculty of Agriculture, Siirt University, Siirt, Turkey

Pradeep Kumar Bhati International Maize and Wheat Improvement Center (CIMMYT), New Delhi, India

Borlaug Institute for South Asia (BISA), New Delhi, India

Melike Biçici Department of Field Crops, Faculty of Agriculture, Çukurova University, Adana, Turkey

Jan Bocianowski Department of Mathematical and Statistical Methods, Poznań University of Life Sciences, Poznań, Poland

Sergio J. Bramardi Centro de Investigaciones en Toxicología Ambiental y Agrobiotecnología del Comahue (CITAAC-CONICET), Universidad Nacional del Comahue, Neuquén, Argentina

Juan Burgueño International Maize and Wheat Improvement Centre (CIMMYT), Texcoco, Mexico

Salvatore Ceccarelli Independent Consultants, Ascoli Piceno, Italy

Marina Ćeran Institute of Field and Vegetable Crops National Institute of the Republic of Serbia, Novi Sad, Republic of Serbia

Jiansheng Chen State Key Laboratory of Wheat Improvement, College of Agronomy, Shandong Agricultural University, Tai'an, Shandong, China

Jairam Choudhary ICAR-Indian Agricultural Research Institute, New Delhi, India

Sabrina Costa-Tártara Departamento de Ciencias Básicas, Departamento de Tecnología, Universidad Nacional de Luján, Buenos Aires, Argentina

Consejo Nacional de Investigaciones Científicas y Técnicas (CONICET), Buenos Aires, Argentina

Jose Crossa International Maize and Wheat Improvement Center (CIMMYT), Texcoco, Edo. de México, Mexico

Colegio de Postgraduados, Montecillos, Edo. de México, Mexico

Ramiro N. Curti Laboratorio de Investigaciones Botánicas (LABIBO), Facultad de Ciencias Naturales, Universidad Nacional de Salta, Salta, Argentina

Consejo Nacional de Investigaciones Científicas y Técnicas (CONICET) CCT Salta-Jujuy, Salta, Argentina

Reza Darvishzadeh Department of Plant Production and Genetics, Faculty of Agriculture, Urmia University, Urmia, Iran

Reshmi Rani Das Global Maize Program, International Maize and Wheat Improvement Center (CIMMYT), Hyderabad, Telangana, India

Vuk Đorđević Institute of Field and Vegetable Crops National Institute of the Republic of Serbia, Novi Sad, Republic of Serbia

D. Dutta ICAR-Indian Agricultural Research Institute, New Delhi, India

Lonishree Dutta Agrotechnology and Rural Development Division, CSIR-North East Institute of Science and Technology, Jorhat, Assam, India

Sara Fatma Department of Botany, Banaras Hindu University, Varanasi, UP, India

Julian Garcia-Abadillo Agronomy Department, University of Florida, Gainesville, USA

Centro de Biotecnología y Genómica de Plantas, Universidad Politécnica de Madrid, Madrid, Spain

P. C. Ghasal ICAR-Vivekananda Parvatiya Krishi Anusandhan Sansthan, Almora, Uttarakhand, India

Anindita Gogoi Agrotechnology and Rural Development Division, CSIR-North East Institute of Science and Technology, Jorhat, Assam, India

K. Gopalareddy ICAR-Sugarcane Breeding Institute, Coimbatore, Tamil Nadu, India

National Institute of Seed Science and Technology, Regional Station, Bengaluru, India

Velu Govindan International Maize and Wheat Improvement Center (CIMMYT), Texcoco, Mexico

P. Govindaraj ICAR-Sugarcane Breeding Institute, Coimbatore, Tamil Nadu, India

Stefania Grando Independent Consultants, Ascoli Piceno, Italy

Tanmita Gupta Agrotechnology and Rural Development Division, CSIR-North East Institute of Science and Technology, Jorhat, Assam, India

Bita Jamshidi Department of Food Security and Public Health, Khabat Technical Institute, Erbil Polytechnic University, Erbil, Iraq

Diego Jarquin Agronomy Department, University of Florida, Gainesville, USA

Arun Kumar Joshi Borlaug Institute for South Asia (BISA), New Delhi, India

Sanaz Khalifani Department of Plant Production and Genetics, Faculty of Agriculture, Urmia University, Urmia, Iran

Praveen Kona ICAR-Sugarcane Breeding Institute, Coimbatore, Tamil Nadu, India

Ankica Kondić-Špika Institute of Field and Vegetable Crops National Institute of the Republic of Serbia, Novi Sad, Republic of Serbia

Prakash H. Kuchanur University of Agricultural Sciences, Raichur, Karnataka, India

Amit Kumar ICAR-Vivekananda Parvatiya Krishi Anusandhan Sansthan, Almora, Uttarakhand, India

Kamlesh Kumar ICAR-Indian Institute of Farming Systems Research, Meerut, Uttar Pradesh, India

T. Lakshmi Pathy ICAR-Sugarcane Breeding Institute, Coimbatore, Tamil Nadu, India

Mohan Lal Agrotechnology and Rural Development Division, CSIR-North East Institute of Science and Technology, Jorhat, Assam, India

H. K. Mahadeva Swamy ICAR-Sugarcane Breeding Institute, Coimbatore, Tamil Nadu, India

Hamid Hatami Maleki Department of Plant Production and Genetics, Faculty of Agriculture, University of Maragheh, Maragheh, Iran

R. T. Maruthi ICAR-Sugarcane Breeding Institute, Coimbatore, Tamil Nadu, India

A. L. Meena ICAR-Indian Institute of Farming Systems Research, Meerut, Uttar Pradesh, India

Dragana Miladinović Institute of Field and Vegetable Crops National Institute of the Republic of Serbia, Novi Sad, Republic of Serbia

Jegor Miladinović Institute of Field and Vegetable Crops National Institute of the Republic of Serbia, Novi Sad, Republic of Serbia

K. Mohanraj ICAR-Sugarcane Breeding Institute, Coimbatore, Tamil Nadu, India

Osval A. Montesinos-López Facultad de Telemática, Universidad de Colima, Colima, Mexico

Ekta Narwal ICAR-Indian Agricultural Research Institute, New Delhi, India

Moyses Nascimento Agronomy Department, University of Florida, Gainesville, USA

Department of Statistics, Federal University of Viçosa, Viçosa, Brazil

Pablo Ortega-Baes Laboratorio de Investigaciones Botánicas (LABIBO), Facultad de Ciencias Naturales, Universidad Nacional de Salta, Salta, Argentina

Consejo Nacional de Investigaciones Científicas y Técnicas (CONICET) CCT Salta-Jujuy, Salta, Argentina

Rodomiro Ortiz Department of Plant Breeding, Swedish University of Agricultural Sciences, Alnarp, Sweden

Hakan Özkan Department of Field Crops, Faculty of Agriculture, Çukurova University, Adana, Turkey

State Key Laboratory of Wheat Improvement, College of Agronomy, Shandong Agricultural University, Tai'an, Shandong, China

Ayyanagouda Patil University of Agricultural Sciences, Raichur, Karnataka, India

Alireza Pour-Aboughadareh Seed and Plant Improvement Institute, Agricultural Research, Education and Extension Organization (AREEO), Karaj, Iran

Liu Qier State Key Laboratory of Wheat Improvement, College of Agronomy, Shandong Agricultural University, Tai'an, Shandong, China

Aleksandra Radanović Institute of Field and Vegetable Crops National Institute of the Republic of Serbia, Novi Sad, Republic of Serbia

Jonatan Rodriguez Laboratorio de Investigaciones Botánicas (LABIBO), Facultad de Ciencias Naturales, Universidad Nacional de Salta, Salta, Argentina

Consejo Nacional de Investigaciones Científicas y Técnicas (CONICET) CCT Salta-Jujuy, Salta, Argentina

Vitor S. Sagae Agronomy Department, University of Florida, Gainesville, USA Department of Statistics, Federal University of Viçosa, Viçosa, Brazil

K. Seetharam Global Maize Program, International Maize and Wheat Improvement Center (CIMMYT), Hyderabad, Telangana, India

Uğur Sesiz Department of Field Crops, Faculty of Agriculture, Şırnak University, Şırnak, Turkey

R. M. Shanthi ICAR-Sugarcane Breeding Institute, Coimbatore, Tamil Nadu, India

Y. S. Shivay ICAR-Indian Agricultural Research Institute, New Delhi, India

Ashok Singamsetti Institute of Agricultural Sciences, Banaras Hindu University, Varanasi, U.P, India

Raghuveer Singh ICAR-Indian Institute of Farming Systems Research, Meerut, Uttar Pradesh, India

Sivakumar Sukumaran Department of Plant and Environmental Sciences, Clemson University, Clemson, SC, USA

Dragana Trkulja Institute of Field and Vegetable Crops National Institute of the Republic of Serbia, Novi Sad, Republic of Serbia

Berta Velásquez Facultad de Ciencias Agrarias, UNJu, SS de Jujuy, Argentina

Vinayaka ICAR-Sugarcane Breeding Institute, Coimbatore, Tamil Nadu, India

M. T. Vinayan Global Maize Program, International Maize and Wheat Improvement Center (CIMMYT), Hyderabad, Telangana, India

Manish Kumar Vishwakarma Borlaug Institute for South Asia (BISA), New Delhi, India

Kanat Yermekbayev Zhetysu University named after I.Zhansugurov, Institute of Biotechnology and Ecology, Taldykorgan, Kazakhstan

Pervez H. Zaidi Global Maize Program, International Maize and Wheat Improvement Center (CIMMYT), Hyderabad, Telangana, India

Genotype × Environment Interaction in Breeding Programs: Wide vs. Specific Adaptation

1

Salvatore Ceccarelli and Stefania Grando

Abstract

The concepts of wide (or broad) and specific adaptation (or narrow) are intimately tied to genotype-by-environment interactions (GEIs) and have been widely, sometimes fiercely, debated among breeders and in the academia. Part of the debate has been fueled by the fact that wide adaptation was at the core of the Green Revolution. GEI is one of the most, if not the most often investigated issue in plant breeding because, when of qualitative type, it affects genetic gains by reducing heritability as shown by the breeder's equation. The two components of GEI, namely Genotype by Locations (GL) and Genotype by Years within Locations (GY_L), should be always estimated independently as they can be managed differently in a breeding program and it is the repeatability of GL that determines whether to breed for wide or for specific adaptation. The chapter argues that to address an heterogenous population of target environments, as it is the case of low input and organic agriculture, and to adequately address issues such as the client profile and the product profile, a breeding program based on decentralized selection, defined as selection in the target geographical and agronomic environment, can exploit GL interactions by subdividing the target population of environments into subgroups characterized by repeatable GL interactions and by selecting for specific spatial adaptation and wide temporal adaptation within each subpopulation. Such a breeding strategy is expected to reverse the trend of modern plant breeding toward a reduction of genetic diversity by increasing agrobiodiversity with beneficial effects both for the planet and for human health.

S. Ceccarelli (✉) · S. Grando
Independent Consultants, Ascoli Piceno, Italy

R. Mohammadi et al. (eds.), *Genotype x Environment Interactions and its Implications for Plant Breeding*, https://doi.org/10.1007/978-981-95-5664-9_1

Keywords

Genotype · Phenotype · Adaptation · Target population of environments · Decentralized selection · Climate change · Selection gain · Adoption

1.1 Introduction

Genotype by environment interaction (GEI) is one of the most, if not the most often investigated issue in plant breeding because it generally reduces the rate of progress in yield improvement in breeding programs (Basford and Cooper 1998), and its analysis is at the base of deciding which breeding strategy to follow, namely whether to select for wide or specific adaptation. The large effect of GEI on genetic gains also affects the accuracy of genomic selection (Bassi et al. 2016). GEI complicates considerably the breeder's work, particularly when it is of qualitative type, namely involves a change in ranking. This type of GEI has been named cross-over interaction by Gail and Simon (1985).

GEI can be defined as the phenomenon in which a single genotype produces different phenotypes depending on the environment (E). Here, E may refer to a physical location (L), a specific year (Y), a combination of locations and years (YL), a type of management practice (M), a social environment (S), or any non-factorial combination of these environmental factors. GEI can be understood as reflecting differences in plasticity—the capacity of a genotype to produce varying phenotypes across different environments (Des Marais et al. 2013; Hufford et al. 2019).

The possibility of obtaining different phenotypes from the same genotype depending on E is interesting because what we harvest, eat, sell, buy, exchange is the phenotype, not the genotype. It is the phenotype that ultimately determines whether the farmers accept (or reject) a new variety (Kholová et al. 2021). Therefore, getting the right phenotype at the end of the selection process in a given E—no matter how defined—determines the success or the failure of the breeding program, if the success is measured by the adoption of the final variety rather than by its release.

Adoption—widely used as a key indicator of breeding program success—remains a persistent challenge, as highlighted by an extensive body of literature. Ceccarelli (2015) provided an early review of this issue, and more recent studies by Kholová et al. (2021, 2024) reaffirm that low adoption rates continue to be a major concern, a finding corroborated by Alary et al. (2020), Thiele et al. (2020), and Itam et al. (2024). This problem extends to the CGIAR system, where numerous crop varieties have been released but seldom adopted by farmers (Kholová et al. 2021, 2024; Ceccarelli and Grando 2021)—a shortfall that may be attributed, at least in part, to genotype-by-environment interactions (GEI). The relevance of this issue to the present chapter lies in the fact that adoption has long served as a benchmark for evaluating the success of breeding efforts aimed at achieving wide adaptation (Morris and Heisey 2003; Pingali 2012).

Related to the issues of adoption and adaptation are two concepts used mostly by breeders in the private sector, namely the *product profile* and the *client profile*. The product profile has been defined by Cobb et al. (2019) as a set of attributes that a new variety should have to be successfully adopted by farmers. The client profile refers to for whom the new variety is intended, namely, for example, the type of farmer or consumer or agri-food chain. GEI is important in the definition of the product profile, which, as stated by Cobb et al. (2019), could be a popular variety that is missing some specific traits, and therefore a product profile is constructed with "must-have" traits and "value-added" traits. Both types of traits should result into a phenotype that ultimately, as said earlier, will depend on GEI.

To these two concepts we may add an additional concept, also related to GEI, namely the *environmen*t profile, defined as the type of environment (rainfed, irrigated, marginal, high input, low input, high elevation, etc.) for which the new variety is being bred. The environment profile is in turn associated with the concepts of wide and specific adaptation because breeding for wide adaptation aims at varieties able of outyielding other varieties regardless of the environment, which in this case includes the agronomic management, in which they are grown.

The concepts of wide and specific adaptation drive the output of a breeding program that ranges from multiple varieties specifically adapted to different customers and different environments to few varieties widely adapted to several customers and several environments as discussed later.

These concepts are still struggling to make their way into public breeding programs, likely because of the reward system, which is often based on the number of varieties released rather than on the number of varieties adopted (Ceccarelli and Grando 2019).

One major challenge in this respect is represented by climate change, which affects E through its complexity (Ceccarelli and Grando 2020). Because of climate change, which includes changes not only in temperature and rainfall but also in the spectrum of insects, fungal diseases, and weeds, by the time the breeding program defines the product profile(s), to the time when the product is very nearly ready, the genotype(s) the breeder assembled may well give different phenotype(s), at least some of them, from those expected. Additionally, as consumer preferences also change rapidly, clients' expectations on which the product profile was developed, may differ considerably thus affecting adoption. Molecular breeding cannot possibly offer solutions to these problems.

1.2 The Wide vs. Specific Adaptation Debate

In plant breeding terminology, a crop variety with wide or good adaptation is one that consistently delivers high yields across a range of environments (Frey 1964). Although the idea of wide adaptation was present in the vocabulary of agricultural scientists as early as the 1960s, it remained peripheral to mainstream agricultural science. This was because the prevailing view in early twentieth-century plant breeding held that selection should take place in the target environment to develop varieties specifically tailored to those local conditions (Baranski 2022, p. 26).

The debate also had a terminology component with terms such as adaptness, which is the state of being adapted, used as a synonym of adaptation, which actually is the process of becoming adapted (van Eeuwijk et al. 2016).

As mentioned earlier, the concept of wide adaptation became very popular when it was adopted as the breeding strategy of the Green Revolution (GR), even though it had been used earlier as a new plant breeding philosophy by Indian scientists (Baranski 2015). The concept of wide adaptation is so embedded in the narrative of the GR, despite being only one component of it, that it is not always clear, in the debate that follows, whether when judgments are made about the GR, these can be extrapolated to the concept of wide adaptation, or vice versa.

Within the breeding strategy centered on the concept of wide adaptation, the ideal variety—as previously noted—is one that delivers both high and stable yields across diverse environments. Stability, a concept frequently misunderstood in plant-breeding literature (Basford and Cooper 1998) and often referred to as broad adaptation or phenotypic stability (Finlay and Wilkinson 1963), was central to this debate. The key question was whether a variety could consistently maintain superior yields across a range of production conditions. Specifically, researchers explored whether it was feasible to combine high yield potential under favorable conditions with drought tolerance (Araus et al. 2002; Blum 2005; Tambussi et al. 2005). Numerous studies have examined stability as a component of genotype-by-environment interaction (GEI) and its relationship with yield levels (Hughes et al. 1987; Adegoke and Frey 1987), yielding conflicting conclusions. Some reported that gains in yield were accompanied by reduced stability (Loffler et al. 1986; Calderini and Slafer 1999), whereas others found a positive association between yield and stability (Haussmann et al. 2000; Mullualem et al. 2024). Eventually, an additional aspect of the debate was about the optimum environment where to select a broadly adapted variety (Allen et al. 1978; Whitehead and Allen 1990).

The debate was between three groups of scientists, namely (1) those in favor of the concept of wide adaptation who cited the benefits of the GR as supporting evidence, (2) those that claimed that the so-called widely adapted varieties were in fact adapted only to high-input agricultural environments and that major gains were expected by selecting for specific adaptation, and (3) those that implicitly agreed with the second group but on the basis of the geographical and social disparity generated by the diffusion of the widely adapted varieties. There are also examples of changing views on the issue of wide vs. specific adaptation (Atlin and Frey 1990; Atlin et al. 2011).

Among the supporters of wide adaptation as a successful breeding strategy were obviously the promoters of the strategy itself, which is based on testing all germplasm under near optimum conditions for its yield potential and disease resistance (Braun et al. 1996, 1997; Rajaram et al. 1997), shuttle breeding, defined as alternate selection in optimal and rainfed environments (Khush 2001), and multi-location testing. Incidentally, Finlay, the co-author of the well-known Finlay and Wilkinson

(1963) regression analysis of adaptation, had some concerns about shuttle breeding. His argument was that such a selection technique "automatically eliminates genotypes with exceptional potential for yield given the correct specific environment" (Baranski 2022, p. 43). A similar concern was expressed by Shorter et al. (1991) as "a potential cost to broad adaptation."

The germplasm was tested under near-optimal conditions based on the rationale that selection in such environments leads to greater yield gains under drought than direct selection in drought-stressed fields (Calhoun et al. 1994). This approach was supported by studies citing benefits like higher genetic variance and heritability, which improves selection precision. However, other research contested this, suggesting that irrigated trials were not statistically superior to dryland trials (Bowman et al. 1993) and that a high-yielding, high-heritability site is not necessarily the most effective for selection (Pederson and Rathjen 1981; Cooper et al. 1997).

Research by Castleberry et al. (1984) in maize and Frey (1964) in oat advocated for selecting cultivars under optimal, high-input conditions, including high fertility, irrigation, and disease control. This strategy of selecting for wide adaptation in favorable environments has been supported by some Australian breeders working with wheat, maize, and sorghum (Brennan et al. 1981; Brennan 1989; Cooper et al. 1995; Chapman et al. 1997, 2000). However, the approach has been challenged. Critics argue that selection in favorable environments inherently biases results toward highly responsive genotypes (Simmonds 1981). Furthermore, Pederson and Rathjen (1981) contend that selection on research stations is only suitable for identifying genotypes for high-yielding environments, whereas selection in farmers' fields is necessary for identifying genotypes suited to low-yielding conditions.

Even in those cases in which the breeding strategy of the GR was considered a "spectacular success," it was also recognized that "it was limited to favoured, usually well-watered areas" (Byerlee and Husain 1993). Doubts have been expressed as to whether the benefits of the GR outweighed the costs in terms of environmental damage (Tilman 1998). Even within Centro Internacional de Mejoramiento de Maíz y Trigo (CIMMYT), some results showed that across sites varieties were not always more stable than specific-site selection (Crossa et al. 1989).

The choice between direct and indirect selection depends on the trait and environment, while the decision to develop specifically or widely adapted varieties must be considered case-by-case (Atlin and Frey 1989a, b). Subsequent research (Atlin and Frey 1990) further suggested that separate breeding programs for low-productivity systems can maximize yield gains. This aligns with the view, supported by researchers like Boyer (1982), that "*considerable success can be expected if plant improvement includes selection under conditions that are often unfavourable for growth*" and adding that with superior genotypes identified under such conditions "*large-scale modification of the environment is less necessary.*"

The emphasis on breeding for specific adaptation rather than broad adaptability has been supported both theoretically—by Ceccarelli (1989) and Simmonds (1991)—and experimentally across a range of crops. Empirical evidence includes studies on barley and lentil (Ceccarelli et al. 1994), pearl millet (Virk and Mangat 1991; van Oosterom et al. 2003), sugarcane (Bull et al. 1992), pearl millet and sorghum (Haussmann et al. 2012), barley (Abay and Bjørnstad 2009; Schneider et al.

2024; van Oosterom et al. 1993), lentil (Erskine et al. 2016), lucerne (Annicchiarico et al. 2020), oat (Yan et al. 2022), and durum wheat (Annicchiarico et al. 2006).

Other research has acknowledged that the GR had significant impact on food security, particularly through increased staple grain production and widespread irrigation adoption, which greatly benefited Asian populations. However, these studies also highlight uneven outcomes, noting adverse effects on human nutrition, reduced crop resilience to environmental stresses, and negative environmental consequences (Bailey-Serres et al. 2019). In the context of this crop, the Green Revolution largely relied on a strong positive GE interaction: the high yield potential of dwarf varieties was fully expressed only under conditions of high soil fertility (Simmonds 1981).

1.3 Target Population of Environments (TPE) and Breeding for Wide vs. Specific Adaptation

A target population of environment (TPE) is commonly intended as the geographical area a breeding program is addressing. The size of a TPE depends on the breeding program and may vary from a State within a country, to a country, to several countries sharing some commonalities such as "the semi-arid tropics," or "the dry areas," or all the areas where a given crop is grown.

A TPE is made of different combinations, not necessarily factorial, of locations, years, management types, and social conditions, which implies that a TPE can have all different degrees of homogeneity/heterogeneity. This in turn, implies the possible need for different product profiles and client profiles within a given TPE. In practical terms, even if a given TPE is characterized by a common constraint, let us say drought, the breeder may need to incorporate drought resistance traits into different genetic backgrounds that correspond to different product profiles.

The definition of a TPE hinges largely on the repeatability of genotype-by-location (GL) interactions (Basford and Cooper 1998). To illustrate this point, Fig. 1.1 presents a simple example: consider a multi-environment trial (MET) involving seven entries evaluated across two locations over 2 years. By comparing the ranking of varieties in locations 1 (Loc 1) and 2 (Loc 2) in Year 1, the presence of a large GL interaction is evident: entry 7, which ranked the last in Loc 1, became the top ranking in Loc 2. Similarly, entry 4, which was the top ranking in Loc 1 became the one before the last in Loc 2. When the trial is repeated in Year 2, the ranking in Loc 1 tends to be similar to the ranking in Loc 2 in Year 1, and vice versa, whereas the ranking in Loc 2 tends to be similar to the ranking of Year 1 in Loc 1.

In each year we observe a large GL interaction, which however is *not repeatable*. In a case like this, the two locations are considered as belonging to the same TPE and we need to breed for varieties adapted to both locations.

Figure 1.2, by contrast, illustrates a case of repeatable GL interaction as shown by the ranking in Loc 1 in Year 1, which is very different from Loc 2 in the same year, and repeats itself in Year 2 as it does in Loc 2.

When the results of a MET are like those in Fig. 1.2, we can consider Loc 1 and Loc 2 as belonging to two different TPE and we can increase the response to

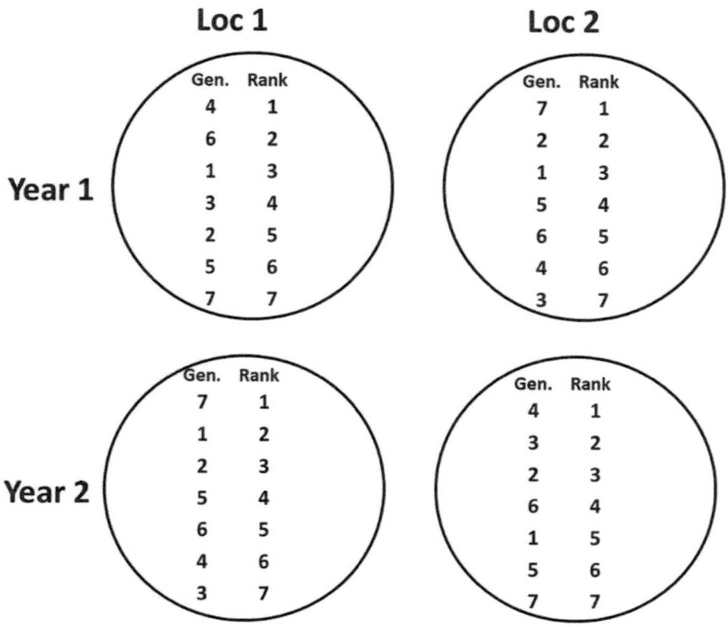

Fig. 1.1 A case of non-repeatable GL interaction

selection by breeding varieties specifically adapted to each location, or, in other words, specifically adapted in space and widely adapted in time.

METs are routine features of a breeding program. From an experimental viewpoint, METs include several locations and are conducted for at least 3 years. The results of a MET, properly analyzed, provide precise indications on which breeding strategies to use.

Significant attention has been devoted to the statistical analysis of datasets generated by METs and to the design of field trials aimed at capturing GEI. Among the most widely used approaches are the Additive Main effects and Multiplicative Interaction (AMMI) models (Gauch 1992) and the Genotype main effect plus Genotype-by-environment Interaction (GGE) biplot method (Yan et al. 2000). Additional techniques include the Site Regression (SREG) model (Cornelius et al. 1996), the Sites Regression and Shifted Multiplicative Model (Crossa and Cornelius 1997), Factor Analytic Models (Burgueño et al. 2008), and Factorial Regression (Voltas et al. 2005). Comprehensive reviews of GEI analysis methods have been provided by Singh et al. (1999) and Malosetti et al. (2013).

In the case of GY, Singh et al. (2006) contended that when locations are sufficiently distant from one another—as is typically the case in METs—the "year" factor (temperature, rainfall and its distribution, sunshine hours, and cloudiness) varies across locations each year. Under these conditions, the LY interaction cannot be meaningfully interpreted. However, the spatial interaction and the genotype with year within location (GY_L) temporal interaction remain interpretable.

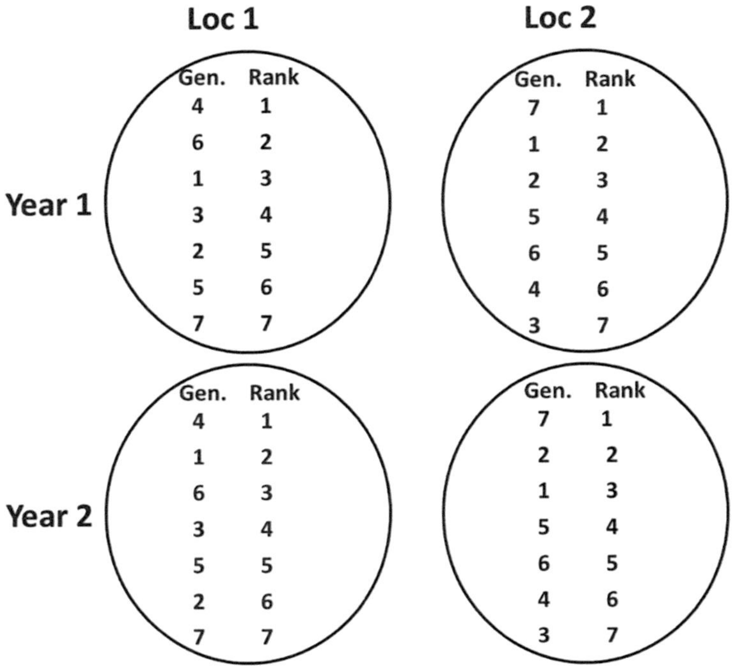

Fig. 1.2 A case of repeatable GL interaction

In terms of breeder's equation (Falconer 1981), the response to selection or genetic gain is equal to:

$$R = i\, h^2\, \sigma_p \tag{1.1}$$

where i is the intensity of selection, h^2 the heritability, and σ_p the phenotypic standard deviation. Considering that the heritability h^2 is equal to:

$$h^2 = \sigma_g^2/\sigma_p^2 = \sigma_g^2/\left(\sigma_g^2 + \sigma_{ge}^2/r + \sigma_e^2/re\right) \tag{1.2}$$

where σ_g^2 is the genetic variance, σ_{ge}^2 is the variance due to GEI, σ_e^2 is the environmental variance, r the number of replications, and e the number of environments, namely locations, years, location–year combination, management types, and social conditions, it becomes obvious that any reduction in σ_{ge}^2 will result in an increase in h^2 and hence in response to selection. This is precisely what the subdivision of a TPE in sub sets, where the GL component of GEI is smaller (like in Fig. 1.2) than in the original set, is achieving and, in terms of breeding strategies, this is implemented in breeding programs based on *decentralized selection*.

1.4 Exploiting GEI: Decentralized Selection

Decentralized selection—specifically selection carried out in the target environment—relies on what has been termed the "extended breeder's equation" (Cooper et al. 2023). In essence, this equation modifies the classic breeder's equation by incorporating the genetic correlation coefficient between the expression of a trait in the selection environment and its expression in the target environment. This approach embodies Falconer's (1990) principle that "when a trait is measured in two environments, the measurements must be treated in a genetic context as two different characters."

Decentralized selection embodies in breeding practice the fundamental distinction between GY and GL interactions—a difference first articulated over 50 years ago by Allard and Hansche (1964) and Allard and Bradshaw (1964). They emphasized that GY and GL cannot be effectively combined, as GY is largely unpredictable, whereas GL exhibits a degree of predictability. Decentralized selection can leverage GL interactions by favoring genotypes with specific local adaptation (Allard and Bradshaw 1964). In contrast, when dealing with the unpredictable nature of GY or GY_L effects described by Singh et al. (2006), the optimal strategy is to develop varieties that are well buffered against environmental fluctuations.

The distinction between GL and GY (or GY_L) is rarely addressed in most GEI literature. Clarifying this difference also enables a deeper interpretation of the underlying causes of GL—specifically, whether they stem from interactions between genotypes and location-specific social factors (S), location-specific management factors (M), or a combination of both.

Decentralized selection is effective to increase selection gains for those traits, particularly yield, influenced by lack of correlation between research stations and on-farm conditions (Cooper et al. 2023; Werner et al. 2025). Decentralized selection allows getting around the problem of low heritability associated with a high GL component of GEI. This implies that for high heritability traits it is still justified to do selection in the research station and later, for low heritability traits, in the TPE. Among other advantages, and in the case of an International Research Organization, decentralized selection allowed to capitalize on their extensive training programs in plant breeding.

One implication of decentralized selection is that it enables the identification of genotypes exhibiting high GL and a low GY_L interaction—characterized by specific spatial adaptation and broad temporal stability. This temporal stability is a key component of resilience, which can be further strengthened by leveraging the benefits of genetically heterogeneous populations, as highlighted by Allard and Hansche (1964) and Allard and Bradshaw (1964).

A practical example of decentralized selection is offered by the barley breeding program at International Center for Agricultural Research in the Dry Areas (ICARDA) between 1984 and 2011. The program was initially based on two research stations in Syria and expanded in the late 1980s to three research stations in Syria, two in Lebanon, and one in Cyprus (Ceccarelli and Grando 1991). At the beginning of the 1990s, the program started using a decentralized selection model

(Ceccarelli et al. 1994), by (1) making targeted crosses for specific countries designed in consultation with national scientists and, in some cases, using parental material supplied by national scientists, (2) distributing segregating populations to the respective countries, (3) ensuring selection is conducted by national programs' breeders (often trained at ICARDA) in the target environments, and (4) using the feedback from local scientists to improve the precision of targeting.

In some cases, the crosses were targeting more than one country, for example Libya and Egypt as early maturity and six-row were the most important traits for both countries.

In other cases, the targeting was for specific areas within a country. For example, in Iraq, in the north of the country, barley, that is mainly grown under rainfed conditions, is mostly two-row with black grains and used as feed for small ruminants, while the barley grown in the irrigated areas surrounding Baghdad is six-row, able to regrow after cutting and used as green fodder to feed dairy cows. Therefore, two different sets of segregating populations were sent to Iraq. Syria was a similar case, as in the drier areas farmers have a strong preference for two-row black seeded barley while in the wetter areas farmers have a strong preference for two-row white seeded barley (Ceccarelli and Grando 2000). Interestingly enough, the farmers in the *Gezira*, namely the area between the Tigris and the Euphrates rivers, which includes the northern part of Iraq and the northeastern part of Syria, have similar preferences concerning barley and suffer from the same major constraint, that is drought. Therefore, this area could be considered as a single TPE even if it encompasses two different countries.

1.5 Conclusions

Agriculture is the second leading cause of biodiversity loss globally, following overexploitation (Maxwell et al. 2016). Historically and increasingly, agricultural practices have diminished both species richness and the genetic diversity within species (Barot et al. 2017). This trend is starkly illustrated by the fact that just nine crop species now account for more than 66% of global crop production by weight, reflecting a dramatic narrowing of the crop base that supplies the majority of the world's food (FAO 2019). Multiple studies suggest that centralized plant breeding programs have contributed to a broad decline in agrobiodiversity, although this issue remains somewhat contested. For instance, Landjeva et al. (2006) reported no loss of genetic diversity in Bulgarian winter wheat, whereas Bonnin et al. (2014) documented a significant reduction in wheat's genetic diversity over the twentieth century—a finding echoed by Reiss and Drinkwater (2018). More recently, Gatto et al. (2021), analyzing varietal release data for 11 major food crops across 44 countries in Asia and Africa, observed a continued decline in crop varietal diversity, largely driven by the replacement and spatial displacement of traditional landraces (Dwivedi et al. 2016).

To reverse the trend toward uniformity, plant breeding should make a combined use of decentralized selection, client participation, and of heterogeneous material such as evolutionary populations, thus reducing the potential negative impact of GEI. The advantages of decentralized selection combined with client participation have been discussed by Grüneberg et al. (2009) and by Christinck and Weltzien (2013). Decentralized selection allows subdividing the TPE in sub regions with more repeatable GL thus contributing to increase the heritability by reducing the variance due to GL. At the same time decentralized selection increases diversity in space, as different varieties are likely to be the winners in different locations. Client participation reduces the possible impact of genotype × management and genotype × social status on the product profile, which may result in a limited adoption; eventually, the use of genetically heterogeneous material such as evolutionary populations, with their demonstrated better stability over time compared to genetically uniform varieties (Ceccarelli and Grando 2020), allows reducing the impact of the unpredictable component of GEI, namely GY. Furthermore, evolutionary populations increase diversity in time as they evolve differently from one year to the next in the same location.

Their ability to adapt over time and across environments in response to both biotic and abiotic stresses is likely the fastest, most cost-effective, and continuously improving solution to the complex challenge of climate change—further enhanced by yield gains achieved through a blend of natural and artificial selection.

References

Abay F, Bjørnstad A (2009) Specific adaptation of barley varieties in different locations in Ethiopia. Euphytica 167:181–195. https://doi.org/10.1007/s10681-008-9858-3

Adegoke AO, Frey KJ (1987) Grain yield response and stability for oat lines with low-, medium-, and high-yielding ability. Euphytica 36:121–127. https://doi.org/10.1007/BF00730655

Alary V, Yigezu YA, Bassi FM (2020) Participatory farmers-weighted selection (PWS) indices to raise adoption of durum cultivars. Crop Breed Genet Genom 2(3):e200014. https://doi.org/10.20900/cbgg20200014

Allard RW, Bradshaw AD (1964) Implications of genotype-environmental interactions in applied plant breeding. Crop Sci 4:503–508. https://doi.org/10.2135/cropsci1964.0011183X000400050021x

Allard RW, Hansche PE (1964) Some parameters of population variability and their implications in plant breeding. Adv Agron 16:281–325. https://doi.org/10.1016/S0065-2113(08)60027-9

Allen FL, Comstock RE, Rasmusson DC (1978) Optimal environments for yield testing. Crop Sci 18(5):747–751. https://doi.org/10.2135/cropsci1978.0011183X001800050013x

Annicchiarico P, Bellah F, Chiari T (2006) Repeatable genotype × location interaction and its exploitation by conventional and GIS-based cultivar recommendation for durum wheat in Algeria. Eur J Agron 24:70–81. https://doi.org/10.1016/j.eja.2005.05.003

Annicchiarico P, Bottazzi P, Ruozzi F, Russi L, Pecetti L (2020) Lucerne cultivar adaptation to Italian geographic areas is affected crucially by the selection environment and encourages the breeding for specific adaptation. Euphytica 216:50. https://doi.org/10.1007/s10681-020-02590-1

Araus JL, Slafer GA, Reynolds MP, Royo C (2002) Plant breeding and drought in C3 cereals: what should we breed for? Ann Bot 89(7):925–940. https://doi.org/10.1093/aob/mcf049

Atlin GN, Frey KJ (1989a) Predicting the relative effectiveness of direct versus indirect selection for oat yield in three types of stress environments. Euphytica 44:137–142. https://doi.org/10.1007/BF00022608

Atlin GN, Frey KJ (1989b) Breeding crop varieties for low-input agriculture. Am J Altern Agric 4(2):53–58. https://doi.org/10.1017/S0889189300002721
Atlin GN, Frey KJ (1990) Selecting oat lines for yield in low-productivity environments. Crop Sci 30:556–561. https://doi.org/10.1007/BF00022608
Atlin GN, Kleinknecht K, Singh KP, Piepho H-P (2011) Managing genotype x environment interaction in plant breeding programs: a selection theory approach. J Indian Soc Agric Stat 65(2):237–247. http://www.isas.org.in/jisas/SearchJournal?volName=65&issue=2
Bailey-Serres J, Parker JE, Ainsworth EA, Oldroyd GED, Schroeder JI (2019) Genetic strategies for improving crop yields. Nature 575:109–118. https://doi.org/10.1038/s41586-019-1679-0
Baranski MR (2015) Wide adaptation of green revolution wheat: international roots and the Indian context of a new plant breeding ideal, 1960-1970. Stud Hist Phil Biol Biomed Sci 50:41–50. https://doi.org/10.1016/j.shpsc.2015.01.004
Baranski MR (2022) The globalization of wheat: a critical history of the green revolution. University of Pittsburgh Press. https://doi.org/10.2307/j.ctv3029w37
Barot S, Allard V, Cantarel A, Enjalbert J, Gauffreteau A, Goldringer I, Lata J-C, Le Roux X, Niboyet A, Porcher E (2017) Designing mixtures of varieties for multifunctional agriculture with the help of ecology. A review. Agron Sustain Dev 37:13. https://doi.org/10.1007/s13593-017-0418-x
Basford KE, Cooper M (1998) Genotype × environment interactions and some considerations of their implications for wheat breeding in Australia. Aust J Agric Res 49(2):153–174. https://doi.org/10.1071/A97035
Bassi FM, Bentley AR, Charmet G, Ortiz R, Crossa J (2016) Breeding schemes for the implementation of genomic selection in wheat (Triticum spp.). Plant Sci 242:23–36. https://doi.org/10.1016/j.plantsci.2015.08.021
Blum A (2005) Drought resistance, water-use efficiency, and yield potential—are they compatible, dissonant, or mutually exclusive? Aust J Agric Res 56:1159–1168. https://doi.org/10.1071/AR05069
Bonnin I, Bonneuil C, Goffaux R, Montalent P, Goldringer I (2014) Explaining the decrease in the genetic diversity of wheat in France over the 20th century. Agric Ecosyst Environ 195:183–192. https://doi.org/10.1016/j.agee.2014.06.003
Bowman D, Raymer P, Dombek D (1993) Crop performance trials under irrigated and dryland conditions. Agron J 85(3):610–614. https://doi.org/10.2134/agronj1993.00021962008500030017x
Boyer JS (1982) Plant productivity and environment. Science 218(4571):443–448. https://doi.org/10.1126/science.218.4571.443
Braun HJ, Rajaram S, van Ginkel M (1996) CIMMYT's approach to breeding for wide adaptation. Euphytica 92:175–183. https://doi.org/10.1007/BF00022843
Braun HJ, Rajaram S, van Ginkel M (1997) CIMMYT'S approach to breeding for wide adaptation. In: Tigerstedt PMA (ed) Adaptation and plant breeding. Kluwer Academic Publishers, Dordrecht, pp 197–205
Brennan JP (1989) Spillover effects of international agricultural research: CIMMYT-based semi-dwarf wheat in Australia. Agric Econ 3:323–332. https://doi.org/10.1016/0169-5150(89)90006-6
Brennan JP, Byth DE, Drake DW, De Lacy IH, Butler DG (1981) Determination of the location and number of test environments for a wheat cultivar evaluation program. Aust J Agric Res 32(2):189–201. https://doi.org/10.1071/AR9810189
Bull JK, Hogarth DM, Basford KE (1992) Impact of genotype × environment interaction on response to selection in sugarcane. Aust J Exp Agric 32:731–735. https://doi.org/10.1071/EA9920731
Burgueño J, Crossa J, Cornelius PL, Yang R-C (2008) Using factor analytic models for joining environments and genotypes without crossover genotype × environment interaction. Crop Sci 48(4):1291–1305. https://doi.org/10.2135/cropsci2007.11.0632

Byerlee D, Husain T (1993) Agricultural research strategies for favoured and marginal areas: the experience of farming system research in Pakistan. Exp Agric 29(2):155–171. https://doi.org/10.1017/S0014479700020603

Calderini DF, Slafer GA (1999) Has yield stability changed with genetic improvement of wheat yield? Euphytica 107:51–59. https://doi.org/10.1023/A:1003579715714

Calhoun DS, Gebeyehu G, Miranda A, Rajaram S, van Ginkel M (1994) Choosing evaluation environments to increase wheat grain yield under drought conditions. Crop Sci 34(3):673–678. https://doi.org/10.2135/cropsci1994.0011183X003400030014x

Castleberry RM, Crum CW, Krull CF (1984) Genetic yield improvement of US cultivars under varying fertility and climatic environments. Crop Sci 24(1):33–36. https://doi.org/10.2135/cropsci1984.0011183X002400010008x

Ceccarelli S (1989) Wide adaptation: how wide? Euphytica 40:197–205. https://doi.org/10.1007/BF00024512

Ceccarelli S (2015) Efficiency of plant breeding. Crop Sci 55:87–97. https://doi.org/10.2135/cropsci2014.02.0158

Ceccarelli S, Grando S (1991) Environment of selection and type of germplasm in barley breeding for low-yielding conditions. Euphytica 57:207–219. https://doi.org/10.1007/BF00039667

Ceccarelli S, Grando S (2000) Barley landraces from the fertile crescent: a lesson for plant breeders. In: Brush SB (ed) Genes in the field: on-farm conservation of crop diversity. IPGRI/IDRC/Lewis Publishers, pp 51–76

Ceccarelli S, Grando S (2019) Participatory plant breeding: who did it, who does it and where? Exp Agric 33:335–344. https://doi.org/10.1017/S0014479719000127

Ceccarelli S, Grando S (2020) Evolutionary plant breeding as a response to the complexity of climate change. iScience 23(12):101815. https://doi.org/10.1016/j.isci.2020.101815

Ceccarelli S, Grando S (2021) Participatory plant breeding. Supplementary material 14 in Kholová et al. (2021) In pursuit of a better world: crop improvement and the CGIAR. J Exp Bot 72(14):5158–5179. https://doi.org/10.1093/jxb/erab226

Ceccarelli S, Erskine W, Grando S, Hamblin J (1994) Genotype x environment interaction and international breeding programmes. Exp Agric 30:177–187. https://doi.org/10.1017/S0014479700024121

Chapman SC, Crossa J, Edmeades GO (1997) Genotype by environment effects and selection for drought tolerance in tropical maize. II Three mode pattern analysis. Euphytica 95:11–20. https://doi.org/10.1023/A:1002922527795

Chapman S, Hammer GL, Butler DG, Cooper M (2000) Genotype by environment interactions affecting grain sorghum. III. Temporal sequences and spatial patterns in the target population of environments. Aust J Agric Res 51:223–233. https://doi.org/10.1071/AR99022

Christinck A, Weltzien E (2013) Plant breeding for nutrition-sensitive agriculture: an appraisal of developments in plant breeding. Food Secur 5:693–707. https://doi.org/10.1007/s12571-013-0288-2

Cobb JN, Juma RU, Biswas PS, Arbelaez JD, Rutkoski J, Atlin G et al (2019) Enhancing the rate of genetic gain in public-sector plant breeding programs: lessons from the breeder's equation. Theor Appl Genet 132:627–645. https://doi.org/10.1007/s00122-019-03317-0

Cooper M, Woodruff DR, Eisemann RL, Brennan PS, DeLacy IH (1995) A selection strategy to accommodate genotype-by-environment interaction for grain yield of wheat: managed-environments for selection among genotypes. Theor Appl Genet 90:492–502. https://doi.org/10.1007/BF00221995

Cooper M, Stucker RE, DeLacy IH, Harch BD (1997) Wheat breeding nurseries, target environments, and indirect selection for grain yield. Crop Sci 37(4):1168–1176. https://doi.org/10.2135/cropsci1997.0011183X003700040024x

Cooper M, Powell O, Gho C, Tang T, Messina C (2023) Extending the breeder's equation to take aim at the target population of environments. Front Plant Sci 14:1129591. https://doi.org/10.3389/fpls.2023.1129591

Cornelius PL, Crossa J, Seyedsadr MS (1996) Statistical tests and estimators of multiplicative models for genotype-by-environment interaction. In: Kang MS, Gauch HG Jr (eds) Genotype-by-environment interaction. CRC Press, Boca Raton, pp 199–234
Crossa J, Cornelius P (1997) Sites regression and shifted multiplicative model clustering of cultivar trial sites under heterogeneity of error variances. Crop Sci 37:406–415. https://doi.org/10.2135/cropsci1997.0011183X003700020017x
Crossa J, Westcott B, Gonzales C (1989) The yield stability of CIMMYT'S maize germplasm. Euphytica 40:245–251. https://doi.org/10.1007/BF00024519
Des Marais DJ, Hernandez KM, Juenger TE (2013) Genotype-by-environment interaction and plasticity: exploring genomic responses of plants to the abiotic environment. Annu Rev Ecol Evol Syst 44:5–29. https://api.semanticscholar.org/CorpusID:85848472
Dwivedi SL, Ceccarelli S, Blair WM, Upadhyaya HD, Are AK, Ortiz R (2016) Landrace germplasm for improving yield and abiotic stress adaptation. Trends Plant Sci 21:31–42. https://doi.org/10.1016/j.tplants.2015.10.012
Erskine W, Sarker A, Agrawal SK (2016) Lentil: breeding. In: Wrigley C, Corke H, Seetharaman K, Faubion J (eds) Encyclopedia of food grains, 2nd edn. Elsevier, Amsterdam, pp 317–323. https://hdl.handle.net/20.500.11766/6344
Falconer DS (1981) Introduction to quantitative genetics, 2nd edn. Longman Group, London
Falconer DS (1990) Selection in different environments: effects on environmental sensitivity (reaction norm) and on mean performance. Genet Res 56(1):57–70. https://doi.org/10.1017/S0016672300028883
FAO (2019) The state of the world's biodiversity for food and agriculture, Bélanger J, Pilling D (eds). FAO Commission on Genetic Resources for Food and Agriculture Assessments, Rome. http://www.fao.org/3/CA3129EN/CA3129EN.pdf. CC BY-NC-SA 3.0 IGO
Finlay KW, Wilkinson GN (1963) The analysis of adaptation in a plant breeding programme. Aust J Agric Res 14:742–754. https://doi.org/10.1071/AR9630742
Frey KJ (1964) Adaptation reaction of oat strains selected under stress and non-stress environmental conditions. Crop Sci 4(1):55–58. https://doi.org/10.2135/cropsci1964.0011183X000400010018x
Gail M, Simon R (1985) Testing for qualitative interactions between treatment effects and patient subsets. Biometrics 41(2):361–372
Gatto M, de Haan S, Laborte A, Bonierbale M, Labarta R, Hareau G (2021) Trends in varietal diversity of main staple crops in Asia and Africa and implications for sustainable food systems. Front Sustain Food Syst 5:626714. https://doi.org/10.3389/fsufs.2021.626714
Gauch HG Jr (1992) Statistical analysis of regional yield trials: AMMI analysis of factorial designs. Elsevier, Amsterdam
Grüneberg WJ, Mwanga R, Andrade M, Espinoza J (2009) Breeding clonally propagated crops. In: Ceccarelli S, Guimaraes EP, Weltzien E (eds) Plant breeding and farmer participation. FAO, Rome, pp 275–321
Haussmann BIG, Obilana AB, Ayiecho PO, Blum A, Schipprack W, Geiger HH (2000) Yield and yield stability of four population types of grain sorghum in a semi-arid area of Kenya. Crop Sci 40(2):319–329. https://doi.org/10.2135/cropsci2000.402319x
Haussmann BIG, Rattunde F, Weltzien-Rattunde E, Traoré PSC, vom Brocke K, Parzies HK (2012) Breeding strategies for adaptation of pearl millet and sorghum to climate variability and change in West Africa. J Agron Crop Sci 198(5):327–339. https://doi.org/10.1111/j.1439-037X.2012.00526.x
Hufford MB, Berny Mier y Teran JC, Gepts P (2019) Crop biodiversity: an unfinished magnum opus of nature. Annu Rev Plant Biol 70:727–751. https://doi.org/10.1146/annurev-arplant-042817-040240
Hughes WG, Westcott B, Sharp PL (1987) Joint selection for high yield and low sensitivity in winter wheat (Triticum aestivum L.). Plant Breed 99(2):107–117. https://doi.org/10.1111/j.1439-0523.1987.tb01159.x

Itam MO, Iohannes SD, Albertsen M, Andrade M, Arap Bor G, Atta-Krah K, Bertram R, Danquah E, Horvath DM, Jones T, Mugehu E, Okwuonu I, Ooko-Ombaka A, Roberts RJ, Slamet-Loedin I, Tripathi L, Ewa Ubi B, Varshney RK, Venturi V, Wagaba H, Zeigler R, Creasey Krainer KM (2024) Demonstrating the benefit of agricultural biotechnology in developing countries by bridging the public and private sectors. Nat Plants 10:2–5. https://doi.org/10.1038/s41477-023-01604-9

Kholová J, Urban MO, Cock J, Arcos J, Arnaud E, Aytekin D et al (2021) In pursuit of a better world: crop improvement and the CGIAR. J Exp Bot 72(14):5158–5179. https://doi.org/10.1093/jxb/erab226

Kholová J, Urban MO, Bavorová M, Ceccarelli S, Grando S, Lensink R, Nchanji E, Pavlík J, Pelaez D, Rubyogo J-C, Sperling L, Thiele G, Ullah A, Voorhaar M, Desczka S, Bulte E (2024) Promoting the adoption of new crop cultivars in low-income countries requires new transdisciplinary approaches that take behavioral factors seriously. Nat Plants 10(11):1610–1613

Khush GS (2001) Green revolution: the way forward. Nat Rev Genet 2:815–822. https://doi.org/10.1038/35093585

Landjeva S, Korzun V, Ganeva G (2006) Evaluation of genetic diversity among Bulgarian winter wheat (Triticum aestivum L.) varieties during the period 1925–2003 using microsatellites. Genet Resour Crop Evol 53:1605–1614. https://doi.org/10.1007/s10722-005-8718-4

Loffler CM, Salaberry MT, Maggio JC (1986) Stability and genetic improvement of maize yield in Argentina. Euphytica 35:449–458. https://doi.org/10.1007/BF00021853

Malosetti M, Ribaut J-M, van Eeuwijk FA (2013) The statistical analysis of multi-environment data: modelling genotype-by-environment interaction and its genetic basis. Front Physiol 4:1–17. https://doi.org/10.3389/fphys.2013.00044

Maxwell SL, Fuller RA, Brooks TM, Watson JEM (2016) Biodiversity: the ravages of guns, nets and bulldozers. Nature 536(7615):143–145. https://doi.org/10.1038/536143a

Morris M, Heisey PW (2003) Estimating the benefits of plant breeding research: methodological issues and practical challenges. Agric Econ 29:241–252

Mullualem D, Tsega A, Mengie T, Fentie D, Kassa Z, Fassil A, Wondaferew D, Gelaw TA, Astatkie T (2024) Genotype-by-environment interaction and stability analysis of grain yield of bread wheat (Triticum aestivum L.) genotypes using AMMI and GGE biplot analyses. Heliyon 10(12):e32918. https://doi.org/10.1016/j.heliyon.2024.e32918

Pederson DG, Rathjen AJ (1981) Choosing trial sites to maximize selection response for grain yield in spring wheat. Aust J Agric Res 32:411–424

Pingali PL (2012) Green revolution: impacts, limits, and the path ahead. Proc Natl Acad Sci USA 109:12302–12308. https://doi.org/10.1073/pnas.0912953109

Rajaram S, Braun HJ, van Ginkel M (1997) CIMMYT's approach to breed for drought tolerance. In: Tigerstedt PMA (ed) Adaptation and plant breeding. Kluwer Academic Publishers, Dordrecht, pp 161–167. https://doi.org/10.1007/978-94-015-8806-5_21

Reiss ER, Drinkwater LE (2018) Cultivar mixtures: a meta-analysis of the effect of intraspecific diversity on crop yield. Ecol Appl 28:62–77. https://doi.org/10.1002/eap.1629

Schneider M, Ballvora A, Léon J (2024) Deep genotyping reveals specific adaptation footprints of conventional and organic farming in barley populations—an evolutionary plant breeding approach. Agron Sustain Dev 44:33. https://doi.org/10.1007/s13593-024-00962-8

Shorter R, Lawn RJ, Hammer GL (1991) Improving genotypic adaptation in crops: a role for breeders, physiologists and modellers. Exp Agric 27:155–175. https://doi.org/10.1017/S0014479700018810

Simmonds NW (1981) Genotype (G), Environment (E) and GE components of crop yields. Exp Agric 17(4):355–362. https://doi.org/10.1017/S0014479700011807

Simmonds NW (1991) Selection for local adaptation in a plant breeding programme. Theor Appl Genet 82:363–367. https://doi.org/10.1007/BF02190624

Singh M, Ceccarelli S, Grando S (1999) Genotype x Environment Interaction of crossover type: detecting its presence and estimating the crossover point. Theor Appl Genet 99:988–995. https://doi.org/10.1007/s001220051406

Singh M, Grando S, Ceccarelli S (2006) Measures of repeatability of genotype by location interactions using data from barley trials in Northern Syria. Exp Agric 42:189–198. https://doi.org/10.1017/S0014479705003364

Tambussi EA, Nogués S, Ferrio P, Voltas J, Araus JL (2005) Does higher yield potential improve barley performance in Mediterranean conditions: a case study. Field Crop Res 91:149–160. https://doi.org/10.1016/j.fcr.2004.06.002

Thiele G, Dufour D, Vernier P, Mwanga ROM, Parker ML, Geldermann ES, Teeken B, Wossen T, Gotor E, Kikulwe E, Tufan H, Sinelle S, Kouakȯu AM, Friedmann M, Polar V, Hershey C (2020) A review of varietal change in roots, tubers and bananas: consumer preferences and other drivers of adoption and implications for breeding. Int J Food Sci Technol 56:1076–1092. https://doi.org/10.1111/ijfs.14684

Tilman D (1998) The greening of the green revolution. Nature 396:211–212. https://doi.org/10.1038/24254

van Eeuwijk FA, Bustos-Korts DV, Malosetti M (2016) What should students in plant breeding know about the statistical aspects of genotype × environment interactions? Crop Sci 56:2119–2140. https://doi.org/10.2135/cropsci2015.06.0375

Van Oosterom EJ, Kleijn D, Ceccarelli S, Nachit MM (1993) Genotype x environment interactions of barley in the Mediterranean region. Crop Sci 33:669–674. https://doi.org/10.2135/cropsci1993.0011183X003300040004x

van Oosterom EJ, Bidinger FR, Weltzien ER (2003) A yield architecture framework to explain adaptation of pearl millet to environmental stress. Field Crop Res 80(1):33–56. https://doi.org/10.1016/S0378-4290(02)00153-3

Virk DS, Mangat BK (1991) Detection of cross over genotype by environment interactions in pearl millet. Euphytica 52:193–199. https://doi.org/10.1007/BF00029396

Voltas J, López-Córcoles H, Borrás G (2005) Use of biplot analysis and factorial regression for the investigation of superior genotypes in multi-environment trials. Eur J Agron 22:309–324. https://doi.org/10.1016/j.eja.2004.04.005

Werner CR, Zaman-Allah M, Assefa T, Cairns JE, Atlin GN (2025) Accelerating genetic gain through early-stage on-farm sparse testing. Trends Plant Sci 30(1):17–20. https://doi.org/10.1016/j.tplants.2024.10.010

Whitehead WF, Allen FL (1990) High- vs low-stress yield test environments for selecting superior soybeans lines. Crop Sci 30(4):912–918. https://doi.org/10.2135/cropsci1990.0011183X003000040031x

Yan W, Hunt LA, Sheng Q, Szlavnics Z (2000) Cultivar evaluation and mega-environment investigation based on GGE biplot. Crop Sci 40:597–605. https://doi.org/10.2135/cropsci2000.403597x

Yan W, Nilsen KT, Beattie A (2022) Mega-environment analysis and breeding for specific adaptation. Crop Sci 63(2):480–494. https://doi.org/10.1002/csc2.20895

Impact of Genotype-By-Environment Interaction on Grain Yield and Related Traits in Multi-Environment Trials

2

Ramiro N. Curti, Jonatan Rodriguez, Berta Velásquez, Sabrina Costa-Tártara, Sergio J. Bramardi, Pablo Ortega-Baes, and Alberto J. Andrade

Abstract

Genotype-by-environment (G × E) interaction is a cornerstone concept in genetic improvement, crucial for enhancing crop yield stability under diverse conditions. This chapter reviews the evolution of G × E research from early agronomic studies and classical statistical models to sophisticated genomic and computational approaches. Major developments include the introduction of bilinear models, the integration of molecular markers, and the use of simulation models to dissect trait components. Key examples illustrate G × E's impact on crops such as wheat, maize, cowpea, and quinoa, addressing traits like grain yield, plant height, flowering time, disease resistance, and grain quality. Despite significant

R. N. Curti (✉) · J. Rodriguez · P. Ortega-Baes
Laboratorio de Investigaciones Botánicas (LABIBO), Facultad de Ciencias Naturales, Universidad Nacional de Salta, Salta, Argentina

Consejo Nacional de Investigaciones Científicas y Técnicas (CONICET) CCT Salta-Jujuy, Salta, Argentina
e-mail: rcurti@agro.uba.ar

B. Velásquez
Facultad de Ciencias Agrarias, UNJu, SS de Jujuy, Argentina

S. Costa-Tártara
Departamento de Ciencias Básicas, Departamento de Tecnología, Universidad Nacional de Luján, Buenos Aires, Argentina

Consejo Nacional de Investigaciones Científicas y Técnicas (CONICET), Buenos Aires, Argentina

S. J. Bramardi
Centro de Investigaciones en Toxicología Ambiental y Agrobiotecnología del Comahue (CITAAC-CONICET), Universidad Nacional del Comahue, Neuquén, Argentina

A. J. Andrade
Instituto de Biología de la Altura, UNJu, SS de Jujuy, Argentina

R. Mohammadi et al. (eds.), *Genotype x Environment Interactions and its Implications for Plant Breeding*, https://doi.org/10.1007/978-981-95-5664-9_2

advances, research gaps persist, notably in understanding abiotic stress-related G × E causes and in the underrepresentation of orphan crops like amaranth and millet. While biotic stress causes have been more thoroughly characterized, abiotic factors demand finer-scale environmental characterization. Furthermore, studies remain heavily skewed toward Old World cereal crops, leaving pulse, oilseed, and pseudocereal native American crops less explored. Looking ahead, the future of G × E research lies in integrating genotyping, phenotyping, and envirotyping with artificial intelligence and predictive models. Emphasis on climate-resilient breeding, targeted multi-environment trials, and broader crop diversification is essential to ensure global food security in a rapidly changing environment.

Keywords

Genotype-by-environment interaction (G × E) · Crop improvement · Grain yield stability · Multi-environment trials (METs) · Abiotic and biotic stress · Genomic selection · Climate-resilient crops · Statistical modeling in plant breeding

2.1 Introduction

Grain yield stands as a cornerstone of global food security and agricultural productivity (Evans 1998). The ability to consistently produce high yields is paramount to feeding an ever-increasing global population and ensuring the sustainability of agricultural systems (Lobell and Burke 2010). Crop performance, however, is not solely determined by the genetic makeup of a plant, known as its genotype (G). The environment (E) in which a crop is grown, encompassing factors such as climate, soil conditions, and biotic stresses, plays an equally critical role (Stamp and Visser 2012). Furthermore, the interplay between a plant's genotype and the environment, termed genotype-by-environment (G × E) interaction, intro-duces a layer of complexity that significantly influences the final expression of traits, particularly grain yield (Langridge et al. 2021). This interaction signifies that different genotypes respond differently to varying environmental conditions, making the selection and development of superior crop varieties a multifaceted challenge (Elias et al. 2016). Understanding the intricacies of G × E interaction is therefore essential for plant breeders, agricultural scientists, and policymakers striving to enhance crop productivity and ensure a stable food supply (de Leon et al. 2016; Bassi et al. 2024).

For a researcher or crop breeder, whether in the public or private sector, as well as for young researchers aspiring to enter this field, a review of the literature on genotype-by-environment (G × E) interactions reveals that the fundamental concepts have undergone few significant changes in recent decades. Key definitions such as genotype, environment, G × E interaction, stability, and selection strategies remain remarkably similar. In the development of this chapter, we do not intend to rewrite the terminology or the theoretical basis, as excellent reviews and compendia address these aspects in detail and can be consulted recurrently. Instead, our primary focus is to evaluate the impact of G × E interactions on the development of new cultivars. To this end, we will analyze the evolution of concepts throughout the extensive period of intense scientific publication on this topic, while also identifying existing gaps in current research. Our primary objective is to guide young researchers initiating themselves in this discipline, providing them with a concise overview of the current state of knowledge on G × E interactions and offering essential references to delve deeper into specific aspects related to their development. Thus, this chapter provides an update analysis of the impact of G × E interaction on grain yield and other related traits, exploring its definition, influencing factors, methodologies for study, breeding strategies for mitigation, and its implications for agriculture and food security, especially in the context of changing environmental conditions.

2.2 Importance of G × E Interactions: A Quantitative Framework

The significance of numerical values in generating a quantitative understanding of a given topic's importance is currently undeniable. In the study of genotype-by-environment (G × E) interactions, some authors reviewing their developments have distinguished three major stages since the origin of agriculture (Ceccarelli 2009). The first encompasses the period between crop domestication and Gregor Mendel's discovery of the laws of heredity; the second, from the rediscovery of Mendel's laws to the Green Revolution; and the third, from this latter event to the present, characterized by the progressive development of genetic and phenotyping tools, as well as artificial intelligence and computational advancement. In this chapter, we have chosen to review scientific articles addressing G × E interactions published since the Green Revolution. This delimitation is due to the scarcity of relevant bibliographic records for crop improvement prior to this stage.

To this end, we conducted a search using different combinations of the title's keywords in the Web of Science[1] scientific repository. This search yielded a total of 5512 articles published between 1970 and 2025. During this period, a progressive increase in the attention dedicated to this topic is observed, evidenced by the rise in the number of publications, a trend that becomes markedly pronounced from 1990 onward (Fig. 2.1).

Subsequently, each article was reviewed to determine the specific aspect of G × E interaction it addressed. Based on the abstracts of the articles, we grouped them according to the following developments: agronomy, statistical models, molecular markers, and computational tools (Fig. 2.1). Throughout this period, we have identified distinct stages dominated by the developments we have classified as follows:

[1] https://clarivate.com/academia-government/scientific-and-academic-research/research-discovery-and-referencing/web-of-science/

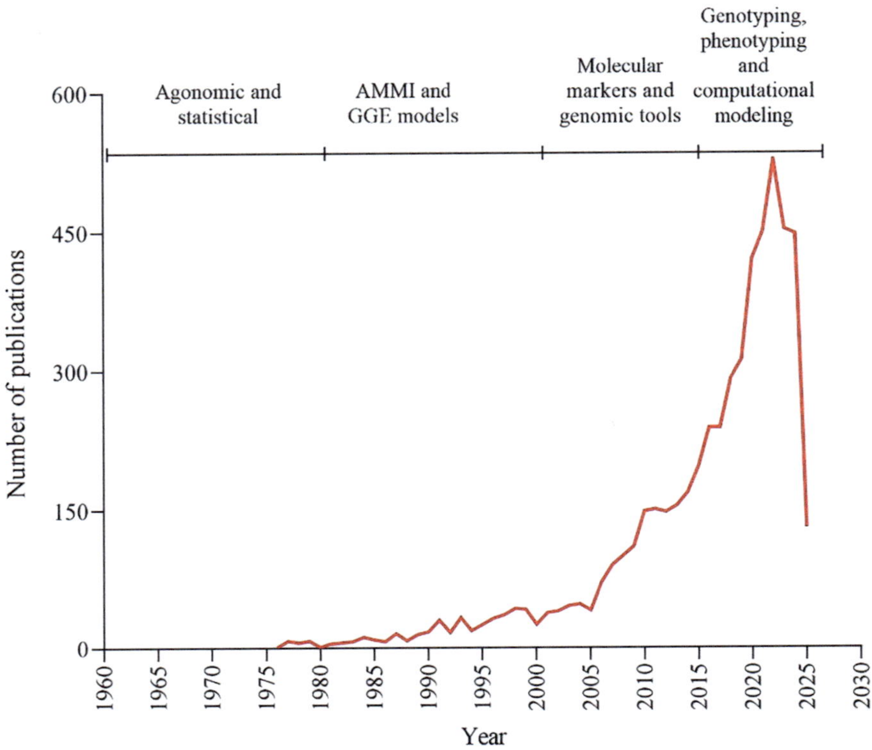

Fig. 2.1 Number of G × E interaction papers categorized according to four broad developments between years 1975 and 2025

The Early Stage (1960s–1980s): Research in this period was primarily limited to focusing on classical agronomic studies and the development of foundational statistical models for the analysis of G × E interaction. During this era, classical regression analyses and pattern analyses were developed, characterized by the complementary use of multivariate ordination and classification methods.

Development of Yield Stability Concept (1980s–1990s): This decade witnessed a progressive development of the concept of yield stability, with its biological and statistical connotations, and the use of uni- and multivariate tools for different stability indices. The hard discussion on selection strategies (broad vs narrow) was also generated during this time.

Sophistication of Statistical Models (1990s–Early 2000s): From the 1990s to the early 2000s, a major and strong development of more sophisticated statistical models, such as linear–bilinear models (AMMI and GGE),[2] occurred, along with the expansion of Biplots, which provided an indisputable advance by facilitating

[2] AMMI: The Additive Main Effects and Multiplicative Interaction
GGE: Genotype plus Genotype-by-Environment interaction

selection strategies in breeding programs. These tools allowed for a better understanding of the multivariate nature of G × E interactions and the determination of their predictability and repeatability. The application of ecophysiological crop models to dissect yield into its determinants and components, with the aim of understanding indirect selection attributes for grain yield (Jackson et al. 1996), is also recognized during this stage. This period culminates with the development of statistical models (factor regression, partial least squares regression [PLSR], structural equation modeling) that incorporate environmental and ecophysiological factors to explain the causes of G × E interactions and evaluate their repeatability in multi-environment trials (METs). The discussion on selection strategies, mainly focusing on breeding for low/high inputs, formal and participatory approaches, was also accentuated during this era (Atlin et al. 2001; Bänziger and Cooper 2001).

Advent of Genomics (2000s): The advent of molecular markers and genomic tools during the 2000s led to a significant increase in the number of publications on G × E interaction, as genetic data (quantitative trait locus, QTL) could be integrated with environmental analyses to understand the variability in yield and related attributes (quality). The development of different, more accessible computational packages and programs (SAS and R) facilitated the application of mixed and generalized linear models to overcome frequent limitations in MET databases (unbalanced) and classical linear models (Basford et al. 2004; Smith et al. 2005). The concept of BLUPs (best linear unbiased predictors), adopted from animal breeding, was introduced during this period and its application proved successful in the context of plant genetic improvement for yield (Piepho 1998; Balzarini 2002; Piepho et al. 2008). There was also a progressive increase in the application of ecophysiological crop simulation models to capture and understand G × E interactions.

Integration and New Frontiers (2010–Present): Since 2010, an unprecedented increase in publications has been observed, primarily driven by genotyping, phenotyping, and envirotyping (Xu 2016; Xu et al. 2023). These new concepts have been strongly linked to the development of novel computational models. Studies have significantly increased in the integration of genomic selection (GS) and environmental covariates to predict genotype performance across different environments. This growing trend reflects the increasing recognition of the importance of G × E interactions in crop improvement and the need to develop cultivars with stable performance across diverse environmental conditions.

In summary, throughout these periods, the study of G × E interactions has regained significant value in recent years. A large part of this resurgence is reflected or accompanied by advances in the development of statistical tools to understand its magnitude and multidimensional nature. While concepts provided by genetics and crop ecology have contributed to interpreting what statistical models quantify, the support and necessity of statistical tools are indisputable in the progress toward a closer understanding of the causes of differential genotype performance across different environments.

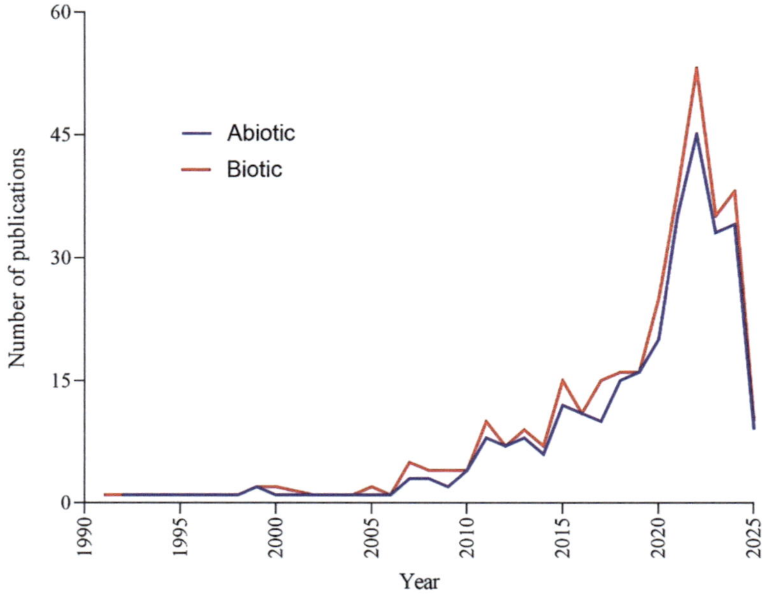

Fig. 2.2 Number of G × E interaction papers categorized according to defined causes (biotic vs. abiotic) between years 1975 and 2025

2.2.1 Defined Causes: Biotic Versus Abiotic Stress

Despite the increasing recognition of the impact of genotype-by-environment (G × E) interactions on the development of high-yielding and stable varieties, our database analysis reveals a limited effort in identifying the probable causes of these interactions (Fig. 2.2). It is well established that the identification of these causes and their repeatability within target population of environments (TPEs) explored in multi-environment trials (METs) will lead to the development of more efficient selection strategies (Basford & Cooper). In this regard, considering that the final expression of a trait such as grain yield is determined throughout the growing season and is exposed to diverse climatic and biotic factors, it is pertinent to investigate the extent to which genotype-by-environment interaction studies have explored how variation in biotic and abiotic stresses determines the differential performance of genotypes and conditions future selection strategies.

The definition of the causes of G × E interactions began to become more evident from 2006 onward, with a notable increase up to the present (Fig. 2.2). The identification of biotic causes has predominated throughout the period (Fig. 2.2), possibly due to their relatively rapid identification. In contrast, the development of varieties resistant to this type of factor does not present the same complexities as addressing abiotic stresses. The latter, on the other hand, require a more detailed characterization of the target populations, necessitating the determination of their variability in frequency at different spatial and temporal scales. This complexity often hinders the integration of G × E interactions within the target population of environments.

2.2.2 The Crop Perspective

Throughout its evolutionary history, humankind has domesticated and utilized a vast array of terrestrial plants (Schaal 2019). However, the sustainability of current human diets is limited to a reduced number of these, which constitute a significant portion of caloric intake (Romero et al. 2025). Many of these cultivated species originated in discrete geographical regions and, with human migrations, became established in diverse environments with novel conditions to which they had to respond (de Casas et al. 2024). Thus, the initial major impact of genotype-by-environment (G × E) interaction can be traced back to the beginnings of early human migrations across continents, carrying seeds of the first domesticated genotypes.

Subsequently, during the Columbian exchange following the colonization of the New World, a substantial interchange of cultivated species occurred between the Old and New Worlds, with a clearly greater migration of grain crops from the Old to the New World (Clement et al. 2015). Currently, the Americas have contributed only a limited number of major grain crops to the world, such as maize, common bean, sunflower, and peanut, in addition to orphan crops like quinoa and amaranth (Khoury et al. 2016). However, at present, only quinoa and amaranth show promising potential for finding niches in the Old World (Kamenya et al. 2021; Hafeez et al. 2022; Kumar et al. 2023), after maize and sunflower.

This asymmetry is reflected in the importance attributed to the impact of G × E interactions among crops from both hemispheres. Our analysis reveals that Old World crops, all monocotyledons, have received a significant impact from public (CGIAR, IRRI, and ICARDA[3]) and private (seed companies) research across all aspects related to the development of G × E interactions. Many of the agronomic, statistical, and computational advancements associated with massive sequencing originate from research conducted on wheat, maize, and rice (Fig. 2.3a); whereas some important, yet non-commodity, crops have received less attention. The contrast is even more pronounced in dicotyledonous cultivated plants. Receiving even less attention, research appears to have primarily focused on pea, followed by soybean, and with some increase in recent years, common bean (Fig. 2.3b). Further behind are orphan crops, which have more recently begun to gain importance for food security. Among these, millet seems to have garnered the most attention; however, quinoa and, more recently, amaranth have increased in relevance (Fig. 2.3c).

[3]Consultative Group on International Agricultural Research (https://www.cgiar.org/)
International Rice Research Institute (https://www.irri.org)
International Center for Agricultural Research in the Dry Areas (https://icarda.org)

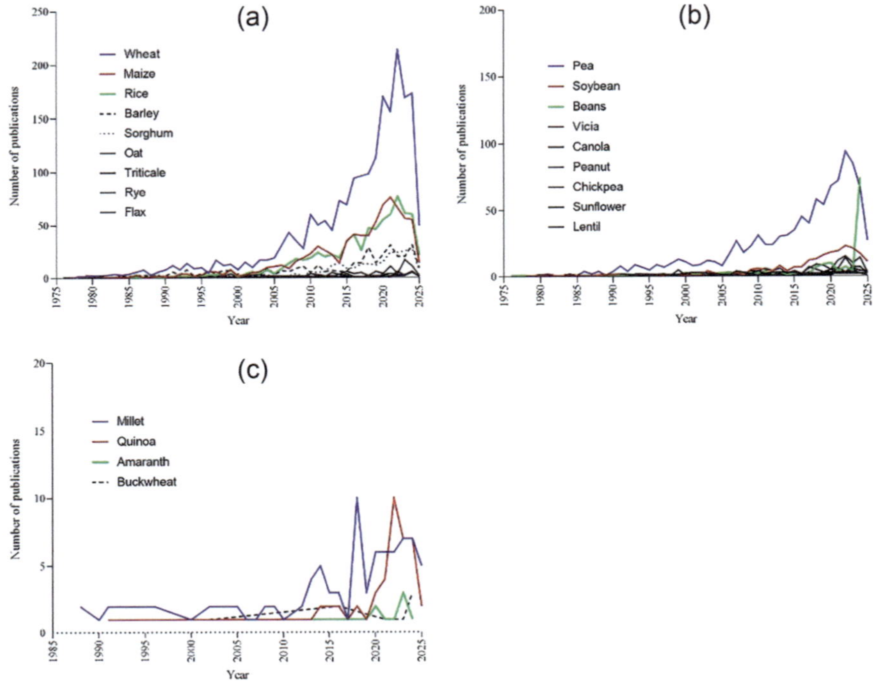

Fig. 2.3 Number of G × E interaction papers categorized according to crops between years 1975 and 2025: (**a**) major and minor cereals crops; (**b**) major pulses and oilseed crops; (**c**) orphan crops

2.3 Defining Genotype-By-Environment Interaction

Genotype-by-environment (G × E) interaction is a fundamental concept in plant breeding and describes the phenomenon where the relative performance of different genotypes varies across different environmental conditions (Cooper and Hammer 1996). This interaction can manifest in several ways, including a change in the ranking order of genotypes from one environment to another, or a change in the magnitude of the differences observed between genotypes without an alteration in their ranking (Ceccarelli et al. 1994; Basford and Cooper 1998).

From a statistical standpoint, G × E can be viewed as genetic variation in trait plasticity, meaning that the extent to which a genotype can alter its phenotype in response to environmental changes is itself genetically determined (Van Eeuwijk et al. 2005; Malosetti et al. 2013). Consequently, populations or traits that lack such inherent genetic variability will not exhibit G × E. Furthermore, G × E signifies a departure from the simple additive effects of heredity and environment, indicating that the combined effect of a particular genotype in a specific environment is not merely the sum of its individual genetic potential and the overall environmental influence (Kang 1998).

Biologically, G × E interactions are evident when the extent to which specific genes contribute to a trait's expression changes depending on the environmental context (Basford and Cooper 1998). Living organisms, including plants, are a product of the continuous interplay between their genetic constitution and the surrounding environment. Phenotypic variations observed in plants are often attributable to the way in which the environment influences the expression and function of the genes that govern various traits (Kang 2002). This results in genotypes exhibiting different levels of phenotypic expression when subjected to different environmental cues.

A crucial aspect of genetic improvement is determining the presence of genotype-by-environment (G × E) interactions, quantifying their magnitude, and categorizing the types of response according to their nature. Furthermore, it is important, once this is accomplished, to ascertain the repeatability of G × E interactions that influence selection decisions. It is well established that G × E interactions are ubiquitous in genotype–environment systems, and their absence is rare. Moreover, the number of G × E interaction types is expected to increase considerably with the number of genotypes and environments evaluated (Allard and Bradshaw 1964).

To exemplify, through a real-world case, the complexity that a target population of environments can exhibit, we present the type of interactions observed when evaluating a significant, albeit not exhaustive, number of genotypes across a limited number of environments. In the following case, 11 genotypes of the species *Amaranthus caudatus* and *Amaranthus mantegazzianum* were evaluated. The *A. caudatus* material consisted of breeding lines (six genotypes), cultivars (three genotypes), and a single landrace, while *A. mantegazzianum* was represented by a single landrace. The evaluation was conducted in two contrasting locations in northwestern Argentina, considering altitude and precipitation and temperature regimes during the crop cycle. The performance graph illustrates the potential existence of a complex network of interactions (Fig. 2.4a), where the different types can vary in their frequency (Fig. 2.4b–d). The absence of interaction was observed, but its frequency was relatively low compared to the present G × E interactions (Fig. 2.4b).

In contrast, interactions exhibiting changes in performance among genotypes were relatively frequent, with heterogeneity of variance across environments being more common in this case than changes in ranking among them (Fig. 2.4c, d). These frequencies were determined for the three cases based on 55 comparisons; however, if the interaction types proposed by Allard and Bradshaw (1964) were considered, the number would increase exponentially.

A major aspect of G × E is its categorization into crossover and non-crossover interactions (Ceccarelli 1989; Cooper and DeLacy 1994; Cooper and Hammer 1996). Crossover interactions occur when the ranking of genotypes changes across different environments, implying that the best genotype in one environment might not be the best in another (Fig. 2.4d). Non-crossover interactions, on the other hand, involve changes in the absolute performance differences between genotypes across environments, but their relative ranking remains consistent (Fig. 2.4c). This distinction is particularly important for plant breeders as crossover interactions often

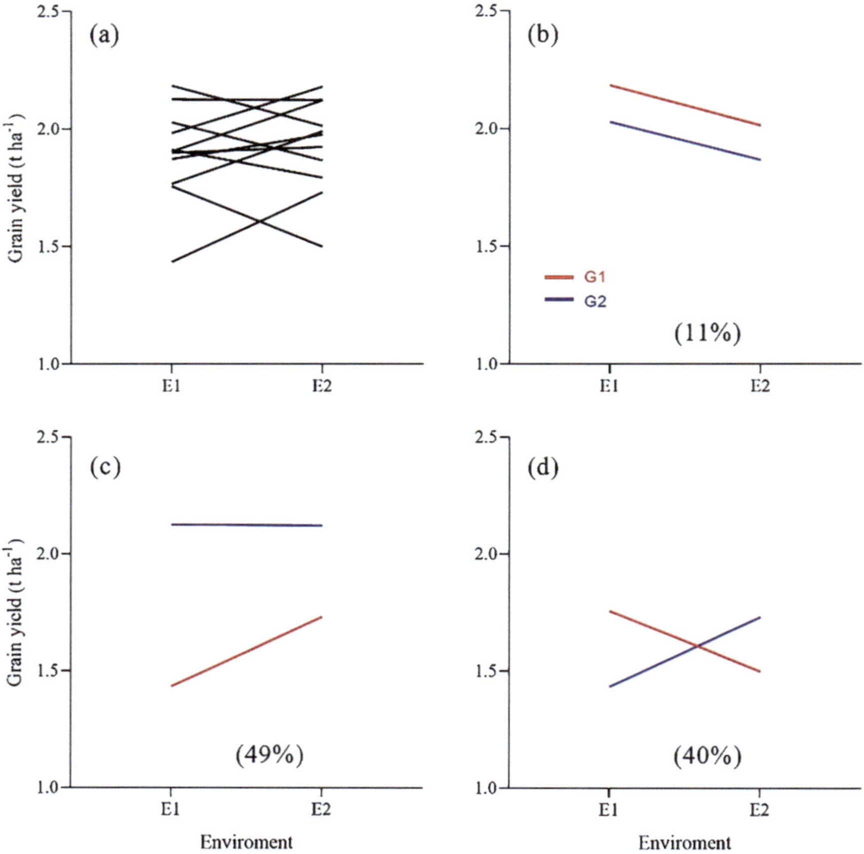

Fig. 2.4 Representation of the performance profiles of *Amaranthus* genotypes evaluated in two types of environments: (**a**) eleven genotypes; (**b**) no G × E interaction; (**c**) G × E interaction due to heterogeneity of variance; (**d**) G × E interaction due to lack of genetic correlation (crossover)

necessitate breeding for specific environments or for broad adaptation strategies, while non-crossover interactions might allow for the identification of genotypes that are generally superior across a wider range of conditions (Annicchiarico 2002, 2007). In essence, G × E interaction highlights the context-dependent nature of genotype performance and poses a significant consideration in the development of improved crop varieties.

2.4 Impact of G × E Interaction on Grain Yield

The interaction between genotype and environment has a profound impact on grain yield, leading to substantial variations in performance across different growing conditions. A particular genotype that demonstrates high grain yield in one environment might exhibit considerably lower yields, or even poor performance, in another (Yan 2016). This phenomenon underscores the challenge faced by plant breeders in identifying genotypes that are consistently high-yielding across a multitude of environments, thereby emphasizing the necessity of conducting multi-environment trials (METs) (Yan 2014).

G × E interaction significantly complicates the selection of genotypes that possess both high yield potential and stable performance (Yue et al. 2025). The presence of G × E often reduces the direct correlation between the observable phenotypic characteristics, such as yield, and the underlying genetic value of a genotype (Bänziger and Cooper 2001). This implies that evaluating a genotype based solely on its performance in a single environment might not accurately reflect its true genetic capacity for yield across a broader range of conditions. To address this complexity, the concepts of stability and adaptability become central in plant breeding programs. Stability, in this context, refers to the ability of a genotype to maintain a relatively consistent level of performance, particularly grain yield, across a diverse array of environmental conditions (Lin et al. 1986; Ceccarelli et al. 1991). Adaptability, on the other hand, describes the capacity of a genotype to perform well in a specific environment or across a range of environments (Basford and Cooper 1998; Annicchiarico 2002). Adaptability can be further categorized as general adaptability, where a genotype performs acceptably well in a wide spectrum of environments, or specific adaptability, where a genotype performs relatively well only in certain, well-defined environments (Ceccarelli 2015). Plant breeders often aim to develop genotypes that exhibit either broad stability, ensuring reliable yields across variable conditions, or specific adaptation to precisely targeted environments, maximizing yield potential in those particular niches (Robles-Zazueta et al. 2024). The analysis of G × E interaction plays a crucial role in identifying genotypes that possess these desirable traits (Bassi et al. 2024).

Several studies across different crops illustrate the significant impact of G × E on grain yield (Kang 2002). In bread wheat, for example, research has consistently shown that grain yield is considerably influenced by the interplay between environments, genotypes, and their interaction (Yue et al. 2025). A recent study identified one genotype as high-yielding in a specific environment, while other demonstrated maximum grain yield when averaged across nine different environments, suggesting its wider adaptability (Mullualem et al. 2024). These findings highlight that different wheat genotypes exhibit varying responses to environmental factors, necessitating comprehensive evaluations across multiple locations and years to pinpoint varieties that are either broadly adaptable and high-yielding or specifically suited to particular conditions, such as rainfed agriculture (Xiong et al. 2021). In rapeseed, stability analysis employing the "additive main effects and multiplicative interaction" (AMMI) model and "genotype main effect and G×E" (GGE) biplots has proven effective in identifying cultivars with consistent high yield across diverse environments, with several genotypes standing out for their stability and high yield (Qasemi et al. 2022). Similarly, in cowpea, significant differences have been observed among genotypes, environments, and G × E for grain yield, with some genotypes showing the most stable performance across different environments (Popoola et al. 2024). This underscores the critical role of G × E in shaping cowpea yields and the importance of selecting stable genotypes for this vital legume, especially in regions characterized by challenging environmental conditions. In summary, G × E interaction is a primary determinant of grain yield in various crops, making the understanding of stability and adaptability in the context of this interaction indispensable for successful plant breeding endeavors.

2.4.1 Influence of G × E on Related Agronomic Traits: Examples from Staple and Minor Crops

Beyond grain yield, genotype-by-environment (G × E) interaction also exerts a significant influence on other crucial agronomic traits that are closely related to grain production. These traits include plant height, flowering time, disease resistance, and grain quality.

Plant height, for instance, is a trait where G × E plays a substantial role. Studies in soybean and maize have shown that G × E contributes significantly to the observed variation in plant height (Azrai et al. 2022; Rani et al. 2023). In wheat, the relationship between plant height and grain yield can be environment-dependent. Research suggests that in rainfed environments, medium-tall wheat genotypes tend to perform better, indicating a positive correlation between plant height and grain yield under such conditions (Singh et al. 2001). This suggests that the optimal plant height for achieving high yields can vary depending on the specific environmental context, particularly concerning water availability. While taller plants might have an advantage in accessing water resources in drier conditions, semi-dwarf varieties might be preferred in other environments due to their increased resistance to lodging and potentially more efficient allocation of resources toward grain development (Singh et al. 2001).

Flowering time, a critical developmental phase in plants, is also subject to phenotypic plasticity and is strongly influenced by the environment (Cooper et al. 1999; Chapman 2008; Zhang et al. 2013; Grogan et al. 2016). In quinoa, for example, studies have indicated that environmental factors explain the largest portion of the total variation in flowering time, with G × E also playing a notable role (Curti et al. 2014). This underscores the sensitivity of flowering time to environmental cues such as temperature and daylength, and how different genotypes can exhibit varying responses to these cues, ultimately affecting their adaptation to specific growing seasons and geographical regions (Curti et al. 2016). While in maize, G × E might have a less pronounced effect on the timing of flowering itself compared to grain yield, research has shown that the genetic relationship, or covariance, between flowering time and grain yield can be significantly altered by the environment (Hudson et al. 2022). This suggests that the optimal flowering time for maximizing yield can be modulated by environmental conditions, potentially revealing environment-specific trade-offs or synergistic effects.

Disease resistance in crops is another important trait where G × E interaction presents a significant challenge in breeding programs. If the relative resistance of different genotypes to a particular disease varies across different environments, it complicates the process of selecting and recommending the most suitable varieties. In rice, while some advanced lines have demonstrated high resistance to blast disease across multiple environments with minimal G × E for blast severity in certain cases (Salleh et al. 2022), other studies have reported significant G × E for disease resistance in various crops (Frahm et al. 2004; Joshi et al. 2007; Sharma et al. 2012; Prasad et al. 2013). This indicates that the effectiveness of disease resistance genes can be environment-dependent. A genotype that exhibits resistance in one location might be susceptible in another due to variations in pathogen races or environmental conditions that favor disease development as was observed in quinoa (Curti et al. 2014). Furthermore, studies on wheat mixtures have shown that genotype diversity can influence disease resistance, but the extent of this effect is often contingent on the specific planting environment (Zhang et al. 2006).

Finally, grain quality is also significantly influenced by G × E interaction, affecting properties such as protein, starch, and oil (Triboi et al. 2006). In maize, for instance, G × E has been shown to significantly impact the protein and fiber content of the grains, with environmental factors often having a more substantial effect than the genetic makeup of the plant (Katsenios et al. 2021). Similarly, in wheat, while genotype might primarily determine grain hardness, environmental factors have been found to be more influential on protein content, wet gluten, and Zeleny sedimentation value (Surma et al. 2012). In sunflower, the magnitude of G × E and G effects on grain were stronger to determine fatty acids composition and oil quality (Aguirrezábal et al. 2015). In soybean, the environment has been identified as the main factor affecting both grain yield and quality traits, although the effects of genotype and G × E are also significant (Monteiro et al. 2021).

2.4.2 Environmental Determinants of G × E Interaction

Genotype-by-environment (G × E) interaction is a complex phenomenon influenced by a multitude of environmental factors. Among the most prominent are climatic conditions, soil properties, biotic stresses, and agricultural management practices. Climatic factors, including temperature, rainfall, humidity, and the duration of sunlight, are significant contributors to G × E in grain yield (Lee et al. 2016; Cooper et al. 2022). For instance, barley has been shown to be particularly sensitive to these conditions during the critical anthesis and grain-filling stages (Bratković et al. 2024). In the North China Plain, winter wheat yield is significantly affected by temperature, relative humidity, and vapor pressure deficit (VPD) (Yue et al. 2025). Similarly, the yield and quality of durum wheat in the Mediterranean basin are influenced by interannual variations in climate (Martínez-Peña et al. 2023).

Soil factors, such as soil type, nutrient availability, and soil water content, also play a crucial role in G × E. Research on maize has demonstrated that low levels of soil nitrogen (N) can significantly reduce grain yield as well as the concentration of iron (Fe) and zinc (Zn) in the grain, with G × E effects being most pronounced under these nutrient-limiting conditions (Akhtar et al. 2023). In wheat, the availability of nutrient contents such as Fe and Zn has been identified as significant contributors to yield variation (Joshi et al. 2010). These findings suggest that the physical and chemical properties of the soil environment interact with the genetic potential of different genotypes to influence their performance.

Biotic factors, including the presence and severity of diseases and pests, can also interact with the genotype to affect grain yield (Newton et al. 2011; Chapman et al. 2012). For example, in Ontario, diseases such as septoria and powdery mildew have been found to be important factors influencing the yield of spring wheat (Gilbert et al. 1998). The effectiveness of a genotype's resistance mechanisms can vary depending on the specific pathogen races present in an environment or the environmental conditions that favor the development and spread of diseases and pests.

Furthermore, agricultural management practices, such as irrigation regimes, sowing dates, and fertilization levels, can also influence G × E (de la Vega and Hall 2002a, b). Studies on durum wheat have included evaluations under different management practices, including irrigation, late sowing, and varying levels of nitrogen fertilization (Cooper et al. 2001). The way a crop is managed can interact with its genetic potential to determine its performance in a given environment (Hatfield and Walthall 2015; Hatfield and Beres 2019). Different genotypes might respond differently to varying levels of inputs or management strategies, highlighting the importance of considering these interactions in optimizing crop production (Duvick and Cassman 1999; Duvick 2005; Rizzo et al. 2022). In summary, understanding these determinants is crucial for developing targeted breeding strategies and optimizing crop management practices for specific environments.

2.5 Methodologies for Studying G × E Interaction

A variety of statistical methods and experimental designs are employed to study and analyze genotype-by-environment (G × E) interaction in plant breeding programs. Among the statistical approaches, analysis of variance (ANOVA) is a foundational technique used to partition the total phenotypic variation observed in multi-environment trials into components attributable to genotype, environment, and their interaction (Crossa 1990). This allows researchers to quantify the relative contribution of each factor to the overall variation. Regression models, such as those proposed by Finlay and Wilkinson (1963) and Eberhart and Russell (1966), are used to assess the stability and adaptability of genotypes by regressing their performance on the mean performance of all genotypes within each environment (Mohammadi et al. 2009, 2021).

The AMMI model is a widely used technique that combines ANOVA with principal component analysis (PCA) to provide a more detailed analysis of G × E (Gauch 1990; Crossa et al. 1991). AMMI utilizes interaction principal component axes (IPCAs) to capture the variance associated with G × E, enabling the identification of genotypes that are stable or specifically adapted to certain environments (Gauch 2006). The GGE biplot is another powerful visualization tool, similar to AMMI, that allows for the joint analysis of the main effects of genotypes and their interaction with environments (Yan et al. 2000; Yan and Kang 2003). GGE biplots are particularly useful for identifying mega-environments and determining which genotypes perform best in which environments, often referred to as "which-won-where" patterns (Yan et al. 2007). More sophisticated methods include "factorial regression," which explicitly models G × E using genotypic and environmental covariates (Voltas et al. 1999a, b; Brancourt-Hulmel et al. 2000; Vargas et al. 2001; Brancourt-Hulmel and Lecomte 2003), and "linear mixed models" (LMM), which can model environment-specific genotype effects and account for the heterogeneity of genetic variances and correlations across different environments (Malosetti et al. 2004; Smith et al. 2005; van Eeuwijk et al. 2016). Partial least squares regression (PLSR) is also employed to identify the most relevant environmental variables that explain the effects of G × E on traits of interest (Vargas et al. 1998; Crossa et al. 1999; Reynolds et al. 2002; Balalić et al. 2012). These statistical methods provide a comprehensive toolkit for dissecting the complex nature of G × E interaction and its impact on crop performance.

The study of G × E interaction heavily relies on well-designed experiments, particularly multi-environment trials (METs) (Basford and Cooper 1998). METs involve evaluating a set of genotypes across multiple locations, years, and under different management conditions. This approach is essential for assessing the stability and wide adaptability of cultivars. In early stages of breeding, Augmented Designs are often used to efficiently screen a large number of genotypes, with standard checks included multiple times for comparison (Ortiz et al. 2023). Within each testing environment, experimental designs such as "randomized complete block design" (RCBD) are commonly used to control for local variability (Damesa et al. 2017). For trials involving a large number of genotypes, incomplete block designs like Alpha-Lattice Designs can be employed (Piepho and Williams 2006). To specifically investigate the response of genotypes to particular environmental stresses, Managed Stress Trials are conducted, where genotypes are exposed to controlled stress conditions such as drought or nutrient deficiency (Bänziger et al. 2006; Weber et al. 2012; Sabadin et al. 2012). Additionally, Common Garden Experiments are used to evaluate phenotypic plasticity and G × E by growing the same genotypes in different environments (Schwinning et al. 2022). The careful selection of appropriate statistical methods and experimental designs is crucial for obtaining meaningful insights into G × E interaction and its implications for plant breeding.

2.5.1 Breeding Strategies for Managing G × E Interaction

Plant breeders employ a variety of strategies to manage the challenges posed by genotype-by-environment (G × E) interaction and to develop stable, high-yielding crop varieties that perform well across diverse environments. One primary strategy is selection for stability, which involves identifying and selecting genotypes that exhibit consistent performance, particularly in terms of grain yield, across a wide range of environmental conditions (Chapman and de la Vega 2002). This often involves the use of stability statistics derived from the analysis of G × E in multi-environment trials. Breeding for stability aims to provide farmers with varieties that offer reliable yields despite environmental fluctuations that may occur from year to year or across different geographical locations (Ceccarelli 2015).

Another key approach is breeding for specific adaptation, which focuses on developing varieties that are optimized for performance in particular target environments or within identified mega-environments (de la Vega et al. 2002). Mega-environments are regions that share similar environmental characteristics. G × E analysis helps in delineating these mega-environments, allowing breeders to tailor their selection and breeding efforts to the specific needs and conditions of these regions, potentially maximizing yield potential in those defined agricultural zones (de la Vega and Chapman 2010).

Cross-breeding and hybridization are traditional yet powerful techniques used to combine desirable traits, including those related to yield, stability, and adaptation to specific environments, from different parent plants (Kang and Priyadarshan 2007). The utilization of indigenous landraces, which have often evolved resilience to local environmental stresses, can be particularly valuable in these programs (Dwivedi et al. 2017). Hybridization can generate new genetic combinations that may exhibit improved performance and stability or enhanced adaptation compared to their parental lines. Modern breeding approaches also leverage marker-assisted selection (MAS) and genomic selection (GS). These techniques use DNA markers that are linked to specific genes or genomic regions associated with traits like yield, stability, and adaptation (Bailey-Serres et al. 2019). By selecting for these markers, breeders can potentially accelerate the breeding process and more efficiently manage G × E by targeting genes that confer stable performance across environments or specific adaptation to certain conditions. This often involves identifying quantitative trait loci (QTLs) that show consistent effects across environments or are specific to particular environmental contexts.

Optimizing multi-environment trial (MET) strategies is crucial for effectively evaluating G × E and identifying superior genotypes. This involves carefully selecting testing locations and years that represent the target growing region and the range of environmental conditions that the crop is likely to encounter. Consideration of both predictable environmental factors, such as soil type, and unpredictable factors, like rainfall patterns, is essential in designing effective METs (Cooper et al. 2022). Breeding for robustness is another strategy that focuses on selecting genotypes that exhibit resilience to environmental fluctuations (Cooper and Messina 2023). This involves identifying traits that allow plants to maintain performance even under variable conditions.

Finally, the use of genotype mixtures, where different genotypes of the same crop are planted together, can potentially improve overall yield and stability at the field level and may also help in reducing disease pressure (Zhang et al. 2024). The idea is that different genotypes within the mixture might perform better under different environmental conditions, leading to a more stable overall yield compared to a monoculture of a single genotype (Moore et al. 2023). In essence, managing G × E interaction and developing improved crop varieties require a multifaceted approach that integrates both traditional and modern breeding techniques, underpinned by a thorough understanding of how genotypes interact with diverse environmental conditions.

2.6 Implications of G × E Interaction for Agriculture and Food Security in a Changing Climate

Genotype-by-environment (G × E) interaction has profound implications for agricultural practices and the overarching goal of ensuring global food security, especially in the face of increasingly volatile and unpredictable environmental conditions brought about by climate change. The inherent inconsistency in the performance of genotypes across different environments, a direct consequence of G × E, presents a significant hurdle for plant breeders in their efforts to recommend and disseminate the most suitable crop varieties (Atlin et al. 2017). The challenge lies in identifying genotypes that not only offer high yield potential but also demonstrate a degree of stability or specific adaptation to the diverse and often fluctuating conditions encountered in agricultural settings.

The escalating challenges posed by climate change, characterized by increased variability in temperature and rainfall patterns, serve to exacerbate the complexities introduced by G × E (Bassi et al. 2024). Unpredictable weather events, such as prolonged droughts, intense heat waves, and severe flooding, can differentially affect the performance of various genotypes (Blum 2018). This increased environmental uncertainty underscores the critical importance of understanding and effectively managing G × E to safeguard crop production and maintain a stable food supply. As climate patterns continue to shift, genotypes that were once well adapted to a particular region may no longer be optimal, necessitating the development and adoption of new varieties that possess either broader adaptability or specific traits that enable them to thrive under the altered environmental conditions.

A thorough understanding of G × E is therefore essential for identifying genotypes that exhibit high levels of adaptability and stability. These traits are paramount for ensuring food security across the diverse agroecological zones that exist worldwide (Wu 2023). Breeding efforts must increasingly focus on developing crops that display resilience to the adverse effects of climate change, including the ability to maintain high yields under various stress conditions. This requires a deep and nuanced understanding of how different genotypes interact with these changing environmental pressures.

Multi-environment trials (METs) play a vital role in this endeavor by providing the necessary data to identify the best-performing genotypes across a range of agroecological locations, ultimately contributing to global food security (Kelly et al. 2007). Furthermore, the concept of mega-environments, which are identified through the analysis of G × E, allows for the development of targeted breeding strategies that are tailored to the specific environmental conditions of particular regions. This approach can significantly enhance the efficiency of breeding programs and improve food security by ensuring that farmers have access to varieties that are best suited to their local environments (Ceccarelli 2015). In conclusion, in an era defined by a rapidly changing climate and increasing environmental variability, the ability to comprehend and effectively manage G × E interaction is more critical than ever for sustaining agricultural productivity and securing the future of global food supplies. Breeding programs must prioritize the development of resilient and adaptable crop varieties that can perform reliably across a wide spectrum of environmental conditions.

2.7 Conclusion and Future Directions

Genotype-by-environment (G × E) interaction stands as a central concept in plant breeding, significantly influencing grain yield and related agronomic traits. The complex interplay between a plant's genetic makeup and the environment in which it grows dictates its performance, posing considerable challenges for developing universally superior crop varieties. Environmental factors such as climate, soil conditions, and biotic stresses play crucial roles in shaping these interactions, and the increasing variability due to climate change further exacerbates the complexities, threatening agricultural productivity and food security.

The methodologies employed to study G × E interaction, including a range of statistical models like ANOVA, regression analysis, AMMI, and GGE biplots, along with carefully designed experiments such as multi-environment trials, pro-vide valuable tools for understanding these intricate relationships. Furthermore, plant breeders utilize diverse strategies, from selecting for stability and specific adaptation to employing modern genomic tools, to mitigate the negative impacts of G × E and develop improved crop varieties. Case studies across major crops like wheat, rice, maize, and soybean underscore the pervasive nature of G × E and the importance of considering it in breeding programs.

Future research should focus on delving deeper into the genetic and physiological mechanisms that underlie G × E for various traits and crops. The development of more sophisticated statistical models capable of predicting G × E under novel and changing environmental conditions is also crucial. Increased integration of genomic tools and high-throughput phenotyping technologies holds immense potential for accelerating the breeding of climate-resilient varieties with stable yields across diverse environments. Additionally, exploring ecological approaches such as genotype mixtures and investigating the interaction between G × E and agricultural management practices offer promising avenues for enhancing crop production in the face of environmental variability. Continued efforts in these areas will be essential to ensure sustainable agriculture and global food security in a rapidly changing world.

References

Aguirrezábal L, Martre P, Pereyra-Irujo G, Echarte MM, Izquierdo N (2015) Chapter 17 – Improving grain quality: ecophysiological and modeling tools to develop management and breeding strategies. In: Sadras VO, Calderini DF (eds) Crop physiology, 2nd edn. Academic, San Diego, pp 423–465

Akhtar S, Mekonnen TW, Osthoff G, Mashingaidz K, Labuschagne M (2023) Genotype by environment interaction in grain iron and zinc concentration and yield of maize hybrids under low nitrogen and optimal conditions. Plants (Basel) 12(7):1463. https://doi.org/10.3390/plants12071463

Allard RW, Bradshaw AD (1964) Implications of genotype-environmental interactions in applied plant breeding. Crop Sci 4(5):cropsci1964.0011183X000400050021x

Annicchiarico P (2002) Defining adaptation strategies and yield-stability targets in breeding programmes. In: Quantitative genetics, genomics and plant breeding, pp 365–383

Annicchiarico P (2007) Wide- versus specific-adaptation strategy for lucerne breeding in northern Italy. Theor Appl Genet 114(4):647–657. https://doi.org/10.1007/s00122-006-0465-1

Atlin GN, Cairns JE, Das B (2017) Rapid breeding and varietal replacement are critical to adaptation of cropping systems in the developing world to climate change. Glob Food Sec 12: 31–37. https://doi.org/10.1016/j.gfs.2017.01.008

Atlin GN, Cooper M, Bjørnstad Å (2001) A comparison of formal and participatory breeding approaches using selection theory. Euphytica 122:463–475

Azrai M, Efendi R, Muliadi A, Aqil M, Suwarti, Zainuddin B, Syam A, Junaedi, Syah UT, Dermail A, Marwiyah S, Suwarno WB (2022) Genotype by environment interaction on tropical maize hybrids under normal irrigation and waterlogging conditions. Front Sustain Food Syst 6. https://doi.org/10.3389/fsufs.2022.913211

Bailey-Serres J, Parker JE, Ainsworth EA, Oldroyd GED, Schroeder JI (2019) Genetic strategies for improving crop yields. Nature 575(7781):109–118. https://doi.org/10.1038/s41586-019-1679-0

Balalić I, Zorić M, Branković G, Terzić S, Crnobarac J (2012) Interpretation of hybrid×sowing date interaction for oil content and oil yield in sunflower. Field Crop Res 137:70–77. https://doi.org/10.1016/j.fcr.2012.08.005

Balzarini M (2002) Applications of mixed models in plant breeding. In: Kang MS (ed) Quantitative genetics, genomics and plant breeding, 1st edn. CABI Publishing, Wallingford, pp 353–363

Bänziger M, Cooper M (2001) Breeding for low input conditions and consequences for participatory plant breeding examples from tropical maize and wheat. Euphytica 122:503–519

Bänziger M, Setimela PS, Hodson D, Vivek B (2006) Breeding for improved abiotic stress tolerance in maize adapted to southern Africa. Agric Water Manag 80(1):212–224. https://doi.org/10.1016/j.agwat.2005.07.014

Basford KE, Cooper M (1998) Genotype×environment interactions and some considerations of their implications for wheat breeding in Australia. Aust J Agric Res 49:153–174

Basford KE, Federer WT, DeLacy IH (2004) Mixed model formulations for multi-environment trials. Agron J 96(1):143–147. https://doi.org/10.2134/agronj2004.1430

Bassi FM, Sanchez-Garcia M, Ortiz R (2024) What plant breeding may (and may not) look like in 2050? Plant Genome 17(1):e20368. https://doi.org/10.1002/tpg2.20368

Blum A (2018) Plant breeding for stress environments, 1st edn. CRC Press

Brancourt-Hulmel M, Denis J-B, Lecomte C (2000) Determining environmental covariates which explain genotype environment interaction in winter wheat through probe genotypes and biadditive factorial regression. Theor Appl Genet 100(2):285–298. https://doi.org/10.1007/s001220050038

Brancourt-Hulmel M, Lecomte C (2003) Effect of environmental variates on genotype environment interaction of winter wheat: a comparison of biadditive factorial regression to AMMI. Crop Sci 43:608–617

Bratković K, Luković K, Perišić V, Savić J, Maksimović J, Adžić S, Rakonjac A, Matković Stojšin M (2024) Interpreting the interaction of genotype with environmental factors in barley using

partial least squares regression model. Agronomy 14(1):194. https://doi.org/10.3390/agronomy14010194
Ceccarelli S (2009) Evolution, plant breeding and biodiversity. J Agric Environ Int Dev 103:131–145
Ceccarelli S (1989) Wide adaptation: how wide? Euphytica 40(3):197–205. https://doi.org/10.1007/BF00024512
Ceccarelli S (2015) Efficiency of plant breeding. Crop Sci 55(1):87–97. https://doi.org/10.2135/cropsci2014.02.0158
Ceccarelli S, Acevedo E, Grando S (1991) Breeding for yield stability in unpredictable environments: single traits, interaction between traits, and architecture of genotypes. Euphytica 56(2):169–185. https://doi.org/10.1007/BF00042061
Ceccarelli S, Erskine W, Hamblin J, Grando S (1994) Genotype by environment interaction and international breeding programmes. Ex Agric 30(02):177. https://doi.org/10.1017/S0014479700024121
Chapman SC (2008) Use of crop models to understand genotype by environment interactions for drought in real-world and simulated plant breeding trials. Euphytica 161(1):195–208. https://doi.org/10.1007/s10681-007-9623-z
Chapman SC, Chakraborty S, Dreccer MF, Howden SM (2012) Plant adaptation to climate change—opportunities and priorities in breeding. Crop Pasture Sci 63(3):251. https://doi.org/10.1071/CP11303
Chapman SC, de la Vega AJ (2002) Spatial and seasonal effects confounding interpretation of sunfower yields in Argentina. Field Crop Res 73:107–120
Clement CR, Denevan WM, Heckenberger MJ, Junqueira AB, Neves EG, Teixeira WG, Woods WI (2015) The domestication of Amazonia before European conquest. Proc R Soc B Biol Sci 282(1812):20150813. https://doi.org/10.1098/rspb.2015.0813
Cooper M, DeLacy IH (1994) Relationships among analytical methods used to study genotypic variation and genotype-by-environment interaction in plant breeding multi-environment experiments. Theor Appl Genet 88(5):561–572. https://doi.org/10.1007/BF01240919
Cooper M, Hammer GL (1996) Plant adaptation and crop improvement. IRRI
Cooper M, Messina CD (2023) Breeding crops for drought-affected environments and improved climate resilience. Plant Cell 35(1):162–186. https://doi.org/10.1093/plcell/koac321
Cooper M, Messina CD, Tang T, Gho C, Powell OM, Podlich DW, Technow F, Hammer GL (2022) Predicting genotype × environment × management (G × E × M) interactions for the design of crop improvement strategies. In: Plant breeding reviews. Wiley, pp 467–585
Cooper M, Rajatasereekul S, Immark S, Fukai S, Basnayake J (1999) Rainfed lowland rice breeding strategies for Northeast Thailand. I. Genotypic variation and genotype environment interactions for grain yield. Field Crop Res 64:131–151
Cooper M, Woodruff DR, Phillips IG, Basford KE, Gilmour AR (2001) Genotype-by-management interactions for grain yield and grain protein concentration of wheat. Field Crop Res 69(1):47–67. https://doi.org/10.1016/S0378-4290(00)00131-3
Crossa J (1990) Statistical analyses of multilocation trials. In: Advances in agronomy. Elsevier, pp 55–85
Crossa J, Fox PN, Pfeiffer WH, Rajaram S, Gauch HG (1991) AMMI adjustment for statistical analysis of an international wheat yield trial. Theor Appl Genet 81(1):27–37. https://doi.org/10.1007/BF00226108
Crossa J, Vargas M, Van Eeuwijk FA, Jiang C, Edmeades GO, Hoisington D (1999) Interpreting genotype × environment interaction in tropical maize using linked molecular markers and environmental covariables. Theor Appl Genet 99(3–4):611–625. https://doi.org/10.1007/s001220051276
Curti RN, de la Vega AJ, Andrade AJ, Bramardi SJ, Bertero HD (2014) Multi-environmental evaluation for grain yield and its physiological determinants of quinoa genotypes across Northwest Argentina. Field Crop Res 166:46–57. https://doi.org/10.1016/j.fcr.2014.06.011

Curti RN, de la Vega AJ, Andrade AJ, Bramardi SJ, Bertero HD (2016) Adaptive responses of quinoa to diverse agro-ecological environments along an altitudinal gradient in North West Argentina. Field Crop Res 189:10–18. https://doi.org/10.1016/j.fcr.2016.01.014

Damesa TM, Möhring J, Worku M, Piepho H (2017) One step at a time: stage-wise analysis of a series of experiments. Agron J 109(3):845–857. https://doi.org/10.2134/agronj2016.07.0395

de Casas RR, Martín-Brull I, Milla R, Ocaña-Calahorro FJ (2024) Selection for domestication favored taxa characterized by fast growth and tolerance of high intraspecific density. PLANTS, PEOPLE, PLANET n/a(n/a). https://doi.org/10.1002/ppp3.10596

de La Vega AJ, Chapman SC (2010) Mega-environment differences affecting genetic progress for yield and relative value of component traits. Crop Sci 50(2):574–583. https://doi.org/10.2135/cropsci2009.04.0209

de la Vega AJ, Hall AJ (2002a) Effects of planting date, genotype, and their interactions on sunflower yield: I. Determinants of oil-corrected grain yield. Crop Sci 42:1191–1201

de la Vega AJ, Hall AJ (2002b) Effects of planting date, genotype, and their interactions on sunflower yield: II. Components of oil yield. Crop Sci 42:1202–1210

de La Vega AJ, Hall AJ, Kroonenberg PM (2002) Investigating the physiological bases of predictable and unpredictable genotype by environment interactions using three-mode pattern analysis. Field Crop Res 78(2–3):165–183. https://doi.org/10.1016/S0378-4290(02)00123-5

Duvick DN (2005) The contribution of breeding to yield advances in maize (Zea mays L.). In: Advances in agronomy. Academic, pp 83–145

Duvick DN, Cassman KG (1999) Post–green revolution trends in yield potential of temperate maize in the North-Central United States. Crop Sci 39(6):1622–1630. https://doi.org/10.2135/cropsci1999.3961622x

Dwivedi SL, van Bueren ETL, Ceccarelli S, Grando S, Upadhyaya HD, Ortiz R (2017) Diversifying food systems in the pursuit of sustainable food production and healthy diets. Trends Plant Sci 22(10):842–856. https://doi.org/10.1016/j.tplants.2017.06.011

Eberhart SA, Russell WA (1966) Stability parameters for comparing varieties. Crop Sci 6(1). https://doi.org/10.2135/cropsci1966.0011183X000600010011x

Elias AA, Robbins KR, Doerge RW, Tuinstra MR (2016) Half a century of studying genotype × environment interactions in plant breeding experiments. Crop Sci 56(5):2090–2105. https://doi.org/10.2135/cropsci2015.01.0061

Evans LT (1998) Feeding the ten billion: plants and population growth. Cambridge University Press

Finlay KW, Wilkinson GN (1963) The analysis and adaptation in a plant-breeding programme. Aust J Agric Res 14:742–754

Frahm MA, Rosas JC, Mayek-Pérez N, López-Salinas E, Acosta-Gallegos JA, Kelly JD (2004) Breeding beans for resistance to terminal drought in the lowland tropics. Euphytica 136(2): 223–232. https://doi.org/10.1023/B:euph.0000030671.03694.bb

Gauch HG (1990) Full and reduced models for yield trials. Theor Appl Genet 80(2):153–160. https://doi.org/10.1007/BF00224379

Gauch HG (2006) Statistical analysis of yield trials by AMMI and GGE. Crop Sci 46(4): 1488–1500. https://doi.org/10.2135/cropsci2005.07-0193

Gilbert J, Woods SM, Tekauz A (1998) Relationship between environmental variables and the prevalence and isolation frequency of leaf-spotting pathogens in spring wheat. Can J Plant Pathol 20(2):158–164. https://doi.org/10.1080/07060669809500421

Grogan SM, Anderson J, Baenziger PS, Frels K, Guttieri MJ, Haley SD, Kim K-S, Liu S, McMaster GS, Newell M, Prasad PVV, Reid SD, Shroyer KJ, Zhang G, Akhunov E, Byrne PF (2016) Phenotypic plasticity of winter wheat heading date and grain yield across the US Great Plains. Crop Sci 56(5):2223–2236. https://doi.org/10.2135/cropsci2015.06.0357

Hafeez MB, Iqbal S, Li Y, Saddiq MS, Basra SMA, Zhang H, Zahra N, Akram MZ, Bertero D, Curti RN (2022) Assessment of phenotypic diversity in the USDA collection of quinoa links genotypic adaptation to germplasm origin. Plants 11(6):738. https://doi.org/10.3390/plants11060738

Hatfield JL, Beres BL (2019) Yield gaps in wheat: path to enhancing productivity. Front Plant Sci 10. https://doi.org/10.3389/fpls.2019.01603

Hatfield JL, Walthall CL (2015) Meeting global food needs: realizing the potential via genetics × environment × management interactions. Agron J 107(4):1215–1226. https://doi.org/10.2134/agronj15.0076

Hudson AI, Odell SG, Dubreuil P, Tixier M-H, Praud S, Runcie DE, Ross-Ibarra J (2022) Analysis of genotype-by-environment interactions in a maize mapping population. G3 (Bethesda) 12(3): jkac013. https://doi.org/10.1093/g3journal/jkac013

Jackson P, Robertson M, Cooper M, Hammer G (1996) The role of physiological understanding in plant breeding; from a breeding perspective. Field Crop Res 49(1):11–37. https://doi.org/10.1016/S0378-4290(96)01012-X

Joshi AK, Crossa J, Arun B, Chand R, Trethowan R, Vargas M, Ortiz-Monasterio I (2010) Genotype×environment interaction for zinc and iron concentration of wheat grain in eastern Gangetic plains of India. Field Crop Res 116(3):268–277. https://doi.org/10.1016/j.fcr.2010.01.004

Joshi AK, Ortiz-Ferrara G, Crossa J, Singh G, Sharma RC, Chand R, Parsad R (2007) Combining superior agronomic performance and terminal heat tolerance with resistance to spot blotch (Bipolaris sorokiniana) of wheat in the warm humid Gangetic Plains of South Asia. Field Crop Res 103(1):53–61. https://doi.org/10.1016/j.fcr.2007.04.010

Kamenya SN, Mikwa EO, Song B, Odeny DA (2021) Genetics and breeding for climate change in orphan crops. Theor Appl Genet 134(6):1787–1815. https://doi.org/10.1007/s00122-020-03755-1

Kang MS (1998) Using genotype-by-environment interaction for crop cultivar development. Adv Agron 62:199–252

Kang MS (ed) (2002) Quantitative genetics, genomics, and plant breeding. CABI Pub, Oxon/New York

Kang MS, Priyadarshan PM (2007) Breeding major food staples, 1st edn. Wiley

Katsenios N, Sparangis P, Chanioti S, Giannoglou M, Leonidakis D, Christopoulos MV, Katsaros G, Efthimiadou A (2021) Genotype × environment interaction of yield and grain quality traits of maize hybrids in Greece. Agronomy 11(2):357. https://doi.org/10.3390/agronomy11020357

Kelly AM, Smith AB, Eccleston JA, Cullis BR (2007) The accuracy of varietal selection using factor analytic models for multi-environment plant breeding trials. Crop Sci 47(3):1063–1070. https://doi.org/10.2135/cropsci2006.08.0540

Khoury CK, Achicanoy HA, Bjorkman AD, Navarro-Racines C, Guarino L, Flores-Palacios X, Engels JMM, Wiersema JH, Dempewolf H, Sotelo S, Ramírez-Villegas J, Castañeda-Álvarez NP, Fowler C, Jarvis A, Rieseberg LH, Struik PC (2016) Origins of food crops connect countries worldwide. Proc R Soc B Biol Sci 283(1832):20160792. https://doi.org/10.1098/rspb.2016.0792

Kumar B, Singh AK, Bahuguna RN, Pareek A, Singla-Pareek SL (2023) Orphan crops: a genetic treasure trove for hunting stress tolerance genes. Food Energy Secur 12(2):e436. https://doi.org/10.1002/fes3.436

Langridge P, Braun H, Hulke B, Ober E, Prasanna BM (2021) Breeding crops for climate resilience. Theor Appl Genet 134(6):1607–1611. https://doi.org/10.1007/s00122-021-03854-7

Lee EA, Deen W, Hooyer ME, Chambers A, Parkin G, Gordon R, Singh AK (2016) Involvement of year-to-year variation in thermal time, solar radiation and soil available moisture in genotype-by-environment effects in maize. Crop Sci 56(5):2180–2192. https://doi.org/10.2135/cropsci2015.04.0231

de Leon N, Jannink J-L, Edwards JW, Kaeppler SM (2016) Introduction to a special issue on genotype by environment interaction. Crop Sci 56(5):2081–2089. https://doi.org/10.2135/cropsci2016.07.0002in

Lin CS, Binns MR, Lefkovitch LP (1986) Stability analysis: where do we stand? Crop Sci 26(5): 894–900. https://doi.org/10.2135/cropsci1986.0011183X002600050012x

Lobell D, Burke M (eds) (2010) Climate change and food security. Springer Netherlands, Dordrecht
Malosetti M, Ribaut J-M, Van Eeuwijk FA (2013) The statistical analysis of multi-environment-data: modeling genotype-by-environment interaction and its genetic basis. Front Physiol 4. https://doi.org/10.3389/fphys.2013.00044
Malosetti M, Voltas J, Romagosa I, Ullrich SE, Eeuwijk FAV (2004) Mixed models including environmental covariables for studying QTL by environment interaction. Euphytica 137(1): 139–145. https://doi.org/10.1023/B:EUPH.0000040511.46388.ef
Martínez-Peña R, Rezzouk FZ, Díez-Fraile M d C, Nieto-Taladriz MT, Araus JL, Aparicio N, Vicente R (2023) Genotype-by-environment interaction for grain yield and quality traits in durum wheat: identification of ideotypes adapted to the Spanish region of Castile and León. Eur J Agron 151:126951. https://doi.org/10.1016/j.eja.2023.126951
Mohammadi R, Haghparast R, Amri A, Ceccarelli S (2009) Yield stability of rainfed durum wheat and GGE biplot analysis of multi-environment trials. Crop Pasture Sci 61(1):92–101. https://doi.org/10.1071/CP09151
Mohammadi R, Sadeghzadeh B, Poursiahbidi MM, Ahmadi MM (2021) Integrating univariate and multivariate statistical models to investigate genotype × environment interaction in durum wheat. Ann Appl Biol 178(3):450–465. https://doi.org/10.1111/aab.12648
Monteiro FF, Bruzi AT, Silva KB, Pulcinelli CE, Soares IO, Carvalho MLM (2021) Breeding for yield and seed quality in soybean. Euphytica 217(12):212. https://doi.org/10.1007/s10681-021-02943-4
Moore VM, Peters T, Schlautman B, Brummer EC (2023) Toward plant breeding for multicrop systems. Proc Natl Acad Sci U S A 120(14):e2205792119. https://doi.org/10.1073/pnas.2205792119
Mullualem D, Tsega A, Mengie T, Fentie D, Kassa Z, Fassil A, Wondaferew D, Gelaw TA, Astatkie T (2024) Genotype-by-environment interaction and stability analysis of grain yield of bread wheat (Triticum aestivum L.) genotypes using AMMI and GGE biplot analyses. Heliyon 10(12):e32918. https://doi.org/10.1016/j.heliyon.2024.e32918
Newton AC, Johnson SN, Gregory PJ (2011) Implications of climate change for diseases, crop yields and food security. Euphytica 179(1):3–18. https://doi.org/10.1007/s10681-011-0359-4
Ortiz R, Reslow F, Huicho J, Vetukuri RR, Crossa J (2023) Optimizing multi-environment testing in potato breeding: using heritability estimates to determine number of replications, sites, and years for field trials. Euphytica 219(4):48. https://doi.org/10.1007/s10681-023-03174-5
Piepho H-P (1998) Empirical best linear unbiased prediction in cultivar trials using factor-analytic variance-covariance structures. Theor Appl Genet 97(1–2):195–201. https://doi.org/10.1007/s001220050885
Piepho HP, Möhring J, Melchinger AE, Büchse A (2008) BLUP for phenotypic selection in plant breeding and variety testing. Euphytica 161(1–2):209–228. https://doi.org/10.1007/s10681-007-9449-8
Piepho HP, Williams ER (2006) A comparison of experimental designs for selection in breeding trials with nested treatment structure. Theor Appl Genet 113(8):1505–1513. https://doi.org/10.1007/s00122-006-0398-8
Popoola BO, Ongom PO, Mohammed SB, Togola A, Ishaya DJ, Bala G, Fatokun C, Boukar O (2024) Assessing the impact of genotype-by-environment interactions on agronomic traits in elite cowpea lines across agro-ecologies in Nigeria. Agronomy 14(2):263. https://doi.org/10.3390/agronomy14020263
Prasad R, Prasad LC, Chand R, Joshi AK (2013) Assessment of diversity for resistance to spot blotch disease and its association with certain phenotypic traits in barely. Field Crop Res 154: 195–200. https://doi.org/10.1016/j.fcr.2013.08.008
Qasemi SH, Mostafavi K, Khosroshahli M, Bihamta MR, Ramshini H (2022) Genotype and environment interaction and stability of grain yield and oil content of rapeseed cultivars. Food Sci Nutr 10(12):4308–4318. https://doi.org/10.1002/fsn3.3023

Rani R, Raza G, Ashfaq H, Rizwan M, Shimelis H, Tung MH, Arif M (2023) Analysis of genotype × environment interactions for agronomic traits of soybean (Glycine max [L.] Merr.) using association mapping. Front Genet 13. https://doi.org/10.3389/fgene.2022.1090994

Reynolds MP, Trethowan R, Crossa J, Vargas M, Sayre KD (2002) Physiological factors associated with genotype by environment interaction in wheat. Field Crop Res 75(2–3):139–160. https://doi.org/10.1016/S0378-4290(02)00023-0

Rizzo G, Monzon JP, Tenorio FA, Howard R, Cassman KG, Grassini P (2022) Climate and agronomy, not genetics, underpin recent maize yield gains in favorable environments. Proc Natl Acad Sci U S A 119(4):e2113629119. https://doi.org/10.1073/pnas.2113629119

Robles-Zazueta CA, Crespo-Herrera LA, Piñera-Chavez FJ, Rivera-Amado C, Aradottir GI (2024) Climate change impacts on crop breeding: targeting interacting biotic and abiotic stresses for wheat improvement. Plant Genome 17(1):e20365. https://doi.org/10.1002/tpg2.20365

Romero AJ, Kolesnikova A, Ezard THG, Charles M, Gutaker RM, Osborne CP, Chapman MA (2025) 'Domesticability': were some species predisposed for domestication? Trends Ecol Evol 40(4):356–363. https://doi.org/10.1016/j.tree.2024.12.007

Sabadin PK, Malosetti M, Boer MP, Tardin FD, Santos FG, Guimarães CT, Gomide RL, Andrade CLT, Albuquerque PEP, Caniato FF, Mollinari M, Margarido GRA, Oliveira BF, Schaffert RE, Garcia AAF, Van Eeuwijk FA, Magalhaes JV (2012) Studying the genetic basis of drought tolerance in sorghum by managed stress trials and adjustments for phenological and plant height differences. Theor Appl Genet 124(8):1389–1402. https://doi.org/10.1007/s00122-012-1795-9

Salleh SB, Rafii MY, Ismail MR, Ramli A, Chukwu SC, Yusuff O, Hasan N (2022) Genotype-by-environment interaction effects on blast disease severity and genetic diversity of advanced blast-resistant rice lines based on quantitative traits. Front Agron 4. https://doi.org/10.3389/fagro.2022.990397

Schaal B (2019) Plants and people: our shared history and future. Plants People Planet 1(1):14–19. https://doi.org/10.1002/ppp3.12

Schwinning S, Lortie CJ, Esque TC, DeFalco LA (2022) What common-garden experiments tell us about climate responses in plants. J Ecol 110(5):986–996. https://doi.org/10.1111/1365-2745.13887

Sharma M, Kiran Babu T, Gaur PM, Ghosh R, Rameshwar T, Chaudhary RG, Upadhyay JP, Gupta O, Saxena DR, Kaur L, Dubey SC, Anandani VP, Harer PN, Rathore A, Pande S (2012) Identification and multi-environment validation of resistance to fusarium oxysporum f. sp. ciceris in chickpea. Field Crop Res 135:82–88. https://doi.org/10.1016/j.fcr.2012.07.004

Singh RP, Huerta-Espino J, Rajaram S, Crossa J (2001) Grain yield and other traits of tall and dwarf Isolines of modern bread and durum wheats. In: Bedö Z, Láng L (eds) Wheat in a global environment. Springer Netherlands, Dordrecht, pp 579–584

Smith AB, Cullis BR, Thompson R (2005) The analysis of crop cultivar breeding and evaluation trials: an overview of current mixed model approaches. J Agric Sci 143(6):449–462. https://doi.org/10.1017/S0021859605005587

Stamp P, Visser R (2012) The twenty-first century, the century of plant breeding. Euphytica 186(3): 585–591. https://doi.org/10.1007/s10681-012-0743-8

Surma M, Adamski, Tadeusz, Banaszak, Zofia, Kaczmarek, Zygmunt, Kuczyńska, Hideyuki, Majcher, Małgorzata, Ługowska, Bogusława, Obuchowskił, Wiktor, Salmanowicz, Bolesław, Krystkowiak K (2012) Effect of genotype, environment and their interaction on quality parameters of wheat breeding lines of diverse grain hardness. Plant Prod Sci 15(3):192–203. https://doi.org/10.1626/pps.15.192

Triboi E, Martre P, Girousse C, Ravel C, Triboi-Blondel A-M (2006) Unravelling environmental and genetic relationships between grain yield and nitrogen concentration for wheat. Eur J Agron 25(2):108–118. https://doi.org/10.1016/j.eja.2006.04.004

van Eeuwijk FA, Bustos-Korts DV, Malosetti M (2016) What should students in plant breeding know about the statistical aspects of genotype × environment interactions? Crop Sci 56(5): 2119–2140. https://doi.org/10.2135/cropsci2015.06.0375

Van Eeuwijk FA, Malosetti M, Yin X, Struik PC, Stam P (2005) Statistical models for genotype by environment data: from conventional ANOVA models to eco-physiological QTL models. Aust J Agric Res 56(9):883. https://doi.org/10.1071/AR05153

Vargas M, Crossa J, Sayre K, Reynolds M, Ramírez ME, Talbot M (1998) Interpreting genotype × environment interaction in wheat by partial least squares regression. Crop Sci 38(3):679–689. https://doi.org/10.2135/cropsci1998.0011183X003800030010x

Vargas M, Crossa J, Van Eeuwijk F, Sayre KD, Reynolds MP (2001) Interpreting treatment × environment interaction in agronomy trials. Agron J 93(4):949–960. https://doi.org/10.2134/agronj2001.934949x

Voltas J, van Eeuwijk FA, Araus JL, Romagosa I (1999a) Integrating statistical and ecophysiological analyses of genotype by environment interaction for grain Ælling of barley II. Grain growth. Field Crops Res 62:75–84

Voltas J, van Eeuwijk FA, Sombrero A, Lafarga A, Igartua E, Romagosa I (1999b) Integrating statistical and ecophysiological analyses of genotype by environment interaction for grain Ælling of barley I. Individual grain weight. Field Crops Res 62:63–74

Weber VS, Melchinger AE, Magorokosho C, Makumbi D, Bänziger M, Atlin GN (2012) Efficiency of managed-stress screening of elite maize hybrids under drought and low nitrogen for yield under rainfed conditions in Southern Africa. Crop Sci 52(3):1011–1020. https://doi.org/10.2135/cropsci2011.09.0486

Wu A (2023) Modelling plants across scales of biological organisation for guiding crop improvement. Funct Plant Biol 50(6):435–454. https://doi.org/10.1071/FP23010

Xiong W, Reynolds MP, Crossa J, Schulthess U, Sonder K, Montes C, Addimando N, Singh RP, Ammar K, Gerard B, Payne T (2021) Increased ranking change in wheat breeding under climate change. Nat Plants 7(9):1207–1212. https://doi.org/10.1038/s41477-021-00988-w

Xu X, Fonseca De Lima CF, Vu LD, De Smet I (2023) When drought meets heat—a plant omics perspective. Front Plant Sci 14:1250878. https://doi.org/10.3389/fpls.2023.1250878

Xu Y (2016) Envirotyping for deciphering environmental impacts on crop plants. Theor Appl Genet 129(4):653–673. https://doi.org/10.1007/s00122-016-2691-5

Yan W (2016) Analysis and handling of G × E in a practical breeding program. Crop Sci 56(5):2106–2118. https://doi.org/10.2135/cropsci2015.06.0336

Yan W (2014) Crop variety trials: data management and analysis, 1st edn. Wiley, Hoboken

Yan W, Hunt LA, Sheng Q, Szlavnics Z (2000) Cultivar evaluation and mega-environment investigation based on the GGE Biplot. Crop Sci 40(3):597–605. https://doi.org/10.2135/cropsci2000.403597x

Yan W, Kang M (2003) GGE biplot analysis: a graphical tool for breeders, geneticists, and agronomists

Yan W, Kang MS, Ma B, Woods S, Cornelius PL (2007) GGE Biplot vs. AMMI analysis of genotype-by-environment data. Crop Sci 47(2):643–653. https://doi.org/10.2135/cropsci2006.06.0374

Yue H, Wang Y, Chen Z, Zhu J, Behera PP, Liu P, Yang H, Wei J, Bu J, Jiang X, Ma W (2025) Assessing the role of genotype by environment interaction of winter wheat cultivars using envirotyping techniques in North China. Front Plant Sci 16:1538661. https://doi.org/10.3389/fpls.2025.1538661

Zhang C, Du Y, Yang T, Schöb C, Li F (2024) Planting environment affects effectiveness of genotype diversity in increasing biomass and yield of a spring wheat mixture. J Plant Ecol 17(6): rtae076. https://doi.org/10.1093/jpe/rtae076

Zhang H, Berger JD, Milroy SP (2013) Genotype×environment interaction studies highlight the role of phenology in specific adaptation of canola (Brassica napus) to contrasting Mediterranean climates. Field Crop Res 144:77–88. https://doi.org/10.1016/j.fcr.2013.01.006

Zhang Y, He Z, Zhang A, Van Ginkel M, Ye G (2006) Pattern analysis on grain yield performance of Chinese and CIMMYT spring wheat cultivars sown in China and CIMMYT. Euphytica 147(3):409–420. https://doi.org/10.1007/s10681-005-9038-7

QTL × Environmental Interactions Underlying Phenotypic Stability or Plasticity in Crops: Which One to Utilize?

3

Sivakumar Sukumaran and Jose Crossa

Abstract

Stability is a key statistical parameter widely used in multi-environment trials to assess the performance consistency of genotypes, particularly for grain yield. While often applied to easily phenotyped traits, their relevance for complex and labor-intensive traits remains limited. A fundamental question in plant breeding is whether the goal of a breeding program should be to develop broadly stable genotypes (mega-varieties) or genotypes specifically adapted to target population of environments (TPEs). Stability plays a critical role in selecting elite lines for varietal release and breeding advancement, yet its biological basis and genetic mechanisms, remain unknown. Moreover, the value of stability is often weighed against phenotypic plasticity, raising questions about which property better supports resilience and productivity under diverse conditions. This chapter explores the continuum of crop performance across varied environments, reasoning why certain genotypes exhibit greater stability while others display higher plasticity, and highlights emerging insights into the genetic underpinnings of these from recent research.

S. Sukumaran (✉)
Department of Plant and Environmental Sciences, Clemson University, Clemson, SC, USA
e-mail: ssukuma@clemson.edu

J. Crossa
International Maize and Wheat Improvement Center (CIMMYT), Texcoco, Edo. de México, Mexico

Colegio de Postgraduados, Montecillos, Edo. de México, Mexico

R. Mohammadi et al. (eds.), *Genotype x Environment Interactions and its Implications for Plant Breeding*, https://doi.org/10.1007/978-981-95-5664-9_3

3.1 Introduction

Why do certain wheat lines and their derivatives consistently outperform others across diverse environments? Why are some lines successfully released as varieties in multiple countries, while others remain confined to the environments where they were developed? These enduring questions lie at the heart of understanding genotype performance, adaptability, and stability in plant breeding.

Examples from CIMMYT's global wheat breeding program illustrate this phenomenon. **Borlaug F2014*, a spring wheat line released by CIMMYT, has repeatedly shown superior grain yield performance across environments globally. It was released in Australia through the evaluations in CAIGE (CIMMYT Australia ICARDA Germplasm Evaluation) program (Trethowan et al. 2024). Similarly, **Sokoll*, a synthetic wheat derivative, has demonstrated outstanding drought and heat tolerance, and served as a parent for several varieties released in stress-prone regions (Reynolds et al. 2017). Landraces—originally adapted to specific locales—and their crosses with synthetic wheats have also shown broad adaptability, leading to successful varietal releases in distant regions (Rosyara et al. 2019).

These cases raise important questions. What enables certain genotypes, developed under specific environmental conditions, to succeed across diverse agroecologies? Is it merely a matter of climatic similarity between target regions, or do these lines carry sets of alleles that confer broad adaptation and resilience? Selection in a specific environment can indeed lead to the accumulation of favorable alleles of target traits, but does this necessarily translate into wide adaptability and long-term phenotypic stability? Are the so-called "mega-varieties" like *Borlaug F2014* truly stable across environments, or are they simply able to cope with environmental diversity through high phenotypic plasticity?

Over the past decades, plant breeders have increasingly used multi-environment trials (METs) not only to evaluate crop performance but also to study stability and genotype × environment (G × E) interactions (GEIs) through advanced statistical models. Linear–bilinear models such as the Additive Main effects and Multiplicative Interaction (AMMI) model (Gauch 1992), Site Regression (SREG) (Crossa and Cornelius 1997), and Genotype plus Genotype-by-Environment (GGE) biplots (Yan et al. 2000) have been widely applied to dissect the components of yield variation across environments. These models help differentiate genotypes with general (wide) adaptability from those with specific adaptation and allow visualization of G × E interaction patterns.

For instance, AMMI combines analysis of variance (ANOVA) for additive effects with principal component analysis (PCA) for interaction effects, allowing breeders to identify genotypes that are both high-yielding and stable. SREG and GGE focus more on the genotype plus interaction term, aligning better with selection objectives in breeding programs. These models have become standard tools for interpreting MET data and guiding cultivar recommendations in target population of environments (TPEs) (Olivoto and Lúcio 2020; Dias et al. 2022).

Grain yield performance in METs is often used as a proxy for stability (Becker and Léon 1988; Finlay and Wilkinson 1963), but it is not the only trait relevant to adaptation. Traits such as flowering time, plant height, root architecture, stay-green, and canopy temperature also contribute significantly and may exhibit distinct stability or plasticity profiles across environments (Yan et al. 2021; Said et al. 2022). Stability is often the outcome of both favorable allelic combinations for key traits and an optimized developmental pattern that aligns with favorable environmental windows (Xiong et al. 2024). For example, major-effect genes may drive performance in optimal conditions (Nadolska-Orczyk et al. 2017), whereas minor-effect modifiers—such as earliness per se (eps) genes—may only manifest after the main flowering genes (VRN [vernalization], PPD [photoperiod sensitivity]) have acted (Sukumaran et al. 2016).

Despite decades of investigation, critical gaps remain. Why do some genotypes express high yield stability, while others show superior plasticity under environmental stress? Which traits—canopy conductance, phenological flexibility, osmotic adjustment—best explain this behavior? Do some genes consistently contribute to stability while others modulate plastic responses? Recent advances in genomic prediction (GP) models, including reaction norm approaches (Jarquín et al. 2014) and deep learning (Montesinos-López et al. 2022), offer promising avenues to integrate genomic and environmental data for improved understanding of G × E interactions and environmental adaptation.

Much remains to be explored. Why do some genotypes exhibit high stability while others show superior plasticity? Which traits—root architecture, canopy temperature, stay-green, phenology—contribute most to consistent yield under stress versus potential environments? Are there “core” genes conferring stability and others enabling plastic responses? As local adaptation continues to be the mainstay of many national breeding programs, insights into the genetic mechanisms of broad adaptation and trait-by-environment interactions will be essential for resource-efficient and climate-resilient crop improvement. This chapter draws on empirical evidence, theoretical insights, and field-level experience to explore the continuum from stability to plasticity. We aim to discuss a conceptual and methodological framework to better understand how stability and plasticity are shaped by selection, genetic architecture, environment, and their interactions—and how this knowledge can be harnessed for climate-resilient, resource-efficient crop improvement.

3.2 Genetic Unit in Genotype × Environment Interactions

3.2.1 Evolution of Genetic Resolution: What is the Fundamental Unit for Studying Genotype × Environment Interactions (GEIs)?

What is the genetic unit that is responsible for its interactions with the environment and management? Is it a quantitative trait loci (QTL), marker trait associations (MTAs), genes, or haplotypes? The first QTL mapping studies in plants, such as the one by Paterson et al. (1988) in tomato, marked the beginning of identifying genomic regions associated with complex traits. These studies relied on biparental populations and low-density marker systems, leading to QTLs with broad confidence intervals, typically ranging from 5 to 20 centimorgans (cM). While fine mapping could improve resolution, the process was time-consuming and often trait specific.

With advances in genotyping and statistical genomics, genome-wide association studies (GWAS) emerged as a powerful alternative. Although the first high-profile GWAS paper was in humans (Ozaki et al. 2002), its influence quickly extended to plants. Early plant GWAS studies had limited resolution due to low marker density, but this changed rapidly (Aranzana et al. 2005). Modern GWAS now use hundreds of thousands to millions of markers, enabling detection of associations at much finer scales (Boatwright et al. 2022; Huang et al. 2010; Tao et al. 2021; Wen et al. 2014). In some cases, candidate genes can be identified directly from initial GWAS peaks, though functional validation remains essential (Chen et al. 2024; Tian et al. 2011).

Most GEI studies to date are based on QTL × environment interactions (Agha et al. 2024; Bustos-Korts et al. 2018; Lowry et al. 2019; Mahmood et al. 2022). Haplotype × environment interaction studies are rare (Han et al. 2025). Gene × environment interaction studiers are rarer but once gene is detected they are specific (Napier et al. 2023). Chen et al. (2023) studied gene-by-environment interactions using CRISPR-BAR to measure the effect of single genetic variants in high-throughput analysis.

As the limitations of single-marker analysis became evident—particularly in capturing the full allelic diversity and complex linkage patterns—haplotypes began to gain traction. Gore et al. (2009) emphasized the value of haplotypes as biologically meaningful units for understanding genetic variation. Over the past decade, haplotype-based approaches have become widely adopted, especially in breeding programs where recombination blocks are often conserved. Recent studies continue to demonstrate their utility for trait dissection and selection (Tong et al. 2024). This progression—from broad QTL intervals to gene-level associations and haplotype-based analyses—has significantly refined how genetic effects are detected and interpreted in the context of environmental variability.

3.2.2 How Many QTLs Control Key Traits

Many statistical approaches have been developed to assess the G × E effects of QTLs, genes, and haplotypes, including mixed models, reaction norm models, and SREG or GGE biplot analyses. However, very few QTLs are consistently discovered across all environments. This is due to the immense complexity of genetic architecture and its environmental sensitivity. The number of gene–environment combinations is astronomically large, making it practically impossible to resolve all contributing loci even with current technologies. The situation calls for future integration of artificial intelligence (AI) and quantum computing to simulate realistic trait architectures. However, in most of the populations where QTLs are discovered, very few QTLs are discovered in every environment. Any study conducted to discover the QTLs do not discover all QTLs in all environments. It is due to the fact that there are numerous genes, and their interactions among themselves and with the environment may require trillions of combinations of genetic materials and infinitely large populations to discover them. In order to test their effects on traits, no known method exists as of now, unless AI and quantum computing could create such simulations to conduct real-life experiments. For any traits, we have not discovered the complete set of major genes or QTLs yet, even after 100 years of Mendel laying the foundation of genetics.

Flowering time remains one of the best-understood and heritable traits, yet it is also highly plastic. Its expression is influenced by photoperiod, temperature, radiation, and water availability (Guo et al. 2024). In crops like sorghum and wheat, several classical genes (e.g., Ma loci in sorghum; VRN, PPD, and EPS in wheat) have been identified, but the complete set of interacting loci remains unknown (Quinby 1967; Rooney and Aydin 1999; Beales et al. 2007; Santra et al. 2009; Mace et al. 2013; Yan et al. 2003; Sukumaran et al. 2016). If flowering time remains genetically unresolved despite its simplicity and high heritability, determining the genetic basis of highly polygenic traits such as yield becomes an even more complex challenge. This has direct implications for breeding strategies that seek stable and broadly adaptive genotypes.

3.2.3 Are QTLs Proxies for Specific Genetic Pathways or Broad-Effect Loci? Does QTLs Work in Tandem?

QTLs can be categorized based on the stability of their effects across environmental conditions. "Constitutive" QTLs are consistently expressed across most environments, whereas "adaptive" QTLs are expressed only under specific conditions or show increased expression in response to environmental stimuli. Adaptive QTLs often show considerable variability between experiments, underscoring the environmental sensitivity of a given trait. This responsiveness may stem from environmentally regulated gene expressions (e.g., transcriptional changes), or may be the result of indirect effects (Monforte et al. 1997).

A key question in understanding QTLs associated with complex traits is whether they act additively or in parallel. A single QTL may encompass thousands of genes, and it should not be considered the basic unit of trait expression. For instance, if 100 QTLs are identified for a given trait, do all associated genes act sequentially in a single pathway, or do they function across multiple, parallel, and perhaps redundant pathways? If alternative pathways exist and can compensate when one fails, this raises questions about the focus of breeding strategies. The fundamental unit is the individual gene, along with its interactions with other genes and environmental factors. Genes do not function in isolation; instead, they operate within networks or pathways, many of which are environmentally dependent. A single trait may be governed by multiple such pathways, each influenced by varying environmental parameters (Cooper et al. 2021).

Should we prioritize the stability of individual QTLs, or rather identify and leverage multiple QTLs that operate in tandem across diverse environments to promote phenotypic plasticity? In other words, should breeding emphasize genotypes that can adapt across a wide range of environmental conditions rather than ones optimized for specific settings? Or do we need all QTLs present in a genotype, ready to be activated when favorable environmental conditions arise? Therefore, the focus of breeding may need to shift from identifying individual stable QTLs to assembling diverse, interacting QTLs that enable plastic responses to environmental change. Multi-environment models like G × E–SREG, factorial regression, and Bayesian reaction norm models are increasingly used to dissect this complexity (Jarquín et al. 2014; Malosetti et al. 2013; Crossa et al. 2021). Such tools allow breeders to predict genotype responses across TPEs, helping bridge the gap between genotype, phenotype, and environment. A key contribution in this direction was provided by Crossa (2012), who reviewed models for GEI and gene × environment interaction, including QTL × E and M × E models, emphasizing the need for biologically meaningful interpretations in breeding contexts.

3.3 Adaptation vs. Stability

3.3.1 Does Adaptation Lead to Stability or Divergence from It?

Cultivated crops diverged from their wild ancestors through selection by early farmers, leading to their adaptation to specific local environments. Over time, these locally adapted populations became known as landraces or heirloom varieties. Landraces often harbor alleles that confer adaptation to regional environmental conditions (Campbell et al. 2025; Lasky et al. 2015; Morris et al. 2013).

In natural systems, local adaptation is shaped by natural selection and leads to adaptive phenotypic divergence along environmental gradients. This often results in a trade-off—enhanced fitness in one environment may come at the cost of reduced fitness in others (Lowry et al. 2019). Understanding these trade-offs is essential for breeding strategies that aim to balance local adaptation with broader environmental resilience.

A key question arises: does local adaptation constrain a genotype's stability across broader environmental contexts? The alleles that drive local adaptation typically enhance performance across a range of environmental parameters, not a single point. Thus, adaptation should be conceptualized not as a fixed relationship to one environment but as a dynamic trait expressed across an environmental gradient. Crossover interactions—where the performance ranking of genotypes changes with the environment—often signal such gradients and may be driven by variables like water availability, photoperiod, and temperature regimes.

3.3.2 The Finlay–Wilkinson Regression

One of the central challenges in understanding genotype-by-environment (G × E) interaction and plasticity is the selection of appropriate genotypes. The Finlay–

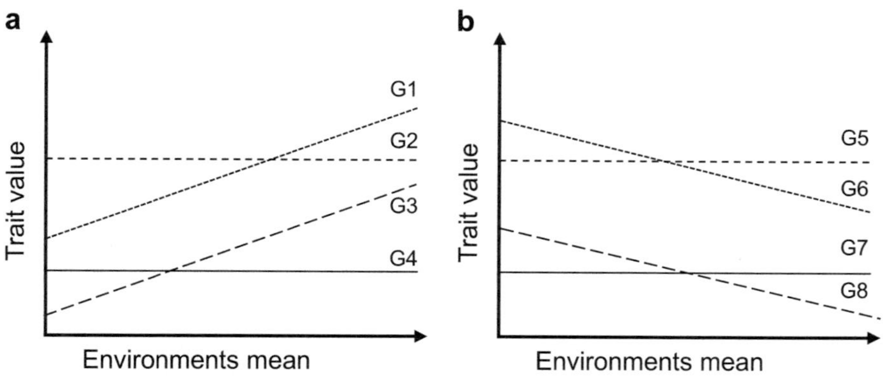

Fig. 3.1 Reaction norms (**a**) of the performance of four varieties over an environmental gradient such as yield, where *x*-axis is the mean yield of the population in an environment in ascending order, G1 is highly plastic variety, G2 is a high-yielding stable variety, G3 is a plastic variety with lower yields, and G4 is low-yielding highly stable variety, and (**b**) shows performance of another four varieties where the weather parameter that adversely affect yield is in ascending order, G5 is stable high-yielding, G6 highly plastic and high-yielding, G7 low-yielding and stable, and G8 low-yielding and plastic

Wilkinson regression remains one of the foundational models in G × E analysis. Finlay and Wilkinson (1963) proposed that, while ideotypes can be designed for individual environments, defining genotypes with broad adaptability is far more complex. Their model regresses genotype yield across environments against environmental means, with the slope (regression coefficient) quantifying the genotype's response. High-yielding genotypes may exhibit either stability or plasticity depending on their slope relative to 1.0.

In this framework, average yield is often used as a surrogate for stability, though this can be misleading—plastic genotypes may outperform others in favorable conditions but underperform in stress-prone environments. Figure 3.1 illustrates an extension of this concept. Traditional breeding prioritizes genotypes with consistent performance across sites, though more recent thinking supports targeted selection for stability within clusters of environments or even exploiting plasticity in variable contexts.

In many cases, linear regression models are used to describe genotype responses to environmental gradients, particularly for traits like yield. However, not all trait–environmental relationships are linear. Non-linear responses are common and biologically meaningful. For instance, grain yield often exhibits a non-linear response to nitrogen inputs—initial increases in nitrogen lead to yield gains, but beyond a certain threshold, additional nitrogen no longer enhances yield and may even lead to negative effects (Puntel et al. 2016; Sinclair and De Wit 1975). These non-linear dynamics underscore the importance of considering both linear and non-linear models when evaluating genotype performance and making selection decisions.

3.3.3 Breeding Goals: Is Broad Stability or Local Adaptation More Desirable?

Should plant breeding prioritize broad adaptability or local optimization? For diseases, which tend to be localized and vary with pathogen races, local adaptation is often more relevant. For abiotic stresses like drought or heat, broader adaptation and plasticity may be preferred (Collins et al. 2008). Most national breeding programs target local environments due to resource constraints, focusing on specific environments for variety selection (Agha et al. 2024). Conversely, international programs aim for wide adaptation to maximize yield and stress tolerance across regions.

For disease resistance, localized adaptation is especially critical. Pathogen populations like yellow rust vary across geographies and seasons, necessitating specific resistance alleles tailored to local contexts. The concept of stable QTLs underpins marker-assisted selection (MAS), which has seen success in both simple and complex traits. For example, MAS has improved disease resistance and even complex traits in crops like chickpea (Varshney et al. 2013, 2014). In this context, identifying QTLs that perform consistently or that respond predictably to environmental cues remains a central goal of molecular breeding.

3.4 What Contributes to Stability?

3.4.1 Stability of Traits Across Environments

A wheat association mapping initiative (WAMI) spring wheat population has been extensively studied for G × E and plasticity (Lopes et al. 2015; Sukumaran et al. 2015, 2017). This population of 288 lines, when grown in several countries, has shown variation in mean yield at the locations and varied responses (Fig. 3.2).

A phenotypic plasticity study was conducted on this population and the effect of environmental parameters on flowering time was established using genomic prediction models and joint genomic regression analysis. Genes related to traits per se and additional genes were found to be associated with stability and plasticity. Since the

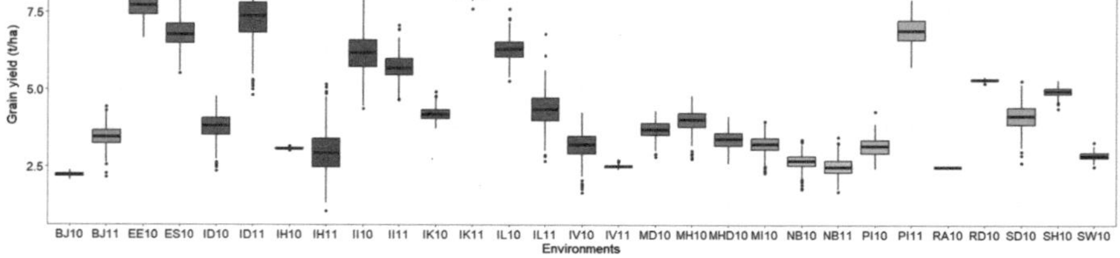

Fig. 3.2 Grain yield performance of WAMI over several environments around the world

traits are the product of GEI continuum, trait per se may not be appropriate term to define QTLs associated with traits in an environment, rather trait is the product of QTL effect in a range of environmental conditions interacting with genotypes and genes are more appropriate (Li et al. 2021).

Genotype–environment interactions are affected by many weather parameters. Recent studies in wheat have shown that climatic factors explained 70% of the year-to-year variability for crossover interactions for yield. Among them, heat and drought stress has contributed to ranking changes of the germplasm (Xiong et al. 2021). Grain hardiness is an important trait for wheat processing and end-product quality. A study detected 153 QTLs, clustered into 12 genomic regions for 13 quality-related traits, out of which, they identified *QGh.cib7D* on wheat grain hardness and further tested its presence in 373 common wheat varieties in different countries. The positive effect of the alleles was detected in 320 out of 373 varieties spreading four continents including Asia, Europe, Oceania, and Africa (Liu et al. 2024).

In a study in maize RIL population, QTLs were mapped and QTL × environmental interactions were studied across seven environments under different water stress scenarios. Among them, 80–6% of the QTLs did not present significant QTL × E interactions in the joint analysis per environment. The stability of the QTLs decreased drastically when the data was combined across water regimes (Messmer et al. 2009). This points to the fact that most of the QTLs detected in some environments could not be detected in another environments. This raises a fundamental question: do the same genes function across different environments, or do each environment activate distinct genetic responses?

3.4.2 Trade-Offs: The Classic Case of Grain Number vs. Grain Weight and Others

Grain number and grain weight are classic traits that are inversely proportional to each other (Calderini et al. 2021; Foulkes et al. 2011; Sukumaran et al. 2018; Vicentin et al. 2024). Compensatory mechanisms may exist in determining the final value of these traits since grain number is decided earlier than grain weight. While in multiple environments where water, photoperiod, and radiation are different, plants might respond through multiple pathways to decide the final grain number and grain weight (Acreche and Slafer 2006). This classic case of thousand-grain weight (TGW) *TaGW2* genes and its interactions in different wheat regions is interesting. Some report the *HapA* allele on chromosome 6A has a positive effect on grain weight; another reported positive effects of *HapG* on grain weight (Liu et al. 2020; Su et al. 2011; Zhang et al. 2018). However, it is one of the genes that is known to have a stable effect on a trait in wheat in addition to the VRN, PPD genes, and grain quality genes.

Spike length is another trait that decides the yield potential and stability in wheat. A stable *QTL qSI-2B* was detected in multiple environments on an RILs' KJ-RIL. This QTL was further tested in another population and a natural mapping population. The results were that the QTL showed inconsistent effects on the two mapping populations. However, selection utilization through *qSI-2B* found positive selection response (Ding et al. 2024). Chromosome substitution lines and RILs allow the effects of the QTLs to be studied individually. One study on wheat chromosome 3A found out QTL × environmental interactions in multiple environments for grain yield and associated traits with pleotropic effects. However, the positive effect of the 3A QTL was larger in high-yielding environments (Campbell et al. 2003). This indicates a range for the QTL effects and that certain environments might have the QTLs expressed while in another it may not be.

3.4.3 The Role of Flowering Time and Thermal Time in Stability

Yield stability across diverse environments is often closely linked to the timing of flowering in relation to local climatic conditions. In many crops, particularly cereals like wheat, aligning the flowering period with the optimal environmental window—when temperature, radiation, and moisture conditions are favorable—is critical to achieving stable yield performance. This concept underpins the rationale for using phenology-based selection in breeding programs. Recent advances in gene-based phenology models, such as those implemented in APSIM, allow for the prediction of wheat heading dates using major genes like *VRN* (vernalization) and *PPD* (photoperiod sensitivity), providing a powerful tool for modeling adaptation and optimizing flowering time under varying environments (Hu et al. 2021). Importantly, it is not just the timing of flowering itself, but also the thermal time and developmental progression from terminal spikelet initiation to anthesis that appears to be crucial. This phase coincides with key processes such as spike development and floret fertility determination, and is highly sensitive to abiotic stresses, especially heat and drought. The availability of water and the absence of stress during this window may ultimately dictate the stability of grain number and, consequently, final yield (Hu et al. 2022).

In addition to wheat, studies in other crops such as *Brassica napus* further underscore the role of flowering time as a stability trait. A genome-wide association study identified 19 stable main-effect quantitative trait nucleotides (QTNs) for flowering time and 32 QTN × environment interactions, highlighting the complexity of flowering regulation under diverse environments (Han et al. 2022). In this study, daily average temperature and precipitation emerged as the most influential environmental covariates associated with variation in flowering time. These findings reflect a broader trend in crop genetics: flowering time is not only a key adaptive trait but also a major integrator of environmental signals. Therefore, its precise control—through marker-assisted or genomic selection targeting phenology genes—offers breeders a strategic lever for improving both adaptation and yield stability. As the frequency and unpredictability of climate stresses increase, particularly during sensitive stages of crop development, understanding and manipulating the interaction between thermal time, phenology genes, and environment will be critical for breeding stable, high-performing cultivars across target environments.

3.5 From Stability to Plasticity: A Continuum of Responses

Plasticity is the ability of single plant genotype to produce different phenotypes in response to environmental stimuli (Pigliucci 2001; Napier et al. 2023). Phenotypic plasticity and *reaction norms* are sometimes used as synonyms. Reaction norm is a fundamental conceptual research tool coined by Woltereck in 1909. It is the function that relates environments to which a particular genotype is exposed and the phenotypes that can be produced by that genotype. Reaction norm models are helpful to map genes and QTLs related to phenotypic plasticity. It has been used successfully in studies to map various traits for phenotypic plasticity (Guo et al. 2024; Mu et al. 2022). However, studies suggest that even the trait like flowering time is determined by many genes with small effects and the effects are below the detection limits of most mapping experiments (Stratton 1998). Large-scale studies with several environments are needed to dissect the phenotypic plasticity in crops. An example is the WAMI population grown at several environments around the world (Fig. 3.3).

Phenotypic plasticity is considered as a major mechanism for evolution rather than a cause of evolution. Some consider phenotypic plasticity as a developmental process facilitating evolution and some consider it an inherent property of genotype (De Jong 2005). Another study suggests that phenotypic plasticity is not a whole plant response, instead a property of individual leaves, meristems, roots, and branches, responding to environmental conditions (De Kroon et al. 2005). It also suggests that the plant's response to the environment is the sum of all responses to local conditions and interaction effects.

Another term that is relevant together with stability, plasticity is *yield reliability* under diverse environments (Sangha et al. 2025). A recent study has shown that it is related to plasticity rather than stability. This proposes several questions such as: Is ideotype breeding—that targets a range of phenotypes (i.e., range of phenotypic values) for breeding—better than stability-based breeding? Do we make more progress by breeding for stability or for more plastic varieties? Do the national agricultural systems and current data analysis approaches support this?

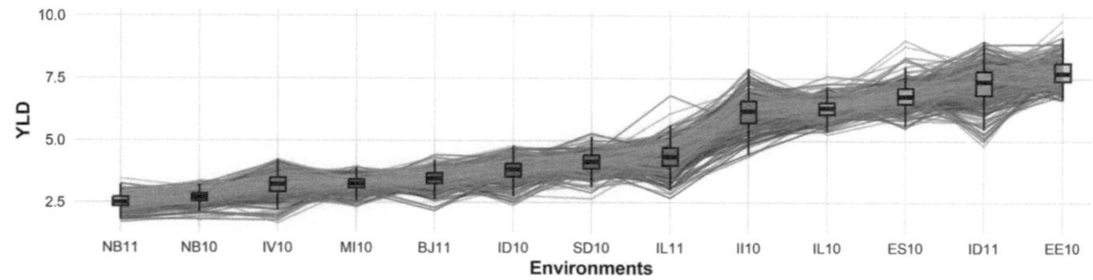

Fig. 3.3 G × E for grain yield for the WAMI population in several environments showing the plasticity of genotypes and ranking change. *YLD* grain yield (t/ha); *x*-axis is different environments

3.5.1 Envirotyping: Utilizing Specific Adaptations for Genetic Studies and Breeding

Understanding how individual genetic loci contribute to trait variation is crucial for studying both local adaptation and broad adaptation across environments. However, loci that are beneficial in one region may have little to no effect in others (Lowry et al. 2019). Envirotyping, when combined with multi-environment trials (MET), has proven effective in identifying stable loci across diverse environments, as demonstrated in rapeseed (Corlouer et al. 2024).

The increasing availability of large-scale phenotypic datasets and geographic information systems (GIS) enhances our ability to identify genes associated with specific environmental parameters—particularly those affecting crop wild relatives and landraces (Xu 2016). A recent study in soybean, conducted across 602 environments spanning from France to Russia, found that envirotyping explained 88% of the genotype-by-environment interactions for yield (Elmerich et al. 2023). This highlights the importance of incorporating both specific and broad adaptations in breeding programs.

Envirotyping is also emerging as a powerful tool for identifying potential parental lines suited to future climates. By leveraging climatic models and genomic prediction, it enables more targeted use of genetic resources conserved in gene banks (Campbell et al. 2025). In grain sorghum, for instance, envirotyping has significantly improved genomic prediction accuracy across mega-environments. One study reported the highest prediction accuracy for sorghum hybrids using envirotyping-assisted genomic prediction (Winans et al. 2024).

To further explore the genetic basis of adaptation using envirotyping, software tools like *EnvRtype* offer valuable support by integrating genomic and environmental data (Costa-Neto et al. 2021b). Our recent studies also reveal that minimum nighttime temperature at different stages of crop growth plays a key role in the stability of spring wheat yield across multiple environments (Fig. 3.4), underscoring

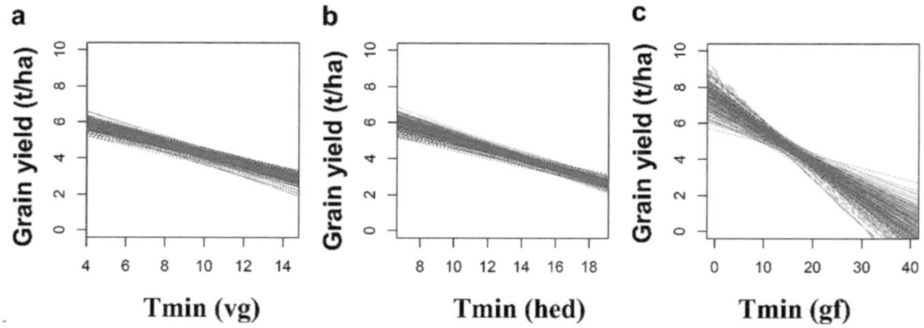

Fig. 3.4 Temperature effect on grain yield in the WAMI population. Each line is the slope of the reaction norm model where temperature is regressed over grain yield (t/ha) in several environments, showing negative effect of weather parameter on the stability of grain yield. *vg* vegetative stage, *hed* for heading, *gf* grain filling

the value of envirotyping in revealing critical environmental drivers of crop performance.

3.5.2 Genetic Models: Allelic Substitution vs. Regulatory Gene Effects

Are there core genes for a specific trait and environment-responsive genes? Phenotypic plasticity is determined by genetic and environmental factors. In a maize study using NAM population grown at 11 environments, using 20 million markers, genetic loci responsible for the plastic response to the identified environmental indices for different traits were identified. The authors proposed two conceptual models: regulatory gene model, and allelic sensitivity model (Tibbs-Cortes et al. 2024). The allelic sensitivity model suggests that phenotypic plasticity is caused by different environmental sensitivities of alleles that control the trait means. It indicates that trait means and trait plasticity are controlled by same genes (Via et al. 1995). Regulatory gene model suggests that different genes and pathways control traits in different environments. In some cases, both models may be present for a trait.

Tibbs-Cortes et al. (2024) identified several candidate genes for 19 traits of which flowering time has the highest number of genes identified. It being an easily observable and highly heritable trait indicates they might be correlated. When it comes to complex traits, our ability to discover QTLs for phenotypic plasticity and G × E interactions is limited. This indicates that the highest variation explained by combined QTL could not reach the heritability and missing heritability will not be able to be accounted for a long time into the future (Manolio et al. 2009).

3.6 Current Challenges and Open Questions

3.6.1 Does Epigenetics and Transcriptional Plasticity Contribute to Stability or Plasticity?

Given that QTLs cannot explain the full heritability of a trait, how much of that could be explained by the epigenetics, such as methylation and post-transcription changes? Why do some QTLs work in certain environmental clues? When does the plant activate the genes? The elucidation of G × E, gene–gene, and QTL × QTL interactions remains a significant challenge, and much progress is still required.

From a quantitative genetics and breeding standpoint, the core objective is to understand and utilize heritable variation to improve crop performance in target environments. Historically, efforts to dissect complex traits have relied on identifying QTLs and estimating additive, dominance, and epistatic variance components. However, a persistent challenge in crop improvement is the “missing heritability” problem—where known genetic markers fail to explain the full phenotypic variance observed, especially across variable environments. This raises a critical question: Are epigenetic regulation and transcriptional plasticity contributing to this unexplained variation, and, if so, how do they influence phenotypic stability or plasticity in breeding populations?

3.6.1.1 QTL Expression and Environmental Sensitivity

It is well recognized that the effect sizes of QTLs are often environment-dependent, with some loci showing strong effects in specific environments while remaining inactive in others. This has direct implications for genotype × environment (G × E) interactions and for decisions on marker-assisted selection or genomic prediction. For breeders, a stable QTL across environments is valuable for broad adaptation, whereas environment-specific QTLs may be leveraged for targeted adaptation.

The activation or silence of QTLs in response to environmental cues points to underlying regulatory mechanisms—likely mediated by epigenetic modifications (e.g., DNA methylation, histone modifications) or transcriptional dynamics. These mechanisms do not alter the DNA sequence but can impact gene expression, adding another layer of complexity in how traits are inherited and expressed. Understanding whether epigenetic variation contributes to heritable phenotypic variation is essential. If stable across generations, epigenetic marks could contribute to broad-sense heritability and thus be useful in selection. However, if they are labile or reset between generations, they may instead contribute to non-genetic plasticity, which is difficult to predict and exploit through standard breeding methods (Richards et al. 2010).

This uncertainty challenges traditional genetic models that rely on additive genetic variance and fixed marker effects. Hierarchical or reaction-norm-based genomic prediction models are emerging as tools to address this, by modeling marker-by-environment interactions and incorporating environmental covariates. These models implicitly accommodate transcriptional plasticity and potential epigenetic effects by allowing allele effects to vary across environments or along continuous environmental gradients (Winans et al. 2024).

3.6.1.2 Gene × Gene and QTL × QTL Interactions

Quantitative geneticists also recognize that epistasis (gene × gene interaction) and QTL × QTL interactions can influence trait expression, particularly under stress conditions or in complex traits like yield, flowering time, or drought tolerance. These interactions may be modulated or mediated by epigenetic regulation, which could explain why some allelic combinations only express under specific conditions. However, such interactions are difficult to detect and validate in traditional biparental populations due to limited recombination and sample size. In contrast, large diversity panels, MAGIC populations, or NAM populations offer greater power to dissect these multi-locus interactions, especially when combined with transcriptomic or methylome data. Integrating these data layers could improve prediction accuracy for traits with high G × E or non-additive effects (Huang et al. 2015).

3.6.1.3 Breeding Utility: Stability vs. Plasticity

For breeders, the ultimate goal is to develop varieties with:

- *High stability*: Consistent performance across environments (e.g., for food security or seed systems with limited input).
- *High plasticity*: Ability to respond favorably to changing conditions (e.g., variable rainfall or temperature).

Epigenetics and transcriptional plasticity might underpin both outcomes depending on the selection regime and trait of interest. For example, a genotype that responds plastically to early-season drought via induced root growth might outperform a stable genotype in erratic rainfall environments. If these regulatory responses are predictable and heritable, they can be modeled and selected. If not, they remain a source of phenotypic noise, potentially reducing the efficiency of genomic selection and breeding progress (Duncan et al. 2014).

For now, breeders should be aware that beyond-DNA mechanisms—though often hidden—may be influencing trait expression, particularly in stress-prone environments. As tools improve, these mechanisms may be harnessed to improve prediction accuracy, better manage G × E, and ultimately deliver varieties that are either more stable or more responsive, depending on breeding goals.

3.6.1.4 Genomic × Environment Interaction and the Role of Environmental Covariables

While epigenetic mechanisms and transcriptional plasticity provide valuable insights into phenotypic expression, an equally critical dimension for modern breeding is the explicit modeling of genomic × environment (G × E) interactions. G × E interaction occurs when the performance of genotypes varies across environments in a non-parallel manner, and it remains a central obstacle—and opportunity—in predictive breeding. These interactions can obscure the true genetic signal if not accounted for, leading to suboptimal selection decisions and reduced genetic gain across target environments.

Recent advances in genomic prediction (GP) have increasingly incorporated environmental covariables (e.g., temperature, precipitation, radiation, soil moisture) to dissect and exploit G × E patterns. Environmental covariables act as quantitative descriptors of environmental conditions and serve as inputs in reaction norm and machine learning models to explain the heterogeneity of allele effects across environments (Jarquín et al. 2014; Cuevas et al. 2017). These models enable environment-specific prediction, improving the accuracy of selection by capturing both main effects and genotype-specific sensitivities to environmental stimuli.

For instance, reaction norm models integrate marker effects and their interaction with continuous environmental gradients, such as temperature or vapor pressure deficit during flowering or grain filling. These models allow the estimation of allele sensitivities to specific environmental conditions, providing a more nuanced and biologically grounded interpretation of G × E (Cooper et al. 2014). Likewise, deep learning models and kernel-based methods offer flexible ways to capture complex, non-linear interactions between genomic features and high-dimensional environmental data (Montesinos-López et al. 2021).

The use of envirotyping—systematic characterization of environmental covariables across sites and seasons—has proven particularly powerful when paired with multi-environment trials (METs). Through envirotyping, researchers can define environmental typologies and use them as meta-predictors, increasing the interpretability and transferability of genomic models (Costa-Neto et al. 2021a, b). This framework also aids in the identification of environmental drivers that underlie QTL × E interactions, bridging the gap between genotype, phenotype, and environment.

In summary, environmental covariables are no longer peripheral—they are central to resolving G × E, increasing prediction accuracy, and enabling climate-resilient breeding. The incorporation of such covariables into GP models is transforming breeding programs from being purely genotype-driven to genotype-environment-informed, where decision-making is tailored to specific environmental scenarios. This paradigm is crucial for developing varieties that can thrive not just in average conditions, but in the specific and variable environments that characterize modern agriculture.

3.6.2 Breeding Pipeline Design: What Works, What Does Not?

In terms of breeding pipeline, most breeding programs that work at the international level across continents would pursue wide adaptation and stability for stress and traits that are complex such as drought tolerance, heat stress tolerance, and salinity tolerance, whereas disease resistance would come under location adaptation or specific environmental adaptation. Traditionally breeders use the Best × Best crossing strategy to advance crops. This has been successful in many cereals, where both the parents may have wide adaptation and are already released varieties (van Ginkel and Ortiz 2018). However, physiological breeding, which emphasizes complementary crossing strategy using source × sink traits and selection based on physiological and agronomic parameters, also gained momentum in the last decade (Reynolds and Langridge 2016; Reynolds et al. 2017, 2021). In addition, use of gene bank accessions in the breeding program enhances the genetic diversity of the breeding populations and increases the genetic gains (Jordan et al. 2011). Accelerating breeding through shuttle breeding was something that did wonders to the wheat breeding; recently, speed breeding is being popular to reduce the generation advancement time (Watson et al. 2018).

Speed breeding together with genomic selection is another approach followed to reduce the time for variety development from crossing (Voss-Fels et al. 2019). Slow rusting genes-based breeding for rust resistance has been the backbone of success of the CIMMYT's breeding program (Huerta-Espino et al. 2020). This approach overcomes the limitation of single gene resistance and breakdown of varieties when a new race of rust evolves. Among these approaches, for different traits all these approaches contribute to developing germplasm with phenotypic stability and plasticity. Locally advantageous alleles could be combined with multiple loci with advantages for wide range of adaptation through traditional and genomic-assisted breeding (Lowry et al. 2019).

3.6.3 G × E × M (Genotype × Environment × Management): A Complex Interplay in Crop Improvement

The performance of a crop genotype is not determined solely by its genetic makeup (G), but also by the intricate interplay between genotype, environment (E), and management (M) practices—collectively referred to as the G × E × M interaction. This multifactorial relationship poses a significant challenge in breeding programs, yet it also offers a unique opportunity to develop more targeted and resilient crop varieties tailored to specific production environments.

Traditionally, breeding efforts have focused on identifying stable quantitative trait loci (QTLs) across a wide range of environments, aiming to ensure broad adaptation. However, the complexity of G × E interactions often makes it difficult to identify QTLs that are consistently expressed across diverse environmental contexts. Instead of solely pursuing stability across environments, a paradigm shift is emerging embracing the variability inherent in G × E × M interactions to breed crops specifically adapted to their target population of environments (TPEs) (Cooper et al. 2022, 2023). This shift reflects a more nuanced understanding of adaptation—prioritizing performance under specific environmental and management scenarios rather than across all possible conditions.

Genomic selection has become a powerful tool in this context, enabling breeders to capture and utilize G × E × M interactions by incorporating large-scale genomic, phenotypic, and environmental data. Rather than treating G × E as noise, modern approaches now model it explicitly. A key innovation in this area is the use of reaction norms—genotype-specific functions that describe trait expression as a function of an environmental gradient. These models enable breeders to quantify and predict how individual genotypes respond to varying environmental conditions, enhancing the precision of selection decisions. For instance, reaction norm-based analyses have been successfully employed in drought tolerance breeding, offering insights into the physiological and genetic mechanisms underlying plasticity (Cooper and Messina 2023).

Building on this, hierarchical genomic prediction models have been proposed to further improve the accuracy of trait prediction across diverse environments. These models account for the non-stationary effects of alleles, acknowledging that the influence of a given allele may change depending on the environment or management context (Powell et al. 2021). By incorporating environmental covariates and interaction terms, these models capture the dynamic nature of gene expression and trait performance under real-world agricultural conditions.

Ultimately, the integration of G × E × M interactions into breeding strategies represents a transformative approach. Rather than seeking universal genotypes, breeders are now empowered to develop environment-specific varieties with tailored management recommendations. This strategy is especially critical in the face of climate change, where the predictability of environments is diminishing, and adaptive capacity becomes essential.

3.7 Conclusions

From a breeding and genetics standpoint, stability is a foundational concept in plant breeding, serving as a key metric for assessing genotype performance across diverse environments. Many marker-assisted applications in plant breeding rely on the assumption of genotypic stability and the consistent effects of genes and QTLs. However, from a practical breeding perspective, highly plastic genotypes can exhibit stability within a defined range of environmental conditions. For crop improvement both must be utilized by understanding the genetic and environmental parameters and its interactions.

References

Acreche MM, Slafer GA (2006) Grain weight response to increases in number of grains in wheat in a Mediterranean area. Field Crop Res 98:52–59

Agha HI, Endelman JB, Chitwood-Brown J, Clough M, Coombs J, De Jong WS, Douches DS, Higgins CR, Holm DG, Novy R, Resende MFR, Sathuvalli V, Thompson AL, Yencho GC, Zotarelli L, Shannon LM (2024) Genotype-by-environment interactions and local adaptation shape selection in the US National Chip Processing Trial. Theor Appl Genet 137:99. https://doi.org/10.1007/s00122-024-04598-w

Aranzana MJ, Kim S, Zhao K, Bakker E, Horton M, Jakob K, Lister C, Molitor J, Shindo C, Tang C, Toomajian C, Traw B, Zheng H, Bergelson J, Dean C, Marjoram P, Nordborg M (2005) Genome-wide association mapping in Arabidopsis identifies previously known flowering time and pathogen resistance genes. PLoS Genet 1:e60. https://doi.org/10.1371/journal.pgen.0010060

Beales J, Turner A, Griffiths S, Snape JW, Laurie DA (2007) A pseudo-response regulator is misexpressed in the photoperiod insensitive Ppd-D1a mutant of wheat (Triticum aestivum L.). Theor Appl Genet 115:721–733

Becker HC, Léon J (1988) Stability analysis in plant breeding. Plant Breed 101:1–23

Boatwright JL, Sapkota S, Jin H, Schnable JC, Brenton Z, Boyles R, Kresovich S (2022) Sorghum Association Panel whole-genome sequencing establishes cornerstone resource for dissecting genomic diversity. Plant J 111:888–904

Bustos-Korts D, Romagosa I, Borràs-Gelonch G, Casas AM, Slafer GA, van Eeuwijk FA (2018) Genotype by environment interaction and adaptation. In: Meyers RA (ed) Encyclopedia of sustainability science and technology. Springer, New York, pp 1–44

Calderini DF, Castillo FM, Arenas-M A, Molero G, Reynolds MP, Craze M, Bowden S, Milner MJ, Wallington EJ, Dowle A, Gomez LD, McQueen-Mason SJ (2021) Overcoming the trade-off between grain weight and number in wheat by the ectopic expression of expansion in developing seeds leads to increased yield potential. New Phytol 230:629–640

Campbell BT, Baenziger PS, Gill KS, Eskridge KM, Budak H, Erayman M, Dweikat I, Yen Y (2003) Identification of QTLs and environmental interactions associated with agronomic traits on chromosome 3A of wheat. Crop Sci 43:1493–1505

Campbell Q, Castañeda-Álvarez N, Domingo R, Bishop-von Wettberg E, Runck B, Nandkangré H, Halpin-McCormick A, Fumia N, Neyhart J, Kilian B, Wambugu P, Nyamongo D, Hübner S,

Sitar S, Thompson A, Rieseberg L, Gore MA, Kantar MB (2025) Prioritizing parents from global genebanks to breed climate-resilient crops. Nat Clim Chang 15:673–681
Chen S-AA, Kern AF, Ang RML, Xie Y, Fraser HB (2023) Gene-by-environment interactions are pervasive among natural genetic variants. Cell Genom 3:100273
Chen W, Li X, Zhang X, Chachar Z, Lu C, Qi Y, Chang H, Wang Q (2024) Genome-wide association study of trace elements in maize kernels. BMC Plant Biol 24:724
Collins NC, Tardieu F, Tuberosa R (2008) Quantitative trait loci and crop performance under abiotic stress: where do we stand? Plant Physiol 147:469–486
Cooper M, Messina CD (2023) Breeding crops for drought-affected environments and improved climate resilience. Plant Cell 35:162–186
Cooper M, Messina CD, Podlich D, Totir LR, Baumgarten A, Hausmann NJ, Wright D, Graham G (2014) Predicting the future of plant breeding: complementing empirical evaluation with genetic prediction. Crop Pasture Sci 65:311–336
Cooper M, Powell OM, Voss-Fels KP, Messina CD, Gho C, Podlich DW, Technow F, Chapman SC, Beveridge CA, Ortiz-Barrientos D, Hammer GL (2021) Modelling selection response in plant-breeding programs using crop models as mechanistic gene-to-phenotype (CGM-G2P) multi-trait link functions. In silico Plants 3:diaa016
Cooper M, Messina CD, Tang T, Gho C, Powell OM, Podlich DW, Technow F, Hammer GL (2022) Predicting genotype × environment × management (G × E × M) interactions for the Design of Crop Improvement Strategies. In: Goldman I (ed) Plant breeding reviews, vol 45. Wiley, pp 467–585
Cooper M, Powell OM, Gho C, Tang T, Messina CD (2023) Extending the breeder's equation to take aim at the target population of environments. Front Plant Sci 14:1180981
Corlouer E, Sauvage C, Leveugle M, Nesi N, Laperche A (2024) Envirotyping within a multi-environment trial allowed identifying genetic determinants of winter oilseed rape yield stability. Theor Appl Genet 137:164. https://doi.org/10.1007/s00122-024-04641-w
Costa-Neto G, Crossa J, Fritsche-Neto R (2021a) Enviromic assembly increases accuracy and reduces costs of the genomic prediction for yield plasticity in maize. Front Plant Sci 12:717552
Costa-Neto G, Galli G, Carvalho HF, Crossa J, Fritsche-Neto R (2021b) EnvRtype: a software to interplay enviromics and quantitative genomics in agriculture. G3 Genes Genomes Genet 11: jkab040
Crossa J (2012) From genotype × environment interaction to gene × environment interaction. Curr Genomics 13:225–244
Crossa J, Cornelius PL (1997) Sites regression and shifted multiplicative model clustering of cultivar trial sites under heterogeneity of error variances. Crop Sci 37(2):406–415
Crossa J, Fritsche-Neto R, Montesinos-Lopez OA, Costa-Neto G, Dreisigacker S, Montesinos-Lopez A, Bentley AR (2021) The modern plant breeding triangle: optimizing the use of genomics, phenomics, and enviromics data. Front Plant Sci 12:651480
Cuevas J, Crossa J, Montesinos-López OA, Burgueño J, Pérez-Rodríguez P, de Los Campos G (2017) Bayesian genomic prediction with genotype × environment interaction kernel models. G3 Genes|Genomes|Genetics 7:41–53
De Jong G (2005) Evolution of phenotypic plasticity: patterns of plasticity and the emergence of ecotypes. New Phytol 166:101–118
De Kroon H, Huber H, Stuefer JF, Van Groenendael JM (2005) A modular concept of phenotypic plasticity in plants. New Phytol 166:73–82
Dias KO, Dos Santos JP, Krause MD, Piepho HP, Guimarães LJ, Pastina MM, Garcia AA (2022) Leveraging probability concepts for cultivar recommendation in multi-environment trials. Theor Appl Genet 135(4):1385–1399
Ding H, Wang C, Cai Y, Yu K, Zhao H, Wang F, Shi X, Cheng J, Sun H, Wu Y, Qin R, Liu C, Zhao C, Sun X, Cui F (2024) Characterization of a wheat stable QTL for spike length and its genetic effects on yield-related traits. BMC Plant Biol 24:292
Duncan EJ, Gluckman PD, Dearden PK (2014) Epigenetics, plasticity, and evolution: how do we link epigenetic change to phenotype? J Exp Zool B Mol Dev Evol 322:208–220

Elmerich C, Faucon M-P, Garcia M, Jeanson P, Boulch G, Lange B (2023) Envirotyping to control genotype x environment interactions for efficient soybean breeding. Field Crop Res 303:109113

Finlay KW, Wilkinson GN (1963) The analysis of adaptation in a plant-breeding programme. Aust J Agric Res 14:742–754

Foulkes MJ, Slafer GA, Davies WJ, Berry PM, Sylvester-Bradley R, Martre P, Calderini DF, Griffiths S, Reynolds MP (2011) Raising yield potential of wheat. III. Optimizing partitioning to grain while maintaining lodging resistance. J Exp Bot 62:469–486

Gauch HG, Jr (1992) Statistical analysis of regional yield trials: AMMI analysis of factorial designs

Gore MA, Chia J-M, Elshire RJ, Sun Q, Ersoz ES, Hurwitz BL, Peiffer JA, McMullen MD, Grills GS, Ross-Ibarra J, Ware DH, Buckler ES (2009) A first-generation haplotype map of maize. Science 326:1115–1117

Guo T, Wei J, Li X, Yu J (2024) Environmental context of phenotypic plasticity in flowering time in sorghum and rice. J Exp Bot 75:1004–1015

Han X, Tang Q, Xu L, Guan Z, Tu J, Yi B, Liu K, Yao X, Lu S, Guo L (2022) Genome-wide detection of genotype environment interactions for flowering time in Brassica napus. Front Plant Sci 13:1059370

Han L, Wang X, Benke R, Tibbs-Cortes LE, Zhao P, Sanguinet KA, Zhang Z, Xu S, Yu J, Li X (2025) Integrated phenomic and genomic analyses unveil modes of altered phenotypic plasticity during wheat improvement. Genome Biol 26(1):1–17

Hu P, Chapman SC, Dreisigacker S, Sukumaran S, Reynolds M, Zheng B (2021) Using a gene-based phenology model to identify optimal flowering periods of spring wheat in irrigated mega-environments. J Exp Bot 72:7203–7218

Hu P, Chapman SC, Sukumaran S, Reynolds M, Zheng B (2022) Phenological optimization of late reproductive phase for raising wheat yield potential in irrigated mega-environments. J Exp Bot 73:4236–4249

Huang X, Wei X, Sang T, Zhao Q, Feng Q, Zhao Y, Li C, Zhu C, Lu T, Zhang Z, Li M, Fan D, Guo Y, Wang A, Wang L, Deng L, Li W, Lu Y, Weng Q, Liu K, Huang T, Zhou T, Jing Y, Li W, Lin Z, Buckler ES, Qian Q, Zhang Q-F, Li J, Han B (2010) Genome-wide association studies of 14 agronomic traits in rice landraces. Nat Genet 42:961–967

Huang BE, Verbyla KL, Verbyla AP, Raghavan C, Singh VK, Gaur P, Leung H, Varshney RK, Cavanagh CR (2015) MAGIC populations in crops: current status and future prospects. Theor Appl Genet 128:999–1017

Huerta-Espino J, Singh R, Crespo-Herrera LA, Villaseñor-Mir HE, Rodriguez-Garcia MF, Dreisigacker S, Barcenas-Santana D, Lagudah E (2020) Adult plant slow rusting genes confer high levels of resistance to rusts in bread wheat cultivars from Mexico. Front Plant Sci 11:824

Jarquín D, Crossa J, Lacaze X, Du Cheyron P, Daucourt J, Lorgeou J, Piraux F, Guerreiro L, Pérez P, Calus MPL, Burgueño J (2014) A reaction norm model for genomic selection using high-dimensional genomic and environmental data. Theor Appl Genet 127:595–607

Jordan DR, Mace ES, Cruickshank AW, Hunt CH, Henzell RG (2011) Exploring and exploiting genetic variation from unadapted sorghum germplasm in a breeding program. Crop Sci 51: 1444–1457

Lasky JR, Upadhyaya HD, Ramu P, Deshpande S, Hash CT, Bonnette J, Juenger TE, Hyma K, Acharya C, Mitchell SE, Buckler ES, Brenton Z, Kresovich S, Morris GP (2015) Genome-environment associations in sorghum landraces predict adaptive traits. Sci Adv 1:e1400218

Li X, Guo T, Wang J, Bekele WA, Sukumaran S, Vanous AE, McNellie JP, Tibbs-Cortes LE, Lopes MS, Lamkey KR, Westgate ME, McKay JK, Archontoulis SV, Reynolds MP, Tinker NA, Schnable PS, Yu J (2021) An integrated framework reinstating the environmental dimension for GWAS and genomic selection in crops. Mol Plant 14:874–887

Liu H, Li H, Hao C, Wang K, Wang Y, Qin L, An D, Li T, Zhang X (2020) TaDA1, a conserved negative regulator of kernel size, has an additive effect with TaGW2 in common wheat (Triticum aestivum L.). Plant Biotechnol J 18:1330–1342

Liu X, Xu Z, Feng B, Zhou Q, Guo S, Liao S, Ou Y, Fan X, Wang T (2024) Dissection of a novel major stable QTL on chromosome 7D for grain hardness and its breeding value estimation in bread wheat. Front Plant Sci 15:1356687

Lopes MS, Dreisigacker S, Peña RJ, Sukumaran S, Reynolds MP (2015) Genetic characterization of the wheat association mapping initiative (WAMI) panel for dissection of complex traits in spring wheat. Theor Appl Genet 128:453–464

Lowry DB, Lovell JT, Zhang L, Bonnette J, Fay PA, Mitchell RB, Lloyd-Reilley J, Boe AR, Wu Y, Rouquette FM, Wynia RL, Weng X, Behrman KD, Healey A, Barry K, Lipzen A, Bauer D, Sharma A, Jenkins J, Schmutz J, Fritschi FB, Juenger TE (2019) QTL × environment interactions underlie adaptive divergence in switchgrass across a large latitudinal gradient. Proc Natl Acad Sci USA 116:12933–12941

Mace ES, Hunt CH, Jordan DR (2013) Supermodels: sorghum and maize provide mutual insight into the genetics of flowering time. Theor Appl Genet 126:1377–1395

Mahmood T, Ahmed T, Trethowan R (2022) Genotype × Environment × Management (GEM) reciprocity and crop productivity. Front Agron 4:826674

Malosetti M, Ribaut JM, van Eeuwijk FA (2013) The statistical analysis of multi-environment data: modeling genotype-by-environment interaction and its genetic basis. Front Physiol 4:44

Manolio TA, Collins FS, Cox NJ, Goldstein DB, Hindorff LA, Hunter DJ, McCarthy MI, Ramos EM, Cardon LR, Chakravarti A, Cho JH, Guttmacher AE, Kong A, Kruglyak L, Mardis E, Rotimi CN, Slatkin M, Valle D, Whittemore AS, Boehnke M, Clark AG, Eichler EE, Gibson G, Haines JL, Mackay TFC, McCarroll SA, Visscher PM (2009) Finding the missing heritability of complex diseases. Nature 461:747–753

Messmer R, Fracheboud Y, Bänziger M, Vargas M, Stamp P, Ribaut J-M (2009) Drought stress and tropical maize: QTL-by-environment interactions and stability of QTLs across environments for yield components and secondary traits. Theor Appl Genet 119:913–930

Monforte AJ, Asins MJ, Carbonell EA (1997) Salt tolerance in Lycopersicon species VI. Genotype-by-salinity interaction in quantitative trait loci detection: constitutive and response QTLs. Theor Appl Genet 95:706–713

Montesinos-López OA, Montesinos-López A, Pérez-Rodríguez P, Barrón-López JA, Martini JWR, Fajardo-Flores SB, Gaytan-Lugo LS, Santana-Mancilla PC, Crossa J (2021) A review of deep learning applications for genomic selection. BMC Genomics 22:19

Montesinos López OA, Montesinos López A, Crossa J (2022) Fundamentals of artificial neural networks and deep learning. In Multivariate statistical machine learning methods for genomic prediction (pp 379–425). Cham: Springer International Publishing

Morris GP, Ramu P, Deshpande SP, Hash CT, Shah T, Upadhyaya HD, Riera-Lizarazu O, Brown PJ, Acharya CB, Mitchell SE, Harriman J, Glaubitz JC, Buckler ES, Kresovich S (2013) Population genomic and genome-wide association studies of agroclimatic traits in sorghum. Proc Natl Acad Sci USA 110:453–458

Mu Q, Guo T, Li X, Yu J (2022) Phenotypic plasticity in plant height shaped by interaction between genetic loci and diurnal temperature range. New Phytol 233:1768–1779

Nadolska-Orczyk A, Rajchel IK, Orczyk W, Gasparis S (2017) Major genes determining yield-related traits in wheat and barley. Theor Appl Genet 130:1081–1098

Napier JD, Heckman RW, Juenger TE (2023) Gene-by-environment interactions in plants: molecular mechanisms, environmental drivers, and adaptive plasticity. Plant Cell 35:109–124

Olivoto T, Lúcio ADC (2020) metan: An R package for multi-environment trial analysis. Methods Ecol Evol 11(6):783–789

Ozaki K, Ohnishi Y, Iida A, Sekine A, Yamada R, Tsunoda T, Sato H, Sato H, Hori M, Nakamura Y, Tanaka T (2002) Functional SNPs in the lymphotoxin-α gene that are associated with susceptibility to myocardial infarction. Nat Genet 32:650–654

Paterson AH, Lander ES, Hewitt JD, Peterson S, Lincoln SE, Tanksley SD (1988) Resolution of quantitative traits into Mendelian factors by using a complete linkage map of restriction fragment length polymorphisms. Nature 335:721–726

Pigliucci M (2001) Phenotypic plasticity: beyond nature and nurture. JHU Press

Powell OM, Voss-Fels KP, Jordan DR, Hammer G, Cooper M (2021) Perspectives on applications of hierarchical gene-to-phenotype (G2P) maps to capture non-stationary effects of alleles in genomic prediction. Front Plant Sci 12:709365

Puntel LA, Sawyer JE, Barker DW, Dietzel R, Poffenbarger H, Castellano MJ, Moore KJ, Thorburn P, Archontoulis SV (2016) Modeling long-term corn yield response to nitrogen rate and crop rotation. Front Plant Sci 7:1630

Quinby JR (1967) The maturity genes of sorghum. In: Norman AG (ed) Advances in agronomy, vol 19. Academic, pp 267–305

Reynolds M, Langridge P (2016) Physiological breeding. Curr Opin Plant Biol 31:162–171

Reynolds MP, Pask AJD, Hoppitt WJE, Sonder K, Sukumaran S, Molero G, Pierre CS, Payne T, Singh RP, Braun H-J, Gonzalez FG, Terrile II, Barma NCD, Hakim A, He Z, Fan Z, Novoselovic D, Maghraby M, Gad KIM, Galal EG, Hagras A, Mohamed MM, Morad AFA, Kumar U, Singh GP, Naik R, Kalappanavar IK, Biradar S, Sai Prasad SV, Chatrath R, Sharma I, Panchabhai K, Sohu VS, Mavi GS, Mishra VK, Balasubramaniam A, Jalal-Kamali MR, Khodarahmi M, Dastfal M, Tabib-Ghaffari SM, Jafarby J, Nikzad AR, Moghaddam HA, Ghojogh H, Mehraban A, Solís-Moya E, Camacho-Casas MA, Figueroa-López P, Ireta-Moreno J, Alvarado-Padilla JI, Borbón-Gracia A, Torres A, Quiche YN, Upadhyay SR, Pandey D, Imtiaz M, Rehman MU, Hussain M, Hussain M, Ud-Din R, Qamar M, Sohail M, Mujahid MY, Ahmad G, Khan AJ, Sial MA, Mustatea P, von Well E, Ncala M, de Groot S, Hussein AHA, Tahir ISA, Idris AAM, Elamein HMM, Manes Y, Joshi AK (2017) Strategic crossing of biomass and harvest index—source and sink—achieves genetic gains in wheat. Euphytica 213:257

Reynolds MP, Langridge P, Lewis JM, Ammar K, Basnet BR, Crespo-Herrera L, Crossa J, Dhugga KS, Dreisigacker S, Juliana P, Karwat H, Kishii M, Krause MR, Lashkari A, Mondal S, Payne T, Pequeno D, Pinto F, Sansaloni C, Schulthess U, Singh RP, Sonder K, Sukumaran S, Xiong W, Braun H-J (2021) Harnessing translational research in wheat for climate resilience. J Exp Bot 72:5134–5157

Richards CL, Bossdorf O, Pigliucci M (2010) What role does heritable epigenetic variation play in phenotypic evolution? Bioscience 60:232–237

Rooney WL, Aydin S (1999) Genetic control of a photoperiod-sensitive response in Sorghum bicolor (L.) Moench. Crop Sci 39:397–400

Rosyara U, Kishii M, Payne T, Sansaloni CP, Singh RP, Braun H-J, Dreisigacker S (2019) Genetic contribution of synthetic Hexaploid wheat to CIMMYT'S spring bread wheat breeding germplasm. Sci Rep 9:12355

Said AA, MacQueen AH, Shawky H, Reynolds M, Juenger TE, El-Soda M (2022) Genome-wide association mapping of genotype-environment interactions affecting yield-related traits of spring wheat grown in three watering regimes. Environ Exp Bot 194:104740

Sangha JS, Wang W, Knox R, Ruan Y, Cuthbert RD, Isidro-Sánchez J, Li L, He Y, De Pauw R, Singh A, Cutler A, Wang H, Selvaraj G (2025) Phenotypic plasticity of bread wheat contributes to yield reliability under heat and drought stress. PLoS One 20:e0312122

Santra DK, Santra M, Allan RE, Campbell KG, Kidwell KK (2009) Genetic and molecular characterization of vernalization genes Vrn-A1, Vrn-B1, and Vrn-D1 in spring wheat germplasm from the Pacific northwest region of the U.S.A. Plant Breed 128:576–584

Sinclair TR, De Wit CT (1975) Photosynthate and nitrogen requirements for seed production by various crops. Science 189:565–567

Stratton DA (1998) Reaction norm functions and QTL–environment interactions for flowering time in Arabidopsis thaliana. Heredity 81:144–155

Su Z, Hao C, Wang L, Dong Y, Zhang X (2011) Identification and development of a functional marker of TaGW2 associated with grain weight in bread wheat (Triticum aestivum L.). Theor Appl Genet 122:211–223

Sukumaran S, Dreisigacker S, Lopes M, Chavez P, Reynolds MP (2015) Genome-wide association study for grain yield and related traits in an elite spring wheat population grown in temperate irrigated environments. Theor Appl Genet 128:353–363

Sukumaran S, Lopes MS, Dreisigacker S, Dixon LE, Zikhali M, Griffiths S, Zheng B, Chapman S, Reynolds MP (2016) Identification of earliness per se flowering time locus in spring wheat through a genome-wide association study. Crop Sci 56:2962–2672

Sukumaran S, Crossa J, Jarquin D, Lopes M, Reynolds MP (2017) Genomic prediction with pedigree and genotype × environment interaction in spring wheat grown in south and West Asia, North Africa, and Mexico. G3 Genes Genomes Genet 7:481–495
Sukumaran S, Lopes MS, Dreisigacker S, Reynolds M (2018) Genetic analysis of multi-environmental spring wheat trials identifies genomic regions for locus-specific trade-offs for grain weight and grain number. Theor Appl Genet 131:985–998
Tao Y, Trusov Y, Zhao X, Wang X, Cruickshank AW, Hunt C, van Oosterom EJ, Hathorn A, Liu G, Godwin ID, Botella JR, Mace ES, Jordan DR (2021) Manipulating assimilate availability provides insight into the genes controlling grain size in sorghum. Plant J 108:231–243
Tian F, Bradbury PJ, Brown PJ, Hung H, Sun Q, Flint-Garcia S, Rocheford TR, McMullen MD, Holland JB, Buckler ES (2011) Genome-wide association study of leaf architecture in the maize nested association mapping population. Nat Genet 43:159–162
Tibbs-Cortes LE, Guo T, Andorf CM, Li X, Yu J (2024) Comprehensive identification of genomic and environmental determinants of phenotypic plasticity in maize. Genome Res 34:1253–1263
Tong J, Tarekegn ZT, Alahmad S, Hickey LT, Periyannan SK, Dinglasan E, Hayes BJ (2024) Stacking haplotypes from the Vavilov wheat collection to accelerate breeding for multiple disease resistance. bioRxiv. https://doi.org/10.1101/2024.04.03.587896
Trethowan RM, Nicol JM, Singh A, Singh RP, Tadesse W, Govidan V, Crespo-Herrera L, Cullis B, Mazur L, Dieters M, Micallef S, Farrell T, Wilson R, Mathews K (2024) The CIMMYT Australia ICARDA germplasm evaluation concept: a model for international cooperation and impact. Front Plant Sci 15:1361780
van Ginkel M, Ortiz R (2018) Cross the best with the best, and select the best: HELP in breeding selfing crops. Crop Sci 58:17–30
Varshney RK, Gaur PM, Chamarthi SK, Krishnamurthy L, Tripathi S, Kashiwagi J, Samineni S, Singh VK, Thudi M, Jaganathan D (2013) Fast-track introgression of "QTL-hotspot" for root traits and other drought tolerance traits in JG 11, an elite and leading variety of chickpea. Plant Genome 6:plantgenome2013.07.0022
Varshney RK, Mohan SM, Gaur PM, Chamarthi SK, Singh VK, Srinivasan S, Swapna N, Sharma M, Pande S, Singh S, Kaur L (2014) Marker-assisted backcrossing to introgress resistance to Fusarium wilt race 1 and Ascochyta blight in C 214, an elite cultivar of chickpea. Plant Genome 7:plantgenome2013.10.0035
Via S, Gomulkiewicz R, De Jong G, Scheiner SM, Schlichting CD, Van Tienderen PH (1995) Adaptive phenotypic plasticity: consensus and controversy. Trends Ecol Evol 10:212–217
Vicentin L, Canales J, Calderini DF (2024) The trade-off between grain weight and grain number in wheat is explained by the overlapping of the key phases determining these major yield components. Front Plant Sci 15:1388382
Voss-Fels KP, Herzog E, Dreisigacker S, Sukumaran S, Watson A, Frisch M, Hayes B, Hickey LT (2019) "SpeedGS" to accelerate genetic gain in spring wheat. In: Miedaner T, Korzun V (eds) Applications of genetic and genomic research in cereals. Woodhead Publishing, pp 303–327
Watson A, Ghosh S, Williams MJ, Cuddy WS, Simmonds J, Rey M-D, Asyraf Md Hatta M, Hinchliffe A, Steed A, Reynolds D, Adamski NM, Breakspear A, Korolev A, Rayner T, Dixon LE, Riaz A, Martin W, Ryan M, Edwards D, Batley J, Raman H, Carter J, Rogers C, Domoney C, Moore G, Harwood W, Nicholson P, Dieters MJ, DeLacy IH, Zhou J, Uauy C, Boden SA, Park RF, Wulff BBH, Hickey LT (2018) Speed breeding is a powerful tool to accelerate crop research and breeding. Nat Plants 4:23–29
Wen W, Li D, Li X, Gao Y, Li W, Li H, Liu J, Liu H, Chen W, Luo J, Yan J (2014) Metabolome-based genome-wide association study of maize kernel leads to novel biochemical insights. Nat Commun 5:3438
Winans ND, Fonseca JMO, Perumal R, Klein PE, Klein RR, Rooney WL (2024) Envirotyping can increase genomic prediction accuracy of new environments in grain sorghum trials depending on mega-environment. Crop Sci 64:2519–2533
Woltereck R (1909) Weitere experimentelle Untersuchungen über Artveränderung, speziell über das Wesen quantitativer Artunterschiede bei Daphniden. Verh Dtsch Zool Ges 1909:110–172
Xiong W, Reynolds MP, Crossa J, Schulthess U, Sonder K, Montes C, Addimando N, Singh RP, Ammar K, Burgueño J, Molero G, Singh S, Bhavani S, Juliana P, He X, Guzman C,

Dreisigacker S, Gerard GS (2021) Increased ranking change in wheat breeding under climate change. Nat Plants 7:1207–1212

Xiong W, Reynolds MP, Montes C, Crossa J, Snapp S, Akin B, Mesut K, Ozdemir F, Li H, He Z, Wang D, Chen F (2024) New wheat breeding paradigms for a warming climate. Nat Clim Chang 14:869–875

Xu Y (2016) Envirotyping for deciphering environmental impacts on crop plants. Theor Appl Genet 129:653–673

Yan L, Loukoianov A, Tranquilli G, Helguera M, Fahima T, Dubcovsky J (2003) Positional cloning of the wheat vernalization gene VRN1. Proc Natl Acad Sci USA 100:6263–6268

Yan W, Hunt LA, Sheng Q, Szlavnics Z (2000) Cultivar evaluation and mega-environment investigation based on the GGE biplot. Crop Sci 40(3):597–605

Yan W, Wang B, Chan E, Mitchell-Olds T (2021) Genetic architecture and adaptation of flowering time among environments. New Phytol 230:1214–1227

Zhang Y, Li D, Zhang D, Zhao X, Cao X, Dong L, Liu J, Chen K, Zhang H, Gao C, Wang D (2018) Analysis of the functions of TaGW2 homoeologs in wheat grain weight and protein content traits. Plant J 94:857–866

Contribution of Yield-Related Traits to Genotype by Environment Interaction for Grain Yield

4

Juan Burgueño

Abstract

This chapter covers the broad subject "grain yield genotype by environment interaction and related traits relationships." The analysis can have diverse objectives, and the subject is complex, involving many data strata, variables, and relationships between all of them. What can you find and not find in this chapter? In this chapter, you will not find a step-by-step guide about data analysis, you will not find theoretical details, and you will not find an exhaustive list of publications. On the contrary, you will find a short list of publications related to the subject, some considerations to take into account before starting the analysis, short codes in R and SAS to do the analysis as well as references to the libraries and procedures, and short interpretation of the results of using some of the methodologies on a real data set, highlighting the main aspects. I intend to introduce the reader to the statistical analysis of the subject, and the methodologies available to do it. The chapter is biased by personal view of the problem and, like in many contracts, it is important to read the fine print amid so much text.

Keywords

Genotype by environment interaction · Statistical procedures · Traits · Yield

J. Burgueño (✉)
International Maize and Wheat Improvement Centre (CIMMYT), Texcoco, Mexico
e-mail: j.burgueno@cgiar.org

R. Mohammadi et al. (eds.), *Genotype x Environment Interactions and its Implications for Plant Breeding*, https://doi.org/10.1007/978-981-95-5664-9_4

4.1 Introduction

Elias et al. (2016) published a review paper covering the last 50 years of genotype by environment interaction analysis, starting with Eberhart and Russel (1966) and Finlay and Wilkinson (1963) regression models up to mixed linear models as the approach but passing through multiplicative models. Most of the models are linear models which do not capture all the genotype by environment interaction; however, they capture the largest proportion of it and the most important components. Close in time, Piepho and Blancon (2023) authored a comprehensive paper focused on the most accepted approach to analyzing multi-environmental trials. He presented several methods and expressed the advantages of a linear mixed model approach. The linear mixed models offer many advantages compared with minimum least square methods or simply linear models. The linear mixed model is able to consider the complexity of multi-environmental trial data. It allows (i) to deal with imbalance data, (ii) to consider other covariance structure for experimental error like spatial models, (iii) to include known relationships between environments and genotypes considering other sources of variation such as environmental variables, and genotypic data, and (iv) to model sub-sampling or longitudinal data. Burgueño et al. (2008) showed the equivalence between the classical additive main effect and multiplicative interaction (AMMI) and site regression (SREG) models using minimum least squares with modeling the genotype by environment interaction or the main effect of genotype plus the genotype by environment interaction effects using a factor analytical covariance structure to modeling those effects in a mixed model framework.

In the literature, we can find different approaches to analyze genotype by environment interaction and the relationships between traits:

1. Using a generalized linear mixed model, including multivariate models, which includes grain yield (GY) genotype by environment interaction (GE) and the related traits effects.
2. Analyzing the genotype by environment interaction of all traits and then comparing the results of different traits.
3. Analyzing the relationship between traits by environment and then comparing the results in different environments.
4. Using specific methodology which deals directly with the problem.

In any case, explicitly (approaches 2, 3, and 4) or implicitly (approach 1), we are analyzing row–column tables in which the rows and columns can be genotypes, environments, or traits. Mandel in 1994 published his book called *Two-Way Table Analysis* with a comprehensive presentation of the possibilities of relationships between rows and columns, and methods of analysis (Mandel 1994). When dealing with three variables, genotypes, environments, and traits, the two-way table can be extended to a three-way table, a cube, in which the three dimensions correspond to the three variables. It is a three-dimensional array, which in mathematics is called three-dimensional tensor. The content of the cells is the genotypic response of one trait in a specific environment. It is important to note that tensors are multidimensional arrays including the two-dimensional arrays which are matrices; all rank-2 tensors are matrices, yet not all matrices are tensors, as tensors offer a flexibility that matrices cannot. A tensor, unlike a matrix, adapts to changes in the coordinate system. This adaptability makes tensors crucial where transformations are common. When coordinates shift, tensors transform accordingly to maintain the same representation in a new system, while matrices cannot automatically adapt to such transformations.

Analytical methods available to work with a three-dimensional array are extensions of two-dimensional analysis, principal components analysis (PCA), factor analysis, and clustering. Among these methods we have the parallel factor analysis proposed by Harshman (1970) and Carroll and Chang (1970), and three-mode factor analysis, proposed by Tucker (1966, 1972) and further elaborated by Kroonenberg and De Leeuw (1980), who renamed the method three-mode principal components analysis. More recently, the latter method has also been called TuckerS analysis (Kiers and van Mechelen 2001).

The analysis of grain yield genotypic by environment interaction and related traits can be formulated as a three-dimensional array problem. The response of several traits in a multi-environmental trial is a three-dimensional array in which the three dimensions correspond to genotypes, environments, and traits. While working with three-dimensional arrays looks as an ideal framework, they have not been extensively used in agronomy. The main use of tensors has been in the implementation of machine learning and deep learning methods. The complexity in a three-dimensional array is difficult to understand and many times it is preferable to split the problem into several two-dimensional ones.

Not so many publications consider three-dimensional array or three-mode analysis to analyze genotype by environment by trait array, but there are many which use the three-mode analysis to analyze arrays in which trait was replaced by management, or environment by trait by year by location (Kroonenberg and Basford 1989; Crossa et al. 1995; Rincón et al. 1997; de la Vega and Chapman 2001; Varela et al. 2006a, b; Varela et al. 2009; Dettori et al. 2011). Vargas, Crossa, and van Eeuwijk published a series of articles from 1995 to 2015 presenting two methodological approaches to incorporate additional information of the genotypes and environments to explain the genotype by environment interaction and to identify the relevant quantitative trait loci (QTL) and environmental covariables (Vargas et al. 1998, 1999a, b, 2006, 2007, 2015; Crossa et al. 1999; van Eeuwijk et al. 1996, 2000, 2002). They used factorial regression and partial least squares (PLS) to analyze the data. Although they did not consider additional traits, it is simple to extend both methodologies, replacing molecular data or environmental covariables by grain yield-related traits like Reynold et al. (2002) did.

Besides, except for the analysis done by Reynolds et al. (2002), in my knowledge only one additional analytical approach was published to explicitly deal with grain yield genotype by environment interaction and related traits analysis (Yan and Frégeau-Reid 2018), which named this approach as "GYT biplot," where GYT means genotype × yield × trait. The method converts the original genotype × trait table into a GYT table, in which each column is the result of yield-trait operation. The operation can be a multiplication or a division depending on the desired direction of the trait to interpret similarly the new variables. When large values of traits are desired, then the multiplication is used while for traits with negative values of interest, division operation is used. The new table, with one less column, is analyzed as usual to generate a biplot or to do a cluster analysis.

In next sections, I will discuss the main concerns when analyzing grain yield genotype by environment interaction and related traits; we will highlight some methods and their application with a real data set.

4.2 Considerations for the Analysis

Selecting a methodology depends, from the most crucial factor to the less, on the objective of the study, the way in which data were collected, and data availability. On the other hand, the last two depend on the first. Additionally, other important considerations are the software availability and computing capacity, as well as the understanding of the methodology to use.

The objective of the study could be only to describe the genotype by environmental interaction and the relationships between several traits. Most of the multivariate methodologies like principal components analysis (PCA), partial least squares (PLS), and cluster analysis, are descriptive and can be used. There have been large efforts to include inference in those methodologies with variable success. The use of a mixed model framework (Piepho and Blancon 2023), Bayesian methodologies, or techniques like bootstrap have been implemented to select the number of dimensions in PCA or PLS (Vargas et al. 1999a), and the number of groups in cluster analysis (Burgueño et al. 2007). For descriptive purposes, it is adequate to explore different methodologies; each one can show and highlight various aspects of the problem. However, when we are trying to do inference, methodology selection is more important. We need to consider cautiously the tested hypothesis, the assumptions behind the tests, and the power to detect significant differences. This does not mean that for descriptive purposes we can dismiss the importance of the methods. We need to understand what the different methods are doing, and how to interpret the results. A frequent question about genotype by environment interaction is: which is better, the AMMI or the SREG model? The question does not make sense because in the former case, we are analyzing exclusively the genotype by environment interaction, while in the last case, we are analyzing the genotypic main effect plus the genotype by environment interaction together (GGE).

Independently of how the data were collected, we can consider, without loss of generality, that the data can be reduced to a three-dimensional array, in which the three dimensions correspond to environment, genotype, and traits. Environments can be various locations, years, or managements, or the combination of them in something called environment, while genotypes correspond to the individual in which some traits were measured.

In another context, environments, genotypes, and traits can be changed for other concepts. For example, in the industry, genotypes could be different devices, different environments, different settings of the devices and traits, and different efficiency measurements of the device. In plant breeding, usually the data come from a multi-environmental trial analysis in which we have replicates for each cell in the three-dimensional array and an initial analysis of the replicated data is done before working with the three-dimensional array. The three-dimensional array can be extended to more dimensions if we consider year, management, and locations like other dimensions. However, the analysis of a dimensional array with more than three dimensions could become so complex. Next in the chapter I will focus on a three-dimensional array of genotypes, environments, and traits, and try to explain and understand grain yield genotype by environment interaction and how other traits contribute to or influence that interaction.

According to the objective of the study, the researcher needs selecting genotypes, environments, and traits to include in his study. Probably, genotype and environment selection, even when cannot be considered a straightforward decision, is usually well defined. The genotypes are those which belong to the population under development or study, environments are selected from the environments to which the results will be extended, but the traits offer more challenges. Most scientists, especially physiologists, know what traits are related to grain yield; however, the relationships are not stable due to changes in environmental conditions. Knowing which traits not to measure is just as important as knowing which traits to measure. It is well known that relationships do not mean causation. However, an omitted trait can be the cause of the response, and we can lose the opportunity to identify it. And more, a trait related to the causal variable but not with grain yield could be identified as probable cause of the result. Additionally, some traits can mask some relationships; hence it is necessary to explore all those situations, at least playing with several kinds of plots. For that reason, it is of paramount importance that a suitable selection of the traits be measured and included in the analysis.

The three-dimensional array has traits that are measured in different scales of measurements, from the weaker to the strongest scale; they are: nominal, ordinal, interval, and ratio. The analytical methods are different for every scale, and there are challenges to analyze variables of different scales with one method. Strongest scales can be transformed to a weaker scale but with losing some information. If possible, it would be good to have all traits measured on a strong scale, but this does not depend on the researcher, instead it primarily depends on the nature of the trait. Additionally, the traits are measured with different units of measurements, which make it difficult to compare the variability between traits and to give a mark of importance to them. Fortunately, a simple solution exists, which is to work with standardized variables, which is equivalent to replacing the variance–covariance matrix with the correlation matrix. But interpretation of the results must be done in terms of standard deviations and not in the original units of measurements. Working with standardized variables could be important also for stabilizing the algorithm of analysis. Some calculi are better done by computer if we are using lower values rather than larger ones.

The last consideration is about missing data. Ideally, we can assume that missing data follows a random pattern and that it is not because of some response of the genotypes to an environmental effect. The first case can be managed with more or less ease using adequate methodologies, while the second case is more complex because missing data cannot be easily managed by analytical methods. In specific cases, missing data can be considered like other results and can be analyzed, but only in extremely specific cases does this make sense. Working with random missing data is easier but we need to be extremely cautious in interpreting results. It is important to consider how much missing data is in the data. Prieto, in his MSc thesis evaluating the genotype by environment interaction, found that more than 30% of missing data has larger effect in the estimation of the genotype by environment variance, and in the estimation of the genotype by environment interaction effects (Prieto Hernández 2014).

While random missing data can be considered as a sampling problem, and it cannot have any additional effect more than affecting the estimation precision, it is common that missing data is not random and therefore affects the genotype by environment interaction analysis. Some descriptive analysis of the missing data and the relationships of missing data with other variables can be done trying to identify if missing data is random or not. Unfortunately, there is not one reliable method to determine randomness of missing data. Generalized linear mixed models and Bayesian methods are useful to manage missing data, always suggesting that missing data is random and that the percentage of missing data is not large. Using those methods, we can get a complete three-dimensional array with complete data but remember that some cells have imputed data. Another approach, usually adopted especially if the percentage of missing data is low, is directly to impute the data in the three-dimensional array. For that, we can use techniques like PCA, PLS, three-mode PCA, or near-neighbor imputation. In my experience it is better to select an imputation method like the method that will be used to analyze the data. In that way, we can expect that the imputation method affects less the results. In some situations, we will work with correlation or distance matrices that can be calculated using pairwise complete data, which is to do the calculus for pair of variables at time omitting individuals with missing data. It works, but in some cases, if the resulting matrix could not be positive definite, and could not have inverse, then, it cannot be used as input for further analysis. It can be converted to a near-positive-definite matrix, but the result must be reviewed with caution to avoid mis results.

4.3 Available Methods for Analysis

4.3.1 Data Set

In the following sections, I will show how the methodologies described previously work using a data set generated by the Global Maize Program of CIMMYT (International Maize and Wheat Improvement Centre). The data set is the result of the evaluation of 18 genotypes in multi-environmental trials run in 14 environments

Table 4.1 List of environments in the data set

Environment	Location	Year	Management
E13O	L1	2013	Optimum
E14O	L1	2014	Optimum
E23I	L2	2013	Intercropping
E23N	L2	2013	Managed low N
E23O	L2	2013	Optimum
E24N	L2	2014	Managed low N
E24O	L2	2014	Optimum
E33I	L3	2013	Intercropping
E33O	L3	2013	Optimum
E44O	L4	2014	Optimum
E53O	L5	2013	Optimum
E54O	L5	2014	Optimum
E63I	L6	2013	Intercropping
E63O	L6	2013	Optimum

as a result of the combination of 6 locations (L1, L2, L3, L4, L5, and L6), 2 years (2013 and 2014), and 3 managements (Optimum, Managed Low Nitrogen, and Intercropping) (Table 4.1). The 18 genotypes are labeled from 1 to 18. The experimental design was a randomized complete block design with three replicates. The following traits were measured in each trial: grain yield (GY, t ha^{-1}), days to anthesis (AD), flowering interval (ASI, days), plant height (PH, cm), ear height (EH), ear position (EPO, ratio [ear height divided plant by height]), grain moisture content (MOI, %), number of ears per plant (EPP), ear aspect (EA, scale from 1 to 5), and ear rot (ER, %). Only ear rot was not measured in Melkassa under optimum management in 2013.

To analyze the grain yield genotype by environment interaction, first, we fitted a linear mixed model (1) using a factor analytical structure (FA) for the environment by genotype interaction effects:

$$y_{ijk} = \text{mu} + \text{environment}_i + \text{block}(\text{environment})_{ij} + \text{environment} \times \text{genotype}_{ik} + \varepsilon_{ijk} \tag{4.1}$$

The overall mean (mu) and the environmental effects were considered as fixed effects. The random effect environment × genotype has a matrix of variance–covariance equals $\Sigma = (\Gamma\Gamma' + \Psi) \otimes \sigma^2 I$, where Γ is the matrix of environmental loadings, $\otimes$ means the Kronecker product, Ψ is a diagonal matrix with specific variances, and σ^2 is a scale parameter. The identity means that genotype effects are independent, there is no relationship between genotypes. The sub-indices i, j, and k identify the individual environments, replicates, and genotypes, respectively. The experimental error matrix of variance–covariances is a diagonal matrix with a different variance in each environment. From this analysis we have a first approximation of the grain yield genotype by environment interaction (Fig. 4.1).

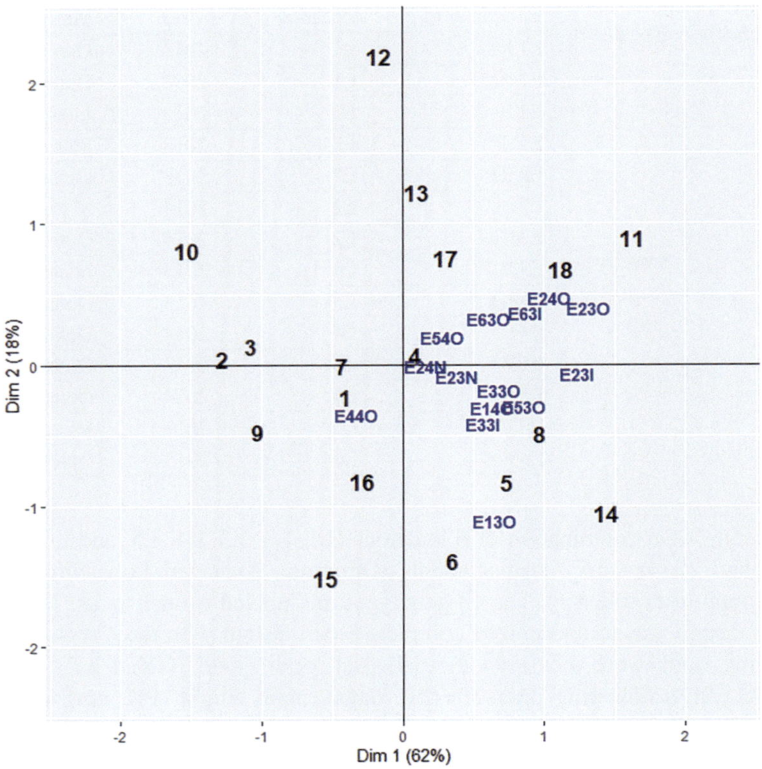

Fig. 4.1 Grain yield biplot obtained using a linear mixed model with a factor analytical structure for the environment effect, Model 1

Because data is complete, for simplicity we will work with the three-dimensional array of adjusted means of genotypes by environments by traits which we got from model 1, with all effects considered as fixed effects. Also, for simplicity, a subset of the data was used as input for the analysis. A usual practice is to not consider the error in the estimation of the adjusted means of genotype by environment table. Adjusted means pass to a second-stage analysis without carrying on their variability. This certainly will affect the results when the data has some degree of imbalance. More formal approaches, like using generalized linear mixed models, multivariate models, and Bayesian methods, are better options. A sub-optimal approach is to weigh the adjusted means by some function of the matrix of variance–covariance of the adjusted means.

The site regression model fitted with model 1 results in the biplot in Fig. 4.1. It shows that except E44O, all the other environments are on the right side of the plot, and all of them have pairwise positive correlation of different degrees. Environment E13O shows the largest absolute value in the second axis, being the environment, which produces the largest genotype by environment interaction. Interpreting similarly the second axis, the genotypes 6, 12, 13, 14, and 15 are those with largest genotype by environment interaction. The most stable environments that do not contribute to discrimination between genotypes are E24N and E23N, located in the middle of the figure (Fig. 4.1). In the case of the site regression model, the first axis explains the overall performance of genotypes; more yielding genotypes are located on the right and less yielding on the left. The second axis measures the genotype by environment interaction; genotypes close to the zero horizontal line have the most stable yield.

Up to here, I analyzed the grain yield genotype by environment interaction, but I did not explain why the genotypes and environments present that pattern. Besides environmental and genotypic characteristics which can explain the results, we are interested in understanding how other traits act to generate this pattern of genotype by environment interaction.

A brief description of the alternative methodologies to explore the grain yield genotype by environment interaction will be provided in next sections using this data set. There are variations of the methodologies as well as several parameters that can be tuned when the methodologies are used; I ask the readers to read the cited literature or any other missing literature in this chapter.

4.3.2 Analysis of Covariance

One of the first used methodologies is the analysis of covariance by Efren and Fisher in 1927, but which was previously published by Fisher in 1925 (Eden and Fisher 1927; Fisher 1958). They described the analysis of oat grain yield under different treatments (environments) using a related trait as straw yield. I reproduce part of the discussion of the paper, which clearly illustrates how the analysis of covariance can explain the grain yield genotype by treatment (environment) interaction based in a related trait as straw:

> It should be observed in passing that with the exception of the large effect ascribable to the first dressing, and to the interaction of quantity and time of application the effect upon the grain is in the six other cases of opposite sign to the effect upon the straw. For each of the four causes of variation into which the total has been divided a correlation between grain and straw may at once be calculated simply by dividing the sum of the products by the geometric mean of the two sums of squares. These correlations are shown in the last column of Table IV. The values found for the blocks and for the treatments being based on only a few degrees of freedom, are of course liable to large sampling errors, and cannot in fact be taken as significantly different from the remaining two better established correlations. If, however, we confine attention to the eight treatments which received single or double dressings, the correlation for the seven comparisons among them is -0.468, which is now significantly different from all of the others. This fact will have an important bearing in the interpretation of the results and serves to illustrate a principle of some importance in the use of the comparatively precise types of experiments now possible.
>
> Nevertheless, the early nitrogenous dressings produced a considerable increase of about 7–3 cwt. of straw to the acre for each cwt. of sulphate of ammonia or its equivalent in muriate. A

> large part of this increase was undoubtedly due to increased tiller formation, and in such, a heavy crop it would be surprising if it should be reflected proportionately in an increased yield of grain. Indeed the additional yield in grain from either the single or the double dressings is only about 3 bushels to the acre. We may conclude that much of the additional tiller formation was superfluous in the sense of not leading to additional ears. With the late dressings in June the condition limiting the effect of the top dressing had become much more stringent; no additional yield of straw was produced by the second cwt. of manure, and no additional grain by either the single or the double dressing. This again fits with the view that such late dressed nitrogen as was taken up by the plant, was utilized only in additional and abortive tillers. The correlation between grain and straw found in Table IV indicates that in the differences between different treatments, though not in the larger differences between blocks of land, some factors were at work acting antagonistically as between grain and straw. Thus the differences in grain yield on the treated plots, though individually insignificant, yet in relation to the straw yields of the same treatments show a significantly subnormal correlation. The same conclusion is suggested, though not so clearly demonstrated, by the low grain-straw correlation in the differential responses of the different treatments on the different blocks of land. The values of Table III show that this effect is not confined to any one manuxial contrast but emerges in considering the aggregate of all. It is not improbable that we have here an indirect effect of the laying of the crop, which has interfered in other ways remarkably little with the success of the experiment, but which may well have hindered grain maturation somewhat more on plots having the greatest abundance of straw. If this is the case the straw yields will afford the better basis for estimating the effects of the manurial treatments on plant nutrition, while the laying of the crop and the small response in additional grain, represent a sporadic risk of loss to which heavy crops are particularly exposed.

In the classical presentation of the analysis of covariance, it is used to control a source of variability originated by a covariable which is not related with treatment effects. The experimental error is reduced, and treatment effects comparison is more accurate. However, like Eden and Fisher (1927) did, and they used first, a covariable related with treatment effects (environments) can be included and it can help to better understand the response in the main variable, grain yield. The analysis of covariance has some limitations, the relationship between the response and the covariable should be linear, and the number of covariables that can be analyzed simultaneously is limited by the degrees of freedom (DF).

A more extensive analysis is the path analysis or as it is known in the economic language, structural equations. Amply used in the past by physiologists, the method analyzes the relationships between many traits with grain yield to identify the grain yield components and to assess their importance. The methodology is not intended to analyze the genotype by environment interaction; however, if the analysis is run for different environments, the paths can be compared and the genotype by environment interaction can be considered.

4.3.3 Path Analysis

The path analysis is more complete than the analysis of covariance since it considers the correlations and partial correlations between all traits. It is possible also to include an error variance for each trait, and covariance estimations between external variables. The analysis starts with a conceptual path between traits. Traits are connected according to researcher knowledge about the relationships between traits. All relationships are written in form of equations and coefficients, which define the degree of relationship between traits are estimated. For each trait direct and indirect effects on the related traits could be estimated, depending on the established relationships, and the null hypothesis of equality to zero is evaluated. Path analysis is useful to analyze the grain yield genotype by environment interaction and related traits because grain yield is the final trait in the path while the other traits build the path to it. Tai et al. (1994) proposed the use of path coefficient analysis to interpret the genotype by environment interaction. The path regression equation is based on a causation model, in which traits are aligned in the order in which they happen in the field, plus an environmental component for each one. In 1990 and 1994, he published about the same topic using the same methodology to describe the genotype by environment interaction in potatoes (Tai et al. 1994). The same model was used by Mohammadi et al. (2016) and by Pour-Aboughadareh et al. (2022), to analyze the genotype by environment interaction in wheat and barley, respectively. A model based on causation looks in some way like an optimal model, but it is difficult in some cases to define the causation relationships. And although some relationships are not causal, they can help interpret the genotype by environment interaction. Ordering the traits and using causal relationships put some limits to this model. Later, Vargas et al. (2007) used a path analysis which is not based on causation, in which the traits are not ordered, and that includes environmental covariables. The flexibility of the model allows a more extensive analysis of the genotype by environment interaction; however, the complexity makes it more difficult to interpret the results.

An illustration of the method will be done applying path analysis to two environments of the example data set in Table 4.1 (E24O and E44O). The analysis was generated using SAS/STAT software, Version 9.4 of the SAS System for Windows, using the capabilities of PROC CALIS of SAS V9.4 (Copyright © 2023 SAS Institute Inc.; SAS and all other SAS Institute Inc. products or service names are registered trademarks or trademarks of SAS Institute Inc., Cary, NC, USA) SAS Institute Inc. (2023a). The input data is the table of adjusted means of genotypes (rows) by environment × trait (columns). The SAS code for the initial complete path is:

```
PROC CALIS data=AME maxiter=5000000;
where env="E24O" or env="E44O";
by env;
path
gy <- moi,
gy <- asi,
gy <- er,
gy <- epp,
gy <- ph,
gy <- ad,
```

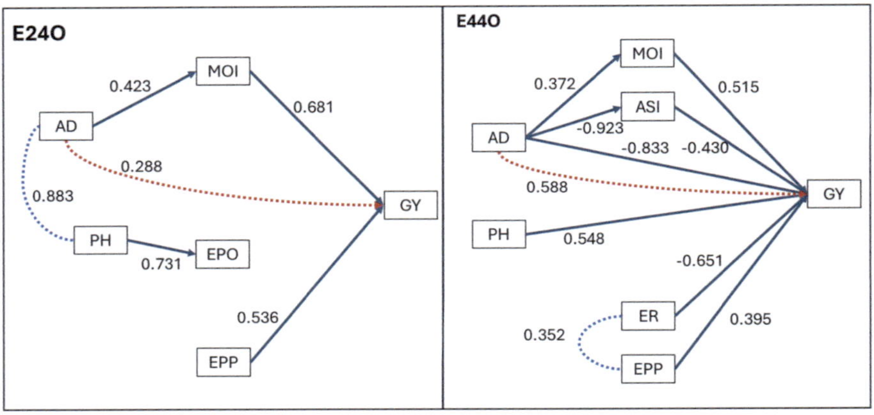

Fig. 4.2 Final fitted path in two environments, E24O and E44O, with standardized coefficients. Black lines correspond to direct effects, dotted red lines to indirect effects, and dotted blue lines to covariances between external traits. Traits are: MOI grain moisture content, AD anthesis date, ASI flowering interval, PH plant height, ER ear rot, EPO ear position, EPP ears per plant, and GY grain yield

```
moi <- ad,
asi <- ad,
epo <- ph,
epp <- ph;
effpart
gy <- ad ph;
run;
```

The first line calls the procedure and defines the number of iterations. Next, we ask to do the analysis by environment, then we define the path. The symbol <– indicates which traits affect which one. The *effpart* is to estimate direct and indirect effects for those traits. All traits with non-significant effects were removed in each environment, and a new path was fitted. The results for the two environments are presented in Fig. 4.2. The first difference is the complexity of both paths; environment E24O presents a simpler path than environment E44O. In E24O, only MOI and EPP affect grain yield while in E44O, ASI, ER, and PH are related with grain yield and all with negative effects.

Looking at the genotype by trait table in each environment, it can be noted that ER is on average 2.8 times larger in E44O than in E24O; if a very susceptible genotype with an average of 55% is not considered, the same ratio falls to 1.6 if it is included in the calculus. Considering other traits, E44O presented 11.9 less days to AD, 3.2 days more in flowering interval, and 44.1 cm less in plant height. Considering the two best and worst grain yield genotypes in each environment, we can see that one of the best in E24O is one of the worst in E44O. That genotype is among the genotypes with lower AD and PH and present the highest ER rate. This is a simple presentation about how the path analysis can be used and results be interpreted to analyze grain yield genotype by environment interaction and the effects of related traits.

4.3.4 Factorial Regression

Analysis of covariance and path analysis do not explicitly analyze the grain yield genotype by environment interaction. The use of factorial regression analysis (Vargas et al. 1999a) is an attempt to explicitly analyze the genotype by environment interaction considering external variables, which can explain the genotype by environment interaction. In factorial regression, external covariables are included in the linear model like covariates, replacing the genotype by environment interaction effect. A regression coefficient is estimated like in the analysis of covariance, but a different one is estimated for each environment. The number of covariables to include is limited by the available degree of freedoms. Vargas et al. (1999a) used factorial regression to explain the genotype by environment interaction association with environmental and molecular data. Despite that, it is easy to include grain yield-related traits instead of using molecular or environmental variables. However, we must consider some points. Grain yield-related traits are result of genotypic and environmental effects, therefore when they are included in the model, they capture part of the genotypic or environmental effects; a solution is to use sum of squares type I, and putting genotype and environment factors first in the model, as well as all design effects. Using sum of squares type I has an additional effect, which is that the covariates should be entered in the model in order of importance. Related traits, with high correlation, generate problems of multi-collinearity. To deal with multi-collinearity problems, we can use sum of squares type I and do a selection procedure of covariables like forward selection, which was the method used with the example. Other two options are to use a method of dimension reduction like PCA to generate new dimensions, which will work as covariate, or to use a kind of penalized regression. See the work of Piepho and Blancon (2023), which discuss the issue in more detail. The next methodology we will describe uses PLS, which is able to deal with the problem of multi-collinearity. PLS was used to relate yearly differences in straw length of barley genotypes to climatic variations over 9 years (Aastveit and Martens 1986). Previously, various authors cited by this reference had published about the use of regression models to analyze the genotype by environment interaction and to consider environmental variables to interpret it.

We used two different approaches to the example data set. In the first, we used as covariates the genotype adjusted means of related traits across environments; this is like using a genotype index, like environmental index used in Eberhart and Russel (1966) or Finlay and Wilinson (1963) methods. For the second case, we used the genotype adjusted means of genotypes in each environment. Therefore, we estimated a regression coefficient for each related trait with grain yield in each environment. The analysis was done using the capabilities of PROC GLMSELECT of SAS (2023c); then we run the final model using PROC GLM capabilities (SAS 2023b). A software to fit factorial regression models among other models to analyze the genotype by environment interaction was developed by CIMMYT and is freely available (Pacheco-Gil et al. 2015). The input for the analysis is the raw data with the additional columns of adjusted means of genotypes, across environments and by environment of the related traits. Adjusted means were standardized before the analysis to avoid the effect of the units of measurements and scales. The SAS code is:

```
* Using the average of each genotype across all environments;
proc glmselect data=era;
class rep env gen;
model gy = env gen rep(env) ADc*env ASIc*env EAc*env EHc*env EPOc*env
EPPc*env ERc*env MOIc*env PHc*env /
selection=forward slentry=0.05 select=SL INCLUDE=3 stb SHOWPVALUES;
run ;

proc glm data=era;
class rep env gen;
model gy = env gen rep(env) ADc*env ASIc*env EPPc*env / SS1 solution;
run;

* Using the average of each genotype in each environment;
proc glmselect data=era;
class rep env gen;
model gy = env gen rep(env) env*AD env*ASI env*EPP env*MOI env*PH
env*EPO env*ER /
selection=forward slentry=0.05 select=SL INCLUDE=3 SHOWPVALUES;
run ;

proc glm data=era;
class rep env gen;
model gy = env gen rep(env) env*AD env*ASI env*EPP env*MOI env*PH env*ER
/ SS1 solution;
run;
```

We used forward selection method for selecting related traits and the criterion to enter the variable in the model was the significance level ($p \leq 0.05$) defined with the instruction select = SL and slentry = 0.05; the three main effects of environment, replicate within environment, and genotype were forced to be included in the model (INCLUDE = 3). Finally, we asked for printing the p-values (SHOWPVALUES).

Table 4.2 shows the analysis of variance (ANOVA) table of the three fitted models. The interaction genotype by environment accounts for 14.4% of the total variability measured as sum of squares, after environment, which explain 59.5%.

Table 4.2 ANOVA table and percentage of sum of squares of the Model 1 and the models that replace the genotype–environment interaction by related traits–environment interaction components

	Source of variation	DF	Type I SS	*F* value	Pr > *F*	%
Model 1	Model	279	3763	8.47	<0.0001	83.2
	Error	476	757.6	–	–	16.8
	Environment (Env)	13	2690	130	<0.0001	59.5
	Replicate/ (Env)	28	94.1	2.11	0.0009	2.1
	Genotype (Gen)	17	328.0	12.12	<0.0001	7.3
	Gen × Env	221	651.0	1.85	<0.0001	14.4
	Corrected total	755	4521	–	–	–
Model with related traits-adjusted means across environments	Model	97	3372	19.92	<0.0001	75.4
	Error	658	1149	–	–	24.6
	ADc × Env	13	151	6.66	<0.0001	23.2
	ASIc × Env	13	49	2.14	0.0107	7.5
	EPPc × Env	13	60	2.66	0.0012	9.3
	Total related traits	39	260	3.61	<0.0001	40.0
Model with related traits-adjusted means by environments	Model	133	3587	19.77	<0.0001	84.9
	Error	562	766	–	–	15.1
	AD × Env	13	103	5.80	<0.0001	15.8
	ASI × Env	13	40	2.26	0.0067	6.2
	EPP × Env	13	120	6.75	<0.0001	18.4
	MOI × Env	13	113	6.36	<0.0001	17.3
	PH × Env	13	85	4.77	<0.0001	13.0
	ER × Env	13	60	3.39	<0.0001	9.2
	Total related traits	78	520	1.60	<0.0001	79.9

Note: The percentages are calculated over the corrected total sum of squares for Model 1, and the Model and Error sources; for the related traits, the percentage is calculated over the genotype by environment interaction sum of squares (SS)

The model which considered the adjusted means of related traits across environments captures 40.0% of the grain yield genotype by environment interaction while the model with adjusted means by environments reaches 79.9%. Three related variables explain this 40.0%: AD, which explains 23.2, EPP with 9.3%, and ASI with 7.5%. The model using the related traits averaged by environments considered the same trait variables, ASI, EPP, and ASI, which explains 15.8, 18.4, and 6.2%, respectively, explaining only a 0.4% additional than the previous model. Additionally, MOI, PH, and ER are the other related traits which explain the difference of 39.5% between models. These analyses allow the researchers to identify those

Table 4.3 Coefficients of regression for related traits and grain yield by environment

Environment	AD	ASI	EPP	ER	MOI	PH
E13O	**−0.180**[a]	0.091	**1.561**	−0.063	−0.021	**0.020**
E14O	−0.022	**0.217**	0.327	−0.026	**−0.258**	**0.026**
E23I	−0.042	0.054	**2.172**	**−0.114**	0.000	**0.025**
E23N	−0.070	−0.004	1.410	−0.008	−0.030	**0.027**
E23O	−0.064	0.170	**2.026**	**−0.079**	−0.087	0.015
E24N	−0.073	−0.178	**2.603**	−0.002	−0.011	0.012
E24O	**0.145**	−0.003	**3.044**	−0.003	0.194	0.008
E33I	−0.125	−0.132	0.516	−0.019	**0.269**	**0.014**
E33O	−0.021	**0.073**	−0.404	**−0.092**	0.124	**0.023**
E44O	**−0.485**	**−0.664**	**1.931**	**−0.082**	**0.373**	**0.035**
E54O	−0.106	0.007	**3.486**	−0.032	0.059	**0.019**
E63I	−0.073	0.001	−0.713	−0.053	0.132	0.016
E63O	−0.077	0.064	1.470	−0.086	0.130	0.013

[a]Coefficients in bold are statistically different of zero; $p \leq 0.05$

related traits that explain the grain yield by environment interaction and to give a grade of the magnitude.

As an example of the additional information we can get from this analysis, Table 4.3 shows the regression coefficient of each related trait with grain yield in each environment. No one trait shows a significant coefficient in all environments. Considering the significant coefficients ($p \leq 0.05$), three traits (EPP, ER, and PH) show all of them positive (EPP and PH) or negative sign (ER), while for the other three traits, the sign of the coefficient changes being positive in some environments and negative in others. Because the coefficients are expressed in t/ha per standard deviations of the related traits, we can see that EPP has the largest effect and PH the lowest. The contribution to explain the grain yield genotype by environment interaction of the related traits cannot be evaluated by the coefficient matrix. It is true that the contribution in terms of sum of squares is related to the presence of significant coefficients with different magnitudes and signs, but the relationships are not strong. An important detail of the analysis is that we are fixing the variability explained by the genotypes and environments as well as for the design factors while the genotype by environment interaction is sent to the residual term. The related traits will extract information of the genotype by environment interaction from the residual of that model; we need to assume that the related traits are not related to the experimental error of grain yield.

Interpreting Table 4.3 could not be straight forward, depending on the size of it and the complexity of the coefficients. Like any two-way table, it can be analyzed to identify patterns using methods like PCA, multidimensional scaling (MDS), or clustering.

4.3.5 Partial Least Square (PLS)

The use of PLS to analyze genotype by environment interaction can be tracked back till 1975. But the methodology takes power with the already cited works of Vargas, Crossa, and van Eeuwijk. Partial least squares take off some limitations of the factorial regression analysis. It allows us to deal with multi-collinearity issues in an effortless way than factorial regression does. In factorial regression we need to use a method for variable selection of which there are many and some of them are not directly associated with our objective. PLS can be considered as a variable selection method that, to some degree, matches better with our objective since it retains those variables that contribute more to explaining the variability. PLS also deals with the usual problem that we have a lower number of "samples" (genotypes) than variables (traits). Besides, it produces new independent dimensions that are easier to interpret because of their orthogonality. The original works considered external variables like molecular data, QTL, and environmental variables; however, we can use related traits with more cautious considerations. Molecular data is a fixed characteristic of the genotypes that do not change by the environment although their effects can change; similarly, environmental covariables have the same values for all genotypes that are in the same environment. On the other hand, related traits are not fixed; they have different values for each genotype in each environment and the relationships with grain yield and between them can change from environment to environment.

The results of the analysis can be represented with the classical biplot for genotype by environment interaction in which the related traits are included (triplot). The general interpretation of the biplot is like the interpretations of the classical biplot. The position of the related traits, length and angle of the vector from the origin, is related to the contribution to the new dimensions and the relationships with the genotypes and environments, while the relative position with genotypes and environments measures the degree of correlation between them.

The input data set to run a PLS analysis is the same as we used for factorial regression with a different format. PLS analyzes the relationships between two matrices: the response matrix, which corresponds to the grain yield genotype by environment interaction, that is, the adjusted means of grain yield of each genotype in each environment; and an explicative matrix, with the adjusted means of the related traits for each genotype across environments or by environment. We did the analysis using the software GEA-R (Pacheco-Gil et al. 2015). For doing it, we need two input files: the raw data in standard format with one observation per row, or the grain yield adjusted means for each genotype in each environment, and one data set with genotypes in the rows and the related traits in the columns.

From the triplot (Fig. 4.3) we can conclude that EA is correlated with ER. Ear rot affected more the environments E44O and E63O and the susceptible genotype 10, while lower effect of the disease was present in environments E24N, E23N, and E23O, affecting the genotypes 1, 2, 11, and 16. Genotypes 12 and 17 presented medium to large values of ER and EA, but they are moved to the left because they presented medium values of ASI and low values of PH, AD, EH, and EPO. Associated with environments E54O and E33I, there are the genotypes 7, 8, and

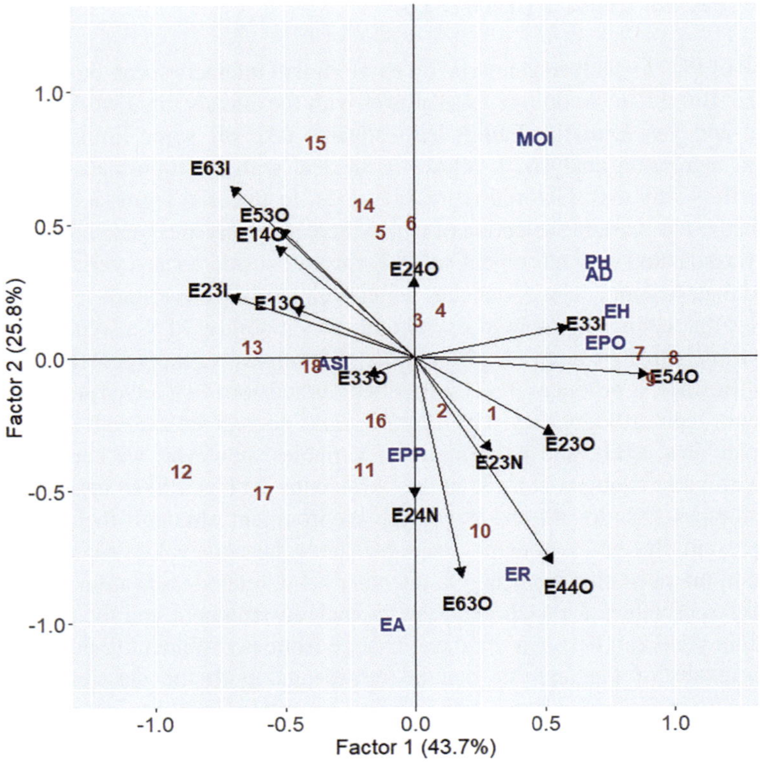

Fig. 4.3 PLS triplot of grain yield genotype by environment interaction including related traits

9 that responded in those environments with larger values of PH, AD, EH, and EPO. Remember that the relationships between genotypes and environments are determined by grain yield. ASI and EPP are traits with low contribution to the grain yield genotype by environment interaction since they are the closest to the origin. More relationships can be figured out from the biplot, I let those for the readers. A paper by Reynolds et al. (2022) presents a detailed analysis of the effect of physiological traits on the grain yield genotype by environment interaction; it is a skilled reading to understand details of the methodology.

4.3.6 Marginal Analysis and Comparison

Although there are methodologies to analyze the grain yield genotype by environment interaction and related traits, many researchers continue using marginal analysis. That is, to analyze two-way tables derived from larger dimensional array.

For example, Birhanu et al. (2024) and Sahmat et al. (2024) analyzed the genotype by environment interaction of wheat genotypes using AMMI and SREG models of several traits; then the observed results on the biplots were compared only by visualization. Jadhav et al. (2019) used principal components analysis and the percentage of contribution of the components for the different traits. The list of publications is extensive, only few more ones will be referenced; for example, Katsenios et al. (2021), Negash and Birr (2022), Martínez-Peña et al. (2023), Kirouani et al. (2023), and Justo et al. (2024). In any case, the analysis considers the comparison of two-way tables, or derived matrices of correlation, distances, or new dimensions obtained using some multivariate method.

Like when we want to know if two variables have a similar pattern, we use some type of correlation. The comparison of two matrices can be done using Mantel correlation (Mantel 1967). The only restriction is that the two matrices need to have the same dimensions. The test is more known in areas like ecology or botany; its use in agronomy is not extensive. Procrustes is another less frequently used methodology in agronomy, developed to compare multidimensional structures. Given two arrays of the same dimensions, the points of one of them are scaled, translated, and rotated to match both arrays. This methodology is frequently used in studies with judges or consumers because each one can use the scales used to measure attributes of the samples in a unique way. In agronomy, for example, different farmers can use the same scale to evaluate a disease but usually use different sectors of the scales; some farmers used to assign largest values than others, for example.

To exemplify the methodology, we started using model 1 to get the FA loadings of environments for five traits (GY, AD, EPP, PH, MOI). Variables were standardized to zero mean and variance 1 across all the data to avoid scale effects. The objective is to compare the distribution of environments in the biplot. The analysis was done in R (R Core Team 2024) using the library ecodist to get Mantel correlation and vegan to do the Procrustes analysis. Both libraries have an instruction Mantel but while ecodist can deal with rectangular matrices, vegan cannot.

The R-code to get Mantel and Procrustes results is:

```
# Mantel correlation
library(ecodist)
xmantel(pher ~ ader, dims=c(14,2))

# Procrustes
library(vegan)
PH_AD <- procrustes(pher,ader))
summary(PH_AD)
plot(PH_AD, kind=1)
plot(PH_AD, kind=2)
```

Where, the objects pher and ader are the rotated environmental loading matrices of PH and AD, respectively. Summary and plot are used to get the results of Procrustes analysis.

Table 4.4 Mantel correlations and p-values of rotated environmental loadings of a factor analytical model 2, between selected traits

Trait	GY	AD	EPP	PH	MOI
GY	–	0.2230	0.4990	0.3740	0.4400
AD	0.5319	–	0.4740	0.0010	0.2290
EPP	0.4102	0.5984	–	0.6690	0.2250
PH	0.4442	0.8816	0.4874	–	0.5030
MOI	0.4360	0.6567	0.6100	0.5344	–

Correlation coefficients in the lower triangle; p-values in the upper diagonal

The first step was to rotate the loadings from the FA model 1 using a varimax rotation; then the first axis explains the maximum variance. Mantel correlation was calculated for all pairs of the 14 × 2 (14 environments, 2 factors) matrices of rotated environmental loadings. Table 4.4 shows the Mantel correlation and the test of a zero-correlation hypothesis. Although correlation coefficients are medium to high, only the correlation between AD and PH is statically significant. The large p-values could happen because of the small number of environments (samples). A significant correlation means that the distributional pattern of environments in the biplot are similar for both traits, however for the other combinations of traits the low correlation indicates different patterns.

To compare the pattern of environments in the biplot, we created the biplots (Fig. 4.4) of the four traits involved in the largest and lowest Mantel correlation, AD–PH (0.8816) and EPP–GY (0.4102). It is not easy to identify if patterns in the biplots are similar or not, and hence the importance of having the Mantel correlation. We can identify some environments that are more or less aligned in the biplots, like E13O, E53O, and E33O, which are located on the left side with similar distribution on the y-axis when comparing PH and AD biplots. Comparing EPP and GY biplots, it was not possible to identify environments with similar relative positions. The Procrustes analysis will help to understand the pattern relationships.

Table 4.5 shows the results of the Procrustes analysis plus the Mantel correlation for reference. The first seven rows of the table show fit statistics and the last six are more interesting. The off-diagonal elements in the matrix rotation indicates how much one pattern must be rotated to match with the other. A value close to zero means less rotation. The translation values indicate a relative difference of the pattern against the origin of the biplot (0,0) since biplots are centered in zero, while the scale indicates how much spread in the biplot are the samples in one array than in the other. The rotation coefficients show that PH–AD comparison (largest Mantel correlation) needs less rotation than the EPP–GY comparison (lower Mantel correlation). Two additional figures can help to understand the differences and similarities between patterns.

Figure 4.5 shows both figures for the two comparisons. In the top figures, both patterns are plotted together, the reference or target pattern and the other pattern after rotation and scaling. The lines indicate the necessary movement of each sample to match the reference pattern. Sometimes it is important to consider the order of the matrices because one can be considered the target and the other the matrix to modify.

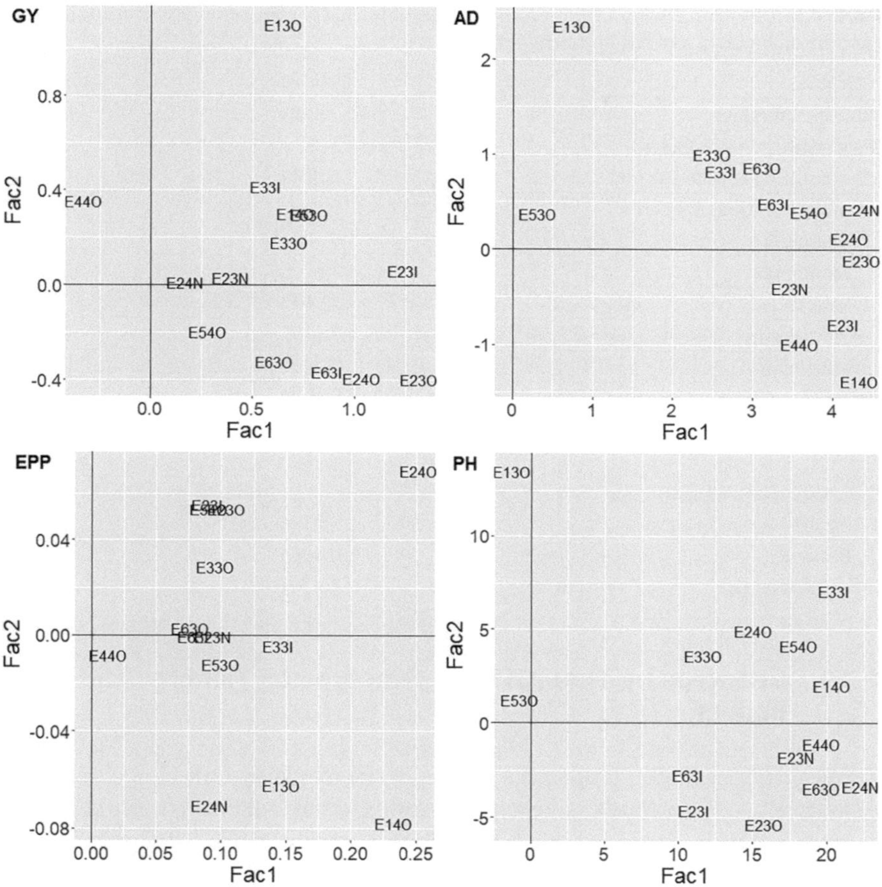

Fig. 4.4 Biplot of rotated environmental loadings of four selected traits: AD days to anthesis, PH plant height, EPP ears per plant, and GY grain yield

The plots also show the original and the rotated axes. The comparison of EPP–GY shows largest lines and more rotation of the axis, indicating a major difference in patterns than for PH-AD comparison. At the bottom of the figure, the residuals for each environment, are plotted in the same order in which they are in the matrix. Largest residual is a sign that it is more difficult to match the environment in the new consensus pattern. Three references with application of Procrustes analysis are Pontes et al. (2020), Yi et al. (2021), and Bramardi et al. (2005). The first uses Procrustes analysis to select traits for coffee phenotyping, the second uses Mantel correlation to analyze the relationship between distance matrices obtained from molecular and genetic data in Bromus, and, finally, the third paper uses both, Mantel correlation and Procrustes analysis, to compare agronomic and molecular characterization of cucumber.

Table 4.5 Results summary of Procrustes analysis plus Mantel correlation of the factor loadings comparison between PH–AD and EPP–GY

Statistics	PH–AD	EPP–GY
Mantel correlation:	0.882	0.410
Procrustes sum of squares:	516	0.0564
Procrustes root mean squared error:	6.0684	0.0635
Quantiles of Procrustes errors:		
Minimum	0.7436	0.0039
1st quartile	3.6206	0.0319
Median	5.8643	0.0428
3rd quartile	7.0292	0.0625
Maximum	9.2282	0.1268
Rotation matrix in three-column format:	0.9995	0.9206
	−0.0317	0.3905
	0.9995	−0.9206
Translation of averages:		
Factor 1	1.0144	0.0705
Factor 2	0.4945	−0.0109
Scaling of target:	4.1702	0.0716

4.3.7 Genotype by Trait (GT) and Genotype by Yield by Trait (GYT) Biplots

Yan and Rajcan (2002) proposed a GT biplot, genotype by trait biplot, generated from the PCA of the matrix of genotypes (rows) by trait (columns). It has been used in several papers (Paramesh et al. 2016; Atnaf et al. 2017). These papers also considered the correlation between traits calculated by environment (Atnaf et al. 2017; Shojaei et al. 2022). Interestingly, no one used Mantel or Procrustes analysis. The analysis of GT biplot by environment can be used to understand, with limitations, the grain yield by environment interaction and related traits relationships. It was in 2018 when a specific analysis of genotypes by environment interaction for grain yield considering grain yield-related traits relationships was proposed by Yan and Frégeau-Reid (2018). Their proposal modifies the input matrix for the singular value decomposition (SVD). GE, GGE, or GT matrices are replaced by GYT matrix. The GT matrix of genotypes by traits is transformed into the GYT matrix. The transformation consists in multiplying each trait by grain yield or its inverse depending on the correlation between grain yield and the trait. If the correlation is positive then the trait is multiplied by grain yield while if the correlation is negative, the trait is divided by grain yield. Therefore, all transformed traits can be analyzed similarly.

After the publication, of GT and GYT biplot, Yan and Rajcan (2002) approaches have been used in many publications for several crops. The use of GT biplot can also be found in other publications (Hassani et al. 2024; Yan et al. 2023; Mohammadi 2019). Some authors like Kendal (2019), Gholizadeh and Ghaffari (2023), and

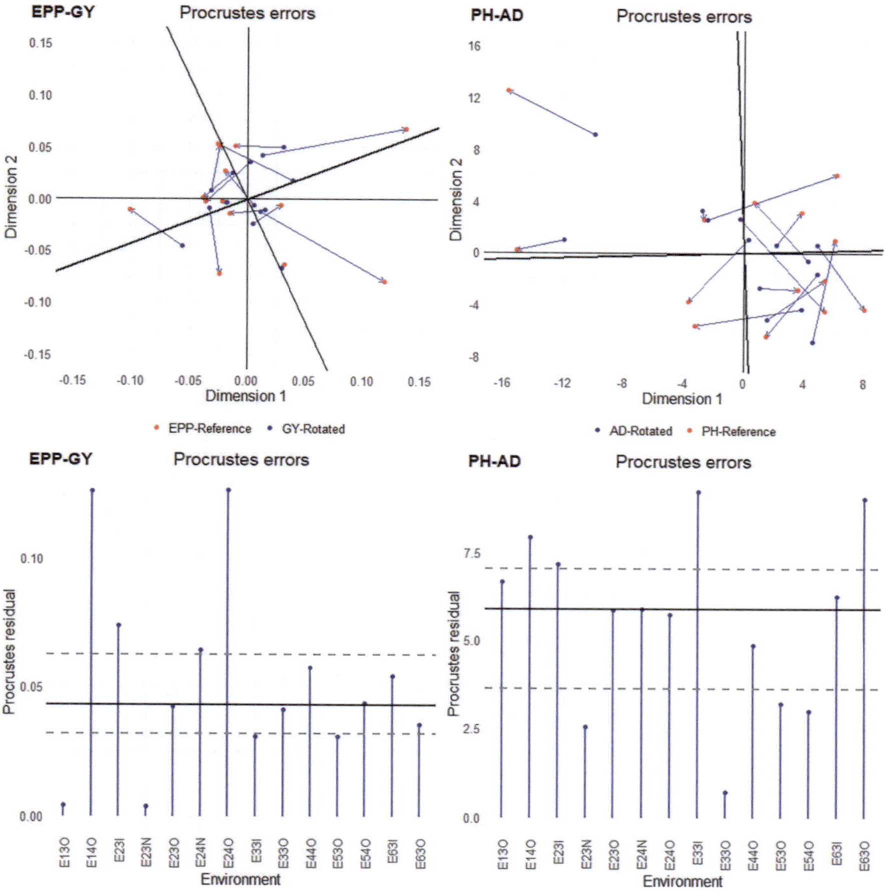

Fig. 4.5 Rotation pattern (top) and residuals (bottom) from Procrustes analysis of environmental loadings of site regression model of four traits. Comparison of ears per plant (EPP) with grain yield (GY) (left), and plant height (PH) with anthesis date (AD) (right)

Shojaei et al. (2023) used both biplots, GT and GYT. Yan et al. (2023) used GYT biplot but also proposed a new biplot that used the GYT index obtained from the GYT biplot calculated for each genotype in each environment, to build a matrix of environments by GYT indices. That matrix is analyzed as usual, by PCA or SVD analysis (Yan et al. 2023). As mentioned previously, most of the available methodologies are descriptive and no unique one will offer the answers we are looking for. It is important to consider more than one methodology that if well interpreted will help us to understand the complex problem we are dealing with.

From our data set, we will work only with two environments E13O and E23O, same year and management, and distinct locations. For each one, we create the GT table. First we calculate the means of all traits for each genotype, and then each trait is multiplied or divided by grain yield according to the direction of selection of the related trait. We look for low values of AD, EA, and MOI, therefore we divide GY by those variables while for EPP we like large values so we multiply GY by EPP. There are some issues in applying this method. For some traits it is not easy to identify the direction because we do not want to change the average value; this can happen with PH and EPO for example. We are dividing by them, but we need to consider in the interpretation of the biplot that we are looking for average values of those traits but large values of grain yield. Another issue is with traits like flowering interval because our interest is a value of zero, complete synchronization of male and female flowering. We can use absolute values; however, in many cases we will have values of zero that produce an infinite value when it divides GY. In our case, ER cannot be used because of the presence of zeros. Traits that can have negative or zero values cannot be analyzed. After those considerations, we continue working with GY/AD, GY/EA, GY/MOI, and GY × EPP. All transformed variables were standardized to zero mean and variance 1.

Individual GYT matrices were analyzed using PCA in R and the scores of genotypes of both environments were compared using Mantel correlation and Procrustes analysis. The code of analysis is:

```
# PCA analysis
E13Op <- princomp(E13O)
E23Op <- princomp(E23O)

# Mantel and Procrustes analysis
xmantel(E13Op$scores[,1:2] ~ E23Op$scores[,1:2], dims=c(18,2))
xmantel(E13Op$loadings[,1:2] ~ E23Op$loadings[,1:2], dims=c(4,2))

proc <- procrustes(E13Op$scores[,1:2], E23Op$scores[,1:2])
```

The objects E13O and E23O are the GYT matrices of each environment.

Figure 4.6 shows both biplots and we can see that the pattern of traits is similar; the Mantel correlation was 0.931 with a p-value of 0.045. The performance of genotypes in both environments is different, with a Mantel correlation of 0.074, which is not statistically significant (p-value $= 0.6990$). Next, we will compare the pattern of genotypes in both biplots using Procrustes analysis. The analysis shows that the rotational coefficient is -0.9729 and the scaling factor is 0.4643. The figures at the bottom of Fig. 4.6 show the large change in the position of genotypes and also large residuals for genotypes 2, 3, 9, 10, 11, and 18. See genotype 18 that is on the right side at the middle of y-axis in environment E13O but almost in the middle of the biplot in environment E23O. The response of genotype 18 on GY × EPP and GY/AD changes in the environments. Like in all these kinds of analyses, we need to take into account that we are representing a multidimensional problem in two dimensions, which produces distortion of the relationships between items, genotypes, environments, and traits. Procrustes rotation typically rotates one matrix to match with a target matrix, and the result is not symmetric, that is, it is important to

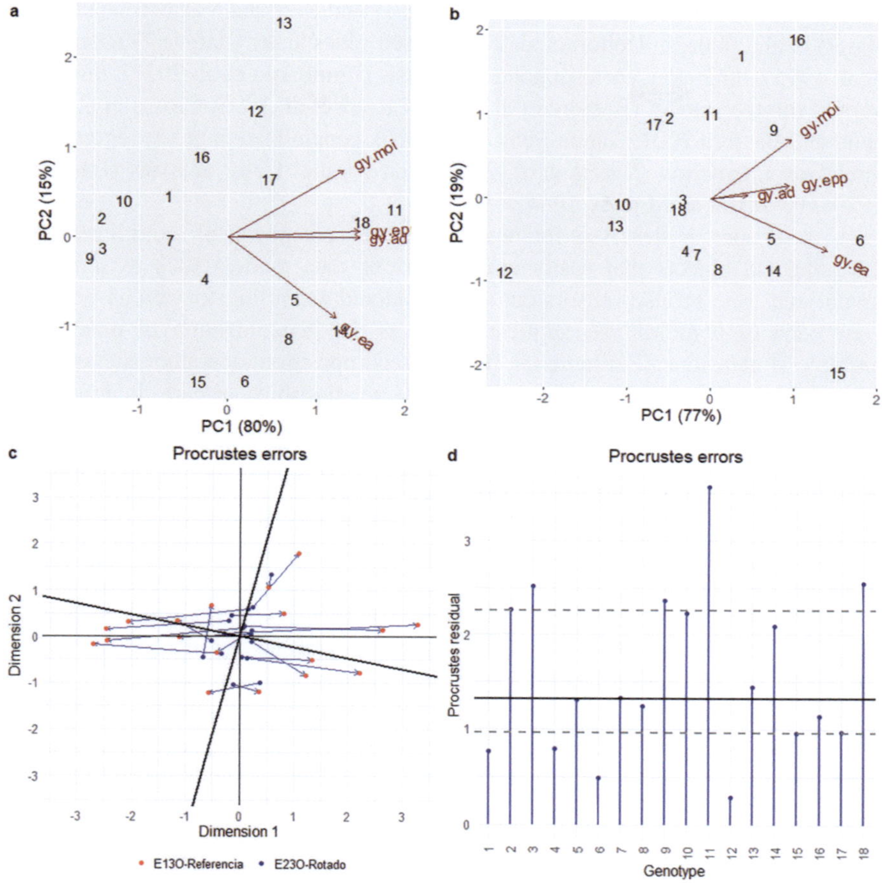

Fig. 4.6 GYT biplot in environments E13O and E23O of selected traits (top); rotation pattern and residual from Procrustes analysis of genotype scores in the biplots of genotypes (bottom)

consider which is the matrix to rotate and which is the target. In many situations this does not matter but it is important to know which matrix is playing what role. It is possible to generate a symmetrical rotation but we need to consider the situations we are working on to decide to do a symmetrical or non-symmetrical rotation.

4.3.8 Three-Way Three-Mode Analysis

Although not frequently used in agronomy or breeding, there are methods that deal directly with the multidimensional problem. Those methods can be found under diverse global names, three-way or three-mode, and with different specific names. Most of them are extensions of two-dimensional methods, which are frequently studied and used. Among those methods, we have the three-mode PCA analysis

(Kroonenberg and Basford 1989; Rincón et al. 1997; Kiers and Van Mechelen 2001; Varela et al. 2006a, b; Dettori et al. 2011), three-way cluster analysis (Kroonenberg et al. 1995), three-way correspondence analysis (Lombardo et al. 2023), three-way normal mixture model (Basford et al. 1991; Crossa et al. 1995; Crossa et al. 2002; Hunt and Basford 2001; García-Peña et al. 2016), generalization of site regression to three-way interaction (Varela et al. 2009), and parallel factor analysis (Harshman 1970; Harshman and Lundy 1984).

A word apart is deserved for tensors. Tensors are powerful multidimensional mathematical objects that easily embed various data models such as relational, graph, and time series. Tensors can be considered according to various points of view such as a multilinear application, or as the generalization of matrices to multiple dimensions. Furthermore, tensor decomposition operators are of great utility to reveal hidden patterns and complex relationships in data. It is used for various purposes such as dimensionality reduction, noise elimination, identification of latent factors, pattern discovery, ranking, and recommendation or data completion. One of these decompositions is Tucker, which factorizes a tensor with N dimensions into a smaller core tensor and a set of N factor matrices (Gillet et al. 2022). Tucker model uses a core tensor to represent the interactions between the component matrices. The core tensor in Tucker decomposition captures the complex relationships between the different modes. The Tucker model offers more flexibility than other methods like three-way PCA in how it represents the data's dimensionality, allowing for different numbers of components in each mode. The Tucker model's core tensor can be more challenging to interpret than the results of other methods. We will illustrate only one methodology which belongs to the category of three-way methods and will refer the readers to the cited literature for others.

4.3.9 Group Factor Analysis

Group factor analysis is an extension of the classical factor analysis in which the variables are grouped by some characteristics. In our case, we can group the variables by environments or by traits, the data are organized with genotypes in the rows, and the variables in the columns are grouped with some criteria. The method looks for latent factors which explain the relationships between samples, sets of variables (sources), and variables within sets (features) (Virtanen et al. 2012; Klami et al. 2014). Using Bayesian methods for the estimation and assuming sparsity on the coefficients of the factors, it is possible to use techniques of biclustering (Bunte et al. 2016). Leppäaho developed an R library, called GFA, to fit this kind of models (Leppäaho et al. 2017).

The methodology has been used in several scientific areas, in medicine (Cheng et al. 2013), crystallography (DeAngelis et al. 1972), and psychology (Merino Soto 2013), but not in agronomy or plant breeding. Farace, in his Doctoral thesis working with maize to produce ethanol, used grouping factor analysis to describe the relationship of several traits including grain yield and other yield-related traits in maize (del Farace 2023). The traits were grouped in four classes: grain chemical quality, grain physical characteristics, biomass chemical quality, and agronomical and yield-related traits.

Following with our example, we selected a subset of environments and traits for which we found interpretable results. We selected the environments E14O, E23O, E24O, E33I, and E53O and the selected variables were GY, AD, EPP, and PH. We considered five sources of data, groups, which correspond to each environment; in each environment we have four features which correspond to the traits. The genotypes are the rows of matrices. The analysis was run in R using the library GFA (Leppäaho et al. 2017). The data were organized in a list with each element of the list being a matrix of genotype by trait response in each environment.

The summary code in R is:

```
library(GFA)

# Normalize the data by feature (trait)
norm <- normalizeData(Y, type="scaleFeatures") # Normalize each
feature

# Fit the model after defining the options.
opts <- getDefaultOpts()
opts[c("iter.burnin", "iter.max", "iter.saved",
"convergenceCheck")] <- c(2500,523000, 100, TRUE)
res <- gfa(norm$train, K=10, opts=opts)

# Generate the results for interpreting and visualization
vis <- visualizeComponents(res, Y, norm)

# Correlation between observed (Yorig) and reconstructed matrix
library(ecodist)
Yh= vis[["Data reconstruction"]]
xmantel(Yorig ~ Yh, dims=c(18,24))

# Components by sources (environments)
CompAct= vis[["Component activities"]]

# Weight of genotypes in each factor
Factor1= vis$`Component 1 effect`
Factor2= vis$`Component 2 effect`
```

Where Y is the list with the data. The figures were produced using the libraries ggplot2 and heatmap.

The GFA converged with a Geweke convergence test of 0.0333; two factors were identified for explaining the data. The Mantel correlation between the original data and the reconstructed data with the three components is 0.918 with a p-value of

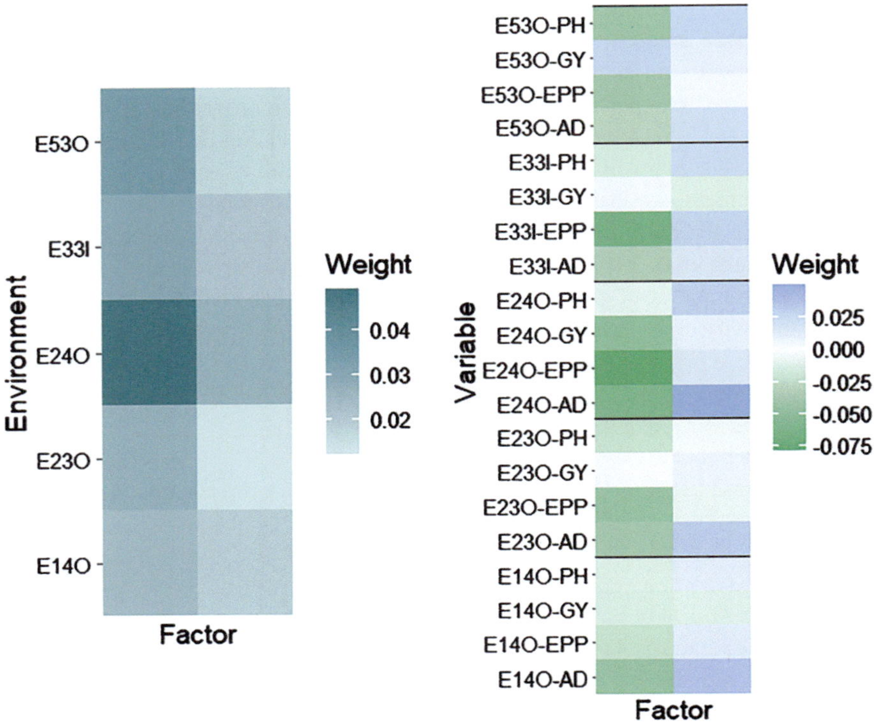

Fig. 4.7 Association of environments (left) and environment by variable combination (right) with the latent factors expressed for the weights from a GFA of 18 genotypes

0.0010. The small number of factors is a signal that the data has a strong structure, and most traits within and between environments are correlated. The data can be reconstructed with high precision. Figure 4.7 shows on the left the weights of the five environments in the two factors; on the right side it shows the weights of each trait–environment combination on the latent factors. The first factor is represented by all the environments with the largest weight of E24O and E53O. Grain yield, EPP, and AD in E24O have the largest negative contribution to the factor, while in E530, the contributing variables are EPP and PH with negative largest weights. In the other environments with lower contribution only one trait in each one has a medium contribution, AD in E14O and EPP in E23O and E33I. The second factor is associated with environments E24O and E33I. While PH and AD explain the contribution of environment E24O, environment E33I is represented by PH and EPP. The other three environments present large weights of AD and EPH in E53O. Figure 4.8 shows the heatmaps and groups of genotypes and environment–trait combinations for factor 1 and factor 2 weights. Clustering used a hierarchical clustering algorithm with the complete method. The first clear result is the

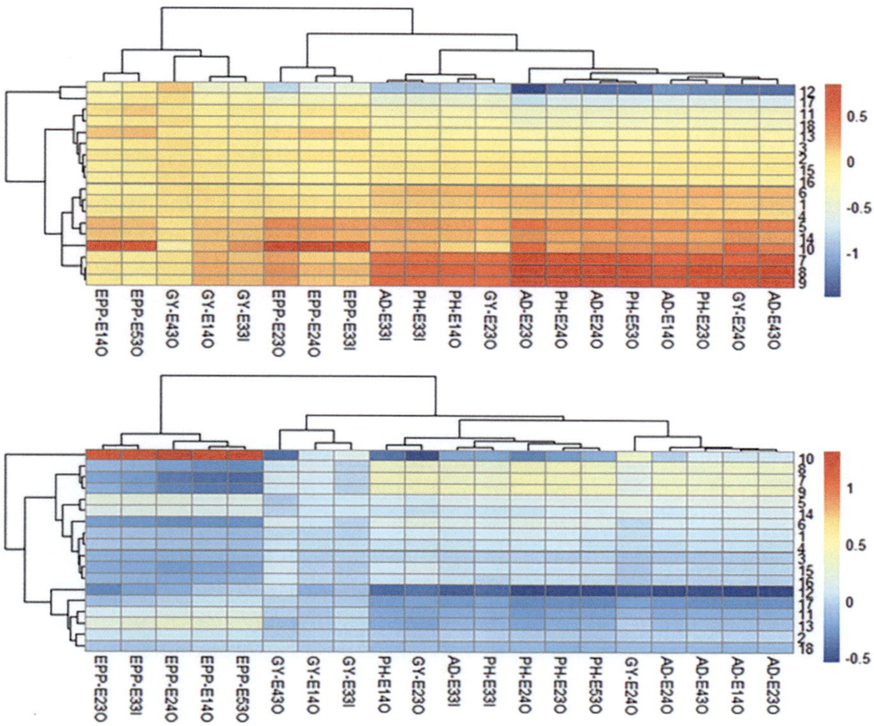

Fig. 4.8 Heatmap and groups of genotypes and environment–trait combinations using factor 1 (top) and factor 2 (bottom) weights. Clustering used a hierarchical clustering algorithm with the complete method

differences in the sign of the weights in factor 1 and factor 2; most weights are positive in the first and negative in the second factor. Although both factors have positive weights on the trait–environments combinations, the response of the genotypes is in opposite direction. The first factor presents four groups of environments–trait combinations, two groups with EPP in all environments and GY in three environments (E43O, E14O, and E33I), and the other two groups with AD and PH mixed in both groups, and GY in two environments (E23O and E24O). The second factor also presents four groups with clear discrimination between traits. All environments of EPP and PH are in one group each, while for AD, four of five environments are in one group. Three environments of grain yield form one group (E43O, E14O, and E33I).

Considering the genotypes for factor 1, three groups can be identified while for factor 2, three or four groups without considering the group with one genotype. The comparison between the groups formed in both factors shows some similitudes; for example, genotypes 7, 8, and 9 are in the same group in both cases. With a deep inspection of the results and the knowledge about genotypes and environments, the group factor analysis can be used to explore the grain yield genotype by environment interaction and related traits (Fig. 4.8).

4.4 Conclusion

The analysis of genotype by environment interaction usually is focused on the main trait, which is grain yield, or another productive trait like biomass in forage crops, or cellulose content in trees. There are several methods to analyze genotype by environment interaction, but a more challenging task is to understand the basis of such interaction in terms of environments, and genotypic characteristics. A two-dimensional problem becomes a multidimensional problem with all its complexity. There are statistical methods to analyze multidimensional problems; however, the ability to disentangle the complexity is limited. Probably, this is the reason most of the methods are not broadly adopted. Particularly, the analysis of the grain yield genotype by environment interaction and related traits has a large descriptive component and many relationships between its components, environments and environmental variables, molecular, morphological, physiological characteristics of the genotypes, and the trait response of the genotypes to the environments. An additional challenge is the data availability; usually, the researchers have few samples (genotypes and environments) and more variables than samples, and all of them are interrelated generating multi-collinearity problems. Selecting the right variables to consider in the study and isolating the specific effects of each one is not easy.

Our major aim in this chapter was to put together most of the alternative methodologies to analyze multidimensional problems and to convey a taste of what can be done with some of them; most details are left unexplained. The readers are referred to other publications where more details can be found.

Acknowledgments I thank Berhanu Tadesse Ertiro (CIMMYT-Kenya) and the CIMMYT Global Maize Programme for providing the sample data.

References

Aastveit AH, Martens H (1986) ANOVA interactions interpreted by partial least squares regression. Biometrics 42:829–844

de la Abelardo Vega J, Chapman SC (2001) Genotype by environment interaction and indirect selection for yield in sunflower: II. Three-mode principal component analysis of oil and biomass yield across environments in Argentina. Field Crop Res 72:163–174. https://doi.org/10.1016/S0378-4290(01)00163-0

Atnaf M, Tesfaye K, Dagne K, Wegary D (2017) Genotype by trait biplot analysis to study associations and profiles of Ethiopian white lupin (*Lupinus albus* L.) landraces. Aust J Crop Sci 11:226–235. https://doi.org/10.21475/ajcs.2017.11.01.pne226

Basford KE, Kroonenberg PM, DeLacy IH (1991) Three-way methods for multiattribute genotype × environment data: an illustrated partial survey. Field Crop Res 27:161–176. https://doi.org/10.1016/0378-4290(91)90027-S

Birhanu C, Mekbib F, Lule D, Bekeko Z, Girma G, Tirfessa A, Ayana G, Nida H, Mengiste T (2024) Genotype by environment interactions and stability for grain yield and other agronomic traits in selected sorghum genotypes in Ethiopia. Agrosyst Geosci Environ 7:e20544. https://doi.org/10.1002/agg2.20544

Bramardi SJ, Bernet GP, Asíns MJ, Carbonell EA (2005) Simultaneous agronomic and molecular characterization of genotypes via the generalized Procrustes analysis: an application to cucumber. Crop Sci 45:1603–1609. https://doi.org/10.2135/cropsci2004.0633

Bunte K, Leppäaho E, Saarinen I, Kaski S (2016) Sparse group factor analysis for biclustering of multiple data sources. Bioinformatics 32:2457–2463. https://doi.org/10.1093/bioinformatics/btw207

Burgueño J, Crossa J, Cornelius PL, Trethowan R, McLaren G, Krishnamachari A (2007) Modeling additive × environment and additive × additive × environment using genetic covariances of relatives of wheat genotypes. Crop Sci 47:311–321

Burgueño J, Crossa J, Cornelius PL, Yang RC (2008) Using factor analytic models for joining environments and genotypes without crossover genotype × environment interaction. Crop Sci 48:1291–1305

Carroll JD, Chang JJ (1970) Analysis of individual differences in multidimensional scaling via an N-way generalization of "Eckart-Young" decomposition. Psychometrika 35:283–319

Cheng WC, Cheng PE, Liou M (2013) Group factor analysis for Alzheimer's disease. Comput Math Methods Med 2013:428385. https://doi.org/10.1155/2013/428385

Crossa J, Basford KE, Taba S, DeLacy IH, Silva E (1995) Three-mode analyses of maize using morphological and agronomic attributes measured in multilocational trials. Crop Sci 35:1361–1370. https://doi.org/10.2135/cropsci1995.0011183X003500050036X

Crossa J, Vargas M, van Eeuwijk FA, Jiang C, Edmeades GO, Hoisington D (1999) Interpreting genotype × environment interaction in tropical maize using linked molecular markers and environmental covariables. Theor Appl Genet 99:611–625

Crossa J, Cornelius PL, Yan W (2002) Biplots of linear-bilinear models for studying crossover genotype × environment interaction. Crop Sci 42:1761–1770. https://doi.org/10.2135/cropsci2002.1761

DeAngelis BA, Newnham RE, White WB (1972) Factor group analysis of the vibrational spectra of crystals: a review and consolidation. Am Mineral 57:255–268

Dettori ML, Crossa J, Ammar K, Peña RJ, Varela M (2011) Three-mode principal component analysis of genotype-by-environment-by-trait data in durum wheat. J Crop Improv 25:592–615. https://doi.org/10.1080/15427528.2011.592930

Eberhart SA, Russell WA (1966) Stability parameters for comparing varieties. Crop Sci 6:36–40

Eden T, Fisher RA (1927) Studies in crop variation. IV. The experimental determination of the value of top dressing with cereals. J Agric Sci 17:548–562. https://doi.org/10.1017/S0021859600018827

van Eeuwijk FA, Denis JB, Kang MS (1996) Incorporating additional information on genotypes and environments in models for two-way genotype by environment tables. In: Kang MS, Gauch HG (eds) Genotype by environment interaction: new perspectives. CRC Press, Boca Raton, pp 173–198

van Eeuwijk FA, Crossa J, Vargas M, Ribaut JM (2000) Modelling QTLs and QTLxE using factorial regression models and partial least squares techniques. In: Zavala G, Treviño H (eds) Simposio Interacción Genotipo × Ambiente, pp 43–60

van Eeuwijk FA, Crossa J, Vargas M, Ribaut JM (2002) Analysing QTL-environment interaction by factorial regression, with an application to the CIMMYT drought and low-nitrogen stress programme in maize. In: Kang MS (ed) Quantitative genetics, genomics and plant breeding. CABI Publishing, Wallingford, pp 245–256

Elias AA, Robbins KR, Doerge RW, Tuinstra MR (2016) Half a century of studying genotype × environment interactions in plant breeding experiments. Crop Sci 56:1419–1432. https://doi.org/10.2135/cropsci2015.01.0061

del Farace ML (2023) Desarrollo de ideotipos de maíz destinados a la producción de bioetanol. Tesis de Doctorado. Facultad de Ciencias Agrarias. Universidad Nacional de Rosario. https://hdl.handle.net/2133/26686

Finlay KW, Wilkinson GN (1963) The analysis of adaptation in a plant-breeding programme. Aust J Agric Res 14:742–754

Fisher RA (1958) The analysis of covariance. In: Statistical methods for research workers, 13th edn. Oliver & Boyd, Edinburgh, pp 161–175

García-Peña M, Arciniegas-Alarcón S, Basford KE, dos Santos Dias CT (2016) Analysis of sunflower data from a multi-attribute genotype × environment trial in Brazil. Euphytica 209: 213–227. https://doi.org/10.1007/s10681-016-1649-7

Gholizadeh A, Ghaffari M (2023) Genotype by yield*trait (GYT) biplot analysis: a novel approach for phenotyping sunflower single cross hybrids based on multiple traits. Food Sci Nutr 11:4567–4580. https://doi.org/10.1002/fsn3.3524

Gillet A, Leclercq E, Sautot L (2022) The Tucker tensor decomposition for data analysis: capabilities and advantages. In: 38ème Conférence sur la Gestion de Données (BDA), Oct 2022, Aubière

Harshman RA (1970) Foundations of the PARAFAC procedure: models and contributions for an "explanatory" multi-modal factor analysis. UCLA Work Pap Phon 16:1–84

Harshman RA, Lundy ME (1984) Data Preprocessing and the Extended PARAFAC Model. In HG Law, CWS Jr, JA Hattie, RP McDonald (eds.), Research Methods for Multimode Data Analysis, pp. 216–284. Praeger, New York

Hassani M, Mahmoudi SB, Saremirad A, Taleghani D (2024) Genotype by environment and genotype by yield*trait interactions in sugar beet: analyzing yield stability and determining key traits association. Sci Rep 14:51061. https://doi.org/10.1038/s41598-023-51061-9

Hunt L, Basford K (2001) Fitting a Mixture Model to Three-mode Three-way Data with Missing Information. J Classif 18:209–226. https://doi.org/10.1007/s00357-001-0016-z

Jadhav S, Balakrishnan D, Shankar V et al (2019) Genotype by environment (G×E) interaction study on yield traits in different maturity groups of rice. J Crop Sci Biotechnol 22:425–449. https://doi.org/10.1007/s12892-018-0082-0

Justo SB, Mkamilo HG, Danga NO, Bakari R, Suleiman F, Mwakapala R, Batare M, Bulegeya V, Kashenge S, Suvi WT, Lee SB, Txd K, Komboka T, Txd I (2024) Genotype and genotype × environment interaction effects on the rice grain yield performance in different agro-ecologies in Tanzania. J Plant Breed Crop Sci 16:36–45. https://doi.org/10.5897/JPBCS2024.1038

Katsenios N, Sparangis P, Chanioti S, Giannoglou M, Leonidakis D, Christopoulos MV, Katsaros G, Efthimiadou A (2021) Genotype × environment interaction of yield and grain quality traits of maize hybrids in Greece. Agronomy 11:357. https://doi.org/10.3390/agronomy11020357

Kendal E (2019) Comparing durum wheat cultivars by genotype × yield × trait and genotype × trait biplot method. Chil J Agric Res 79:512–521. https://doi.org/10.4067/S0718-58392019000400512

Kiers HAL, Van Mechelen I (2001) Three-way component analysis: principles and illustrative application. Psychol Methods 6:84–110. https://doi.org/10.1037/1082-989X.6.1.84

Kirouani A, Boukhalfoun L, Ouldkiar R, Bouzerzour H (2023) Analysis of the effect of GE interaction on the grain yield and its related traits in rain-fed Algerian durum wheat (*Triticum turgidum* L. var. durum) grown in contrasting environments. Rev Fac Nal Agron Medellín 76: 10297–10308. https://doi.org/10.15446/rfnam.v76n2.102517

Klami A, Virtanen S, Leppäaho E, Kaski S (2014) Group factor analysis. arXiv. https://doi.org/10.48550/arXiv.1411.5799

Kroonenberg PM, Basford KE (1989) An investigation of multi-attribute genotype response across environments using three-mode principal component analysis. Euphytica 44:235–244. https://doi.org/10.1007/BF00022605

Kroonenberg PM, De Leeuw J (1980) Principal component analysis of three-mode data by means of alternating least squares algorithms. Psychometrika 45:69–97

Kroonenberg PM, Basford KE, Ebskamp AGM (1995) Three-way cluster and component analysis of maize variety trials. Euphytica 84:159–172. https://doi.org/10.1007/BF01677554

Leppäaho E, Ammad-ud-din M, Kaski S (2017) GFA: exploratory analysis of multiple data sources with group factor analysis. J Mach Learn Res 18:1–5

Lombardo R, van de Velden M, Beh EJ (2023) Three-Way Correspondence Analysis in R. The R Journal 15(2):237–262. https://doi.org/10.32614/RJ-2023-049

Mandel J (1994) Analysis of two-way layouts. Chapman & Hall, New York

Mantel N (1967) The detection of disease clustering and a generalized regression approach. Cancer Res 27:209–220

Martínez-Peña RF, Rezzouk FZ, Díez-Fraile MC, Nieto-Taladriz MT, Araus JL, Gutiérrez NA, Vicente R (2023) Genotype-by-environment interaction for grain yield and quality traits in durum wheat: identification of ideotypes adapted to the Spanish region of Castile and León. Eur J Agron 151:126951. https://doi.org/10.1016/j.eja.2023.126951

Merino Soto C (2013) Análisis factorial de grupo múltiple de una versión muy breve del cuestionario revisado de personalidad de Eysenck (EPQR-BV). Int J Psychol Res 6:79–88

Mohammadi R (2019) Genotype by yield*trait biplot for genotype evaluation and trait profiles in durum wheat. Cereal Res Commun 47:479–491. https://doi.org/10.1556/0806.47.2019.32

Mohammadi R, Farshadfar E, Amri A (2016) Path analysis of genotype × environment interactions in rainfed durum wheat. Plant Prod Sci 19:123–132. https://doi.org/10.1080/1343943X.2015.1128100

Negash G, Birr A (2022) Adaptability study of yield and yield related trait performance of improved bread wheat (*Triticum aestivum* L.) varieties in North Shewa Zone Oromia, Ethiopia. South Asian Res J Agric Fish 4:1–12. https://doi.org/10.36346/sarjaf.2022.v04i06.001

Pacheco-Gil RA, Vargas M, Alvarado G, Rodríguez F, Crossa J, Burgueño J (2015) GEA-R (genotype x environment analysis with R for windows) version 4.1. CIMMYT Research Data & Software Repository Network V16. https://hdl.handle.net/11529/10203

Paramesh M, Mohan Reddy D, Shanthi Priya M, Sumathi P, Sudhakar P, Reddy KHP (2016) GT biplot analysis for yield and drought related traits in mung bean (*Vigna radiata* L. Wilczek). Electron J Plant Breed 7:69–77. https://doi.org/10.5958/0975-928X.2016.00069.7

Piepho HP, Blancon J (2023) Extending Finlay–Wilkinson regression with environmental covariates. Plant Breed 142:621–631. https://doi.org/10.1111/pbr.13130

Pontes DS, Rosado RDS, Cruz CD, Nascimento M, Oliveira AMC, Pensky SM (2020) Trait selection using Procrustes analysis for the study of genetic diversity in Conilon coffee. Acta Sci Agron 42:e43195. https://doi.org/10.4025/actasciagron.v42i1.43195

Pour-Aboughadareh A, Poodineh O, Marzooghian A (2022) Interaction effects of genotype × environment using path analysis and mixed models in barley superior lines. Iran J Field Crop Sci 53:177–191. https://doi.org/10.22059/IJFCS.2021.323545.654826

Prieto Hernández VM (2014) Análisis de la estructura de covarianza para la interpretación de la interacción genotipo-ambiente con datos desbalanceados. MSc Thesis, Facultad de Agronomía, Universidad de la República, Uruguay. https://www.colibri.udelar.edu.uy/jspui/bitstream/20.500.12008/24147/1/PrietoHern%C3%A1ndezV%C3%ADctorManuel.pdf

R Core Team (2024) R: a language and environment for statistical computing. R Foundation for Statistical Computing, Vienna. https://www.R-project.org/

Reynolds MP, Trethowan R, Crossa J, Vargas M, Sayre KD (2002) Physiological factors associated with genotype by environment interaction in wheat. Field Crop Res 75:139–160

Reynolds MP, Slafer GA, Foulkes JM, Griffiths S, Murchie EH, Carmo-Silva E, Asseng S, Chapman SC, Sawkins M, Gwyn J, Bailey Flavell R (2022) A wiring diagram to integrate physiological traits of wheat yield potential. Nat Food 3:318–324. https://doi.org/10.1038/s43016-022-00512-z

Rincón F, Johnson B, Crossa J, Taba S (1997) Identifying subsets of maize accessions by three-mode principal component analysis. Crop Sci 37:1944–1951. https://doi.org/10.2135/cropsci1997.0011183X003700060044X

Sahmat SS, Rafii MY, Oladosu Y, Jusoh M, Hakiman M, Mohidin H (2024) Unravelling the dynamics of genotype and environment interactions on chilli (*Capsicum annuum* L.) yield-related attributes in soilless planting systems. Sci Rep 14:1698. https://doi.org/10.1038/s41598-023-50381-0

SAS Institute Inc (2023a) The CALIS procedure in SAS/STAT® 15.3 user's guide. Cary

SAS Institute Inc (2023b) The GLM procedure in SAS/STAT® 15.3 user's guide. Cary
SAS Institute Inc (2023c) The GLMSELECT procedure in SAS/STAT® 15.3 user's guide. Cary
Shojaei SH, Ansarifard I, Mostafavi K, Bihamta MR, Zabet M (2022) GT biplot analysis for yield and related traits in some sunflower (*Helianthus annuus* L.) genotypes. J Agric Food Res 10. https://doi.org/10.1016/j.jafr.2022.100370
Shojaei SH, Mostafavi K, Ansarifard I, Bihamta MR, Zeinalzadeh-Tabrizi H, Omrani A, Gore M, Mousavi SMN (2023) Comparison of genotype × trait and genotype × yield-trait biplots in Sunflower cultivars. Internat J Agricult Environ Food Sci 7(1):136–147. https://doi.org/10.31015/jaefs.2023.1.17
Tai GCC, Levy DA, Coleman WK (1994) Path analysis of genotype-environment interactions of potatoes exposed to increasing warm-climate constraints. Euphytica 78:193–201. https://doi.org/10.1007/BF00024531
Tucker LR (1966) Some mathematical notes on three-mode factor analysis. Psychometrika 31:279–311
Tucker LR (1972) Relations between multidimensional scaling and three-mode factor analysis. Psychometrika 37:3–27
Varela M, Crossa J, Rane J, Joshi AK, Trethowan R (2006a) Analysis of a three-way interaction including multi-attributes. Aust J Agric Res 57:1185–1193
Varela M, Crossa J, Onathan HC, Rane J, Joshi AK, Trethowan R (2006b) Analysis of a three-way interaction including multi-attributes. Crop Pasture Sci 57:1185–1193. https://doi.org/10.1071/AR06081
Varela M, Crossa J, Joshi AK, Cornelius PL, Manès Y (2009) Generalizing the sites regression model to three-way interaction including multi-attributes. Crop Sci 49:1783–1794. https://doi.org/10.2135/cropsci2008.12.0702
Vargas M, Crossa J, Sayre K, Reynolds MP, Ramírez ME, Talbot M (1998) Interpreting genotype × environment interaction in wheat by partial least squares regression. Crop Sci 38:679–689
Vargas M, Crossa J, van Eeuwijk FA, Ramírez ME, Sayre K (1999a) Using AMMI, factorial regression and partial least squares regression models for interpreting genotype × environment interaction. Crop Sci 39:955–967
Vargas M, Crossa J, van Eeuwijk FA, Ramírez ME, Sayre K (1999b) Using partial least squares regression, factorial regression, and AMMI models for interpreting genotype × environment interaction. Crop Sci 39:955–967
Vargas M, van Eeuwijk FA, Crossa J, Ribaut JM (2006) Mapping QTLs and QTL × environment interaction for CIMMYT maize drought stress program using factorial regression and partial least squares methods. Theor Appl Genet 112:1009–1023
Vargas M, Crossa J, Reynolds MP, Dhungana P, Eskridge KM (2007) Structural equation modelling for studying genotype × environment interactions of physiological traits affecting yield in wheat. J Agric Sci 145:151–161. https://doi.org/10.1017/S0021859607006806
Vargas M, Glaz B, Alvarado G, Pietragalla J, Morgounov A, Zelenskiy Y, Crossa J (2015) Analysis and interpretation of interactions in agricultural research. Agron J 107:1–12. https://doi.org/10.2134/agronj13.0405
Virtanen S, Klami A, Khan SA, Kaski S (2012) Proceedings of the 15th International Conference on Artificial Intelligence and Statistics (AISTATS). La Palma, Canary Islands
Yan W, Frégeau-Reid J (2018) Genotype by yield*trait (GYT) biplot: a novel approach for genotype selection based on multiple traits. Sci Rep 8:8863. https://doi.org/10.1038/s41598-018-26688-8
Yan W, Rajcan I (2002) Biplot Analysis of Test Sites and Trait Relations of Soybean in Ontario. Crop Sci 42(1):11–20. https://doi.org/10.2135/cropsci2002.1100
Yan W, Hadinezhad M, DeHaan B et al (2023) Exploring the trait-yield association patterns in different oat mega-environments of Canada. Crop Sci 63:2255–2273. https://doi.org/10.1002/csc2.21106
Yi L, Dong Z, Lei Y, Zhao J, Xiong Y, Yang J, Xiong Y, Gou W, Ma X (2021) Genetic diversity and molecular characterization of worldwide prairie grass (*Bromus catharticus* Vahl) accessions using SRAP markers. Agronomy 11:2054. https://doi.org/10.3390/agronomy11102054

Does Modern Plant Breeding Techniques Increase Crop Stability Performance?

5

Marina Ćeran, Vuk Đorđević, Jegor Miladinović, Dragana Miladinović, Dragana Trkulja, Aleksandra Radanović, and Ankica Kondić-Špika

Abstract

Global food security is increasingly threatened by climate change, population growth, limited arable land, and the growing impacts of abiotic and biotic stresses on crop production. Yield projections for major crops indicate declines of up to 25% by the century's end, highlighting the urgent need for resilient, high-yielding crop varieties. Traditional breeding, while historically successful, is labor-intensive and insufficient to meet current and future demands. Advances in molecular biology, bioinformatics, and computational tools have transformed plant breeding, introducing innovative techniques such as marker-assisted selection, genomic selection, genome editing, epigenomics, high-throughput phenotyping, speed breeding, and artificial intelligence. These tools enhance breeding efficiency, accelerate genetic gain, and enable the rapid development of crops with improved yield stability, resilience to climate variability, and superior nutritional profiles. Integrating modern breeding techniques with traditional approaches addresses critical agricultural challenges by enabling the creation of adaptive and sustainable agricultural systems. However, the adoption of advanced techniques requires skilled expertise, interdisciplinary collaboration, and substantial investment, posing challenges for breeders in developing regions. Future breakthroughs will likely involve the convergence of existing and emerging technologies, driving the development of adaptive and stable crop varieties essential for mitigating the impacts of climate change and ensuring global food security.

M. Ćeran (✉) · V. Đorđević · J. Miladinović · D. Miladinović · D. Trkulja · A. Radanović · A. Kondić-Špika
Institute of Field and Vegetable Crops National Institute of the Republic of Serbia, Novi Sad, Republic of Serbia
e-mail: marina.ceran@ifvcns.ns.ac.rs

R. Mohammadi et al. (eds.), *Genotype x Environment Interactions and its Implications for Plant Breeding*, https://doi.org/10.1007/978-981-95-5664-9_5

5.1 Introduction

Climate change and population growth are among the primary challenges to global food security. Abiotic stresses such as drought, salinity, changes in annual rainfall, temperature fluctuations, and increased atmospheric CO_2 levels, along with biotic stresses such as pests and diseases, significantly affect crop development, agricultural productivity, and the global food supply. Additionally, the constrained availability of arable land has contributed to a stagnation in agricultural productivity, leading to concerns about meeting future food demands (Yu et al. 2021). Projections indicate that the yields of the world's top ten crops, including barley, cassava, maize, oil palm, rapeseed, rice, sorghum, soybean, sugarcane, and wheat, could decrease by 3–12% by the middle of the century and by 11–25% by its end (Ray et al. 2019). As the global population grows, food security concerns are becoming even more urgent. By 2050, it is projected that global crop yields must double to sustain a population of ten billion people (Campbell et al. 2023). In addition to meeting quantity demands, there is an increasing need for crops with enhanced nutritional profiles to address dietary deficiencies and malnutrition and to adapt to shifting dietary preferences (Fróna et al. 2019; Miladinović et al. 2019). Furthermore, recent global events, such as the coronavirus disease 2019 (COVID-19) pandemic and geopolitical conflicts, have underscored the importance of self-sustainability, particularly in reducing reliance on imports for agricultural needs. To address these multifaceted challenges, crop breeders must develop varieties that combine high and stable productivity under unpredictable conditions with resistance or resilience to emerging biotic and increasingly extreme abiotic stresses, meeting consumer demands for safer and more nutritious food.

Modern plant breeding can generate elite varieties that will address ongoing challenges effectively and enable stable crop production (Miladinovic et al. 2021). Striking a balance between maximizing yield under specific conditions and reducing vulnerability to changing environments is critical to establishing practical and sustainable agricultural systems. Yield and most of the agronomically important traits are complex, influenced by many genes with small effects and by interactions between genes and the environment. Genotype-by-environment interaction (GEI) plays a key role in determining a plant's performance under different conditions (Laitinen and Nikoloski 2018) and a major challenge in plant breeding is predicting how genotypes will perform across locations and growing seasons. Identification of stable and high-performing lines in plant breeding relies heavily on multi-year testing of genotypes in field trials across multiple environments (Voss-Fels et al. 2019). Stability, or the consistency of a trait over time and space, is crucial for ensuring reliable crop production, particularly in the face of climate change, as climate variability increasingly affects agricultural productivity (Olesen et al. 2011; Müller et al. 2018). For plant breeders, selecting genotypes with stable performance across environments ensures adaptability to changing conditions (Mühleisen et al. 2014). For farmers, stable yield reduces economic risks and improves predictability. Cropping system diversification, often associated with improved yield stability, provides further resilience to environmental fluctuations

(Marini et al. 2020). On a larger scale, yield stability has national and global implications for food security (Kalkuhl et al. 2016).

Traditional breeding achieved remarkable success in the last century, increasing annual genetic gains by 0.5–1% for the most important crops (Hristov et al. 2009; Mladenov et al. 2011; Achilli et al. 2022; Abdala et al. 2024; Mirosavljević et al. 2024). However, it mainly focuses on improving traits through phenotypic selection and these breeding methods remain labor-intensive, time-consuming, and are increasingly inadequate in addressing the current and future demands. Recent scientific and technological advancements have significantly improved the development of high-yield, stable, resilient crop varieties. High-throughput phenotyping systems facilitate the rapid screening of large populations (Araus et al. 2018), while new sequencing technologies have enabled a deeper understanding of crop genetics (Bassi et al. 2016). Investigating the genetic architecture underlying stability and employing genomics-assisted breeding has proven effective in enhancing trait stability (Wang et al. 2015b; Liu et al. 2017a). The integration of genomics-assisted breeding with high-throughput phenomics has allowed for precise gene pyramiding, resulting in crop varieties with stable performance under a range of abiotic and biotic stress conditions (Kondić-Špika et al. 2023). Additionally, speed breeding has addressed the challenges of lengthy generation times, enabling faster trait fixation and generation advancement through specialized protocols across various crop species (Watson et al. 2018). The adoption of innovative technologies is imperative to accelerating plant breeding, increasing crop productivity, and addressing the challenges of global food security. These tools can help redesign existing breeding pipelines or can be integrated into conventional breeding programs to meet the demands of sustainable agriculture (Crossa et al. 2021; Kondić-Špika et al. 2022; Cvejić et al. 2023; Potts et al. 2023).

5.2 Concept of Crop Stability in Plant Breeding

Crop cultivars and hybrids have been primarily developed by traditional plant breeding methods based on the selection of desirable genotypes, genetic recombination of the selections through intermating, and then reselection. Selection techniques and intermating systems are numerous and differ depending on plant species and breeding goals (Burton 2011). However, the selection and recommendation of ideal genotypes with high performance and stability across multiple environments based on multiple traits have always been a difficult task (Li and Wu 2023). The assessment of phenological development as a function of specific environmental variables is a basic piece of information necessary for any attempt at modeling plant growth, adaptation, and productivity as a dynamic process (Wang et al. 1987; Colson et al. 1995). Genotype-by-environment interactions are common for most quantitative traits of economic importance. Different responses of a genotype to environmental changes are the cause of the GEI (Becker and Leon 1988). GEI occurs when differences between genotypes are not the same in all locations within and across the years (Edmeades et al. 1989). It represents the inconsistency of the relative

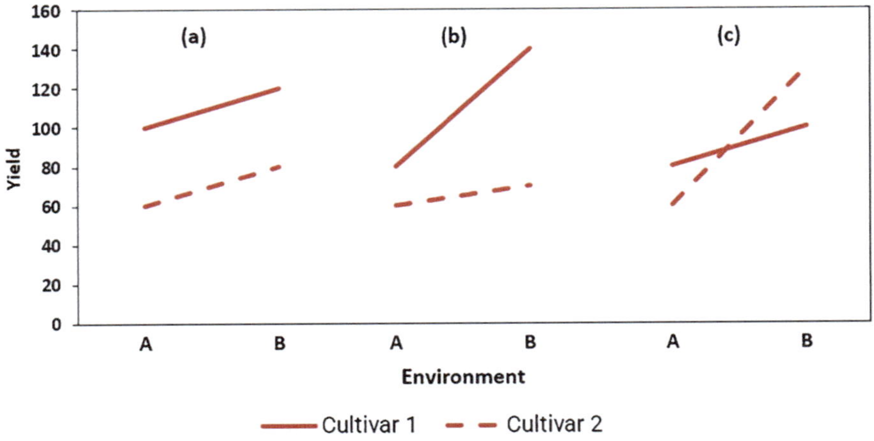

Fig. 5.1 Types of genotype-by-environment interactions: (**a**) no interactions; (**b**) interaction but no rank changes; (**c**) interaction and rank changes

performance of genotypes over environments (Hill et al. 1998). If cultivars 1 and 2 are evaluated in environments A and B, three types of GEI can be observed (Fig. 5.1). According to Allard and Bradshaw (1964), the two varieties may show similar responses (Fig. 5.1a) which indicate the independence of the genotype performances over the environments. The non-intersecting but non-parallel lines explain the same relative ranking of varieties, but their absolute differences vary across the environments (Fig. 5.1b). The GEI is believed to be a crossover or qualitative if it leads to the changes in relative ranks of genotypes across different environments (Fig. 5.1c). Crossover interactions are important in plant breeding since they affect the selection of genotypes in a given environment and suggest that genotypes are specifically adapted to some particular environments. In practice, the situation is immensely more complicated, because many genotypes and environments must be considered and there is a great number of possible types of interaction.

Genotype-by-environment interaction plays a significant role in crop improvement, making it essential for plant breeders to identify high-yielding genotypes that are both adaptable and stable across diverse environmental conditions before their release as varieties. Adaptability, a result of GEI is generally categorized into two types: (1) general adaptability, or the ability of a genotype to perform at an acceptable level across a wide range of environments, and (2) specific adaptability, or the ability to perform well in favorable environments (Begna 2020). Adaptability refers to the better survival and performance of a genotype in a specific environment, while stability indicates its suitability across diverse environments. Stability is used to select genotypes that remain relatively unaffected by environmental fluctuations (Moorthy et al. 2012; Chandrakanth et al. 2016). A stable genotype is characterized by a consistent yield across different environmental conditions, with minimal variance in performance. Due to this criterion, genotypes with a low yield variance across environments are considered stable (Miladinović et al. 2003; Miladinovic et al. 2006). Yield stability is directly influenced by GEI, as differences in stability among genotypes arise from the widespread occurrence of these interactions (Rani et al. 2023b). A variety or genotype is considered adaptive or stable if it exhibits high mean yield and low fluctuation in yield when grown across diverse environments. There are two primary concepts of phenotypic stability: the biological (static) and agronomic (dynamic) concepts. The biological concept emphasizes constant performance across a wide range of environments, whereas the agronomic concept allows for responsive performance under favorable conditions (Begna 2020).

It is important to emphasize that the stability with which we are concerned in crop improvement does not imply a general constancy of phenotype across varying environments. Instead, it focuses on stability in economically important traits, such as yield and quality. Achieving such stability often requires maintaining some morphological and physiological traits in a steady state while allowing others to vary. Therefore, ideal varieties exhibit low GEI for agriculturally important traits, particularly yield, without necessarily showing stability in other characteristics. A variety that adjusts its genotypic or phenotypic state in response to transient environmental fluctuations, while maintaining high and stable economic returns for specific locations and years, can be described as "well-buffered" (Allard and Bradshaw 1964).

Li and Wu (2023) emphasize that the recommendation of ideal genotypes for widespread cultivation should be based on their overall performance across multiple traits, rather than isolated performance of individual traits (Gollob 1968; Gauch 2013; Olivoto et al. 2019b). However, the evaluation of yield stability is often prioritized over mean yield performance under diverse environmental conditions (Tariku et al. 2013; Islam et al. 2016). Genotype recommendation based on multiple traits across multi-environment trials poses significant challenges for crop breeders and agronomists. This complexity arises from the additive effects of genotype (G) and environment (E), along with their interaction (GEI) (Fischer 2011; Crossa et al. 2017). GEI reflects the variation in genotype responses across environments and changes in their rank order under different conditions (Yan and Kang 2003). Accurate estimation of G, E, and GEI is essential for making valid recommendations of suitable varieties capable of withstanding variable conditions (Egea-Gilabert et al. 2021).

Genotype-by-environment interaction has limitations in the study of important agronomic traits like yield and its components, as it complicates the understanding of genetic experimentations and restricts the selection of varieties adaptive to specific conditions (Farshadfar and Sutka 2003). Normally, in plant breeding programs, the selection of genotypes for a specific environment is conducted by multi-environmental trials for the evaluation of genotypes based on their performance across environments (Li et al. 2020). Numerous research studies have been conducted using several statistical modeling approaches for checking the effect of GEI on yield and other agronomic traits (Grüneberg et al. 2005). These approaches mainly utilize a generalized linear model (GLM) to measure the variation caused by genotype, environment, and GEI for each variable by linear regression and joint analysis of variance (ANOVA) (Arif et al. 2021). GLMs lower the supposition of dependent variables (Olsson 2002). To overcome the difficulty of estimating and quantifying the GEI, some models with developing procedures, namely stability analyses, have been exploited (Gollob 1968; Olivoto et al. 2019a). Additive main effects and multiplicative interaction (AMMI) have been widely applied in GEI analysis because they provide a convenient estimate in comparison with the traditional analysis of variance (ANOVA) (Gauch 2013). In the AMMI method, ANOVA is used to access G, E, and GEI to keep the genotype as fixed and the environment as a random effect, as described by Olivoto and Lúcio (2020). This method is further divided into interaction principal component analysis (IPCA) and AMMI main effect biplot analysis, where GEI was plotted on the x-axis and IPCA values on the y-axis, while the second method is G and GEI biplot (Yan and Tinker 2006). However, the AMMI analysis mainly depends on the linear–bilinear model, which is not generally more accurate than those obtained by the best linear unbiased prediction (BLUP) through the linear mixed-effect model (van Eeuwijk et al. 2016; Olivoto et al. 2019a). A recent efficient method for quantifying the genotypic stability of single traits has been developed using the weighted average of absolute scores from the singular value decomposition (SVD) of the matrix through the BLUP framework for the GEI effects (WAASB) index. Subsequently, a superiority index, WAASBY (weighted average of WAASB and response variable), was introduced to integrate the mean performance and stability of single traits (Olivoto et al. 2019a). Recently genomic-based stability metrics were introduced to estimate GEI in unbalanced designs and predict GE metrics for new genotypes based on multi-environment genomic prediction model (Carvalho et al. 2024). Sparse testing, where only a subset of lines is tested in specific environments, aims to estimate the genetic performance of all genotypes across all environments and adjust the observed phenotypic data, accounting for genomic information (Montesinos-López et al. 2024). The availability of accurate GE metric estimates provided by new estimators should help improve the identification of high performance and stable genotypes in a range of target environments.

In addition, the second difficulty for crop breeders in genotype recommendation is that stability analysis and mean performance evaluation across multiple environments are generally unfeasible for multiple traits, although being feasible for single parameters, such as yield (Nowosad et al. 2016; Mohammadi et al. 2018). The reliability and accuracy of genotype recommendations can be significantly improved when multiple traits are simultaneously considered (Nduwumuremyi et al. 2017; Koundinya et al. 2019). Thus, ideal genotype selection in combination with high performance and stability across multi-environment trials based on multiple traits has always been a challenging task. Recently, there has been a powerful and comprehensive method, the multi-trait stability index (MTSI), which can be used in multi-environment trials to combine the simultaneous selection of multiple traits into a single and easy-to-interpret indicator obtained by the framework of the linear mixed-effect model (Olivoto et al. 2019b).

5.3 Modern Breeding Techniques in Achieving Crop Stability

In the face of numerous challenges, breeders must identify traits that ensure higher and more stable yields under diverse conditions, provide resistance or resilience to emerging biotic and abiotic stresses, and meet evolving consumer demands. Over the past two decades, significant advancements in molecular biology, bioinformatics, and computational tools have transformed plant breeding, leading to the widespread adoption of modern breeding techniques.

Techniques such as marker-assisted selection, genomic selection, high-throughput phenotyping, genome editing, genetic mapping, epigenetic modifications, speed breeding, and artificial intelligence (AI) have greatly enhanced the precision and efficiency of crop breeding for more resilient crops (Fig. 5.2). These advancements have enabled the discovery and generation of novel genetic variations, improved the scale and precision of phenotyping, and provided deeper insights into the genetic architecture of simple and complex traits. They have also facilitated the efficient introgression of traits related to growth, life cycle, stress resistance, and metabolism into elite cultivars.

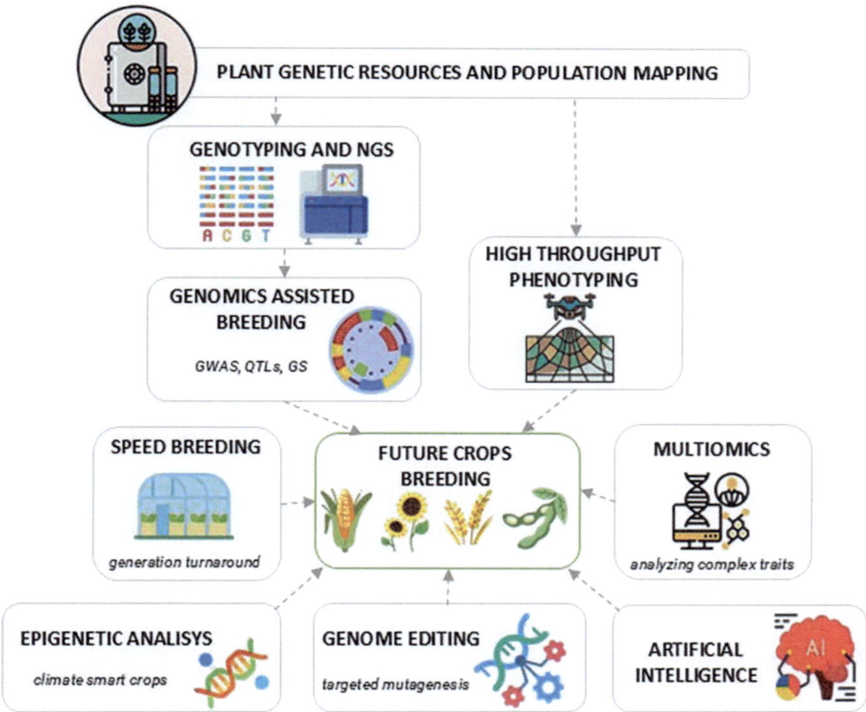

Fig. 5.2 Overview of modern plant breeding techniques for designing superior genotypes. (This figure has been designed using icons from Flaticon.com)

Furthermore, rapid advancements in next-generation sequencing (NGS) and phenotyping technologies have exponentially increased the volume of genetic, phenotypic, and environmental data available for crop breeding. The integration of information technology, AI, and biotechnology is now driving a transition from traditional breeding and field screening methods to a data-driven, intelligent breeding paradigm.

5.3.1 Molecular Markers, Gene Mapping and Marker-Assisted Selection (MAS)

Advancements in molecular biology over the last four decades, from the development of different molecular marker technologies in the 1980s to the recent breakthroughs in genomics and next-generation sequencing technologies, have dramatically transformed crop breeding (Fig. 5.3). Particularly over the past 20 years, there has been a rapid expansion in our knowledge of plant and pathogen genome sequences, as well as the physiological and molecular roles of various plant genes. This progress has revolutionized molecular genetics, significantly enhancing its efficiency in plant breeding programs (Collard et al. 2005).

Molecular markers offer several advantages over morphological and biochemical markers, making them particularly valuable in crop breeding. These advantages include a high level of polymorphism, widespread abundance within the plant genome, independence from environmental conditions, and the ability to be detected at any tissue type or stage of plant growth. A molecular marker is associated with a specific gene or quantitative trait locus (QTL) of interest, and its polymorphism can be easily tracked and quantified using hybridization techniques (e.g., RFLP—restriction fragment length polymorphism) or PCR-based methods (e.g., RAPD—random amplified polymorphic DNA, AFLP—amplified fragment length polymorphism, microsatellites or SSRs, SNP—single nucleotide polymorphism, SCAR—sequence-characterized amplified region, EST—expressed sequence tags) (Nadeem et al. 2018). The most common applications of molecular markers include:

- Detection of genetic diversity and identification of sources of new variability (Glogovac et al. 2013; Mikić et al. 2017; Miladinović et al. 2018b; Uhlarik et al. 2022);

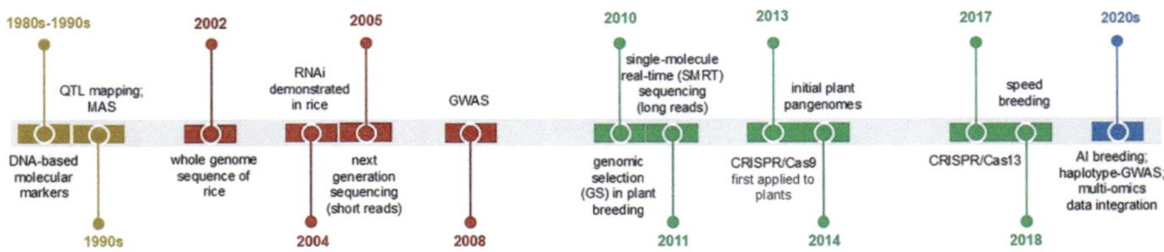

Fig. 5.3 A timeline of key advancements in modern plant breeding

- DNA profiling of crop varieties (Mangini et al. 2018);
- Studying genetic distances between populations, inbred lines, and other breeding materials (Mikić et al. 2018; Trkulja et al. 2019; Brbaklić et al. 2021)
- Mapping major genes and QTLs (Trkulja et al. 2012, 2016; Brdar-Jokanović et al. 2013; Imerovski et al. 2013, 2014, 2016; Brbaklić et al. 2015; Mikić et al. 2016; Mladenov et al. 2019);
- Tracking selection footprints (Tomičić et al. 2015);
- Marker-assisted and genomic selection (Dimitrijević et al. 2010; Kurasch et al. 2017; Miladinović et al. 2018a; Đorđević et al. 2019; Radanović et al. 2022; Ćeran et al. 2024a).

All these aspects of using molecular markers in breeding facilitated the identification and introgression of beneficial alleles for specific traits that improve crop stability, particularly under environmental stress. MAS, by combining specific genetic markers linked to desirable traits, allows breeders to target resilience traits while saving time and effort involved in traditional phenotypic selection, particularly in complex G × E interactions. This includes identifying new alleles associated with yield and its components, improving tolerance to drought, one of the most significant effects of climate change, or resistance to emerging races of pathogens. Notably, the QTL mapping made the most remarkable breakthrough in dissecting the complex nature of yield (Cui et al. 2014; Dargahi et al. 2014; Yang et al. 2020b), as well as abiotic (Gupta et al. 2017; Kumar et al. 2022) and biotic stress tolerance traits (Davar et al. 2011; Poole et al. 2012; Qi et al. 2016) all of which significantly impact crop yield potential. It was anticipated that application of molecular markers would strengthen and accelerate breeding for complex traits through gene pyramiding. To date, MAS has proven to be highly effective for relatively simple traits that are controlled by a few genes (Barloy et al. 2007; Ramalingam et al. 2020). However, the QTLs detected using the linkage mapping approach are often challenging to implement in breeding programs due to the association of large genomic regions with linkage drag or undesirable genes. Additionally, the complex genetic architecture of yield involves multiple small-effect loci with significant GEI (Hasan et al. 2021). Early attempts to dissect yield QTLs using bi-parental populations and interval mapping (IM) often resulted in an overestimation of QTL effects due to limited recombination events. To overcome these limitations, the association mapping (AM) or linkage disequilibrium (LD) approach has been introduced. This approach relies on the use of a set of genotypes of known/unknown ancestry, designed to carry most of the genetic variability for the trait of interest representing the products of hundreds of historic recombination cycles, thus providing higher resolution during QTL mapping (Breseghello and Sorrells 2006). The progress in genome sequencing over the past 20 years, along with the development of various high-throughput genotyping techniques such as SNP Arrays (Single Nucleotide Polymorphism Arrays), GBS (Genotyping by Sequencing), and DArT (Diversity Arrays Technology), has been a significant leap forward in the dissection of the genetic architecture of complex traits. This advancement introduced the strategy of genome-wide association studies (GWAS). The approach relies on the principle of genotyping a sufficient number of markers across the genome, ensuring that functional alleles are likely in linkage disequilibrium with at least one of the genotyped markers (Nordborg and Weigel 2008). The GWAS is one of the leading strategies for complex trait dissection and identification of novel and superior alleles using existing germplasm. There are numerous studies where the GWAS was used to identify new genes and QTLs for yield-related traits (Schierenbeck et al. 2021, 2023; Rani et al. 2023a), drought (Pranneshraj et al. 2022; Aleem et al. 2024) and frost resistance (Visioni et al. 2013; Soleimani et al. 2022), as well as tolerance to prevalent diseases (Guidini et al. 2023). In addition, the GWAS was performed to understand how stability is genetically controlled and to identify genomic regions influencing phenotypic consistency across environments. In the GWAS conducted for both yield genotype-environment interaction (GEI) and one stability measure, six significant QTLs for GxE (variance explained 20–35%) and one significant QTL associated with the stability measure (variance explained 1%) were identified (Xavier et al. 2018). The majority of stability QTL discovered contained nearby genes generally known to play a role in stress response like water, salinity, flooding, photo-oxidative, and development under stress. This is expected because the ability for a genotype to maintain a consistent phenotype across environments can depend on its response in low-yielding environments. Additionally, certain genomic regions were identified as regulating both trait values and stability indices, indicating pleiotropic effects or close linkage, thus allowing for the simultaneous selection of both trait sets (Lozada and Carter 2020a, b). Moreover, a combination of four markers that explained 20% of variance for yield stability coefficient in wheat was identified, describing the effect of epistasis on yield stability (Sehgal et al. 2017). Identified QTL for yield stability coefficient shared the same genomic regions as those reported for yield in drought and heat stress environments. Also, half of the markers associated with yield stability coefficient were involved in significant epistatic effects causing significant differences in stability. Moreover, recent advancements in whole-genome sequencing, transcriptomics, the availability of numerous DNA polymorphic markers, and improved methodologies for the GWAS have enabled the dissection of GEI into components such as quantitative trait nucleotide × environment (QEI) or Gene × Environment interactions, representing a major step forward in understanding complex traits (Laitinen and Nikoloski 2018; Li et al. 2022; Jin et al. 2023; Napier et al. 2023). Thus, integrated with high-throughput phenotyping, the GWAS can become one of the most powerful tools to help facilitate the process of selecting stable and high-yielding genotypes.

5.3.2 Genomic Selection (GS)

In contrast to conventional MAS, which typically focus on a limited set of markers associated with major QTLs, genomic selection (GS) is a genomics-assisted approach that selects superior individuals for breeding by simultaneously using genome-wide DNA data to predict their genetic potential (Meuwissen et al. 2001;

Heffner et al. 2009). As a result, this breeding technique is particularly suited for quantitative traits influenced by a large number of loci with a small effect. The individual effects of genome-wide markers on quantitative traits are estimated using genotyped and phenotyped training populations (Fig. 5.4a), by establishing an advanced prediction model. The cumulative sum of marker effects is employed to calculate the genomic-estimated breeding values (GEBV) for a selection of candidates based only on their genotypic profiles. The accuracy of GS is quantified by the Pearson correlation between GEBV and true breeding values (Isidro et al. 2015), providing an assessment of selection accuracy that correlates directly with genetic gain (Merrick et al. 2022). Genomic prediction can be implemented at different stages of breeding. The first is prebreeding, either to look for interesting germplasm or to choose elite parents for crossing. It can also be applied to select promising germplasm early in the variety development process usually predicting additive values, before the time-consuming and resource-intensive field phenotyping (Crossa et al. 2017; Varshney et al. 2021). Besides reduced time and resource consumption, additional benefits of GS include increased accuracy of estimated breeding values and selection intensity (Crossa et al. 2017; Đorđević et al. 2019; Bonnett et al. 2022). A number of factors may affect the efficiency and practical application of genomic selection, such as trait architecture and heritability, training set size and composition, genomic relationship between the training and test set, marker density, genotype-by-environment interaction, linkage disequilibrium (LD) between markers, and model for estimation of marker effects (Fig. 5.4b) (Jarquín et al. 2014; Isidro et al. 2015; Crossa et al. 2017; Akdemir and Isidro-Sánchez 2019; Ćeran et al. 2024a).

One of the major advantages of GS is its ability to improve crop stability by predicting how a genotype will perform across diverse environments. Genomic selection could be useful for the improvement in stability traits by allowing for prediction and selection of stability at a stage before phenotypic stability assessment is possible. Even though the inconsistent performance of a model across multiple environments can make GS prediction more difficult, it is crucial to incorporate genotype-by-environment interactions into GS models and selection process allowing breeders to choose cultivars that exhibit consistent performance in a range of environments by forecasting crop performance under variable conditions (Heslot et al. 2014). Several studies have suggested that enhanced phenotyping can improve genomic prediction models by better capturing GEI across diverse target environments (Bao 2014; Lado et al. 2016; Đorđević et al. 2019). The phenotyping depth of the training population should account for the heterogeneity of the target environment, particularly about climatological variables. Additionally, GS models could integrate extra variables to capture the variability associated with GEI, thereby improving the accuracy of genomic predictions (Varshney et al. 2021). The accuracy of single-environment studies in bread wheat, maize, and legumes improved by an average of 10–40% in all models that incorporated GEI (Burgueño et al. 2012; Jarquín et al. 2014, 2016, 2017; Saint-Pierre et al. 2016; Jarquín et al. 2017). The marker and pedigree-based G-BLUP prediction model was successfully applied by Burgueño et al. (2012), with crop growth model (CGM) outputs used to evaluate

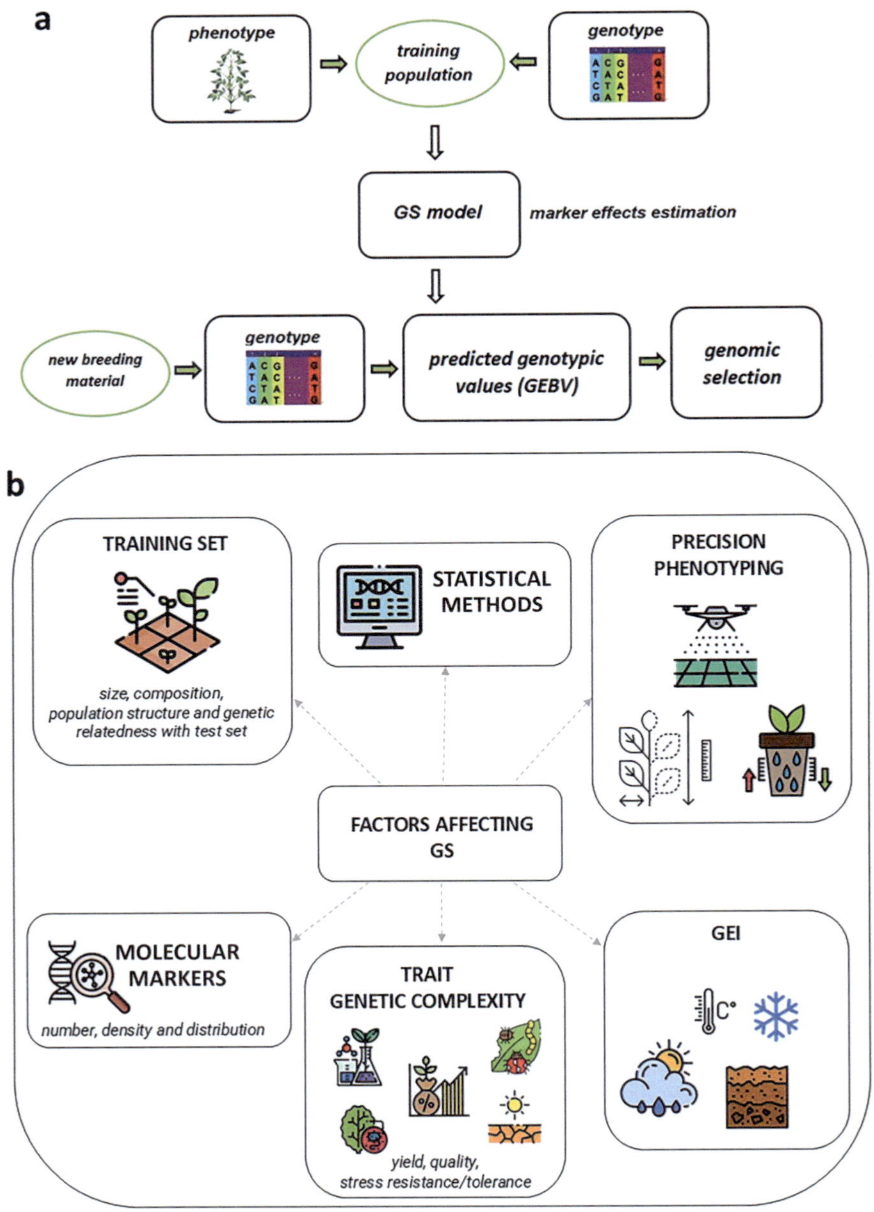

Fig. 5.4 Schematic overview of: (**a**) genomic selection (GS) model development and (**b**) major factors affecting GS. GEBV, genomic estimated breeding value. (This figure has been designed using icons from Flaticon.com)

GEI (Heslot et al. 2014; Rincent et al. 2019). This approach was further integrated with GS (Messina et al. 2018; Robert et al. 2020). Additionally, a method for multi-environment trial analysis proposed by de los Campos et al. (2020) accelerated the evaluation of grain yield stability and adaptation. It assessed GEI patterns learned from field trials, predicted cultivar performance in specific environments, and evaluated performance under various weather conditions.

Furthermore, the main and interaction effects of markers and environmental covariates are incorporated into reaction norm model, an extension of the G-BLUP model (Jarquín et al. 2014). This model has been successfully applied in multi-environment studies with added environmental covariates, including rice and maize (Millet et al. 2019; Monteverde et al. 2019), wheat gene bank accessions (Crossa et al. 2016b), lines evaluated across diverse agroecological zones (Jarquín et al. 2016; Saint-Pierre et al. 2016), and different continents (Pérez-Rodríguez et al. 2017; Sukumaran et al. 2017). The reaction norm model, which incorporates GEI, demonstrated a 7–20% increase in GS accuracy compared to models that exclude GEI (Crossa et al. 2017). The importance of integrating high-throughput environmental data into genomic selection models is emphasized, as this allows predictions to be made in new contexts with similar environmental features (Rogers et al. 2021).

The marker × environment (MxE) interaction model introduced by Lopez-Cruz et al. (2015) partitions marker effects into two components: those that are consistent across environments (stability) and those that vary by environment (interaction), allowing marker effects to change across environments. The MxE model is best applied when studying environments with positive correlations (Lopez-Cruz et al. 2015). To identify genomic regions responsible for GEI and those with stable effects across environments, this approach can be enhanced using shrinkage and variable selection techniques. The group of authors successfully applied the MxE GS model to predict untested individuals and to dissect genomic regions with both environment-specific and stable effects across environments (Crossa et al. 2016a).

In practical applications, GS has been successfully applied to select for abiotic and biotic stress tolerance, with the resulting crops demonstrating stable yields across different environments. This approach has proved its role in improving tolerance against biotic stresses which are quantitatively controlled in wheat (Sarinelli et al. 2019), rice (Huang et al. 2019), maize (Badu-Apraku et al. 2019), and barley (Lorenz et al. 2012). Furthermore, the usefulness of GS has been shown for abiotic stress tolerance (drought and heat) in wheat (Bhandari et al. 2019), maize (Beyene et al. 2015), and rice (Sukumaran et al. 2018). Genomic prediction accuracies using multi-environment models for drought stress tolerance were higher than those using single-environment models in rice and wheat. Prediction accuracies were higher for heat and drought stress in case of wheat when secondary traits contributing to yield were considered under stress rather than yield per se using genomic prediction (Rutkoski et al. 2016).

In addition, different studies developed genomic prediction models for distinct stability measures (AMMI stability index, Eberhart and Russel regression) of different traits, such as yield, test weight, plant height, and heading date in wheat breeding program across two years and 12 environments (Huang et al. 2016), yield, heading date, and plant height in diverse winter wheat lines evaluated in 9 environments (Lozada and Carter 2020a, b), yield stability in Australian macadamia tree production (O'Connor et al. 2021), and stability metrics in coffee evaluated for 20 traits across multiple years and locations (Adunola et al. 2023). As genomic prediction models consider genetic effects, the accuracy of GS for trait stability demonstrates that genetic control of trait stability metrices is predictable using molecular information. Moreover, observed independent relationship between trait and trait stability, such as for yield, indicated that GS could be a valuable tool to predict and select genotypes that are both high yielding and stable across different environments. Furthermore, in soybean was evaluated the efficacy of genomic prediction for yield, protein and oil content stability (Harms 2023). Assessed predictive abilities were too low for use in a breeding program, but rank coincidence gave more promising results for applied genomic selection for stability traits, allowing selections to be made from individuals in the top half of predicted value rankings. It was concluded that observed lower prediction accuracy in different studies can result from experimental design. Also, some studies have reported higher prediction accuracies for static stability measures in seed quality traits, so it is possible that a different metrics can better capture the GEI pattern across environments for stability of different traits (Wang et al. 2015b). By improving prediction models through the selection of model type, SNP type, stability measures, and increased sampling of training and validation environments, there is potential for genomic selection to effectively improve stability in breeding programs, allowing selection at an early stage even when phenotypic data are insufficient to directly select for stability. Selecting for stability traits at earlier stages of the breeding process would help enhance lines for better yield stability in later stages, ultimately leading to improved cultivars and hybrids.

5.3.3 Epigenetic Modifications

Epigenetics provides information on changes occurring not on the DNA-sequence level but on gene expression and genome stability modulation. Epigenetic modifications are heritable and include DNA methylation, histone modifications, chromatin remodeling, and activities of non-coding RNA (ncRNA) molecules. As crops, and plants in general, are sessile organisms, and cannot move as animals can, response to outside stimuli must be quick to ensure survival in newly emerging conditions. Bearing in mind that changes at the genomic level (DNA sequences) are slow and in plants can take years for desirable alleles to evolve, epigenetic modifications offer a quicker and more flexible option and thus may be crucial in crop development and adaptation to environmental conditions.

5.3.3.1 DNA Methylation

DNA methylation, essential for epigenetic control, influences plant growth, development, stress response, and adaptability. Plant genomes exhibit a high level of nuclear DNA methylation, targeting CG, CHG, and CHH (H = A, T, C) sequences.

Over 30% of the total 5-methylcytosine (m5C) in plant DNA is found in non-CG sites (Vanyushin and Ashapkin 2011). DNA methylation in plants is specific to species, tissues, organelles, and developmental stages.

In crops, DNA methylation can be significantly altered during exposure to stress which can be utilized for enhancing crop resilience. Several studies in rice, maize, and wheat showed that drought stress causes significant changes in DNA methylation across different developmental stages, which can influence gene expression, drought memory, and the epigenetic landscape, contributing to the domestication and stress resilience of crops (Gahlaut et al. 2020; Auler et al. 2021; Wang et al. 2021). Similarly, salinity stress has been shown to cause tissue-specific changes in DNA methylation in rice. While methylation levels significantly decreased in root tissues under salt stress, only minor changes were observed in leaf tissues (Wang et al. 2015a).

5.3.3.2 Histone Modifications

Most histone modification sites are on the N-terminal tails and sometimes in nucleosome core regions (Zhao et al. 2019). These modifications, known as the "histone code," specify chromatin functions and transcriptional activities (Jenuwein and Allis 2001). They primarily include methylation, acetylation, ubiquitination, and phosphorylation. Specific histone modification patterns are linked to distinct transcriptional processes. H3K4me (methylation at Lys4 of H3) and H3K36me indicate gene activation, while H3 deacetylation (H3ac), H3K9me, and H3K27me are generally linked to transcriptional repression (Liu et al. 2010). Acetylation of conserved lysine residues neutralizes their positive charges, decreasing their affinity for negatively charged DNA, thereby promoting chromatin accessibility to transcription factors and enhancing transcription. These processes have a pivotal role in developing resistance to abiotic stress in crops (Li et al. 2014a; Feng et al. 2022).

5.3.3.3 Chromatin Remodeling

Chromatin remodeling involves disrupting nucleosome structure and loosening histone–DNA contacts to regulate gene expression by enabling transcription factor binding. ATP-dependent chromatin remodelers (e.g., SWI/SNF, ISWI, NURD/Mi-2/CHD, INO80, SWR1) catalyze this process and exchange histone variants. High mobility group (HMG) proteins (HMGA, HMGB) and topoisomerases also play roles in chromatin remodeling (Song et al. 2021b). Besides having a significant role in adaptation to stress, it has been shown that chromatin modifications also have an important role in developmental processes. Interestingly, it has been reported that flowering in rice can be promoted by CHD3/Mi-2 chromatin remodeling factor regardless of photoperiod (Yoon et al. 2021).

5.3.3.4 Activities of Non-coding RNA Molecules

Various types of ncRNAs, such as small RNAs (siRNAs), circular RNAs (circRNAs) and long non-coding RNAs (lncRNAs), can regulate gene expression at transcriptional, post-transcriptional, and translational levels. lncRNAs also influence target genes epigenetically by modifying histone status and chromatin organization (Hazra et al. 2023). These RNA molecules are involved in plant growth and development as well as response to biotic and abiotic stress. Three lncRNAs (COOLAIR, COLDAIR, and COLDWRAP) regulate FLC (Flowering Locus C) repression via chromatin modifications and RNA–DNA interactions. COOLAIR, an antisense transcript from the 3' region of FLC, initiates repression earlier than COLDAIR and COLDWRAP, reducing FLC mRNA expression transiently (Swiezewski et al. 2009). In rapeseed, it was found that overexpression of MSTRG.22563 reduces seed oil content by 3.1–3.9%, while overexpression of MSTRG.86004 increases it by around 2% (Li et al. 2023).

As can be seen, unravelling different epigenetic mechanisms leads to a better understanding of the important mechanisms allowing further boosting of crop yield, resilience, and overall fitness.

5.3.4 Genome Editing

Genome or gene editing (GE) includes a wide range of innovative techniques that enable scientists to accurately modify genome sequences and adjust gene expression regulation in specific regions, providing valuable insights into an organism's functional genomics (Abdallah et al. 2015). Genome editing tools utilize either site-specific recombinases (SSRs) or site-specific nucleases (SSNs). Both systems rely on recognizing a known sequence. SSR technologies, like Cre/loxP and Flp/FRT systems, can knockdown or knockin genes in eukaryotic genomes, depending on the orientation of the specific sites (loxP, FLP, etc.) surrounding the target site. Recombination-based genome engineering relies on the specific recombinase (sub-) family and the target site, leading to high rates of homologous recombination (Abdallah et al. 2015). The SSN system induces single or double-strand DNA breaks, which then trigger the activation of endogenous DNA repair mechanisms. Four main classes of SSNs have been developed to cleave genomic sequences: mega-nucleases (homing endonucleases), zinc finger nucleases (ZFNs), transcriptional activator-like effector nucleases (TALENs), and the CRISPR/Cas system (clustered regularly interspaced short palindromic repeats/CRISPR-associated proteins).

The CRISPR/Cas9 genome-editing method has been applied to over 20 crop species, offering a precise method to directly modify genes involved in stress responses, nutrient uptake, disease resistance, and pest resistance, providing a new avenue for developing crops that perform more consistently with enhanced resilience to specific environmental conditions, improving both yield and stability in diverse environments (Hamdan et al. 2022). Due to its simplicity, precision, and effectiveness, genome editing opens up new possibilities for developing improved crop varieties by precisely adding beneficial traits or removing undesirable ones, making it one of the best methods for creating crops that can withstand pests and flourish in harsh environments, in addition to improving the nutritional value of crops (for a review, see Yıldırım et al. 2023). Ongoing research focuses on enhancing crop varieties to achieve higher yields, better stress tolerance, improved disease and pest resistance, reduced input costs, and greater nutritional value—traits that directly impact crop productivity and stability (Abdallah et al. 2015). Below are some examples of concrete application of genome editing in improvement in traits contributing to stability of crop production.

5.3.4.1 Disease and Pest Resistance

In order to achieve crop stability, breeders aim to create new crop varieties that are resilient to biotic stresses by developing novel breeding techniques and incorporating genetic traits for long-lasting resistance. However, new diseases and the quick evolution of pathogens brought on by climate change pose a threat to their efficacy (Hussain 2015; Bhoi et al. 2022). To improve crop resilience to diseases and provide greater food security within more sustainable agricultural systems, genome editing technology has brought new tools and techniques for identifying genes implicated in crop defense (Gosavi et al. 2020). Improved pest and disease resistance by genome editing may result in more resilient crops, which would help manage pests and diseases sustainably. More robust resistance mechanisms can be created by downregulating susceptibility (S) genes and upregulating genes that identify pathogens, prevent infection, and lessen pathogenicity. Together with controlling pest and pathogen exposure levels in IPM (Integrated Pest Management) strategies, these developments—made possible by genetic modifications, gene editing, RNA interference (RNAi), and gene drives—can significantly increase the sustainability of agricultural production, especially in light of climate change. One notable example is application of CRISPR/Cas9 for improvement in blast resilience in rice, where mutant lines performed well in the field trials and did not differ from the wild-type plants with regard to the agronomic traits tested, thus proving that gene modification via CRISPR/Cas9 is a useful approach for enhancing disease resilience in crops (Wang et al. 2016). Also, genome editing achieved significant success in wheat and tomato in enhancing resistance against different pathogens including wheat powdery mildew and bacterial wilt in tomatoes (Nekrasov et al. 2017; Li et al. 2018).

5.3.4.2 Abiotic Stress Tolerance

Historically, traditional marker-assisted breeding has produced a large number of crops resistant to abiotic stress. However, it takes around ten years to successfully produce abiotic stress-resilient crop plants using classical breeding techniques, because of backcrossing practices, and intensive screening (Hoang et al. 2016; Mishra et al. 2022; Yadav et al. 2022). Hence, CRISPR/Cas-based genome-editing technology could be of use for targeting complex quantitative genes that are either directly or indirectly related to abiotic stresses (Yadav et al. 2023).

The use of CRISPR/Cas as an additional tool in crop breeding programs aimed at producing more resilient cultivars in response to various abiotic conditions was recently covered by Nascimento et al. (2023). Through a methodical database search on the use of the CRISPR/Cas strategy for abiotic stress tolerance, the review provides a comprehensive and objective compilation of relevant studies (Nascimento et al. 2023).

The genetic mechanism behind tolerance against a variety of abiotic stresses, such as drought, salt, heat, and nutritional values in different agricultural plants, has currently been successfully understood through the application of CRISPR/Cas-based genome engineering (Shi et al. 2017; Li et al. 2019). Shi et al. utilized CRISPR/Cas9 to create new variants of ARGOS8. It was demonstrated before that transgenic plants with constitutive overexpression of ARGOS8 exhibit reduced sensitivity to ethylene and enhanced grain yield under drought stress conditions. Furthermore, by targeting OsRR22 gene in rice, improvement in rice's ability to withstand salinity was reported (Zhang et al. 2019a). In addition, the majority of possible uses of the CRISPR/Cas9-mediated genome editing technique in crop plants to control abiotic stressors like heat, salinity, and drought are outlined in review by Kumar et al. (2023), going over the potential uses of CRISPR/Cas-based systems in the future to create crop varieties that can withstand stress. Promising results were obtained in breeding climate-smart wheat to assure global food and nutritional security (Afroz et al. 2023).

5.3.4.3 Yield Improvement Through Genome Editing

Complex interactions between soil and climate conditions, biotic and abiotic stressors, and crop management techniques have a significant impact on crop yield and, in turn, food production. The effects of these interactions on plant growth and productivity can be either positive or detrimental (Jovičić et al. 2019; Kondić-Špika et al. 2019; Marjanović Jeromela et al. 2019). Thus, a great deal of research has been done on the combined effects of biotic and abiotic stresses (Pautasso et al. 2012; Pandey et al. 2015), as well as other abiotic stresses (Song et al. 2014; Paul et al. 2019), on plant physiology and development. Studies using disease forecasting models for various regional climatic scenarios have been conducted because biotic stresses play a significant role in potential yield loss and account for 15% of global declines in food production (Oerke 2005; Oldenburg et al. 2009; Caffarraa et al. 2012; Jevtić et al. 2017). Modern breeding techniques, which may produce desirable features more quickly and precisely than traditional breeding procedures, are crucial to addressing the challenges brought by climate change. Among these, genome editing based on CRISPR/Cas9 has become a potent technology and has been effectively implemented in many crops (Ricroch et al. 2017; Miladinovic et al. 2021). By deleting the genes that adversely affect yield-related variables, CRISPR/Cas9 has been shown in numerous studies to be an effective method for increasing agricultural output (Li et al. 2016; Liu et al. 2017b; Lu et al. 2018; Zhang et al. 2018). The following outcomes have been achieved by the use of a multiplexed GE method for trait pyramiding: higher kernel weight in wheat (Zhang et al. 2016), early heading in rice (Li et al. 2017), and improved grain size and weight in rice (Xu et al. 2018). In a recent study, Huang et al. (2018) used genome sequencing and the CRISPR/Cas9 method to identify 57 genes that affect yield-related features in 30 kinds of the Green Revolution phenotype known as "miracle rice." They then created knockout mutants of those 57 genes. These mutants' phenotyping made it possible to identify a number of genes that are essential for controlling rice yield-related characteristics. When analyzing intricate quantitative characteristics, such as yield, this novel method can be quite helpful.

5.3.5 Pangenomics

With the advancements in next-generation sequencing technology and the availability of more sequence data, it becomes clear that the reliance on a single reference genome does not properly reflect the species' diversity (Golicz et al. 2016). This becomes especially clear when analyzing biotic stress defense response, as gene duplication and differentiation play a vital role in sustaining pest defense responses. The pangenome concept was introduced to fully encompass the diversity in gene content within a species by sequencing multiple accessions representing that species (Tettelin et al. 2005). Furthermore, *de novo* sequencing of wild species accessions has enabled the development of superpangenome (Khan et al. 2020).

Pangenomics identifies species' whole gene pool that is present in different individuals, including the core genes present in all individuals and variable genes that are either partially shared or unique to each individual. The core genome represents the minimum genome that an individual requires for survival and basic, important functions (Wang et al. 2018), whereas the variable (dispensable) genome provides a greater diversity and is associated with crop growth and biotic and abiotic environmental adaptation (Golicz et al. 2016; Hoopes et al. 2019). The information of dispensable genome regulates the phenotypic variation related to agronomic traits, and is crucial for increasing agricultural productivity. It is also a major contributor to the domestication and adaptive evolution of species (Jayakodi et al. 2021). A vast amount of genetic variations was lost as a result of the genetic bottleneck during domestication, which may include the favorable variations that allow the plants to adapt to various environmental conditions. Thus, wild crop relatives serve as a reservoir of genetic diversity, offering new allelic variations essential for breeding programs aimed at improving the adaptive capacity of agricultural systems. The crop wild relatives have proven to be highly beneficial in the environment adaptation of crop varieties, high yield potential, disease challenges, and market needs. In recent decades, there has been a notable rise in the incorporation of traits from wild species into cultivated crops, primarily aimed at addressing biotic and abiotic stresses (Naqvi et al. 2022). The reintroduction of variable genes into elite germplasm led to improved varieties by constructing superior gene combinations, enabling plant breeders to shift from useful individuals to useful haplotypes and useful genes when planning breeding programs (Varshney et al. 2021). Moreover, identification of variable genes across wild and domesticated lines helps to retrace the impact of domestication and breeding bottlenecks on the pangenome (Varshney et al. 2021). Utilizing pangenomes can help identify and characterize valuable genes retained or lost due to the artificial selection and breeding of crop cultivars that can support their reintroduction into breeding programs. Pangenomics expedites the use of wild crop relatives and cultivated species to enhance tolerance to biotic and abiotic stresses, improve nutritional characteristics, and enhance other economically valuable traits in commonly grown cultivars (Aziz and Masmoudi 2024).

With the advancement in NGS technologies, a huge wave of pangenomic studies has been launched initially involving restricted number of accessions, relying on short-read sequencing (Golicz et al. 2016). However, recently long-read sequencing and hybrid methodologies, which combine sequences of long and short reads, enabled the reconstruction of pangenome assemblies with a larger set of accessions, including cultivated, landraces, or wild progenitors. In recent years, pangenomes for different major crops and closely related species have been developed, including rice (Zhao et al. 2018), tomato (Gao et al. 2019), soybean (Li et al. 2014b; Liu et al. 2020b), sunflower (Hübner et al. 2019), maize (Hirsch et al. 2014), *Brassica napus* (Song et al. 2020), and wheat (Montenegro et al. 2017). In general, a better understanding of the genetic diversity of crop stability and resilience or adaptability to various environmental stresses enables selecting crops that are better suited to withstand changing climates and can enable better-targeted breeding strategies that enhance food security and agricultural sustainability. (Tao et al. 2019). Moreover, pangenomics helps to pinpoint regional variations in crops and better understand how different varieties perform in various climates. This knowledge can lead to the optimization of crop varieties tailored for specific environmental conditions, leading to more stable yields.

5.3.6 Haplotype-Based Breeding

The accessibility of whole genome sequencing data of a large number of lines for a given crop and extensive phenotypic data enabled identification of haplotypes, assessment of their diversity, and validation of their phenotypic effects (Bevan et al. 2017). A haplotype represents a group of polymorphic markers that are inherited together in progeny with minimum recombination (Garg 2021). Haplotype breeding can be achieved by combining desirable haplotypes with well-known phenotypic effects into the genetic background of crops. Haplotype-based breeding has shown potential for the improvement in several crops including rice (Abbai et al. 2019), wheat (Brinton et al. 2020), and pigeon pea (Sinha et al. 2020). Given the strong genome structure present in the breeding pools of the majority of crops, the usage of haplotypes may be a powerful supplementary technique to increase the precision and effectiveness of both MAS and GS. Haplotypes can enhance GS models and improve the prediction of stable-performing genotypes since they offer more genetic information than individual markers. By utilizing advantageous genetic variations and their interactions, haplotype-based breeding can increase crop stability, enabling breeders to monitor and preserve entire genomic regions linked to complex stability traits, such as yield stability and stress tolerance, which are usually polygenic. To preserve and further accumulate the advantageous haplotypes in subsequent breeding cycles, lines with a greater number of complimentary and favorable haplotypes can be found and used as parents in breeding programs, hence promoting phenotypic stability.

Recent studies have highlighted the potential of haplotype-based approaches in improving yield stability in wheat and soybean. Sehgal et al. (2020) conducted a haplotype-based genome-wide association study (GWAS) on 6461 advanced breeding lines of CIMMYT spring bread wheat, identifying stable genomic regions associated with grain yield across various environments. Similarly, Liu et al. (2024) discovered a novel quantitative trait locus (QTL) for wheat yield stability under drought stress, with haplotype analysis revealing that the allele of a specific gene was favorable for maintaining yield stability in drought conditions. In soybean, the GWAS was combined with haplotype analysis to identify stable QTLs and putative genes linked to yield, highlighting various allelic combinations that regulate different phenotypes and underscoring the usefulness of haplotype-based approaches yield improvement (Bhat et al. 2022).

Haplotype introgression might take a long time due to the requirement of more breeding cycles and long generation time for the crop. However, an integrated approach that combines haplotype breeding and speed breeding and genomic selection has the potential to accelerate genetic gain (Ćeran et al. 2024b). Bevan et al. (2017) propose prospective methods for haplotype utilization in genomics-based crop improvement through resequencing of large ancestral and wild relative populations of a given crop species, to find haplotypes with a wider range of genetic variation than those found in elite breeding pools. The goal is to identify novel haplotypes, and subsequently use molecular markers to define haplotype blocks which can be included into breeding programs. Varshney et al. (2005) acknowledged the great potential of haplotypes for genomic selection, using thorough haplotype maps to find and use genome regions linked to desired trait in populations with pronounced LD structures.

5.3.7 High-Throughput Phenotyping (HTPP)

Plant phenotyping involves a comprehensive assessment of complex plant traits, such as height, phenological characteristics, yield components, and tolerance to various abiotic and biotic stresses. It was one of the earliest methods used in plant selection and the development of new varieties. However, plant phenotypic variation results from the interaction between genotype and environmental conditions (Romagosa et al. 2013), requiring experiments to be conducted across multiple environments with replicated trials. Early measurements of crop phenotypic features, which relied on traditional methods, were costly, labor-intensive, destructive, and prone to significant errors. With rapid advancements in genomics, proteomics, metabolomics, bioinformatics, and big data computing, these traditional methods have become insufficient for extracting meaningful insights from today's vast data sets. To fully exploit current genome sequence information and improve the understanding of complex characteristics governed by multiple genes, critical for future agricultural development, there is a growing need for efficient, automated, and accurate technologies and platforms (Song et al. 2021a). These innovations capable of capturing phenotypic data across all growth stages are now considered as important as high-throughput genotyping.

Over the past 20 years, the rapid development of non-destructive sensing and imaging techniques has significantly improved the measurement of crop phenotypic traits in both controlled environments and field settings. Collaboration among scientists and engineers from diverse fields—such as crop breeding, agronomy, electronic engineering, molecular biology, and computer image processing, mathematics, and statistics—has led to the creation of various imaging techniques. These techniques include visible, thermal infrared, fluorescence, three-dimensional (3D), multi- or hyperspectral imaging, and tomographic imaging using magnetic resonance imaging (MRI) or X-ray computed tomography (CT) (Jin et al. 2021). This field of research, known as "phenomics," is defined as the high-throughput, accurate acquisition and analysis of multi-dimensional phenotypes across crop growth stages. It encompasses the organism level, including cell, tissue, organ, individual plant, plot, and field levels (Yang et al. 2020a, b). With the advancements in novel sensors, imaging technologies, and analytical methods, numerous infrastructure platforms have been developed for phenotyping that can be divided in three group:

- Microscopic phenotyping—It is used for imaging various tissue morphologies and micro-phenotypes (Beck et al. 2012; Hughes et al. 2017)
- Ground-based phenotyping— It used for crop phenotyping at the individual and field-plot scales and includes portable phenotyping systems, as well as stationary and movable platforms deployed in glasshouses or field trials (Jin et al. 2021; Vollmann et al. 2022)
- Aerial phenotyping—It includes satellites and unmanned aerial vehicles (UAVs), widely used for crop phenotyping in the field, mostly in large-scale crop breeding (Jang et al. 2020).

High-throughput (HTP) platforms offer numerous advantages, such as efficiency, non-destructiveness, and full automation. The phenotypic data obtained through HTP techniques possess new features that are unattainable through traditional research methods. These features include convex hull area, convex hull vertices, compactness, grayscale values, top/side projection areas, top/side contour areas, and more (He et al. 2024). All these, manual, semi-autonomous, or autonomous platforms, along with UAVs and portable devices equipped with single or multiple sensors, have generated an enormous volume and variety of data collected at multiple time points, leading to significant big data issues. Until a few years ago, storing and processing such vast amounts of data were quite challenging. However, advancements in storage capacities and the development of powerful computer processors have made it possible (Gill et al. 2022).

The next major step forward in high-throughput phenotyping is closely related to the development of AI and machine learning (ML). ML focuses on creating systems that can learn from data and make decisions or predictions without explicit programming. Instead of following predefined instructions, ML models identify patterns and relationships within data, improving their performance over time (Sandhu et al.

2021). There are numerous studies available where HTTP and ML approaches have been used for processing images to identify different types and levels of stresses, both abiotic (Atieno et al. 2017; Correia et al. 2022) and biotic stresses (Abdalla et al. 2019; Lin et al. 2019; Zhang et al. 2019b; Prabhakar et al. 2020), yield-related traits (Ranđelović et al. 2020, 2023; Koh et al. 2021; Wang et al. 2023) and to predict crop yield (Rajković et al. 2022, 2023).

HTTP with or without combination with ML has many practical applications, such as a wide range of research studies, through plant breeding to precision agriculture. For different purposes there are different aims, needs and demands, requiring different approaches, monitoring methods, sensors, and carrier vehicles for plant assessment. Consequently, the technology development and methodologies used are very diverse and should be carefully chosen and adjusted to the specific needs (Chawade et al. 2019). Each research study has its own aim but in general they are focused on identifying processes and mechanisms in plants that contribute to stable yields, crop quality, abiotic and biotic stresses tolerance. Also, they are usually small-scale experiments and conducted in controlled conditions or fields. In plant breeding, HTTP aids in the selection of relevant genotypes, requiring measurement of thousands of field plots at multiple environments and providing possibilities for increasing selection intensity, and improving selection accuracy and the decision support system (DSS) (Araus et al. 2018). In precision agriculture, HTTP is used to improve management practices, by monitoring irrigation, fertilization, and pest management and it usually needs the assessment of large arable areas.

There are several challenges remaining to be solved in application of these technologies in research and practice. A key challenge is related to the data management and establishment of a common set of guidelines and standards for protocol optimization and reporting, required for proper interpretation and reproducibility. Minimum Information About a Plant Phenotyping Experiment (MIAPPE) is a set of guidelines that considers a wide range of attributes for samples, experiment, environment, and the results. Additional challenges are: a) identification of cheap, robust, easy-to-use, rapid, and automated phenotyping methods that can feed into DSS; b) real-time analyses, necessary for timely and appropriate intervention of results; c) development of phenotyping systems that could be easily upgraded with the new sensors. Despite all these challenges, HTTP technologies are widely present in the agricultural research, while applications in breeding and production should be significantly improved.

5.3.8 Speed Breeding

Long generation times are the main breeding bottleneck. From the initial cross until the final release of an improved variety, traditional methods for the majority of crops take over ten years. To increase productivity and stability of crops to meet the changing climatic conditions, there is need to increase the rate of cultivar development. Given these challenges, breeders must integrate several methods to accelerate crop development. Off-season nursery/shuttle breeding, *in vitro*/embryo culture, and double haploid technology have been used to shorten generational interval time in various crops. Speed breeding (SB) is a promising technique that allows rapid generation advancement in controlled environments by manipulating the conditions to accelerate plant growth processes, such as photoperiod, light quality and intensity, temperature, humidity, carbon dioxide level, and nutrient supply (Watson et al. 2018; Bhatta et al. 2021). In comparison to typical greenhouse and field conditions, this results in a reduction in generation time from 2.5 to 5, which lowers generation time and resources while increasing the number of generations annually. Achieving rapid generation turnover requires an understanding of the various growth conditions required by different crops, and protocol optimization at the crop and variety levels (Jähne et al. 2020; Pandey et al. 2022). The rapid generation of homozygous and stable genotypes, rapid gene identification, crossing, backcrossing, trait pyramiding, and the creation of mapping populations are just a few of the breeding activities that can be accelerated by speed breeding (Hickey et al. 2019; Bhatta et al. 2021).

Compared to one to two generations per year using traditional selection methods, speed breeding produces between three to nine generations annually, which accelerate the development and release of new crop cultivars (Ghosh et al. 2018). For example, speed breeding made it possible to produce four to six generations in a year in barely, chickpea, canola, and wheat (Watson et al. 2018; Voss-Fels et al. 2019), whereas in soybeans, up to five generations annually have been achieved (Jähne et al. 2020). Lentils and faba beans can produce about eight generations annually, by using plant growth regulators and adjusting the temperature and photoperiod (Mobini and Warkentin 2016). Other crops where speed breeding has been successfully used are rice, sorghum, millets, rapeseed, sugarcane, tomato, and potato (Hickey et al. 2019).

Enhanced breeding efficiency achieved through SB ensures the timely development of climate-resilient, adaptive, and stable crop varieties (Gudi et al. 2022). With an integrated genomics assisted speed breeding (GASB) method that maximizes the advantages of each technique, genetic gain might be further accelerated and stable and high-performing crop varieties could be developed to meet global challenges like food security and climate change (Ćeran et al. 2024b). For example, combining speed breeding with genomic selection (GS) allows breeders to predict and select for yield stability across diverse environments before field testing. Also, rapid breeding cycles facilitate the pyramiding of multiple genes related to stability, such as disease resistance, drought tolerance, and nutrient efficiency. Moreover, the discovery of candidate genes using pan-genome assembly, GS, gene editing, haplotype breeding, or metabolic pathway editing for desired traits may be part of an integrated GASB method (Razzaq et al. 2021).

5.4 Challenges of Applying Modern Breeding Techniques for Crop Stability

While addressing major problems of conventional breeding, there are still some obstacles to using modern plant breeding techniques in increasing crop stability. Their use usually requires skilled manpower and knowledge across multiple biology, agronomy, and computer science branches. Additionally, modern breeding uses expensive, high-tech machinery that can, at times, be hard to operate and unaffordable for researchers and breeders working in developing countries.

One of the more accessible and widely used modern breeding techniques, the use of molecular markers in MAS has been giving good results in accelerating breeding and enabling desirable gene pyramiding. However, not all markers are equally effective. Significant efforts need to be put in evaluating the correlation between the QTL and the marker used for its selection. Failing to accurately measure and account for this correlation is a primary cause of unclear MAS results in contemporary breeding programs (Cobb et al. 2019). Creating functional markers is the most effective as it targets the desirable gene and is usually independent on the genetic background; however, the majority of markers used in MAS are not functional but linked to a desirable QTL. This linkage can vary between in different crosses depending on the genetic background and this is one of big obstacles in MAS. Choosing the best strategy for MAS is crucial in the successful application of this modern breeding tool. Some QTLs, like small-effect QTLs, are better addressed through phenotypic or genomic selection. Due to technological advancements made from the beginning of the twenty-first century, molecular breeders now have high-throughput systems at their disposal. The GWAS, for example, can uncover associations between genomic variants and important agronomical traits. Despite its positive sides, the GWAS faces limitations such as sample size, phenotype quality, and genotype quality. It also struggles to identify rare or multiple independent alleles (Kaňovská et al. 2024). Post-GWAS approaches, like fine mapping, integration of omics data, and machine learning, help narrow down associated variant positions. These methods, also applicable to post-QTL analysis, require experimental confirmation to verify causality.

Furthermore, the genetic control of yield stability is not straightforward and considering interaction between genotype and environment (GEI), there are challenges when utilizing genomic tools for selection in a breeding programs. Stability traits have a complex genetic architecture, lacking stable and large-effect QTLs (Wang et al. 2015b; Lozada and Carter 2020a, b). Moreover, the inconsistency and the variety of stability measures add to the complexity of understanding genetic control of stability traits and reliable QTL discovery and genome-wide association studies (GWAS). This complexity was demonstrated by identifying different QTLs by the GWAS when mapping genomic regions contributing to AMMI stability, conventional stability, and explicit modeling of GEI, with only few QTLs shared across stability categories (Happ et al. 2021). Thus, the absence of stable QTL across models and the absence of a standardized method for analyzing stability make marker-assisted selection (MAS) difficult in practical application for improving yield stability. Genomic selection can be used to overcome the complex genetic architecture of stability traits and accelerate the rate of genetic gain where MAS is ineffective.

Genomic prediction might face a limitation in assessing stability traits due to the lack of phenotype data from a greater number of environments. The large amount of phenotypic data across multiple environments and years for each genotype would have improved stability estimation and possibly prediction accuracy. In addition, if genomic selection for yield stability is to be applied in a breeding program, there are certain elements of genomic selection methods that must be first assessed and optimized by examining the effect of factors such as SNP density, SNP type, and stability measure.

Speed breeding, on the other hand, faces challenges in controlling growing conditions (photoperiod, temperature, humidity, and pest infestations) and requires specialist equipment for trait selection. Breeding programs often lack reliable field trial facilities and skilled workers, with limited opportunities for professional development. Significant costs, mainly from lighting and temperature control, can be reduced using energy-efficient systems and solar power. While speed breeding is costly and infrastructure-limited, combining it with techniques like genomics-assisted breeding can optimize resource use. This method accelerates breeding in crops with long juvenile periods and produces elite inbred lines, making it practical for under-resourced breeding programs. Connecting these programs to fast commercialization systems helps recover extra breeding costs (Sharma et al. 2023).

Introducing innovative approaches in breeding for improved crop stability such as epigenetics requires identifying stable and heritable variations for transmission to progeny. So far, examination of epigenetics shows the most promise in tracking stress resilience and controlling some developmental stages, such as flowering. Most stress-induced epigenetic modifications revert to initial levels post-stress, but some remain stable and inheritable. Tools like mathematical models and epigenotyping procedures can help enhance breeding programs by identifying heritable epigenetic phenotypes and understanding epiallele inheritance (Varotto et al. 2020). Efficient integration of epigenetics into breeding requires new methods to correlate epigenetic variation with desired traits, enabling targeted modifications. Stress memory's impact on yield must also be considered. Predictive models can assess the impact of epigenetic variations on plant performance and improvement.

Combining new breeding techniques could even be more advantageous in crop breeding. Combining speed breeding and epigenomics with genome editing has sparked an innovative revolution in breeding, but still faces significant practical challenges. One is transformation efficiency which is crucial for advancing crop GE techniques, while another is regeneration which still presents a major limitation in obtaining stable edited plants. CRISPR/Cas9 genome editing faces challenges with off-target effects, causing unwanted genome cleavage and chromosomal rearrangements, which can be harmful to plants (Kweon et al. 2013). Cas9's specificity is influenced by the sgRNA guide sequence and PAM sequence, with mismatches causing off-target effects. Strategies to reduce off-target events include manipulating sgRNA composition and structure, such as altering GC content or truncating the sgRNA (Tsai et al. 2015; Zhang et al. 2015). Various methods, like Digenome-seq and GUIDE-seq, help identify off-target events (Ansari et al. 2020). To reduce off-target effects, improved Cas9 variants like fCas9 and modified Cas9 with amino acid substitutions have been developed, offering increased specificity and reduced off-target events (Guilinger et al. 2014; Kleinstiver et al. 2016). In addition, finding the genes that provide resistance to the pathogen is essential to the effectiveness of GE-mediated disease resistance. Single genes can occasionally provide significant resistance, which makes them perfect targets for editing. In some cases, resistance is controlled by several genes or quantitative trait loci (QTLs), necessitating the use of more sophisticated editing techniques such gene stacking or regulatory element change (Zaidi et al. 2018). Another obstacle in complete acceptance of GE in agriculture is the public opinion about GE as well as current regulative on national levels in many countries. Regulatory and ethical discussions focus on how to manage gene-edited products, including whether to adjust current genetically modified organism (GMO) risk regulations. Concerns include gaps in information about target and off-target impacts on genomes and the need for environment management approaches. Addressing these complex issues involves collaboration among researchers, regulators, and stakeholders. At this moment in the European Union (EU), regulatory guidelines for gene-edited plants are based on GMO guidelines, with considerations for human, animal, and ecological health. National and pan-European regulations assess risks case-by-case, with the European Commission responsible for risk management decisions (Bortesi et al. 2016; Stazi and Jovine 2022). Debates continue on whether new gene-editing techniques should be exempt from existing GMO guidelines, considering their similarity to traditional methods and associated risks.

5.5 Future Outlook

Future of agricultural production and global food security will greatly depend on breeding success, novel strategies employed, and further improvement in disciplines related to the development of stable and adaptive crop varieties and hybrids. In order to facilitate this success, classical and modern breeding approaches must be combined with the aim to exploit the best of each technology and to complement their potentials. In this chapter, most of the currently available methodologies are discussed and their advantages as well as weaknesses in supporting the improvement of crop stability are described. These methods enable the development of crops that are not only high-yielding but also stable in a variety of environmental conditions. They are more resistant to biotic (like pests and diseases) and abiotic (like drought, heat, and salinity) stresses, resulting in more consistent yields. Furthermore, by using these methods, breeders may better account for GEIs, guaranteeing that the crops produced are appropriate for particular geographical areas and agricultural methods. Having in mind a rapid scientific development of many technologies in the last few decades, it is highly difficult to predict what would be the next breakthrough that will address most of the today's challenges and boundaries. For sure, the science will try firstly to solve the challenges mentioned above by different approaches. Also, the precision of each technology to target specific traits related to crop stability (stress tolerance, resistance to prevalent and newly appeared diseases, climate change resilience, etc.) will be very important and should be further improved in order to develop an adequate tools for fast and successful adaptation to new challenges and limitations.

There are important results from genomic prediction studies that can be translated to new methods to predict stability traits. These studies inform future research looking to improve prediction for stability traits and develop methods for effective use in real-world breeding programs. While heritability and genetic architecture are specific to the trait of interest, experimental design, training population design, the method of calculating stability, marker density and type, and statistical methods are elements that can be optimized to increase prediction accuracy and are crucial for the application in breeding programs. The independent relationship between trait and trait stability indicated that genomic selection could be a valuable tool for selecting both high-yielding and stable lines. Genomic prediction for stability traits would be particularly beneficial during the early yield trial phase of the breeding process, when there is not enough data from testing environments to select based on phenotypic selection. Furthermore, an assessment of the genetic control of agronomically important traits stability and its consistency across populations grown in different environments would also be helpful for understanding the future potential and application of MAS and genomic prediction for stability. The stable QTLs as well as the desirable SNP alleles and haplotype alleles underlying these stable QTLs identified for the agronomically important traits are potential resource for the future crop improvement.

Another promising approach is the combination of various technologies to overcome the limitations each one has, while simultaneously improving them. For example, integrating molecular techniques with speed breeding or combining different omics methodologies could lead to the development of innovative breeding strategies.

Additionally, harnessing the full potential of artificial intelligence (AI) could further enhance these efforts.

Recent advances in high-throughput technologies including genomics, epigenomics, transcriptomics, proteomics, and metabolomics facilitate analysis of global genetic diversity. By integrating data from different omics layers, multi-omics approaches enable understanding of the genetic and biochemical mechanisms underlying stress resilience and complex traits involved in stability (Akiyama 2021). This integration helps bridge the gap between genotype and phenotype, facilitates targeted breeding strategies optimized for specific growing conditions and environmental stresses to improve crop performance, yield, stability, and quality (Khan et al. 2024). There are several successful examples of the integration of multi-omics data in crop improvement. For example, in rice a combination of transcriptomics and proteomics was used to identify the genes and proteins involved in disease resistance that helped in developing new varieties more resistant to a range of diseases (Scossa et al. 2021). In tomatoes, a combination of metabolomics and genetics data was used to identify the genes and metabolic pathways involved in the production of pigment lycopene (Derbyshire et al. 2022). Multi-omics-based genomic prediction has been implemented successfully, improving prediction accuracy in different crops such as maize (Schrag et al. 2018), wheat (Zhao et al. 2015), oats (Hu et al. 2021), barley (Wu et al. 2022), and rapeseed (Knoch et al. 2021).

The integration of advanced AI technologies allows breeders to analyze and interpret extensive data sets from different sources, including genotypic and phenotypic information, facilitating informed decision-making, reducing the time and resources required to develop new plant varieties, and enhancing the success rate of breeding programs. Machine learning is a sub-class of AI that uses statistical and computational methods to analyze large data sets characterized by significant noise, dimensionality, and/or incompleteness (Liu et al. 2020a; Mahood et al. 2020). ML efficiently identifies patterns, structures, and relationships in existing data, enabling accurate description, categorization, and prediction of new data (Singh et al. 2016). Three key components involve the use of AI in breeding: using algorithms and programs to transform sensory data into phenotypic information; model development to understand genotype–phenotype association with environmental factors; and database management to facilitate the exchange of data and resources (Khan et al. 2022). Moreover, omics-based data evaluation has become too large and complicated for statistical analysis. Among other fields of large-scale data processing, modern ML tools and AI approaches systematically integrate and address complex biological issues in metabolomics, genomics, proteomics, transcriptomics, and systems biology (Xu and Jackson 2019). Examples include identifying regulatory and non-regulatory elements and evaluating gene expression, determination of the specificity of transcription factor DNA binding, identification of protein-coding genes, interaction between proteins, tertiary structure of proteins, gene ontology, and subcellular localization (Naqvi et al. 2022). In addition to its function in genotype–phenotype relationships, the effectiveness of ML methods has been evaluated for genomic prediction. In scenarios where non-additive variances had a substantial impact, "deepGS" performed better than the conventional genomic selection (GS) models (Abdollahi-Arpanahi et al. 2020). Additionally, ML helps to analyze the complex function of gene activity in determining plant phenotypes in response to environmental fluctuations as well as accurate forecasting of crop stability under different conditions, including soil type, temperature, and humidity (Großkinsky et al. 2018; Harfouche et al. 2019). Modeling GEI helps identify the optimal ecological range for achieving the highest yield productivity and estimates yield stability across different ecological zones. If more complicated climate factors are considered, the model also helps estimate the influence of climate change on yield performance and quality and identifies the optimal genotypes adapted to climate change. Most methods for modeling GEI are based on linear regression algorithms to infer correlations between yield performance and a few environmental factors. Statistical models have become unsuitable for modeling increasingly complicated genotypic, phenotypic, environmental, and climate data sets, prompting the need for ML methods (Cheng and Wang 2024). Furthermore, ML techniques will be used more frequently for high-throughput crop stress phenotyping, crop yield prediction, and climate change effect assessment (Singh et al. 2016; Esposito et al. 2019; Tong and Nikoloski 2021). These technologies reduce the need for extensive manual labor and enable more frequent crop monitoring, which will lead to earlier detection of issues and better informed decision-making that will ultimately result in higher stability of crop yields and reduced losses from pests, diseases, or environmental factors. In addition, open science and FAIR (Findable, Accessible, Interoperable, and Reusable) data principles will enable re-use of existing experimental data, an integration of data from a large number of different sources, different angles and models of analyses, as well as more efficient use of resources. Application of all above-mentioned complex statistical and AI methodologies in analyzing such a huge amount of data obtained worldwide will improve our knowledge about crop stability in different environments and enable mapping of the best performing areas for each crop or cultivar.

5.6 Conclusion

Accurately assessing the stability of genotypes across diverse environments is crucial for developing crops that perform well under varying conditions. Traditional breeding methods, while successful, require significant time and resources to obtain reliable stability data. The complexity of traits like yield stability, affected by both genetic and environmental factors, makes prediction and progress increasingly difficult. These challenges highlight the limitations of conventional methods, which are often too slow to address the demands of modern agriculture.

In the past two decades, molecular techniques have made huge strides, and high-throughput phenotyping (HTP) methodologies have significantly advanced, enabling a deeper understanding of important processes related to abiotic stress tolerance and biotic resistance in crop species. These innovations, in conjunction with modern genomic techniques, have contributed to the development of more adaptive and stable crop varieties across diverse environmental conditions. Speed breeding, which accelerates crop development while minimizing time, space, and resources, has further enhanced the ability to create high-performance varieties with desirable traits. Modern approaches, such as genomic selection and marker-assisted selection (MAS), offer efficient ways to select for stability traits and stress tolerance, even at early breeding stages. By identifying and manipulating genetic factors responsible for stress resilience and yield stability, these technologies enable breeders to expedite crop development. Advances in gene editing technologies like CRISPR/Cas9, as well as the exploration of epigenetic modifications, improve the ability to design crops resilient to environmental stresses. The integration of genomics, multi-omics approaches, and artificial intelligence (AI) offers additional opportunities to refine breeding strategies.

Modern breeding techniques not only help address genotype-by-environment ($G \times E$) interactions but also empower breeders to develop crops that are resilient to the growing challenges of climate change, pests, and diseases. The integration of classical and modern approaches is key to producing varieties that are not only high-yielding but also stable and resilient across diverse environments. These innovations are vital for ensuring food security, as well as the stability and sustainability of crop production under increasing environmental variability. As breeding technologies continue to evolve, new strategies will emerge by combining existing and novel approaches to meet future agricultural demands. Collectively, these advancements will play a crucial role in optimizing crop performance and tackling the global challenges of food production in the coming decades.

Competing Interests This study was funded by the Ministry of Science, Technological Development and Innovation of the Republic of Serbia, grant number: 451-03-66/2024-03/200032, by the Science Fund of the Republic of Serbia through IDEAS project "Creating Climate Smart Sunflower for Future Challenges" (SMARTSUN), grant number 7732457, by the Science Fund of the Republic of Serbia through PRISMA project "Soybean Yield Prediction Using Multi-omics Data Integration" (SoyPredict), grant number 6788, and by the European Commission through Twinning Western Balkans project CROPINNO, grant number 101059784, by Center of Excellence for Innovations in Breeding of Climate-Resilient Crops—Climate Crops and Center of Excellence for Legumes, Institute of Field and Vegetable Crops, Novi Sad, Serbia. The authors have no conflicts of interest to declare that are relevant to the content of this chapter.

References

Abbai R, Singh VK, Nachimuthu VV, Sinha P, Selvaraj R, Vipparla AK, Singh AK, Singh UM, Varshney RK, Kumar A (2019) Haplotype analysis of key genes governing grain yield and quality traits across 3K RG panel reveals scope for the development of tailor-made rice with enhanced genetic gains. Plant Biotechnol J 17:1612–1622

Abdala LJ, Otegui ME, Di Mauro G (2024) On-farm soybean genetic progress and yield stability during the early 21st century (2024): A case study of a commercial breeding program in Argentina and Brazil. Field Crops Res 308:109277. https://doi.org/10.1016/j.fcr.2024.109277

Abdalla A, Cen H, Wan L, Rashid R, Weng H, Zhou W, He Y (2019) Fine-tuning convolutional neural network with transfer learning for semantic segmentation of ground-level oilseed rape images in a field with high weed pressure. Comput Electron Agric 167:105091

Abdallah NA, Prakash CS, McHughen AG (2015) Genome editing for crop improvement: challenges and opportunities. GM Crops Food 6(4):183–205. https://doi.org/10.1080/21645698.2015.1129937

Abdollahi-Arpanahi R, Gianola D, Peñagaricano F (2020) Deep learning versus parametric and ensemble methods for genomic prediction of complex phenotypes. Genet Sel Evol 52(1):12. https://doi.org/10.1186/s12711-020-00531-z

Achilli AL, Roncallo PF, Echenique V (2022) Genetic gains in grain yield and agronomic traits of Argentinian durum wheat from 1934 to 2015. Agronomy 12:2151. https://doi.org/10.3390/agronomy12092151

Adunola P, Ferrão MAG, Ferrão RG, Fonseca AFA, Volpi PS, Comério M, Filho ACV, Munoz PR, Ferrão LFV (2023) Genomic selection for genotype performance and environmental stability in Coffea canephora. G3 13(6):jkad062. https://doi.org/10.1093/g3journal/jkad062

Afroz N, Ansary MWR, Islam T (2023) CRISPR-cas genome editing for the development of abiotic stress-tolerant wheat. In: Abiotic stresses wheat, vol 195–207. Academic. https://doi.org/10.1016/B978-0-323-95368-9.00014-X

Akdemir D, Isidro-Sánchez J (2019) Design of training populations for selective phenotyping in genomic prediction. Sci Rep 9:1446. https://doi.org/10.1038/s41598-018-38081-6

Akiyama M (2021) Multi-omics study for interpretation of genomewide association study. J Hum Genet 66:3–10

Aleem M, Razzaq MK, Aleem M et al (2024) Genome-wide association study provides new insight into the underlying mechanism of drought tolerance during seed germination stage in soybean. Sci Rep 14:20765. https://doi.org/10.1038/s41598-024-71357-8

Allard RW, Bradshaw AD (1964) Implication of genotype–environmental interaction in applied plant breeding. Crop Sci 5:503–506

Ansari WA, Chandanshive SU, Bhatt V, Nadaf AB, Vats S, Katara JL, Sonah H, Deshmukh R (2020) Genome editing in cereals: approaches, applications and challenges. Int J Mol Sci 21(11):4040

Araus JL, Kefauver SC, Zaman-Allah M, Olsen MS, Cairns JE (2018) Translating high-throughput phenotyping into genetic gain. Trends Plant Sci. 23:451–466. https://doi.org/10.1016/j.tplants.2018.02.001

Arif A, Parveen N, Waheed MQ, Atif RM, Waqar I, Shah TM (2021) A comparative study for assessing the drought-tolerance of chickpea under varying natural growth environments. Front Plant Sci 11:607869

Atieno J, Li Y, Langridge P et al (2017) Exploring genetic variation for salinity tolerance in chickpea using image-based phenotyping. Sci Rep 7:1300. https://doi.org/10.1038/s41598-017-01211-7

Auler PA, Nogueira do Amaral M, Bolacel Braga EJ, Maserti B (2021) Drought stress memory in rice guard cells: proteome changes and genomic stability of DNA. Plant Physiol Biochem 169: 49–62

Aziz AM, Masmoudi K (2024) Molecular breakthroughs in modern plant breeding techniques. Hortic Plant J 11(1):15–41. https://doi.org/10.1016/j.hpj.2024.01.004

Badu-Apraku B, Talabi AO, Fakorede MAB, Fasanmade Y, Gedil M, Magorokosho C et al (2019) Yield gains and associated changes in an early yellow bi-parental maize population following genomic selection for Striga resistance and drought tolerance. BMC Plant Biol 19:129

Bao Y (2014) Genome wide association mapping and genomic selection for agronomic and disease traits in soybean. Dissertation, University of Minnesota

Barloy D, Lemoine J, Abelard P et al (2007) Marker-assisted pyramiding of two cereal cyst nematode resistance genes from Aegilops variabilis in wheat. Mol Breed 20:31–40. https://doi.org/10.1007/s11032-006-9070-x

Bassi FM, Bentley AR, Charmet G, Ortiz R, Crossa J (2016) Breeding schemes for the implementation of genomic selection in wheat (*Triticum* spp.). Plant Sci 242:23–36. https://doi.org/10.1016/j.plantsci.2015.08.021

Beck M, Zhou J, Faulkner C, MacLean D, Robatzek S (2012) Spatio-temporal cellular dynamics of the Arabidopsis flagellin receptor reveal activation status-dependent endosomal sorting. Plant Cell 24(10):4205–4219. https://doi.org/10.1105/tpc.112.100263

Becker HC, Leon J (1988) Stability analysis in plant breeding. Plant Breed 101(1):1–23

Begna T (2020) The role of genotype by environmental interaction in plant breeding. Int J Agri Biosci 9(5):209–215

Bevan MW et al (2017) Genomic innovation for crop improvement. Nature 543:346–354

Beyene Y, Semagn K, Mugo S, Tarekegne A, Babu R, Meisel B et al (2015) Genetic gains in grain yield through genomic selection in eight bi-parental maize populations under drought stress. Crop Sci 55:154–163. https://doi.org/10.2135/CROPSCI2014.07.0460

Bhandari A, Bartholomé J, Cao-Hamadoun T-V, Kumari N, Frouin J, Kumar A et al (2019) Selection of trait-specific markers and multi-environment models improve genomic predictive ability in rice. PLoS One 14:e0208871. https://doi.org/10.1371/JOURNAL.PONE.0208871

Bhat JA, Adeboye KA, Ganie SA, Barmukh R, Hu D, Varshney RK, Yu D (2022) Genome-wide association study, haplotype analysis, and genomic prediction reveal the genetic basis of yield-related traits in soybean (Glycine max L.). Front Genet 13:953833. https://doi.org/10.3389/fgene.2022.953833

Bhatta M, Sandro P, Smith MR, Delaney O, Voss-Fels KP, Gutierrez L et al (2021) Need for speed: manipulating plant growth to accelerate breeding cycles. Curr Opin Plant Biol 60:101986. https://doi.org/10.1016/j.pbi.2020.101986

Bhoi A, Yadu B, Chandra J, Keshavkant S (2022) Mutagenesis: a coherent technique to develop biotic stress resistant plants. Plant Stress 3:100053. https://doi.org/10.1016/j.stress.2021.100053

Bonnett D, Li Y, Crossa J, Dreisigacker S, Basnet B, Pérez-Rodríguez P, Alvarado G, Jannink JL, Poland J, Sorrells M (2022) Response to early generation genomic selection for yield in wheat. Front Plant Sci 12:718611. https://doi.org/10.3389/fpls.2021.718611

Bortesi L, Zhu C, Zischewski J, Perez L, Bassié L, Nadi R, Forni G, Lade SB, Soto E, Jin X et al (2016) Patterns of CRISPR/Cas9 activity in plants, animals and microbes. Plant Biotechnol J 14: 2203–2216

Brbaklić L, Trkulja D, Kondić-Špika A, Hristov N, Denčić S, Mikić S, Tomičić M, Kobiljski B (2015) Genetic associations in the detection of QTLs for wheat spike-related traits. Pesq Agropec Bras 50(2):149–159

Brbaklić L, Trkulja D, Mikić S, Mirosavljević M, Momčilović V, Dudić B, Procházková L, Aćin V (2021) Genetic diversity and population structure of Serbian barley (*Hordeum vulgare* L.) collection during a 40-year long breeding period. Agronomy 11(1):118. https://doi.org/10.3390/agronomy11010118

Brdar-Jokanović M, Trkulja D, Nikolić-Đorić E, Kondić-Špika A, Kobiljski B (2013) The possibilities of applying marker assisted selection in breeding boron tolerant wheat. Genetika-Belgrade 45(3):769–776. https://doi.org/10.2298/GENSR1303769J

Breseghello F, Sorrells ME (2006) Association analysis as a strategy for improvement of quantitative traits in plants. Crop Sci 46:1323–1330

Brinton J et al (2020) A haplotype-led approach to increase the precision of wheat breeding. Commun Biol 3:712

Burgueño J et al (2012) Genomic prediction of breeding values when modeling genotype_environment interaction using pedigree and dense molecular markers. Crop Sci 52: 707–719

Burton J (2011) Quantitative genetics: results in soybean breeding. In: Miladinović J, Hrustić M, Vidić M (eds) Soybean. AMB Grafika, Novi Sad, pp 137–183

Caffarraa A, Rinaldia M, Eccela E, Rossib V, Pertota I (2012) Modelling the impact of climate change on the interaction between grapevine and its pests and pathogens: European grapevine moth and powdery mildew. Agric Ecosyst Environ 148:89–101. https://doi.org/10.1016/j.agee.2011.11.017

Campbell B, Thornton P, Loboguerrero A, Dinesh D, Nowak A (2023) Transforming food systems under climate change through innovation. Cambridge University Press, Cambridge, UK

Carvalho HF, Rio S, García-Abadillo J et al (2024) Revisiting superiority and stability metrics of cultivar performances using genomic data: derivations of new estimators. Plant Methods 20:85

Ćeran M, Đorđević V, Miladinović J, Vasiljević M, Đukić V, Ranđelović P, Jaćimović S (2024a) Selective genotyping and phenotyping for optimization of genomic prediction models for populations with different diversity. Plants 13:975. https://doi.org/10.3390/plants13070975

Ćeran M, Miladinović D, Đorđević V, Trkulja D, Radanović A, Glogovac S, Kondić-Špika A (2024b) Genomics-assisted speed breeding for crop improvement: present and future. Front Sustain Food Syst 8:1383302. https://doi.org/10.3389/fsufs.2024.1383302

Chandrakanth N, Moorthy SM, Rekha M, Sivaprasad V (2016) Stability and path analysis for yield and related traits in silkworm, (*Bombyx mori* L.) reared under stress conditions. Genetika 48(1): 271–284

Chawade A, van Ham J, Blomquist H, Bagge O, Alexandersson E, Ortiz R (2019) High-throughput field-phenotyping tools for plant breeding and precision agriculture. Agronomy 9:258. https://doi.org/10.3390/agronomy9050258

Cheng Q, Wang X (2024) Machine learning for AI breeding in plants. Genomics Proteomics Bioinformatics 22(4):qzae051. https://doi.org/10.1093/gpbjnl/qzae051

Cobb JN, Biswas PS, Platten JD (2019) Back to the future: revisiting MAS as a tool for modern plant breeding. Theor Appl Genet 132:647–667

Collard BC, Jahufer MZ, Brouwer JB et al (2005) An introduction to markers, quantitative trait loci (QTL) mapping and marker-assisted selection for crop improvement: the basic concepts. Euphytica 142(1–2):169–196

Colson J, Bouniols A, Jones JW (1995) Soybean reproductive development: adapting a model for European cultivars. Agron J 87:1129–1139

Correia PMP, Westergaard JC, da Silva AB, Roitsch T, Carmo-Silva E, da Silva JM (2022) High-throughput phenotyping of physiological traits for wheat resilience to high temperature and drought stress. J Exp Bot 73(15):5235–5251. https://doi.org/10.1093/jxb/erac160

Crossa J et al (2016a) Extending the Marker × Environment interaction model for genomic-enabled prediction and genome-wide association analyses in durum wheat. Crop Sci 56:1–17

Crossa J, Jarquín D, Franco J, Pérez-Rodríguez P, Burgueńo J, Saint-Pierre C, Vikram P, Sansaloni C, Petroli C, Akdemir D et al (2016b) Genomic prediction of gene bank wheat landraces. G3-Genes Genom Genet 6:1819–1834. https://doi.org/10.1534/g3.116.029637

Crossa J, Pérez-Rodríguez P, Cuevas J, Montesinos-López O, Jarquín D, de Los CG, Burgueño J, González-Camacho JM, Pérez-Elizalde S, Beyene Y, Dreisigacker S, Singh R, Zhang X, Gowda M, Roorkiwal M, Rutkoski J, Varshney RK (2017) Genomic selection in plant breeding: methods, models, and perspectives. Trends Plant Sci 22(11):961–975

Crossa J, Fritsche-Neto R, Montesinos-Lopez OA, Costa-Neto G, Dreisigacker S, Montesinos-Lopez A, Bentley AR (2021) The modern plant breeding triangle: optimizing the use of genomics, phenomics, and enviromics data. Front Plant Sci 12:651480. https://doi.org/10.3389/fpls.2021.651480

Cui F, Zhao C, Ding A et al (2014) Construction of an integrative linkage map and QTL mapping of grain yield-related traits using three related wheat RIL populations. Theor Appl Genet 127:659–675. https://doi.org/10.1007/s00122-013-2249-8

Cvejić S, Jocić S, Mitrović B, Bekavac G, Mirosavljević M, Marjanović Jeromela A et al (2023) Innovative approaches in breeding of climate-resilient crops. In: Benkeblia N (ed) Climate change and agriculture: perspectives, sustainability and resilience. Wiley, New Jersey, pp 111–156

Dargahi H, Tanya P, Somta P, Abe J, Srinives P (2014) Mapping quantitative trait loci for yield-related traits in soybean (*Glycine max* L.). Breed Sci 64(4):282–290. https://doi.org/10.1270/jsbbs.64.282

Davar R, Darvishzadeh R, Majd A, Ghosta Y, Sarrafi A (2011) QTL mapping of partial resistance to basal stem rot in sunflower using recombinant inbred lines. Phyto-pathologia Mediterranea 49(3):330–341. https://doi.org/10.14601/Phytopathol_Mediterr-8374

de los Campos G, Pérez-Rodríguez P, Bogard M, Gouache D, Crossa J (2020) A data-driven simulation platform to predict cultivars' performances under uncertain weather conditions. Nat Commun 11:4876. https://doi.org/10.1038/s41467-020-18480-y

Derbyshire MC, Batley J, Edwards D (2022) Use of multiple 'omics techniques to accelerate the breeding of abiotic stress tolerant crops. Curr Plant Biol 32:100262

Dimitrijević A, Imerovski I, Miladinović D, Tančić S, Dušanić N, Jocić S, Miklič V (2010) Use of SSR markers in identification of sunflower isogenic lines in late generations of backcrossing. Helia 33(53):191–198. https://doi.org/10.2298/HEL1053191D

Đorđević V, Ćeran M, Miladinović J, Balešević-Tubić S, Petrović K, Miladinov Z, Marinković J (2019) Exploring the performance of genomic prediction models for soybean yield using different validation approaches. Mol Breeding 39:74. https://doi.org/10.1007/s11032-019-0983-6

Edmeades GO, Bolanos J, Lafitte HR, Rajaram S, Pfeiffer WH, Fisher RA (1989) Traditional approaches to breeding for drought resistance in cereals. In: Baker WG (ed) Drought resistance in cereals. ICSU press, Paris, pp 27–52

Egea-Gilabert C, Pagnotta MA, Tripodi P (2021) Genotype × Environment Interactions in Crop Breeding. Agronomy 11(8):1644

Esposito S, Carputo D, Cardi T, Tripodi P (2019) Applications and trends of machine learning in genomics and phenomics for next-generation breeding. Plants 9:34. https://doi.org/10.3390/plants9010034

Farshadfar E, Sutka J (2003) Locating QTLs controlling adaptation in wheat using AMMI model. Cereal Res Commun 31:249–256

Feng P, Sun X, Liu X, Li Y, Sun Q, Lu H, Li M, Ding X, Dong Y (2022) Epigenetic regulation of plant tolerance to salt stress by histone acetyltransferase GsMYST1 from wild soybean. Front Plant Sci 13:860056

Fischer RA (2011) Wheat physiology: a review of recent developments. Crop Pasture Sci 62(2): 95–114

Fróna D, Szenderák J, Harangi-Rákos M (2019) The challenge of feeding the world. Sustainability 11:5816. https://doi.org/10.3390/su11205816

Gahlaut V, Samtani H, Khurana P (2020) Genome-wide identification and expression profiling of cytosine-5 DNA methyl-transferases during drought and heat stress in wheat (*Triticum aestivum*). Genomics 112:4796–4807

Gao L, Gonda I, Sun H, Ma Q, Bao K, Tieman DM et al (2019) The tomato pan-genome uncovers new genes and a rare allele regulating fruit flavor. Nat Genet 51:1044–1051. https://doi.org/10.1038/s41588-019-0410-2

Garg S (2021) Computational methods for chromosome-scale haplotype reconstruction. Genome Biol 22:101. https://doi.org/10.1186/s13059-021-02328-9

Gauch HG Jr (2013) A simple protocol for AMMI analysis of yield trials. Crop Sci 53:1860–1869

Ghosh S, Watson A, Gonzalez-Navarro OE, Ramirez-Gonzalez RH, Yanes L, Mendoza-Suárez M et al (2018) Speed breeding in growth chambers and glasshouses for crop breeding and model plant research. Nat Protocol 13:2944–2963. https://doi.org/10.1038/s41596-018-0072-z

Gill T, Gill SK, Saini DK et al (2022) A comprehensive review of high throughput phenotyping and machine learning for plant stress phenotyping. Phenomics 2:156–183. https://doi.org/10.1007/s43657-022-00048-z

Glogovac S, Takač A, Lj B, Trkulja D, Červenski J, Gvozdanović-Varga J, Popović V (2013) Molecular evaluation of genetic variability in tomato (*Lycopersicon esculentum* Mill.) genotypes by microsatellite markers. Field Veg Crops Res 50(3):1–5

Golicz AA, Batley J, Edwards D (2016) Towards plant pangenomics. Plant Biotechnol J 14:1099–1105. https://doi.org/10.1111/pbi.12499

Gollob HF (1968) A statistical model which combines features of factor analytic and analysis of variance techniques. Psychometrika 33:73–115

Gosavi G, Yan F, Ren B, Kuang Y, Yan D, Zhou X et al (2020) Applications of CRISPR technology in studying plant-pathogen interactions: overview and perspective. Phytopathol Res 2:21. https://doi.org/10.1186/s42483-020-00060-z

Großkinsky DK, Syaifullah SJ, Roitsch T (2018) Integration of multiomics techniques and physiological phenotyping within a holistic phenomics approach to study senescence in model and crop plants. J Exp Bot 69:825–844. https://doi.org/10.1093/jxb/erx333

Grüneberg WJ, Manrique K, Zhang D, Hermann M (2005) Genotype × Environment interactions for a diverse set of sweet potato clones evaluated across varying ecogeographic conditions in Peru. Crop Sci 45:2160–2171

Gudi S, Kumar P, Singh S, Tanin MJ, Sharma A (2022) Strategies for accelerating genetic gains in crop plants: special focus on speed breeding. Physiol Mol Biol Plants 28:1921–1938. https://doi.org/10.1007/s12298-022-01247-8

Guidini R, Jahani M, Huang K, Rieseberg L, Mathew FM (2023) Genome-wide association mapping in sunflower (*Helianthus annuus*) reveals common Loci and putative candidate genes for resistance to *Diaporthe gulyae* and *D. helianthi* Causing Phomopsis stem canker. Plant Disease 107(3):667–674

Guilinger JP, Thompson DB, Liu DR (2014) Fusion of catalytically inactive Cas9 to FokI nuclease improves the specificity of genome modification. Nat Biotechnol 32:577

Gupta PK, Balyan HS, Gahlaut V (2017) QTL analysis for drought tolerance in wheat: present status and future possibilities. Agronomy 7(1):5. https://doi.org/10.3390/agronomy7010005

Hamdan MF, Karlson CKS, Teoh EY, Lau SE, Tan BC (2022) Genome editing for sustainable crop improvement and mitigation of biotic and abiotic stresses. Plants 11:2625

Happ MM, Graef GL, Wang H, Howard R, Posadas L, Hyten DL (2021) Comparing a mixed model approach to traditional stability estimators for mapping genotype by environment interactions and yield stability in Soybean [*Glycine max* (L.) Merr.]. Front Plant Sci 12:1–12. https://doi.org/10.3389/fpls.2021.630175

Harfouche AL, Jacobson DA, Kainer D, Romero JC, Harfouche AH, Mugnozza GS et al (2019) Accelerating climate resilient plant breeding by applying next-generation artificial intelligence. Trends Biotechnol 37:1217–1235. https://doi.org/10.1016/j.tibtech.2019.05.007

Harms B (2023) Genomic selection for yield and seed composition stability in an applied soybean breeding program. Master Thesis, University of Nebraska – Lincoln. https://digitalcommons.unl.edu/agronhortdiss/245

Hasan N, Choudhary S, Naaz N, Sharma N, Laskar RA (2021) Recent advancements in molecular marker-assisted selection and applications in plant breeding programmes. J Genet Eng Biotechnol 19(1):128. https://doi.org/10.1186/s43141-021-00231-1

Hazra S, Moulick D, Mukherjee A, Sahib S, Chowardhara B, Majumdar A, Upadhyay MK, Yadav P, Roy P, Santra SC, Mandal S (2023) Evaluation of efficacy of non-coding RNA in abiotic stress management of field crops: current status and future prospective. Plant Physiol Bioch 203:107940

He S, Li X, Chen M, Xu X, Tang F, Gong T, Xu M, Yang W, Liu W (2024) Crop HTP technologies: applications and prospects. Agriculture 14:723. https://doi.org/10.3390/agriculture14050723

Heffner EL, Sorrells ME, Jannink JL (2009) Genomic selection for crop improvement. Crop Sci 49: 1–12. https://doi.org/10.2135/cropsci2008.08.0512

Heslot N et al (2014) Integrating environmental covariates and crop modeling into the genomic selection framework to predict genotype by environment interactions. Theor Appl Genet 127: 463–480

Hickey LT, Hafeez AN, Robinson H, Jackson SA, Leal-Bertioli SCM, Tester M et al (2019) Breeding crops to feed 10 billion. Nat Biotechnol 37:744–754. https://doi.org/10.1038/s41587-019-0152-9

Hill J, Becker HC, Tigerstedt PMA (1998) Quantitative and ecological aspects of plant breeding. Chapman and Hall, London

Hirsch CN, Foerster JM, Johnson JM, Sekhon RS, Muttoni G, Vaillancourt B et al (2014) Insights into the maize pan-genome and pantranscriptome. Plant Cell 26:121–135. https://doi.org/10.1105/tpc.113.119982

Hoang TML, Tran TN, Nguyen TKT, Williams B, Wurm P, Bellairs S, Mundree S (2016) Improvement of salinity stress tolerance in rice: challenges and opportunities. Agronomy 6:54

Hoopes GM, Hamilton JP, Wood JC, Esteban E, Pasha A, Vaillancourt B et al (2019) An updated gene atlas for maize reveals organ-specific and stress-induced genes. Plant J 97:1154–1167. https://doi.org/10.1111/tpj.14184

Hristov N, Mladenov N, Đuric V, Kondic-Spika A, Marjanovic-Jeromela A (2009) Improvement of wheat quality in cultivars released in Serbia during the 20th century. Cereal Res Commun 38(1): 111–121. https://doi.org/10.1556/CRC.37.2009.4.9

Hu H, Campbell MT, Yeats TH, Zheng X, Runcie DE, Covarrubias-Pazaran G, Broeckling C, Yao L, Caffe-Treml M, Gutiérrez LA et al (2021) Multi-omics prediction of oat agronomic and seed nutritional traits across environments and in distantly related populations. Theor Appl Genet 134:4043–4054. https://doi.org/10.1007/s00122-021-03946-4

Huang M, Cabrera A, Hoffstetter A, Griffey C, Van Sanford D, Costa J, McKendry A, Chao S, Sneller C (2016) Genomic selection for wheat traits and trait stability. Theor Appl Genet 129: 1697–1710. https://doi.org/10.1007/s00122-016-2733-z

Huang J, Li J, Zhou J, Wang L, Yang S, Hurst LD, Li WH, Tian D (2018) Identifying a large number of high-yield genes in rice by pedigree analysis, whole-genome sequencing, and

CRISPR-Cas9 gene knockout. PNAS 115:E7559–E7567. https://doi.org/10.1073/pnas.1806110115

Huang M, Balimponya EG, Mgonja EM, McHale LK, Luzi-Kihupi A, Wang G-L et al (2019) Use of genomic selection in breeding rice (*Oryza sativa* L.) for resistance to rice blast (*Magnaporthe oryzae*). Mol. Breed 39:1–16. https://doi.org/10.1007/S11032-019-1023-2

Hübner S, Bercovich N, Todesco M, Mandel JR, Odenheimer J, Ziegler E et al (2019) Sunflower pan-genome analysis shows that hybridization altered gene content and disease resistance. Nat Plants 5:54–62. https://doi.org/10.1038/s41477-018-0329-0

Hughes A, Askew K, Scotson CP, Williams K, Sauze C, Corke F, Doonan JH, Nibau C (2017) Non-destructive, high-content analysis of wheat grain traits using X-ray micro computed tomography. Plant Methods 13:76. https://doi.org/10.1186/s13007-017-0229-8

Hussain B (2015) Modernization in plant breeding approaches for improving biotic stress resistance in crop plants. Turkish J Agric Forest 39:515–530. https://doi.org/10.3906/tar-1406-176

Imerovski I, Dimitrijević A, Miladinović D, Dedić B, Jocić S, Kovačević B, Obreht D (2013) Identification of PCR markers linked to different Or genes in sunflower. Plant Breed 132(1): 115–120. https://doi.org/10.1111/pbr.12022

Imerovski I, Dimitrijević A, Miladinović D, Jocić S, Dedić B, Cvejić S, Šurlan-Momirović G (2014) Identification and validation of breeder-friendly DNA markers for Pl(arg) gene in sunflower. Mol Breed 34(3):779–788. https://doi.org/10.1007/s11032-014-0074-7

Imerovski I, Dimitrijević A, Miladinović D, Dedić B, Jocić S, Kočiš-Tubić N, Cvejić S (2016) Mapping of a new gene for resistance to broomrape races higher than F. Euphytica 209(2): 281–289. https://doi.org/10.1007/s10681-015-1597-7

Isidro J, Jannink J-L, Akdemir D, Poland J, Heslot N, Sorrells ME (2015) Training set optimization under population structure in genomic selection. Theor Appl Genet 128:145–158. https://doi.org/10.1007/s00122-014-2418-4

Islam M, Sarker M, Sharma N, Rahman M, Collard B, Gregorio G, Ismail AM (2016) Assessment of adaptability of recently released salt tolerant rice varieties in coastal regions of South Bangladesh. Field Crops Res 190:34–43

Jähne F, Hahn V, Würschum T, Leiser WL (2020) Speed breeding short-day crops by LED-controlled light schemes. Theor Appl Genet 133:2335–2342. https://doi.org/10.1007/s00122-020-03601-4

Jang G, Kim J, Yu J-K, Kim H-J, Kim Y, Kim D-W, Kim KH, Lee CW, Chung YS (2020) Review: cost-effective Unmanned Aerial Vehicle (UAV) platform for field plant breeding application. Remote Sens 12:998. https://doi.org/10.3390/rs12060998

Jarquín D et al (2014) A reaction norm model for genomic selection using high-dimensional genomic and environmental data. Theor Appl Genet 127:595–607

Jarquín D et al (2016) Genotyping by sequencing for genomic prediction in a soybean breeding population. BMC Genomics 15:740

Jarquín D et al (2017) Increasing genomic-enabled prediction accuracy by modeling genotype_environment interaction in Kansas wheat. Plant Genome 10(2):plantgenome2016-12. https://doi.org/10.3835/plantgenome2016.12.0130

Jayakodi M, Schreiber M, Stein N, Mascher M (2021) Building pangenome infrastructures for crop plants and their use in association genetics. DNA Res 28:dsaa030. https://doi.org/10.1093/dnares/dsaa030

Jenuwein T, Allis CD (2001) Translating the histone code. Science 293:1074–1080

Jevtić R, Župunski V, Lalošević M, Župunski L (2017) Predicting potential winter wheat yield losses caused by multiple disease systems and climatic conditions. Crop Protect 99:17–25. https://doi.org/10.1016/j.cropro.2017.05.005

Jin X, Zarco-Tejada PJ, Schmidhalter U, Reynolds MP, Hawkesford MJ, Varshney RK, Yang T, Nie C, Li Z, Ming B et al (2021) High-throughput estimation of crop traits: a review of ground and aerial phenotyping platforms. IEEE Geosci Rem Sen M 9:200–231

Jin M, Liu H, Liu X, Guo T, Guo J, Yin Y, Ji Y, Li Z, Zhang J, Wang X, Qiao F, Xiao Y, Zan Y, Yan J (2023) Complex genetic architecture underlying the plasticity of maize agronomic traits. Plant Commun 4(3):100473. https://doi.org/10.1016/j.xplc.2022.100473

Jovičić D, Popović MB, Marjanović Jeromela A, Nikolić Z, Ignjatov M, Milošević D (2019) The interaction between salinity stress and seed ageing during germination of Brassica napus seeds. Seed Sci Technol 47:47–52

Kalkuhl M, von Braun J, Torero M (2016) Volatile and extreme food prices, food security, and policy: an overview. In: Kalkuhl M, von Braun J, Torero M (eds) Food price volatility and its implications for food security and policy. Springer Open, New York City, pp 3–31

Kaňovská I, Biová J, Škrabišová M (2024) New perspectives of post-GWAS analyses: from markers to causal genes for more precise crop breeding. Curr Opin Plant Biol 82:102658

Khan AW et al (2020) Super-pangenome by integrating the wild side of a species for accelerated crop improvement. Trends Plant Sci 25:148–158

Khan MHU, Wang S, Wang J, Ahmar S, Saeed S, Khan SU, Xu X, Chen H, Bhat JA, Feng X (2022) Applications of artificial intelligence in climate-resilient smart-crop breeding. Int J Mol Sci 23: 11156

Khan AA, Iqbal B, Jalal A et al (2024) Advanced molecular approaches for improving crop yield and quality: a review. J Plant Growth Regul 43:2091–2103. https://doi.org/10.1007/s00344-024-11253-7

Kleinstiver BP, Pattanayak V, Prew MS, Tsai SQ, Nguyen NT, Zheng Z, Joung JK (2016) High-fidelity CRISPR–Cas9 nucleases with no detectable genome-wide off-target effects. Nature 529: 490

Knoch D, Werner CR, Meyer RC, Riewe D, Abbadi A, Lücke S, Snowdon RJ, Altmann T (2021) Multi-omics-based prediction of hybrid performance in canola. Theor Appl Genet 134:1147–1165. https://doi.org/10.1007/s00122-020-03759-x

Koh JCO, Spangenberg G, Kant S (2021) Automated machine learning for high-throughput image-based plant phenotyping. Remote Sens 13:858

Kondić-Špika A, Mladenov N, Grahovac N, Zorić M, Mikić S, Trkulja D, Marjanović-Jeromela A, Miladinović D, Hristov N (2019) Biometric analyses of yield, oil and protein contents of wheat (*Triticum aestivum* L.) genotypes in different environments. Agronomy 9:270. https://doi.org/10.3390/agronomy9060270

Kondić-Špika A, Mikić S, Mirosavljević M, Trkulja D, Marjanović Jeromela A, Rajković D et al (2022) Crop breeding for a changing climate in the Pannonian region: towards integration of modern phenotyping tools. J Exp Bot 73:5089–5110. https://doi.org/10.1093/jxb/erac181

Kondić-Špika A, Trkulja D, Brbaklić L, Mikić S, Glogovac S, Johansson E et al (2023) Marker-assisted selection for the improvement of cereals and pseudocereals. In: Rakszegi M, Papageorgiou M, Rocha JM (eds) Developing sustainable and health promoting cereals and pseudocereals. Academic, Cambridge, MA, pp 253–283

Koundinya AVV, Pandit MK, Ramesh D, Mishra P (2019) Phenotypic stability of eggplant for yield and quality through AMMI, GGE and cluster analyses. Sci Hortic (Amsterdam) 247:216–223

Kumar J, Mir RR, Shafi S, Sen Gupta D, Djalovic I, Miladinovic J, Kumar R, Kumar S, Kumar R (2022) Genomics associated interventions for heat stress tolerance in cool season adapted grain legumes. Int J Mol Sci 23(1):399. https://doi.org/10.3390/ijms23010399

Kumar M, Prusty MR, Pandey MK, Singh PK, Bohra A, Guo B, Varshney RK (2023) Application of CRISPR/Cas9-mediated gene editing for abiotic stress management in crop plants. Front Plant Sci 14:1157678. https://doi.org/10.3389/fpls.2023.1157678

Kurasch AK, Hahn V, Leiser WL, Vollmann J, Schori A, Bétrix CA, Mayr B, Winkler J, Mechtler K, Aper J, Sudaric A, Pejic I, Sarcevic H, Jeanson P, Balko C, Signor M, Miceli F, Strijk P, Rietman H, Muresanu E, Djordjevic V, Pospišil A, Barion G, Weigold P, Streng S, Krön M, Würschum T (2017) Identification of mega-environments in Europe and effect of allelic variation at maturity E loci on adaptation of European soybean. Plant Cell Environ 40(5): 765–778. https://doi.org/10.1111/pce.12896

Kweon J, Kim A, Chon JK, Yoo JY, Kim HJ, Kim S, Lee C, Jeong E, Chung E (2013) A library of TAL effector nucleases spanning the human genome. Nat Biotechnol 31(3):251–258

Lado B, González Barrios P, Quincke M, Silva P, Gutiérrez L (2016) Modeling genotype × environment interaction for genomic selection with unbalanced data from a wheat breeding program. Crop Sci 56:2165–2179. https://doi.org/10.2135/cropsci2015.04.0207

Laitinen RAE, Nikoloski Z (2018) Genetic basis of plasticity in plants. J Exp Bot 70:739–745

Li Z, Wu W (2023) Genotype recommendations for high performance and stability based on multiple traits selection across a multi-environment in rapeseed. Eur J Agron 145:126787

Li H, Yan S, Zhao L, Tan J, Zhang Q, Gao F, Wang P, Hou H, Li L (2014a) Histone acetylation associated up-regulation of the cell wall related genes is involved in salt stress induced maize root swelling. BMC Plant Biol 14:1–14

Li Y-H, Zhou G, Ma J, Jiang W, Jin L-G, Zhang Z et al (2014b) De novo assembly of soybean wild relatives for pan-genome analysis of diversity and agronomic traits. Nat Biotechnol 32:1045–1052. https://doi.org/10.1038/nbt.2979

Li M, Li X, Zhou Z, Wu P, Fang M, Pan X, Lin Q, Luo W, Wu G, Li H (2016) Reassessment of the four yield-related genes Gn1a, DEP1, GS3, and IPA1 in rice using a CRISPR/Cas9 system. Front Plant Sci 7:377. https://doi.org/10.3389/fpls.2016.00377

Li C, Unver T, Zhang BH (2017) A high-efficiency CRISPR/Cas9 system for targeted mutagenesis in cotton (Gossypium hirsutum L.). Sci Rep 7:43902. https://doi.org/10.1038/srep43902

Li C, Zong Y, Wang Y, Jin S, Zhang D, Song Q et al (2018) Expanded base editing in rice and wheat using a Cas9-adenosine deaminase fusion. Genome Biol 19:59–59. https://doi.org/10.1186/s13059-018-1443-z

Li R, Liu C, Zhao R, Wang L, Chen L, Yu W et al (2019) CRISPR/Cas9-mediated SlNPR1 mutagenesis reduces tomato plant drought tolerance. BMC Plant Biol 19:1–13. https://doi.org/10.1186/s12870-018-1627-4

Li M, Liu Y, Wang C, Yang X, Li D, Zhang X, Xu C, Zhang Y, Li W, Zhao L (2020) Identification of traits contributing to high and stable yields in different soybean varieties across three Chinese latitudes. Front Plant Sci 10:1642

Li M, Zhang YW, Zhang ZC, Xiang Y, Liu MH, Zhou YH et al (2022) A compressed variance component mixed model for detecting QTNs, and QTN-by-environment and QTN-by-QTN interactions in genome-wide association studies. Mol Plant 15(4):630–650. https://doi.org/10.1016/j.molp.2022.02.012

Li Y, Tan Z, Zeng C, Xiao M, Lin S, Yao W (2023) Regulation of seed oil accumulation by lncRNAs in *Brassica napus*. Biotechnol Biofuels Bioprod 16:22–40. https://doi.org/10.1186/s13068-022-02256-1

Lin K, Gong L, Huang Y et al (2019) Deep learning-based segmentation and quantification of cucumber powdery mildew using convolutional neural network. Front Plant Sci 10:155. https://doi.org/10.3389/fpls.2019.00155

Liu C, Lu F, Cui X, Cao X (2010) Histone methylation in higher plants. Annu Rev Plant Biol 61: 395–420

Liu G, Zhao Y, Mirdita V, Reif JC (2017a) Efficient strategies to assess yield stability in winter wheat. Theor Appl Genet 130:1587–1599

Liu X, Surui W, Jiao X, Sui C, Wei J (2017b) Application of CRISPR/Cas9 in plant biology. Acta Pharm Sinica B 7:292–302. https://doi.org/10.1016/j.apsb.2017.01.002

Liu J, Li J, Wang H, Yan J (2020a) Application of deep learning in genomics. Sci China Life Sci 63: 1860–1878. https://doi.org/10.1007/s11427-020-1804-5

Liu Y, Du H, Li P, Shen Y, Peng H, Liu S et al (2020b) Pan-genome of wild and cultivated soybeans. Cell 182:162–176. https://doi.org/10.1016/j.cell.2020.05.023

Liu X, Yang Z, Hu W, Liu S, Sun R, Jin S, Nergui K, Zhao G, Gao L, Liu Y, Deng X (2024) A genome-wide association study identifies novel QTL for wheat yield stability under drought stress, Current Plant Biology 37:100326. https://doi.org/10.1016/j.cpb.2024.100326

Lopez-Cruz M, Crossa J, Bonnett D, Dreisigacker S, Poland J, Jannink J-L, Singh RP, Autrique E, de los Campos G (2015) Increased prediction accuracy in wheat breeding trials using a marker ×

environment interaction genomic selection model. G3 5:569–582. https://doi.org/10.1534/g3.114.016097
Lorenz AJ, Smith KP, Jannink JL (2012) Potential and optimization of genomic selection for Fusarium head blight resistance in six-row barley. Crop Sci 52:1609–1621. https://doi.org/10.2135/cropsci2011.09.0503
Lozada DN, Carter AH (2020a) Genomic selection in winter wheat breeding using a recommender approach. Genes 11(7):779. https://doi.org/10.3390/genes11070779
Lozada DN, Carter AH (2020b) Insights into the genetic architecture of phenotypic stability traits in winter wheat. Agronomy 10(3):368. https://doi.org/10.3390/agronomy10030368
Lu K, Wu B, Wang J, Zhu W, Nie H, Qian J, Huang W, Fang Z (2018) Blocking amino acid transporter OsAAP3 improves grain yield by promoting outgrowth buds and increasing tiller number in rice. Plant Biotechnol J 16:1710–1722. https://doi.org/10.1111/pbi.12907
Mahood EH, Kruse LH, Moghe GD (2020) Machine learning: a powerful tool for gene function prediction in plants. Appl Plant Sci 8:e11376. https://doi.org/10.1002/aps3.11376
Mangini G, Nigro D, Margiotta B, De Vita P, Gadaleta A, Simeone R et al (2018) Exploring SNP diversity in wheat landraces germplasm and setting of a molecular barcode for fingerprinting. Cereal Res Commun 46(3):377–387
Marini L, St-Martin A, Vico G, Baldoni G, Berti A, Blecharczyk A, Małecka-Jankowiak I, Morari F, Sawinska Z, Bommarco R (2020) Crop rotations sustain cereal yields under a changing climate. Environ Res Lett 15:124011. https://doi.org/10.1088/1748-9326/abc651
Marjanović Jeromela A, Terzić S, Jankulovska M, Zorić M, Kondić-Špika A, Jocković M, Hristov N, Crnobarac J, Nagl N (2019) Dissection of year related climatic variables and their effect on winter rapeseed (*Brassica napus* L.) development and yield. Agronomy 9:517
Merrick LF, Herr AW, Sandhu KS, Lozada DN, Carter AH (2022) Optimizing plant breeding programs for genomic selection. Agronomy 12:714. https://doi.org/10.3390/agronomy12030714
Messina CD, Technow F, Tang T, Totir R, Gho C, Cooper M (2018) Leveraging biological insight and environmental variation to improve phenotypic prediction: integrating crop growth models (CGM) with whole genome prediction (WGP). Eur J Agron 100:151–162. https://doi.org/10.1016/j.eja.2018.01.007
Meuwissen THE, Hayes BJ, Goddard ME (2001) Prediction of total genetic value using genome-wide dense marker maps. Genetics 157:1819–1829. https://doi.org/10.1093/genetics/157.4.1819
Mikić S, Kondić-Špika A, Lj B, Stanisavljević D, Trkulja D, Tomičić M, Nastasić A, Kobiljski B, Prodanović S, Šurlan-Momirović G (2016) Multiple marker-traits associations for maize agronomic traits. Chil J Agric Res 76(3):300–306
Mikić S, Kondić-Špika A, Lj B, Stanisavljević D, Ćeran M, Trkulja D, Mitrović B (2017) Molecular and phenotypic characterisation of diverse temperate maize inbred lines in Southeast Europe. Zemdirbyste 104(1):31–40. https://doi.org/10.13080/z-a.2017.104.005
Mikić S, Brbaklić L, Stanisavljević D, Kondić-Špika A, Bekavac G, Mitrović B, Trkulja D, Mirosavljević M (2018) Molecular diversity and microsatellite polymorphism of modern maize hybrids. Field Veg Crops Res 55(2):95–102. https://doi.org/10.5937/ratpov55-17535
Miladinović J, Hrustić M, Vidić M, Balešević-Tubić S, Tatić M (2003) Adaptability and stability of new soybean genotypes. Sel Seed Prod IX(1–4):51–55
Miladinovic J, Kurosaki H, Burton JW, Hrustic M, Miladinovic D (2006) The adaptability of short season soybean genotypes to varying longitudinal regions. Eur J Agron 25(3):243–249
Miladinović D, Miler M, Marjanović-Jeromela A, Imerovski I, Dimitrijević A, Kovačević B, Jocić S, Cvejić S, Hladni N, Obreht-Vidaković D (2018a) Evaluation of RAPD markers as a marker-assisted selection tool for variety type and erucic acid content in rapeseed. Genetika-Belgrade 50(2):421–430. https://doi.org/10.2298/GENSR1802421M
Miladinović J, Ćeran M, Đorđević V, Balešević-Tubić S, Petrović K, Đukić V, Miladinović D (2018b) Allelic variation and distribution of the major maturity genes in different soybean collections. Front Plant Sci 9:1286. https://doi.org/10.3389/fpls.2018.01286

Miladinović J, Đorđević V, Balešević-Tubić S, Petrović K, Ćeran M, Cvejić J, Bursać M, Miladinović D (2019) Increase of isoflavones in the aglycone form in soybeans by targeted crossings of cultivated breeding material. Sci Rep 9:10341

Miladinovic D, Antunes D, Yildirim K et al (2021) Targeted plant improvement through genome editing: from laboratory to field. Plant Cell Rep 40:935–951. https://doi.org/10.1007/s00299-020-02655-4

Millet EJ, Kruijer W, Coupel-Ledru A, Alvarez Prado S, Cabrera-Bosquet L, Lacube S et al (2019) Genomic prediction of maize yield across European environmental conditions. Nat Genet 51: 952–956. https://doi.org/10.1038/s41588-019-0414-y

Mirosavljević M, Momčilović V, Dražić T, Aćin V, Jocković B, Mikić S, Brbaklić L, Živančev D, Zorić M, Pržulj N (2024) Genetic progress in grain yield and associated changes in spikelet and grain traits in historical set of Pannonian wheat cultivars. Euphytica 220:10. https://doi.org/10.1007/s10681-023-03262-6

Mishra N, Tripathi MK, Tiwari S, Tripathi N, Gupta N, Sharma A, Solanki RS, Tiwari S (2022) Characterization of soybean genotypes on the basis of yield attributing traits and SSR molecular markers. In: Innovations in science and technology, vol 3. BP International, Hong Kong, pp 87–106

Mladenov N, Hristov N, Kondic-Spika A, Djuric V, Jevtic RM, Mladenov V (2011) Breeding progress in grain yield of winter wheat cultivars grown at different nitrogen levels in semiarid conditions. Breed Sci 61(3):260–268. https://doi.org/10.1270/jsbbs.61.260

Mladenov V, Dimitrijević M, Petrović S, Boćanski J, Banjac B, Kondić-Špika A, Trkulja D (2019) Genetic analysis of spike length in wheat. Genetika 51(1):167–178

Mobini SH, Warkentin TD (2016) A simple and efficient method of in vivo rapid generation technology in pea (*Pisum sativum* L.). In Vitro Cell Dev Biol Plant 52:530–536. https://doi.org/10.1007/s11627-016-9772-7

Mohammadi R, Armion M, Zadhasan E, Ahmadi MM, Amri A (2018) The use of AMMI model for interpreting genotype × environment interaction in durum wheat. Exp Agric 54(5):670–683

Montenegro JD, Golicz AA, Bayer PE, Hurgobin B, Lee H, Chan CKK et al (2017) The pangenome of hexaploid bread wheat. Plant J 90:1007–1013. https://doi.org/10.1111/tpj.13515

Montesinos-López OA, Vitale P, Gerard G, Crespo-Herrera L, Saint Pierre C, Montesinos-López A, Crossa J (2024) Genotype performance estimation in targeted production environments by using sparse genomic prediction. Plants 13(21):3059. https://doi.org/10.3390/plants13213059

Monteverde E, Gutierrez L, Blanco P, Pérez de Vida F, Rosas JE, Bonnecarrère V et al (2019) Integrating molecular markers and environmental covariates to interpret genotype by environment interaction in rice (*Oryza sativa* L.) grown in subtropical areas. G3 9:1519–1531. https://doi.org/10.1534/g3.119.400064

Moorthy SM, Das NK, Mandal K (2012) Genotype × Environment interaction and stability analysis in bivoltine silkworm genotypes of *Bombyx mori* L. IJBPAS 1(10):1443–1449

Mühleisen J, Piepho H-P, Maurer HP, Longin CFH, Reif JC (2014) Yield stability of hybrids versus lines in wheat, barley, and triticale. Theor Appl Genet 127(2):309–316. https://doi.org/10.1007/s00122-013-2219-1

Müller C, Elliott J, Pugh TAM, Ruane AC, Ciais P, Balkovic J, Deryng D, Folberth C, Cesar Izaurralde R, Jones CD, Khabarov N, Lawrence P, Liu W, Reddy AD, Schmid E, Wang X (2018) Global patterns of crop yield stability under additional nutrient and water inputs. PLoS One 13(6):e0198748

Nadeem MA, Nawaz MA, Shahid MQ, Doğan Y, Comertpay G, Yıldız M, Hatipoğlu R, Ahmad F, Alsaleh A, Labhane N, Özkan H, Chung G, Baloch FS (2018) DNA molecular markers in plant breeding: current status and recent advancements in genomic selection and genome editing. Biotechnol Biotechnol Equip 32(2):261–285. https://doi.org/10.1080/13102818.2017.1400401

Napier JD, Heckman RW, Juenger TE (2023) Gene-by-environment interactions in plants: molecular mechanisms, environmental drivers, and adaptive plasticity. Plant Cell 35:109–124. https://doi.org/10.1093/plcell/koac322

Naqvi RZ, Siddiqui HA, Mahmood MA, Najeebullah S, Ehsan A, Azhar M, Farooq M, Amin I, Asad S, Mukhtar Z, Mansoor S, Asif M (2022) Smart breeding approaches in post-genomics era for developing climate-resilient food crops. Front Plant Sci 13:972164. https://doi.org/10.3389/fpls.2022.972164

Nascimento FDS, Rocha AJ, Soares JMDS, Mascarenhas MS, Ferreira MDS, Morais Lino LS, Ramos APS, Diniz LEC, Mendes TAO, Ferreira CF, Santos-Serejo JAD, Amorim EP (2023) Gene editing for plant resistance to abiotic factors: a systematic review. Plants 12(2):305. https://doi.org/10.3390/plants12020305

Nduwumuremyi A, Melis R, Shanahan P, Theodore A (2017) Interaction of genotype and environment effects on important traits of cassava (*Manihot escu*lenta Crantz). Crop J 5(5):373–386

Nekrasov V, Wang C, Win J, Lanz C, Weigel D, Kamoun S (2017) Rapid generation of a transgene-free powdery mildew resistant tomato by genome deletion. Sci Rep 7(1):482. https://doi.org/10.1038/s41598-017-00578-x

Nordborg M, Weigel D (2008) Next-generation genetics in plants. Nature 456(7223):720–723. https://doi.org/10.1038/nature07629. PMID: 19079047

Nowosad K, Liersch A, Popławska W, Bocianowski J (2016) Genotype by environment interaction for seed yield in rapeseed (*Brassica napus* L.) using additive main effects and multiplicative interaction model. Euphytica 208:187–194

O'Connor KM, Hayes BJ, Hardner CM, Alam M, Henry RJ, Topp BL (2021) Genomic selection and genetic gain for nut yield in an Australian macadamia breeding population. BMC Genomics 22(1):1–12

Oerke EC (2005) Crop losses to pests. J Agric Sci 144:31–43. https://doi.org/10.1017/s0021859605005708

Oldenburg E, Manderscheid R, Erbs M, Weigel HJ (2009) Interaction of free air carbon dioxide enrichment (FACE) and controlled summer drought on fungal infections of maize. In: Feldmann F, Alford DV, Furk C (eds) Crop plant resistance to biotic and abiotic factors: current potential and future demands. Julius Kühn-Institut, Berlin, pp 75–83

Olesen JE, Trnka M, Kersebaum KC, Skjelvåg AO, Seguin B, Peltonen-Sainio P, Rossi F, Kozyra J, Micale F (2011) Impacts and adaptation of European crop production systems to climate change. Eur J Agron 34(2):96–112. https://doi.org/10.1016/j.eja.2010.11.003

Olivoto T, Lúcio ADC (2020) metan: an R package for multi-environment trial analysis. Methods Ecol Evol 11:783–789

Olivoto T, Lúcio ADC, da Silva JAG, Marchioro VS, de Souza VQ, Jost E (2019a) Mean performance and stability in multi-environment trials I: combining features of AMMI and BLUP techniques. Agron J 111:2949–2960

Olivoto T, Lúcio ADC, da Silva JAG, Sari BG, Diel MI (2019b) Mean performance and stability in multi-environment trials II: selection based on multiple traits. Agron J 111:2961–2969

Olsson U (2002) Generalized linear models. An applied approach. Studentlitteratur, Lund

Pandey P, Ramegowda V, Senthil-Kumar M (2015) Shared and unique responses of plants to multiple individual stresses and stress combinations: physiological and molecular mechanisms. Front Plant Sci 6:723. https://doi.org/10.3389/fpls.2015.00723

Pandey S, Singh A, Parida SK, Prasad M (2022) Combining speed breeding with traditional and genomics-assisted breeding for crop improvement. Plant Breed 141:301–313. https://doi.org/10.1111/pbr.13012

Paul K, Pauk J, Kondic-Spika A, Grausgruber H, Allahverdiyev T, Sass L et al (2019) Co-occurrence of mild salinity and drought synergistically enhances biomass and grain retardation in wheat. Front Plant Sci 10:501. https://doi.org/10.3389/fpls.2019.00501

Pautasso M, Döring TF, Garbelotto M, Pellis L, Jeger MJ (2012) Impacts of climate change on plant diseases–opinions and trends. Eur J Plant Pathol 133:295–313. https://doi.org/10.1007/s10658-012-9936-1

Pérez-Rodríguez P, Crossa J, Rutkoski J, Poland J, Singh R, Legarra A, Autrique E, Campos GL, Burgueño J, Dreisigacker S (2017) Single-step genomic and pedigree genotype × environment interaction models for predicting wheat lines in international environments. Plant Genome 10(2):plantgenome2016.09.0089. https://doi.org/10.3835/plantgenome2016.09.0089

Poole GJ, Smiley RW, Paulitz TC, Walker CA, Carter AH, See DR, Garland-Campbell K (2012) Identification of quantitative trait loci (QTL) for resistance to Fusarium crown rot (*Fusarium pseudograminearum*) in multiple assay environments in the Pacific Northwestern US. Theor Appl Genet 125(1):91–107. https://doi.org/10.1007/s00122-012-1818-6

Potts J, Jangra S, Michael VN, Wu X (2023) Speed breeding for crop improvement and food security. Crops 3:276–291. https://doi.org/10.3390/crops3040025

Prabhakar M, Purushothaman R, Awasthi DP (2020) Deep learning based assessment of disease severity for early blight in tomato crop. Multimedia Tools Appl 79:28773–28784. https://doi.org/10.1007/s11042-020-09461-w

Pranneshraj V, Sangha MK, Djalovic I, Miladinovic J, Djanaguiraman M (2022) Lipidomics-assisted GWAS (lGWAS) approach for improving high-temperature stress tolerance of crops. Int J Mol Sci 23(16):9389. https://doi.org/10.3390/ijms23169389

Qi A-Y, Zhang P-P, Zhou Y, Yao Z-J, Li Z-F, Liu D-Q (2016) Mapping of QTL conferring leaf rust resistance in Chinese wheat lines W014204 and Fuyu 3 at adult plant stage. J Integr Agric 15(1): 18–28. https://doi.org/10.1016/S2095-3119(14)60974-6

Radanović A, Sprycha Y, Jocković M, Sundt M, Miladinović D, Jansen C, Horn R (2022) KASP markers specific for the fertility restorer locus Rf1 and application for genetic purity testing in sunflowers (*Helianthus annuus* L.). Genes 13(3):465. https://doi.org/10.3390/genes13030465

Rajković D, Marjanović Jeromela A, Pezo L, Lončar B, Zanetti F, Monti A, Kondić Špika A (2022) Yield and quality prediction of winter rapeseed—artificial neural network and random forest models. Agronomy 12:58. https://doi.org/10.3390/agronomy12010058

Rajković D, Marjanović Jeromela A, Pezo L, Lončar B, Grahovac N, Kondić Špika A (2023) Artificial neural network and random forest regression models for modelling fatty acid and tocopherol content in oil of winter rapeseed. J Food Compos Anal 115:105020. https://doi.org/10.1016/j.jfca.2022.105020

Ramalingam J, Raveendra C, Savitha P, Vidya V, Chaithra TL, Velprabakaran S, Saraswa-thi R, Ramanathan A, Arumugam Pillai MP, Arumugachamy S, Vanniarajan C (2020) Gene pyramiding for achieving enhanced resistance to bacterial blight, blast, and sheath blight diseases in rice. Front Plant Sci 11:591457. https://doi.org/10.3389/fpls.2020.591457

Ranđelović P, Đorđević V, Milić S, Balešević-Tubić S, Petrović K, Miladinović J, Đukić V (2020) Prediction of soybean plant density using a machine learning model and vegetation indices extracted from RGB images taken with a UAV. Agronomy 10(8):1108. https://doi.org/10.3390/agronomy10081108

Ranđelović P, Đorđević V, Miladinović J, Prodanović S, Ćeran M, Vollmann J (2023) High-throughput phenotyping for non-destructive estimation of soybean fresh biomass using a machine learning model and temporal UAV data. Plant Methods 19:89. https://doi.org/10.1186/s13007-023-01054-6

Rani R, Raza G, Ashfaq H, Rizwan M, Razzaq MK, Waheed MQ, Shimelis H, Babar AD, Arif M (2023a) Genome-wide association study of soybean (Glycine max [L.] Merr.) germplasm for dissecting the quantitative trait nucleotides and candidate genes underlying yield-related traits. Front Plant Sci 14:1229495. https://doi.org/10.3389/fpls.2023.1229495

Rani R, Raza G, Ashfaq H, Rizwan M, Shimelis H, Tung MH, Arif M (2023b) Analysis of genotype × environment interactions for agronomic traits of soybean (*Glycine max* [L.] Merr.) using association mapping. Front Genet 13:1090994

Ray DK, West PC, Clark M, Gerber JS, Prishchepov AV, Chatterjee S (2019) Climate change has likely already affected global food production. PLoS One 14:e0217148. https://doi.org/10.1371/journal.pone.0217148

Razzaq A, Kaur P, Akhter N, Wani SH, Saleem F (2021) Next-generation breeding strategies for climate-ready crops. Front Plant Sci 12:620420. https://doi.org/10.3389/fpls.2021.620420

Ricroch A, Clairand P, Harwood W (2017) Use of CRISPR systems in plant genome editing: toward new opportunities in agriculture. Emerg Top Life Sci 1:169–182. https://doi.org/10.1042/etls20170085

Rincent R, Malosetti M, Ababaei B, Touzy G, Mini A, Bogard M et al (2019) Using crop growth model stress covariates and AMMI decomposition to better predict genotype-by-environment interactions. Theor Appl Genet 132:3399–3411. https://doi.org/10.1007/s00122-019-03432-y

Robert P, Le Gouis J, Rincent R (2020) Combining crop growth modeling with trait-assisted prediction improved the prediction of genotype by environment interactions. Front Plant Sci 11:1–11. https://doi.org/10.3389/fpls.2020.00827

Rogers AR, Dunne JC, Romay MC, Bohn M, Buckler ES, Ciampitti IA et al (2021) The importance of dominance and genotype-by-environment interactions on grain yield variation in a large-scale public cooperative maize. G3 1:jkaa050. https://doi.org/10.1093/g3journal/jkaa050

Romagosa I, Borràs-Gelonch G, Slafer G, van Eeuwijk F (2013) Genotype by environment interaction and adaptation. In: Christou P, Savin R, Costa-Pierce BA, Misztal I, Whitelaw CBA (eds) Sustainable food production. Springer, New York. https://doi.org/10.1007/978-1-4614-5797-8_199

Rutkoski J, Poland J, Mondal S, Autrique E, Pérez LG, Crossa J et al (2016) Canopy temperature and vegetation indices from high-throughput phenotyping improve accuracy of pedigree and genomic selection for grain yield in wheat. G3 6:2799–2808. https://doi.org/10.1534/G3.116.032888

Saint-Pierre C et al (2016) Genomic prediction models for grain yield of spring bread wheat in diverse agro-ecological zones. Sci Rep 6:27312

Sandhu KS, Patil SS, Pumphrey MO, Carter AH (2021) Multi-trait machine and deep learning models for genomic selection using spectral information in a wheat breeding program. Plant Genome 14(3):e20119. https://doi.org/10.1002/TPG2.20119

Sarinelli JM, Murphy JP, Tyagi P, Holland JB, Johnson JW, Mergoum M et al (2019) Training population selection and use of fixed effects to optimize genomic predictions in a historical USA winter wheat panel. Theor Appl Genet 132:1247–1261. https://doi.org/10.1007/S00122-019-03276-6

Schierenbeck M, Alqudah AM, Lohwasser U, Tarawneh RA, Simón MR, Börner A (2021) Genetic dissection of grain architecture-related traits in a winter wheat population. BMC Plant Biol 21(1):1–14. https://doi.org/10.1186/s12870-021-03183-3

Schierenbeck M, Alqudah AM, Thabet SG, Lohwasser U, Simón MR, Börner A (2023) Association mapping unravels the genetics controlling seedling drought stress tolerance in winter wheat. Front Plant Sci 14:1061845. https://doi.org/10.3389/fpls.2023.1061845

Schrag TA, Westhues M, Schipprack W, Seifert F, Thiemann A, Scholten S, Melchinger AE (2018) Beyond genomic prediction: combining different types of omics data can improve prediction of hybrid performance in maize. Genetics 208:1373–1385. https://doi.org/10.1534/genetics.117.300374

Scossa F, Alseekh S, Fernie AR (2021) Integrating multi-omics data for crop improvement. J Plant Physiol 257:153352. https://doi.org/10.1016/j.jplph.2020.153352

Sehgal D, Autrique E, Singh R, Ellis M, Singh S, Dreisigacker S (2017) Identification of genomic regions for grain yield and yield stability and their epistatic interactions. Sci Rep 7:41578. https://doi.org/10.1038/srep41578

Sehgal D, Mondal S, Crespo-Herrera L, Velu G, Juliana P, Huerta-Espino J, Shrestha S, Poland J, Singh R and Dreisigacker S (2020) Haplotype-Based, Genome-Wide Association Study Reveals Stable Genomic Regions for Grain Yield in CIMMYT Spring Bread Wheat. Front. Genet 11:589490. https://doi.org/10.3389/fgene.2020.589490

Sharma S, Kumar A, Dhakte P, Raturi G, Vishwakarma G, Barbadikar KM, Das BK, Shivaraj SM, Sonah H, Deshmukh R (2023) Speed breeding opportunities and challenges for crop improvement. J Plant Growth Regul 42:46–59

Shi J, Gao H, Wang H, Lafitte HR, Archibald RL, Yang M et al (2017) ARGOS8 variants generated by CRISPR-Cas9 improve maize grain yield under field drought stress conditions. Plant Biotechnol J 15:207–216. https://doi.org/10.1111/pbi.12603

Singh A, Ganapathysubramanian B, Singh AK, Sarkar S (2016) Machine learning for high-throughput stress phenotyping in plants. Trends Plant Sci 21:110–124. https://doi.org/10.1016/j.tplants.2015.10.015

Sinha P et al (2020) Superior haplotypes for haplotype-based breeding for drought tolerance in pigeonpea (*Cajanus cajan* L.). Plant Biotechnol J 18:2482–2490

Soleimani B, Lehnert H, Babben S et al (2022) Genome wide association study of frost tolerance in wheat. Sci Rep 12:5275. https://doi.org/10.1038/s41598-022-08706-y

Song Y, Yu J, Huang B (2014) Elevated CO2-mitigation of high temperature stress associated with maintenance of positive carbon balance and carbohydrate accumulation in kentucky bluegrass. PLoS One 9:e89725. https://doi.org/10.1371/journal.pone.0089725

Song J-M, Guan Z, Hu J, Guo C, Yang Z, Wang S et al (2020) Eight high-quality genomes reveal pan-genome architecture and ecotype differentiation of *Brassica napus*. Nat Plants 6:34–45. https://doi.org/10.1038/s41477-019-0577-7

Song P, Wang J, Guo X, Yang W, Zhao C (2021a) High-throughput phenotyping: breaking through the bottleneck in future crop breeding. Crop J 3:633–645. https://doi.org/10.1016/j.cj.2021.03.015

Song ZT, Liu JX, Han JJ (2021b) Chromatin remodeling factors regulate environmental stress responses in plants. J Integr Plant Biol 63(3):438–450

Stazi A, Jovine R (2022) Food traceability in Europe, the US and China: comparative law and regulatory technology. BioLaw J 2:399–425

Sukumaran S et al (2017) Genomic prediction with pedigree and genotype × environment interaction in spring wheat grown in South and West Asia, North Africa, and Mexico. G3 7:481–197

Sukumaran S, Jarquin D, Crossa J, Reynolds M (2018) Genomic-enabled prediction accuracies increased by modeling genotype × environment interaction in durum wheat. Plant Genome 11:170112. https://doi.org/10.3835/PLANTGENOME2017.12.0112

Swiezewski S, Liu F, Magusin A, Dean C (2009) Cold-induced silencing by long antisense transcripts of an Arabidopsis Polycomb target. Nature 462(7274):799–802

Tao Y, Zhao X, Mace E, Henry R, Jordan D (2019) Exploring and exploiting pan-genomics for crop improvement. Mol Plant 12:156–169

Tariku S, Lakew T, Bitew M, Asfaw M (2013) Genotype by environment interaction and grain yield stability analysis of rice (*Oryza sativa* L.) genotypes evaluated in north Western Ethiopia. Ethiopia Net J Agri Sci 1:10–16

Tettelin H, Masignani V, Cieslewicz MJ, Donati C, Medini D, Ward NL et al (2005) Genome analysis of multiple pathogenic isolates of *Streptococcus agalactiae*: implications for the microbial "pan-genome". Proc Natl Acad Sci USA 102:13950–13955. https://doi.org/10.1073/pnas.0506758102

Tomičić M, Djordjević V, Obreht D, Miladinović J, Brbaklić L, Mikić A, Mikić S (2015) Tracking footprints of selection associated with soybean adaptation to Central-East Europe environments. Euphytica 203:701–713. https://doi.org/10.1007/s10681-014-1313-z

Tong H, Nikoloski Z (2021) Machine learning approaches for crop improvement: leveraging phenotypic and genotypic big data. J Plant Physiol 257:153354. https://doi.org/10.1016/j.jplph.2020.153354

Trkulja D, Kondić-Špika A, Brbaklić L, Kobiljski B, Hristov N (2012) Marker-trait associations for spike-related characters in a doubled haploid population of wheat. Rom Agric Res 29:9–16

Trkulja D, Chhetry M, Griffiths S, Leverington-Waite M, Kondić-Špika A, Brbaklić L, Kobiljski B (2016) Genotyping and linkage analysis of wheat experimental populations using KASP platform. In: Proceedings of the 16th International EWAC Conference, Lublin, Poland, 24–29 May 2015, pp 155–157

Trkulja D, Kondić-Špika A, Brbaklić L, Kobiljski B, Mikić S, Mirosavljević M, Glogovac S, Šurlan-Momirović G (2019) Genetic structure and allelic richness of the wheat core collection for association mapping of yield. Zemdirbyste 106(3):257–264. https://doi.org/10.13080/z-a.2019.106.033

Tsai SQ, Zheng Z, Nguyen NT, Liebers M, Topkar VV, Thapar V, Aryee MJ (2015) GUIDE-seq enables genome-wide profiling of off-target cleavage by CRISPR-Cas nucleases. Nat Bitechnol 33:187

Uhlarik A, Ćeran M, Živanov D, Grumeza R, Skøt L, Sizer-Coverdale E, Lloyd D (2022) Phenotypic and genotypic characterization and correlation analysis of pea (*Pisum* sativum L.) diversity panel. Plants 11(10):1321. https://doi.org/10.3390/plants11101321

van Eeuwijk FA, Bustos-Korts DV, Malosetti M (2016) What should students in plant breeding know about the statistical aspects of genotype × environment interactions? Crop Sci 56:2119–2140

Vanyushin BF, Ashapkin VV (2011) DNA methylation in higher plants: past, present and future. BBA-Gene Regul Mech 1809(8):360–368

Varotto S, Tani E, Abraham E, Krugman T, Kapazoglou A, Melzer R, Radanović A, Miladinović D (2020) Epigenetics: possible applications in climate-smart crop breeding. J Exp Bot 71(17):5223–5236

Varshney RK, Graner A, Sorrels M (2005) Genomics-assisted breeding for crop improvement. Trends Plant Sci 10:621–630. https://doi.org/10.1016/j.tplants.2005.10.004

Varshney RK, Bohra A, Yu J, Graner A, Zhang Q, Sorrells ME (2021) Designing future crops: genomics-assisted breeding comes of age. Trends Plant Sci 26:631–649. https://doi.org/10.1016/j.tplants.2021.03.010

Visioni A, Tondelli A, Francia E et al (2013) Genome-wide association mapping of frost tolerance in barley (*Hordeum vulgare* L.). BMC Genomics 14:424. https://doi.org/10.1186/1471-2164-14-424

Vollmann J, Rischbeck P, Pachner M, Đorđević V, Manschadi AM (2022) High-throughput screening of soybean di-nitrogen fixation and seed nitrogen content using spectral sensing. Comput Electron Agric 199:107169. https://doi.org/10.1016/j.compag.2022.107169

Voss-Fels KP, Stahl A, Wittkop B, Lichthardt C, Nagler S, Rose T et al (2019) Breeding improves wheat productivity under contrasting agrochemical input levels. Nat Plants 5:706–714. https://doi.org/10.1038/s41477-019-0445-5

Wang J, McBlain BA, Wooley JT, Hesketh JD, Bernard RL (1987) A soybean phenology model and associated data base. Biotronics 16:25–38

Wang W, Huang F, Qin Q, Zhao X, Li Z, Fu B (2015a) Comparative analysis of DNA methylation changes in two rice genotypes under salt stress and subsequent recovery. Biochem Biophys Res Commun 465:790–796

Wang Y, Mette MF, Miedaner T, Wilde P, Reif JC, Zhao Y (2015b) First insights into the genotype–phenotype map of phenotypic stability in rye. J Exp Bot 66:3275–3284

Wang F, Wang C, Liu P, Lei C, Hao W, Gao Y, Liu YG, Zhao K (2016) Enhanced rice blast resistance by CRISPR/Cas9-targeted mutagenesis of the ERF transcription factor gene OsERF922. PLoS One 11(4):e0154027. https://doi.org/10.1371/journal.pone.0154027

Wang W, Mauleon R, Hu Z, Chebotarov D, Tai S, Wu Z et al (2018) Genomic variation in 3,010 diverse accessions of Asian cultivated rice. Nature 557:43–49. https://doi.org/10.1038/s41586-018-0063-9

Wang Q, Xu J, Pu X, Lv H, Liu Y, Ma H, Wu F, Wang Q, Feng X, Liu T, Tang Q (2021) Maize DNA methylation in response to drought stress is involved in target gene expression and alternative splicing. Int J Mol Sci 22(15):8285

Wang W, Guo W, Le L, Yu J, Wu Y, Li D, Wang Y, Wang H, Lu X, Qiao H, Gu X, Tian J, Zhang C, Pu L (2023) Integration of high-throughput phenotyping, GWAS, and predictive models reveals the genetic architecture of plant height in maize. Mol Plant 16(2):354–373. https://doi.org/10.1016/j.molp.2022.11.016

Watson A, Ghosh S, Williams MJ, Cuddy WS, Simmonds J, Rey MD et al (2018) Speed breeding is a powerful tool to accelerate crop research and breeding. Nat Plants 4:23–29. https://doi.org/10.1038/s41477-017-0083-8

Wu PY, Stich B, Weisweiler M, Shrestha A, Erban A, Westhoff P, Inghelandt DV (2022) Improvement of prediction ability by integrating multi-omic datasets in barley. BMC Genomics 23:200. https://doi.org/10.1186/s12864-022-08337-7

Xavier A, Jarquin D, Howard R, Ramasubramanian V, Specht JE, Graef GL, Beavis WD, Diers BW, Song Q, Cregan PB, Nelson R, Mian R, Grover Shannon J, McHale L, Wang D, Schapaugh W, Lorenz AJ, Xu S, Muir WM, Rainey KM (2018) Genome-wide analysis of grain yield stability and environmental interactions in a multiparental soybean population. G3 8(2):519–529. https://doi.org/10.1534/g3.117.300300

Xu C, Jackson SA (2019) Machine learning and complex biological data. Springer, Berlin. https://doi.org/10.1186/s13059-019-1689-0

Xu R, Duan P, Yu H, Zhou Z, Zhang B, Wang R, Li J, Zhang G, Zhuang S, Lyu J, Li N, Chai T, Tian Z, Yao S, Li Y (2018) Control of grain size and weight by the OsMKKK10-OsMKK4-OsMAPK6 signaling pathway in rice. Mol Plant 11(6):860–873. https://doi.org/10.1016/j.molp.2018.04.004

Yadav PK, Singh AK, Tripathi MK, Tiwari S, Yadav SK, Tripathi N (2022) Morphophysiological and molecular characterization of maize (*Zea mays* L.) genotypes for drought tolerance. Eur J Appl Sci 10:65–87

Yadav RK, Tripathi MK, Tiwari S, Tripathi N, Asati R, Chauhan S, Tiwari PN, Payasi DK (2023) Genome editing and improvement of abiotic stress tolerance in crop plants. Life 13:1456. https://doi.org/10.3390/life13071456

Yan W, Kang M (2003) GGE biplot analysis: a graphical tool for breeders, geneticists, and agronomists. CRC Press, Boca Raton

Yan W, Tinker NA (2006) Biplot analysis of multi-environment trial data: principles and applications. Can J Plant Sci 86:623–645

Yang W, Feng H, Zhang X, Zhang J, Doonan JH, Batchelor WD, Xiong L, Yan J (2020a) Crop phenomics and high-throughput phenotyping: past decades, current challenges, and future perspectives. Mol Plant 13(2):187–214. https://doi.org/10.1016/j.molp.2020.01.008

Yang J, Liu Z, Chen Q et al (2020b) Mapping of QTL for grain yield components based on a DH population in maize. Sci Rep 10:7086. https://doi.org/10.1038/s41598-020-63960-2

Yıldırım K, Miladinović D, Sweet J, Akin M, Galović V, Kavas M, Zlatković M, de Andrade E (2023) Genome editing for healthy crops: traits, tools and impacts. Front Plant Sci 14:1231013. https://doi.org/10.3389/fpls.2023.1231013

Yoon H, Shim Y, Yoo SC, Kang K, Paek NC (2021) The rice CHD3/Mi-2 chromatin remodeling factor rolled fine striped promotes flowering independent of photoperiod. Int J Mol Sci 22(3):1303

Yu C, Miao R, Khanna M (2021) Maladaptation of U.S. corn and soybeans to a changing climate. Sci Rep 11:12351. https://doi.org/10.1038/s41598-021-91192-5

Zaidi SSEA, Mukhtar MS, Mansoor S (2018) Genome editing: targeting susceptibility genes for plant disease resistance. Trends Biotechnol 36:898–906. https://doi.org/10.1016/j.tibtech.2018.04.005

Zhang XH, Tee LY, Wang XG, Huang QS, Yang SH (2015) Off-target effects in CRISPR/Cas9-mediated genome engineering. Mol Ther Nucleic Acids 4:264

Zhang Y, Liang Z, Zong Y, Wang Y, Liu J, Chen K, Qiu JL, Gao C (2016) Efficient and transgene-free genome editing in wheat through transient expression of CRISPR/Cas9 DNA or RNA. Nat Commun 7:12617. https://doi.org/10.1038/ncomms12617

Zhang Y, Li D, Zhang D, Zhao X, Cao X, Dong L, Liu J, Chen K, Zhang H, Gao C, Wang D (2018) Analysis of the functions of TaGW2 homoeologs in wheat grain weight and protein content traits. Plant J 94:857–866. https://doi.org/10.1111/tpj.13903

Zhang A, Liu Y, Wang F, Li T, Chen Z, Kong D, Bi J, Zhang F, Luo X, Wang J, Tang J, Yu X, Liu G, Luo L (2019a) Enhanced rice salinity tolerance via CRISPR/Cas9-targeted mutagenesis of the *OsRR22* gene. Mol Breed 39:47. https://doi.org/10.1007/s11032-019-0954-y

Zhang X, Han L, Dong Y, Shi Y, Huang W, Han L et al (2019b) A deep learning-based approach for automated yellow rust disease detection from high-resolution hyperspectral UAV images. Remote Sens 11:1554. https://doi.org/10.3390/rs11131554

Zhao Y, Li Z, Liu G, Jiang Y, Maurer HP, Würschum T, Mock H-P, Matros A, Ebmeyer E, Schachschneider R et al (2015) Genome-based establishment of a high-yielding heterotic pattern for hybrid wheat breeding. Proc Natl Acad Sci USA 112:15624–15629. https://doi.org/10.1073/pnas.1514547112

Zhao Q, Feng Q, Lu H, Li Y, Wang A, Tian Q et al (2018) Pan-genome analysis highlights the extent of genomic variation in cultivated and wild rice. Nat Genet 50:278–284. https://doi.org/10.1038/s41588-018-0041-z

Zhao T, Zhan Z, Jiang D (2019) Histone modifications and their regulatory roles in plant development and environmental memory. J Genet Genomics 46(10):467–476

Does Breeding for Abiotic Stress Tolerance Mitigate G × E Interactions Under a Changing Climate?

6

Pervez H. Zaidi, Reshmi Rani Das, Ashok Singamsetti, K. Seetharam, Sara Fatma, Prakash H. Kuchanur, M. T. Vinayan, and Ayyanagouda Patil

Abstract

Escalating climate variability, compounded by intensifying abiotic stresses, is amplifying genotype-by-environment interaction (G × E) effects and threatening the sustainability of global food systems. The dynamic nature of G × E demands a paradigm shift toward breeding for high and stable yields across diverse, often suboptimal environments. Conventional breeding approaches emphasizing yield maximization under optimal conditions inadequately capture the spatiotemporal complexity of G × E, necessitating a shift toward breeding for yield stability and resilience across diverse and frequently suboptimal environments. This chapter outlines an integrated, G × E-informed breeding framework incorporating target population of environments (TPE) redefinition through genetic correlations and environmental covariates, genomic-assisted selection, enviromics, multi-environment trials (METs), and predictive modeling. Advances in high-throughput precision phenotyping (HTPP) enable field-based quantification of dynamic stress responses, enhancing the linkage of phenotypic, genomic, and environmental data. Evidence from maize and rice programs (e.g., CIMMYT's climate-resilient breeding strategy; IRRI's OneRice Framework) demonstrates

P. H. Zaidi (✉) · R. R. Das · K. Seetharam · M. T. Vinayan
Global Maize Program, International Maize and Wheat Improvement Center (CIMMYT), Hyderabad, Telangana, India
e-mail: phzaidi@cgiar.org

A. Singamsetti
Institute of Agricultural Sciences, Banaras Hindu University, Varanasi, U.P, India

S. Fatma
Department of Botany, Banaras Hindu University, Varanasi, UP, India

P. H. Kuchanur · A. Patil
University of Agricultural Sciences, Raichur, Karnataka, India

R. Mohammadi et al. (eds.), *Genotype x Environment Interactions and its Implications for Plant Breeding*, https://doi.org/10.1007/978-981-95-5664-9_6

that genetic gains under both stressed and unstressed conditions can be simultaneously achieved through balanced selection for plasticity, resilience, and high yield potential. Operationalizing such pipelines requires institutional investment in distributed phenotyping networks, data integration platforms, and enabling policy frameworks to sustain long-term adaptive breeding capacity. By integrating modern statistical modeling and application of contemporary systems-based tools—such as crop simulation models (CSMs), AI- and ML-driven predictive frameworks, digital decision support systems (DSS), and leveraging dynamic G × E, modern breeding programs can accelerate genetic gain and deliver cultivars with broad or environment-specific adaptation suited to current and projected climatic regimes.

6.1 Introduction

Climate change-induced variability in weather is one of the major factors contributing to the year-to-year variation in global crop yields (Vogel et al. 2019; Zaidi et al. 2020). The global climate has changed relative to the pre-industrial period, and there are multiple lines of evidence that these changes have had impacts on organisms and ecosystems, as well as human well-being (Hoegh-Guldberg et al. 2018). It poses a significant challenge to global food security through its impact on yield potential, stability, and predictability of crop performance across environments (Xiong et al. 2024; Kopeć 2024; Prabhu et al. 2023). The predicted reduction in yields results from increases in temperature, alterations in rainfall patterns, and elevated surface carbon dioxide levels attributable to anthropogenic greenhouse gas emissions (Mersmann et al. 2021). A study by the National Aeronautics and Space Administration (NASA) under the Agricultural Model Intercomparison Project (AgMIP) projected that in the current trend of climate, average global maize yields may record a 24% decrease by the late century, whereas wheat may uptick by about 17% (Fig. 6.1). Li et al. (2025), however, projected that under the business-as-usual scenario an estimated yield response of −22% (maize), −15% (soybean), −14% (wheat) and −9% (rice) from 2015 to 2080–2100. Rising global temperatures are linked to changes in rainfall patterns and increased frequency and duration of heatwaves and droughts (Subba Rao and Jägermeyr 2022). Growing-season climate factors account for 20–49% of yield anomaly variance across staple crops, with 18–43% attributable to climate extremes (Vogel et al. 2019). In addition to its effects on crop yields, climate change also impacts natural resources, primarily land and water, which are crucial for agricultural productivity. Water availability is expected to decline due to climate change, while agrarian water consumption is projected to increase by 19% by 2050 (UN-Water 2013). Climate-induced stressors, particularly drought, salinity, and heat, affect a significant portion of global cropland, and this footprint is expanding (Lesk et al. 2016; Neupane et al. 2022; Jamil et al. 2011). Projections indicate that staple crop yields in sub-Saharan Africa may decline by 10–20% by 2050, with maize yields in Ethiopia alone potentially decreasing by 15% due to increased temperatures and erratic rainfall (Abebaw 2025).

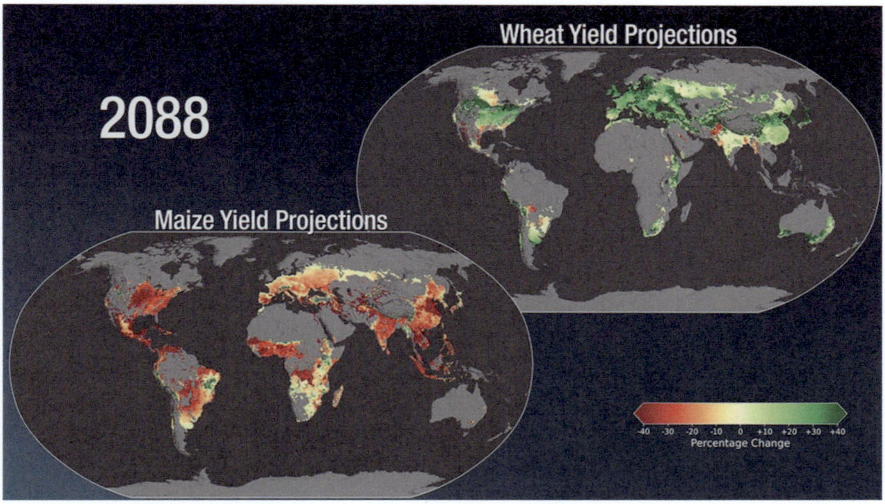

Fig. 6.1 Predicted changes in yields of maize and wheat based on an ensemble of crop and climate models (https://svs.gsfc.nasa.gov/4974)

In general, crop farming is one of the riskiest businesses due to its high dependence on prevailing climatic conditions, which often vary with space and time. Despite variability in crop management and input utilization, climate change-mediated weather extremes exert detrimental effects on crops, though the level of sensitivity might vary with different crops and with different growth stages of the crop (Arshad et al. 2017). The disparities in micro- and mega-environments across locations and years contribute to the differential responses of cultivars, which are categorized as G × E interaction effects (Vinayan et al. 2020). A complex interplay between genetic makeup and environmental conditions results in G × E effects, leading to variability in phenotypic expression. It is expressed in terms of variable performance and/or rankings of genotypes under different environments, complicating the selection of superior cultivars across multiple target environments (Vinayan et al. 2020; Cooper and Messina 2023). It is particularly problematic in regions characterized by high climatic variability and low-input systems, where year-to-year and site-to-site stress profiles fluctuate significantly. The climate-induced abiotic stresses further undermine genotype stability and reduce the efficiency of breeding programs that rely on consistent phenotypic selection across environments (Ceccarelli 2015). Classical breeding approaches often fall short in unravelling the complexity of G × E effects, especially in heterogeneous and shifting agro-ecological zones with dynamic climate scenarios and associated weather extremes (Zaidi et al. 2020). Therefore, addressing the increasing contribution of G × E in the backdrop of climate change-mediated weather variability and associated abiotic stresses has thus emerged as a priority for crop breeding programs, especially in lowland tropics and sub-tropical regions.

Breeding for abiotic stress tolerance in field crops is no longer a niche endeavor but a priority for food and nutritional security, which not only can help in yield stability but also improves system resilience, farmer income, and dietary outcomes (Prasanna et al. 2021). Stress-resilient crop varieties offer a vital option for risk mitigation and livelihood sustainability for millions of smallholder farmers living in marginal, stress-prone ecologies (Thomas et al. 2019; Dar et al. 2021). Mainstreaming these traits into breeding pipelines is thus a priority adaptation strategy under the United Nations Sustainable Development Goals (SDGs), particularly SDG2 (Zero Hunger) and SDG13 (Climate Action). Recent advances in novel genomic tools, coupled with multi-environment trials (METs), including managed stress phenotyping for key stresses, have enhanced the ability to dissect and model G × E, enabling environment-specific prediction and selection strategies (Jarquín et al. 2014). However, the success of these strategies remains contingent upon an integrated understanding of the target population of the environment (TPE), stress-responsive traits, and adaptive mechanisms. In this chapter, we reviewed recent advances in breeding strategies for tolerance/resilience to climate change-induced abiotic stresses in major field crops, and their contribution in minimizing G × E and associated yield losses. Of course, the renewed approach is based on a different paradigm, that is, focus on high-stable yields rather than just high-yielding (though risky) cultivars. This chapter synthesizes the conceptual and analytical framework for understanding G × E, highlighting its impact on realized genetic gains and effectiveness in robust selections. It also delves into modern statistical models and further explores how contemporary system approaches, including crop simulation models (CSMs), novel prediction models using artificial intelligence (AI) and machine learning (ML) approaches, and digital decision support systems (DSS), can be leveraged to manage G × E under projected climate changes.

6.2 Conceptual and Analytical Framework of G × E Effects

Development of high-performing and stable cultivars that can thrive across diverse conditions, including stress-prone agricultural systems, remains a key objective of modern plant breeding programs (Ceccarelli and Grando 2020; Varshney et al. 2021). To ensure broad adaptability or climate and region-specific suitability, breeders perform METs for evaluating genotypes across different locations/years/seasons. The data from METs helps in partitioning of the observed phenotypic variation into its main components, viz., genotype (G), environment (E), and G × E (Yan and Kang 2002; Crossa et al. 2017). Precise calibration of G × E is very critical for identifying genotypes with either wide or specific adaptation, and performing under specific environmental conditions (Cooper et al. 2006; Hickey et al. 2019). Understanding the nature and magnitude of G × E enhances the precision in cultivar recommendations or product placement under variable climate and site-specific constraints (Egea-Gilabert 2021).

6.2.1 Defining and Classifying G × E

G × E refers to the differential performance of genotypes across diverse environments, where the relative ranking and magnitude of genotype responses vary depending on environmental conditions (Falconer and Mackay 1996; Crossa et al. 2017). This suggests that the phenotypic expression of a genotype is not solely determined by its genetic constitution, but also by the environmental context in which it is expressed (El-Soda et al. 2014). A statistically significant G × E indicates that genotypes may not respond uniformly when tested across distinct agro-ecoregions (Fig. 6.2), resulting in changes in either rank (crossover/qualitative interaction) or a shift in performance magnitude without altering ranks (non-crossover/quantitative interaction) (Cooper et al. 2006). The crossover G × E is particularly significant in plant breeding because it implies that no single genotype performs consistently across all environments. Although non-crossover interactions may still impact yield magnitude and variability, they pose less of a challenge for breeders compared to crossover interactions, as selection decisions remain relatively consistent across environments.

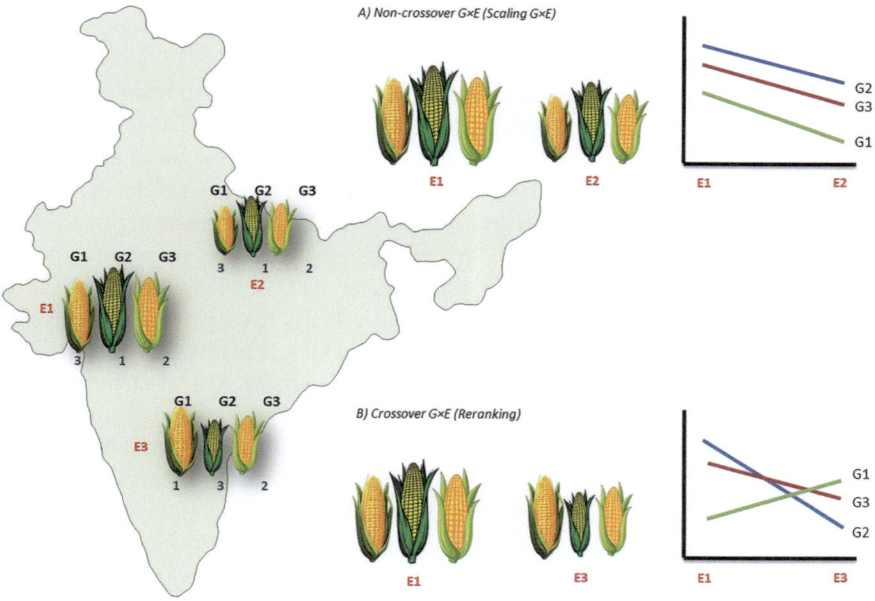

Fig. 6.2 Illustration of non-crossover (scaling) and crossover (reranking) G × E interactions in maize hybrids across three environments, distinct agro-ecoregion: E1 = Hot semi-arid ecoregion, E2 = Hot sub-humid dry region, and E3 = Hot sub-humid to semi-arid ecoregion. (**a**) In non-crossover G × E, genotypes maintain consistent rank order across E1 and E2, despite environmental differences—indicative of broad adaptability. (**b**) In crossover G × E, genotype rankings change drastically between E1 and E3—highlighting specific adaptation and the need for location-specific recommendations

When G × E is non-significant, selection is relatively straightforward, while significant G × E necessitates evaluating genotypes in multiple representative environments to identify high-stable performers or those with specific adaptation (Vinayan et al. 2020; Zaidi et al. 2023). There are two key concepts of stability: biological (static) stability, where genotypes maintain constant performance regardless of environmental variation, and agronomic (dynamic) stability, where genotypes respond proportionally to environmental improvements (Becker and Leon 1988). Selection for high mean yield with low variance across environments is central to breeding stable, high-performing cultivars under climate variability and diverse production systems. In brief, distinguishing between crossover and non-crossover interactions allows breeders to refine selection strategies: opting for broadly adapted genotypes in the case of minimal crossover, or developing environment-specific cultivars when crossover G × E is prominent. Moreover, understanding and quantifying G × E is increasingly critical in modern genomic prediction frameworks, where the inclusion of the G × E term enhances the accuracy of genomic estimated breeding values (GEBVs) (Crossa et al. 2017). This integration supports more precise selection, improved targeting of genotypes to specific regions, and ultimately, higher genetic gains.

6.2.2 Dissecting G × E Effects

One of the earliest robust tools for dissecting G × E is the additive main effects and multiplicative interaction (AMMI) model, which combines ANOVA for additive effects (G and E) with principal component analysis (PCA) for the multiplicative G × E component (Zobel et al. 1988). The model is instrumental when significant crossover interactions exist and helps to identify stable genotypes using interaction principal component axes (IPCAs). In contrast, the genotype plus genotype by environment (GGE) biplot model (Yan and Kang 2002) focuses solely on the G and G × E, removing the environmental main effects. It employs singular value decomposition (SVD) of the environment-centered G × E matrix and is visualized using a biplot. More recently, multivariate indices such as AMMI stability value (ASV) and weighted average of absolute scores (WAAS) have gained popularity due to their integrative capacity to combine multiple IPCAs into a single measure (Olivoto et al. 2019). With advances in genomics, genomic selection (GS) has been expanded to account for G × E through genomic-enabled prediction models, such as genomic best linear unbiased prediction (gBLUP) with interaction terms (Crossa et al. 2017). These models extend traditional BLUP by incorporating marker data and allow breeders to predict untested G × E combinations with greater accuracy.

Though the classical models offer valuable visual insights into G × E, linear mixed models (LMMs) provide a more flexible and statistically rigorous framework for analyzing MET data. LMMs are particularly well suited for handling unbalanced data, heterogeneous error structures, and genetic correlations across environments, making them indispensable for advanced plant breeding programs under diverse and evolving agro-ecologies. A powerful extension of LMMs is the factor analytic (FA) model, which offers a biologically intuitive and computationally efficient way to model the G × E covariance matrix. In the face of climate change, environments are becoming increasingly variable and less predictable, making the modeling of heterogeneity and genotype-specific responses a breeding imperative (Bustos-Korts et al. 2019; Argaw et al. 2025). The LMMs and FA structures not only accommodate such heterogeneity but also exploit it to enhance prediction accuracy and selection decisions, even in unbalanced or incomplete trials typical of real-world breeding pipelines (even for early generation testing). Despite their historical utility, traditional ANOVA-based methods for MET analysis face critical limitations (Smith et al. 2001; Kelly et al. 2007; Piepho 2022). Therefore, LMM-based approaches, particularly in a one-stage framework, have emerged as the standard for MET data analysis.

6.2.3 Climate Factors Drive G × E Effects

Environmental heterogeneity and climatic variability represent major drivers of G × E, which are due to spatial and temporal variability in factors such as temperature, rainfall, soil conditions, and crop management practices (Challinor et al. 2014). These sources of variation lead to changes in genotype performance across space and time, often resulting in crossover interactions that complicate selection. Understanding and modeling these interactions are critical for breeding stable and climate-resilient varieties. Temporal environmental variation drives G × E, which is especially relevant under current unpredictable weather patterns due to climate change effects. Envirotyping, which integrates environmental covariates such as growing degree days (GDD), vapor pressure deficit (VPD), relative humidity (RH), and photoperiod into regression analysis, allows researchers to better explain the sources of G × E (Xu 2016). In a multi-season maize trial, Vinayan et al. (2020) found that incorporating vapor pressure deficit (VPD) and relative humidity as environmental covariates enhanced the identification of hybrids exhibiting stable performance under escalating heat stress. Egea-Gilabert et al. (2021) applied envirotyping alongside GGE biplot analysis to categorize 19 Indian soybean trial locations into four distinct mega-environments. This environmental stratification enabled the development of zone-specific genotype recommendations, thereby improving the effectiveness of targeted breeding strategies. Similarly, in maize METs conducted across China's Huanghuaihai region, Yue et al. (2021) applied multi-trait stability indices to delineate macro-environments. They identified climate-resilient hybrids using historical temperature and rainfall data. Messina et al. (2018) demonstrated that integrating crop growth models with whole-genome prediction (CGM-WGP) enables a more accurate assessment of genotype performance by accounting for the temporal dynamics of environmental stressors such as heat and drought. Abebe et al. (2024) applied FA models in soybean METs across Nigeria to partition G × E variance and identified genotypes with broad adaptability across different agro-ecologies. This approach is especially useful when dealing with large, unbalanced datasets typical of modern breeding programs. Identification of mega-environments, the subsets of environments where genotype performance is consistent, is a strategic breeding objective (Gauch and Zobel 1997).

6.2.4 Implications of G × E in Breeding and Selection Decisions

6.2.4.1 Trait Architecture

In the context of climate-adaptive breeding, trait architecture is rarely static, and fluctuating environmental signals profoundly influence it. G × E emerges when the genetic variance and the genetic rank of genotypes vary across environments, as seen with traits such as flowering time, grain yield, or quality (Malosetti et al. 2013; van Eeuwijk et al. 2016). For instance, QTL-by-environment (Q × E) interactions have been observed in maize and wheat for yield under drought and heat stress, implying that specific alleles are conditionally neutral or even antagonistic depending on climatic context (Jarquín et al. 2014; Cooper and Messina 2021).

6.2.4.2 Selection Response

Selection response ($R = i \times \sigma_p \times h^2$, where R is the response to selection, i is the selection intensity, σ_p is the phenotypic standard deviation, and h^2 is heritability) is maximized when heritability and accuracy of selection are high. This is particularly problematic in climate-unstable systems, where genotype performance may shift unpredictably across years due to weather extremes such as intermittent drought, waterlogging, and/or heat stress. By modeling G × E, breeders can partition genetic variance into main effect and interaction components and obtain BLUPs that are environment-specific or environment-aggregated (Smith et al. 2001; Kelly et al. 2007). Furthermore, type-B genetic correlations quantify the stability of genetic effects across environments, guiding the identification of broadly vs. specifically adapted genotypes (Piepho and Möhring 2005).

6.2.4.3 Genetic Gain and Breeding Progress

The ultimate objective of G × E modeling is to develop highly stable yielding cultivars and enhance genetic gain per unit time. Modern breeding pipelines aim to increase selection accuracy by incorporating genomic prediction, G × E-aware models, and genotype-specific crop models (Messina et al. 2018). Abebe et al. (2024) demonstrated that linear mixed models facilitated the identification of widely adapted high-yielding soybean cultivars in METs by integrating spatial and climatic heterogeneity. For instance, by integrating high-throughput phenotyping with genomic and environmental data, breeding programs are achieving sustained genetic gains much faster, more accurately, and with greater climate adaptation, forming the foundation for resilient maize production under changing agro-climatic conditions (Prasanna et al. 2021; Badu-Apraku et al. 2023). Integration of analytical insights with practical breeding decisions, with G × E and climate variable analysis, can aid in trait optimization and accelerated gains.

6.3 Exacerbated G × E Under Abiotic Stresses

Climate change has intensified the occurrence, severity, and overlap of abiotic stresses, directly magnifying G × E effects and challenging the selection of superior genotypes for the target population of environments (TPEs). Increase in inter-annual climatic variability, manifesting through unpredictable rainfall patterns, rising temperatures, and increased frequency of extreme weather events, has exacerbated G × E in major cropping systems (Challinor et al. 2014). The intensified G × E affects both performance and ranking of genotypes across locations and years, and thereby slows down the breeding progress (Mittler 2006). Extreme weather episodes cause instability in phenotypic expression and reduce the predictability of breeding outcomes (Chenu et al. 2011; Jiang et al. 2025). Major abiotic stresses, such as drought, heat, and waterlogging, rarely occur in isolation; instead, they often manifest as compound or sequential stress episodes, further amplifying G × E and thereby reducing selection efficiency across locations (Mickelbart et al. 2015; Jiang et al. 2025). The exacerbated G × E under the climate-induced abiotic stresses affects the breeding efforts in many ways, because:

- *It narrows down the available genetic variation as high selection pressure leads to the dominance of a few resilient genotypes* (Cortés 2024). *In essence, the inherent genetic superiority of a specific variety to others may become less evident under conditions of severe stress.*
- *It increase inconsistency in genotypic performance across years/locations, thereby complicating performance predictions of test genotypes* (Lobell et al. 2015).
- *It results in significant variation in the environmental mean across locations/ years and therefore challenges a robust selection decision* (Kumar et al. 2023).
- *It affects the prediction accuracy and results in a less reliable selection process across years and locations* (Zhang et al. 2015). *This decline in accuracy was attributed to the poor heritability of targeted traits under a stressed environment.*

6.3.1 Drought Stress

Agricultural drought is one of the key drivers of G × E rainfed agro-ecosystems. It affects key plant functions such as stomatal conductance, photosynthetic rate, and reproductive development, and thereby reduces yield potential and stability (Blum 2011). The impact of drought varies considerably across environments, resulting in pronounced G × E effects (Passioura 2012). In rice, yield reduction under reproductive-stage drought can range from 30 to 80%, and genotypic response varies sharply between lowland and upland ecologies (Kumar et al. 2014). In wheat, terminal drought influences kernel filling and hastens maturity, often reversing genotypic rankings across trials (Reynolds et al. 2007). Similarly, the maize crop that is predominantly grown in rainfed lowland tropics shows high G × E for flowering synchrony and stay-green traits under drought (Cairns et al. 2013). Most of the pulses (e.g., chickpea, pigeon pea, and lentil) are extremely drought-sensitive, particularly at flowering and pod-filling. Studies show that early-maturing pigeon pea varieties may outperform long-duration ones in terminal drought years but underperform in normal years, a classic example of crossover G × E (Upadhyaya et al. 2012). Similarly, in oilseeds, drought-induced reduction in seed filling and oil content in groundnut and mustard shows significant environmental dependence and G × E in multi-location evaluations (Hayat et al. 2025). Breeding programs attempt to address this challenge through targeted drought screening under managed stress environments and the use of drought-prone TPEs (Reynolds et al. 2020). However, unpredictable onset and escape dynamics of drought make it difficult to ensure the stable performance of genotypes in MLTs.

6.3.2 Heat Stress

Climate change-induced heat stress combines two challenges: high day- and night temperatures, and physiological water deficit caused by an increase in VPD at high temperatures (Zaidi et al. 2023). Under heat stress, both cross-over and non-crossover G × E effects were observed, as maize genotypes exposed to similar high temperatures (Tmax) across different locations responded differently, depending upon the level of VPD at Tmax. Heat-induced oxidative stress responses vary by genotype and growth stage in tropical maize, creating high temporal and spatial G × E (Bänziger 2000). Given the diverse topography of TPEs and the disparities due to climatic, biotic, and abiotic factors, augmented the difficulties in developing and identifying suitable cultivars with high and stable performance across locations/years (Vinayan et al. 2020). These disparities contributed to the differential responses of cultivars and were partitioned as G × E component (Vargas et al. 2015). Heat stress, especially during reproductive stages, can cause pollen sterility, accelerated senescence, and reduced grain filling period, and eventually limit the yield potential of the maize crop (Hatfield and Prueger 2015; Noor et al. 2019). It imposes strong G × E effects by shortening phenophases and disrupting reproductive development in maize.

In wheat, exposure to temperatures above 34 °C during flowering led to accelerated senescence and loss in yield attributing traits that show wide genotype variability across environments (Lobell et al. 2012). Rice genotypes exhibited spikelet sterility at high nighttime temperatures, with variation in thermotolerance mechanisms contributing to G × E (Jagadish et al. 2015). The flower buds were unable to mobilize essential assimilates, carbohydrates, and derived products when subjected to extreme heat stress. G × E substantially contributed to genotypic variation in thermotolerance (Wang et al. 2019; Fábián et al. 2019). Physiological, biochemical, and molecular responses to heat stress underpin genotype-specific tolerance and contribute to the complex G × E observed in rice breeding programs targeting climate resilience cultivars (Wang et al. 2008; Wu et al. 2017). In chickpea and lentil, heat stress during flowering drastically reduces pod-set and seed viability, and genotypes differ significantly in stress escape and/or tolerance strategies, leading to an unstable performance across planting dates (Gaur et al. 2019). High temperatures during flowering affected oil biosynthesis and seed set in mustard, which often interact with photoperiod sensitivity and lead to unpredictable performance across locations (Pillai and Walia 2024). These differential genotype responses to heat across seasons and sites amplify crossover interactions and challenge genomic prediction models unless G × E under different thermal regimes are accurately characterized and incorporated.

6.3.3 Excessive Moisture/Waterlogging Stress

Soil waterlogging reduces gas exchange and mineral nutrient homeostasis in the plants, while prolonged waterlogging leads to hypoxia in the rhizosphere, reduced tillering, and delayed flowering in rice (Colmer and Voesenek 2009). In lowland rice, waterlogging or submergence frequently co-occurs with salinity or intermittent drought stress due to erratic rainfall, leading to significant G × E and eventually a genotype-specific performance. The presence or absence of the Sub1A gene explains much of the G × E under flash-flood conditions, but variation in recovery traits adds further interaction complexity (Ismail et al. 2013). Crop like maize is highly susceptible to oxygen deprivation in the root zone and therefore expresses significant G × E-specific variation under waterlogged conditions (Zaidi et al. 2003). In wheat, waterlogging induces strong crossover interactions, particularly in poorly drained soils (Setter and Waters 2003). Oilseeds such as rapeseed and soybean also experience severe yield instability due to high moisture stress, with G × E often confounded by soil texture (Wollmer et al. 2018).

6.3.4 Salt Stress

Soil-based abiotic stresses such as salinity, sodicity (alkalinity), boron toxicity, and iron toxicity, exert localized yet severe effects on crop performance, resulting in high location-based G × E effects. In rice, salinity tolerance at the seedling stage (e.g., via Saltol) does not always translate into yield stability under field conditions, as genotypic response is modulated by season, soil pH, and crop management (Thomson et al. 2010). Chickpea and lentil cultivars show variable tolerance to boron and salinity across agro-ecologies, contributing to crossover G × E effects (Rameshwaran et al. 2016).

6.3.5 Multiple Stresses

While a substantial amount of research has focused on individual stress factors, most crops face compound or sequential stress episodes, which result in nonadditive effects on plant functions and yield (Mittler 2006; Zaidi et al. 2020). These combined stresses tend to magnify G × E because the ranking and expression of genotypes vary under single versus multiple stress conditions, which is found to have non-additive effects on phenotype and further exacerbate G × E effects. Predicted climate change scenarios are likely to increase the combination of stresses that plants must face in the field. Climate change projections suggest massive yield losses in many regions, underlining the high priority that needs to be accorded to developing climate-resilient germplasm to offset these losses (Cairns et al. 2012). Frequent spells of high temperatures (often ≥35 °C) in conjunction with moisture stress are a common phenomenon in most tropical arid/semi-arid regions (Zaidi et al. 2023). In the absence of frequent irrigation (to maintain high humidity in the crop canopy), the impact of physiological drought due to high VPD on crop plants is exacerbated, resulting in significant yield losses. Combined high temperature and (physiological) drought stress causes reproductive failure in wheat and maize, destabilizing genotype rankings even in advanced lines (Novick et al. 2024). Heat stress and drought are principal abiotic stresses that lead to significant crop losses and compromises in the quality of rice (Yang et al. 2022). In coastal rice-growing regions, simultaneous exposure to salinity and submergence is frequent, which disrupts ionic balance and depletes carbohydrate reserves, thereby requiring genotypes to cope simultaneously with osmotic stress and hypoxic conditions (Ismail et al. 2013). Breeding for tolerance to compound stresses is inherently more complex, requiring trait pyramiding, improved phenotyping protocols, and genomic prediction models that explicitly incorporate G × E effects (Shahi et al. 2022).

6.4 Approaches to Manage G × E Effects

A better understanding of G × E is a must for breeding genotypes for stress-prone ecologies (Amin et al. 2025). There are various options (Table 6.1) to define G × E and account for it while breeding new cultivars for wider adaptability.

6.4.1 Mapping Mega-Environments

A mega-environment (ME) can be defined as a large, repeatable geographical area or group of environments that consistently share similar climatic conditions, soil types, cropping systems, and G × E patterns, such that a set of genotypes can be commonly targeted across locations within a ME (Yan et al. 2023). The concept, introduced by CIMMYT in the context of wheat breeding, has since been widely adopted across crops like maize, rice, and sorghum to stratify global or regional breeding targets. Mega-environment analysis involves evaluating MET data to quantify G × E and define breeding strategies that target specific adaptive zones (Cooper et al. 2006; Xu 2016). In contrast to agro-ecological zones (AEZs), which are based on broad environmental criteria (e.g., latitude, altitude, soil, climate) and are generally non-crop-specific, MEs are crop-specific and defined by genotypic main effects (G) and G × E patterns (Gauch and Zobel 1997). It represents patterns of crop performance variability across diverse environmental settings (Yan et al. 2023). In

Table 6.1 Differences in aspects of mega-environment (ME), envirotyping, and climate typing

S. No.	Aspects	Mega-environment	Envirotyping	Climate typing
1.	Definition	It's a macro-level classification of regions or testing sites based on long-term climatic means, such as temperature, humidity, and rainfall, among others, often used for zoning and identification of test sites.	A broader and detailed characterization of the environment, often dynamic, including weather data (temporal variation), soil status, and stress indicators during the crop life cycle. It's just like phenotyping the environment.	Grouping environments based on long-term climatic patterns
2.	Temporal resolution	Multi-year stability (repeatable G × E)	Daily to seasonal (within-season dynamics)	Long-term averages (≥30 years)
3.	Spatial scale	Large (regional, national, or global)	Field to regional (can be fine-scale)	Large-scale (continental, global zoning)
4.	Data sources	MET data, long-term yield data	Real-time or historical weather, soil, and crop data	Historical climate records (temperature, rainfall, etc.)
5.	Tools used	AMMI, GGE biplots, clustering based on yield/ G × E	Crop models (e.g., *APSIM*, *DSSAT*), sensors, remote sensing	Köppen classification, GIS, FAO AEZs
6.	Purpose in breeding	Group locations for efficient testing & genotype targeting	Model G × E interactions, refine prediction & trait selection	Define climatic boundaries for breeding strategy
7	Output	ME maps and genotype clusters (GGE biplots: *Which-won-where,* etc.)	Environmental similarity matrices, stress profiles	Climate zone maps (e.g., humid tropics, arid subtropics)

close view, AEZs and MEs often overlap, but the latter offer a more functional basis for breeding decisions by focusing on crop performance and G × E structure.

6.4.2 Climate Typing and Envirotyping

Climate typing is the method of classifying environments primarily based on long-term climatic patterns often spanning a few decades, such as average rainfall, temperature, radiation, and evapotranspiration (Xu 2016; Amin et al. 2025). It includes the analysis of the historical weather datasets over multi-year and multi-season periods to identify recurring climatic regimes or dominant stress patterns (e.g., drought-prone, heat-prone, sub-humid, or optimal conditions). The prime objective of climate typing is to delineate macro-environmental categories or mega climate-zones. These categories serve as the basis for selecting test sites and designing environment-specific breeding strategies. On the other hand, *envirotyping* is a more comprehensive and high-resolution approach that goes beyond long-term climate changes. It describes the process of systematically characterizing environments using high-resolution environmental covariates, such as weather parameters (e.g., temperature, rainfall, humidity, radiation), soil properties (e.g., texture, structure, available organic matter, pH, water-holding ability), and agronomic practices (e.g., sowing window, irrigation, fertilization) (Xu 2016). The phenomenon focuses on the actual situations experienced by a crop during its growth cycle, often down to the daily or weekly level. The required data are collected via remote sensing, meteorological or weather stations, farmer management records, and historical data on cropping patterns. Envirotyping converts this raw data into eco-physiological variables (e.g., GDDs, VPD, and abiotic stress indices) that are biologically meaningful and can be integrated into complex G × E analyses and crop modeling for predictions. Climate typing and envirotyping both aim to define and categorize environments; they vary in scope, resolution of data, and application, each offering unique insights into breeding for climate-resilience.

6.4.3 Defining the Target Population of Environment (TPE)

The TPE encompasses both the representative biophysical conditions and the temporal variability (Piepho 2022). A TPE analysis involves aggregating potential crop environments into clusters where test genotypes show minimal G × E across conditions over multiple years. In practice, the TPEs enable environment-specific breeding strategies. For instance, CIMMYT's wheat breeding program in South Asia stratified the region into distinct TPEs, such as fully irrigated, heat-stressed, and drought-prone environments, to optimize trialing sites and selection pressure (Singh et al. 2021). TPE-based frameworks also improve the accuracy of genomic selection (GS) by more precisely modeling the G × E patterns. In rice and wheat, such stratification has enhanced prediction accuracies for complex traits like yield under drought, particularly under variable stress scenarios (Heslot et al. 2015; Millet et al. 2019). Overall, integrating TPE in the breeding pipeline supports strategic trial design, improves the efficiency of selection, and strengthens the deployment of climate-resilient cultivars.

The two complementary approaches—climate typing and envirotyping play a pivotal role in understanding the performance and ranking patterns of test entries in TPEs. While climate typing is typically used in the early stages of breeding program design to define representative TPEs, envirotyping comes into play during phenotyping, G × E analysis, and model-based prediction, offering a granular understanding of the environmental drivers behind genotype performance. Aligning TPE delineation with envirotyping enables dynamic breeding pipelines that can adapt to both current and projected climate scenarios (Cooper et al. 2006; Xu 2016). Integration of envirotyping with current genomic selection models (known as "*genomic-environmental prediction*") enhances predictive accuracy across environments. This approach is particularly beneficial in scenarios where new genotypes are evaluated in previously unobserved environments, such as a cross-validation scheme, where genomic estimated breeding values (GEBVs) are predicted for untested genotypes under unobserved environmental conditions (Alemu et al. 2024).

6.4.4 Multi-Environment Trials (METs)

The METs help in identifying genotypes with either broad adaptation (stable performance across environments) or specific adaptation (outperforming in targeted conditions) (Argaw et al. 2024). METs enable breeders to—(a) quantify environmental variance, which often exceeds genotypic variance under stress (Badu-Apraku et al. 2023), (b) stratify target environments into mega-environments using GGE or AMMI, improving selection accuracy and deployment strategies (Badu-Apraku et al. 2012; Vinayan et al. 2020), (c) identify stable, drought-resilient genotypes (Dias et al. 2018), and (d) enhance selection efficiency through secondary traits (Yue et al. 2021; Singamsetti et al. 2023b; Mekonnen et al. 2023). Through replicated trials across variable environments, METs capture both main and interaction effects essential for genomic prediction models, clustering of testing zones, and incorporating physiological traits. Integrating MET-derived phenotypes with genomic and envirotyping information thus provides a comprehensive framework enabling breeders to predict performance in untested environments and accelerate development of climate-smart maize cultivars. The advent of linear mixed models (LMMs) revolutionized MET analysis by modeling genotype and G × E as random effects, enabling the estimation of genetic merit even when working with incomplete or unbalanced datasets (Meuwissen et al. 2001; Smith et al. 2001, 2005; Piepho et al. 2008). When extended with appropriate covariance structures, such as factor analysis (FA) variance structures, LMMs can capture genotype correlations across environments, improving both model fit and selection decisions (Kelly et al. 2007). The FA models can also incorporate pedigree-based (A matrix) or genomic-based (G matrix) relationship information to improve prediction accuracy and interpretability further.

6.4.5 Integrating Crop Simulation Models with Genomic Tools

Crop simulation models (CSMs) are employed to simulate crop growth and developmental processes by integrating various biological, physical, and chemical mechanisms, including key physiological pathways (Boote et al. 2021). Integration of CSMs with genomic tools represents a transformative step toward predictive, climate-resilient, and resource-efficient breeding (Onogi et al. 2016). Integrated approaches of CSMs with genomic prediction strategies improve accuracy under stress environments and across TPEs and enable environment-specific breeding that

Table 6.2 Case studies on recent crop models integrated with genomics

S. No.	Crops	Modeling approach	Input data	Result & impact	References
1.	Maize	CGM–WGP (APSIM + WGP)	106 DH maize lines, managed drought environments	Enhanced yield prediction under drought; validated in real breeding trials	Cooper et al. (2016)
2.	Alfalfa	Enviromics + GBLUP	Multi-harvest trials with envirotyping	Reduced trial needs; higher genomic model accuracy	Filho et al. (2023)
3.	Sorghum	Envirotyping + genomic prediction	Sorghum MET across mega-environments	Improved accuracy when accounting for environment-specific effects	Winans et al. (2024)
4.	Maize	ML + APSIM simulation features	Corn Belt data; APSIM variables	7–20% reduction in error by leveraging simulation variables	Shahhosseini et al. (2020)
5.	Wheat	Global sensitivity analysis (APSIM-wheat)	4 sites × 125 years; 90 traits across sowing/N/future CO_2	Identified critical traits for ideotypes under stress	Casadebaig et al. (2016)

supports simulating performance under defined climate/soil/management scenarios (Table 6.2). Traditional genomic prediction methods often treat environments as statistical noise or fixed effects, limiting their ability to model G × E × M interactions explicitly. The trait-based parameters can be linked to genetic markers or GEBVs allow breeders to move from phenotype-based to gene-to-phenotype (*G2P*) modeling. By Trait Parameter Estimation (*Trait-to-Gene Integration*), key physiological traits used in CSMs are estimated for multiple genotypes (Poudel et al. 2023). Genomic predictions are used to associate these traits with molecular markers. Messina et al. (2018) introduced frameworks that compute envirotyping and enviromics matrices, combining genetic relationship matrices (GRM) with environmental kinships that enable mixed models to account simultaneously for genotype main effects and G × E in a unified, scalable manner. Moreover, coupling CSM outputs with machine learning (ML) algorithms, such as the integration of APSIM-simulated water stress metrics into *LightGBM* or Random Forest (RF) models, can predict 7–20% reductions in yield prediction error (Shahhosseini et al. 2020; Xie et al. 2025).

6.4.6 Integrating G × E in Genomic Selection

In METs, classical GS models such as ridge regression BLUP (rrBLUP) and genomic BLUP (gBLUP) are widely used, particularly when extended to account for G × E, a significant source of prediction inaccuracy (Heslot et al. 2012; Jarquín et al. 2014). However, the predictive accuracy of GS is often hindered by G × E, where the effects of genetic variants on phenotypic expression vary across environments. Ignoring G × E leads to biased GEBVs and reduced selection efficiency under climate variability. To address this, reaction norm models, multi-environment gBLUPs, and structured covariance models have been developed to improve predictive power in site-year-specific conditions (Jarquín et al. 2014). Recent advances incorporate non-linear models and ML algorithms such as Random Forests, Artificial Neural Networks (ANN), and Deep Learning (DL), better capture complex marker-trait-environment interactions and enhance multi-trait, multi-environment prediction accuracy (Montesinos-López et al. 2016; Millet et al. 2019).

6.4.7 Support of Novel Computational Tools in Genomic Predictions

By modeling complex, unbalanced, non-linear relationships in genomic, phenotypic, and environmental data, artificial intelligence (AI) and machine learning (ML) are transforming genomic predictions in plant breeding. Unlike traditional BLUP or gBLUP models, AI models can handle interactions, collinearity, missing data, and hierarchical covariates, making them ideal for multi-environment genomic prediction under G × E. Coupling APSIM-derived drought indices with ML algorithms reduced error in maize yield prediction by 20% across the US Corn Belt (Shahhosseini et al. 2021). Montesinos-López et al. (2016) reported higher prediction accuracy in wheat using deep learning (DL) models versus standard gBLUP in METs.

6.4.8 Participatory and Inclusive Breeding (PIB)

As a system-level response to G × E, PIB empowers farmers to participate throughout the breeding cycle, starting from design, evaluation, advancement, and dissemination (Ceccarelli 2015). It also improves heritability (h^2) and selection intensity (i) by capturing real-world variability via farmer-managed trials, refining selection indices using contextualized data (e.g., local stress markers, farmer scores) (van Etten et al. 2019). A systematic implementation of PIB strengthens climate-resilient breeding pipelines, particularly under unbalanced datasets and high G × E scenarios, by blending genomic tools with socially grounded design, ensuring both biological impact and community relevance (Steinke and van Etten 2017).

6.4.9 Decision Support Systems

Decision support systems (DSS) are emerging as critical components in climate-resilient breeding pipelines by integrating diverse datasets that include genomic, environmental, and socio-economic information into actionable tools for plant breeders. It enables what-if scenario analysis, environment-specific trait prioritization, and genotype deployment strategies tailored to specific TPEs. DSS platforms increasingly incorporate satellite-derived evapotranspiration maps, probabilistic weather forecasting, and poverty and nutrition overlays to guide breeding decisions that are context-specific and impact-driven (Shahhosseini et al. 2020; van Etten et al. 2019; Xie et al. 2025). Modern DSS now harnesses ML and hybrid AI-mechanistic models, improving prediction accuracy across environments by dissecting predictable and stochastic components of G × E (Montesinos-López et al. 2016; Crossa et al. 2017). Such systems enable in silico ideotype design, reduce breeding cycles, and improve varietal targeting to AEZs. The synergy between DSS, envirotyping, and trait optimization under stress conditions can significantly enhance selection efficiency and impact potential.

6.5 New Breeding Paradigm: From High-Yield to High-Stable-Yield

Climate change implies incremental pressure on the agrarian community and its dependents through the year-on-year unpredictable change in climatic patterns and the increased frequency of weather vagaries like drought, heat, and water-logging stresses (Zaidi et al. 2020; Monteleone et al. 2023; Yang et al. 2023). These stresses impair crop growing conditions, nutrient uptake, and deviate the flowering periods and grain filling duration, which have a direct impact on crop yield and quality, and an indirect effect on ecosystem changes, food security issues, etc. (Yuan et al. 2024). Therefore, developing climate-resilient crop varieties to cope with variable and extreme weather conditions requires a deliberate attempt on a strategic breeding approach, from selecting the right genetic resources to designing precise testing environments and integrating novel predictive tools (Cooper et al. 2016; Varshney et al. 2021). Developing climate-smart crop varieties through next-generation breeding approaches can be a shield for overcoming the multidimensional, spatial, and temporal abiotic and biotic stresses coupled with exacerbated G × E effects.

Conventional plant breeding programs have developed high-yielding crop varieties through hybridization and phenotypic selection, usually under well-managed conditions. From the farmer's point of view, yield stability is equally essential to high yields to reduce the risk of investment by resource-constrained farm families. To select superior genotypes with high agronomic performance across diverse environments, breeders must test numerous candidates through multi-year, multi-location phenotypic trials, including managed stress phenotyping. However, many agronomic traits are complex, influenced by multiple small-effect QTL, and their expression is heavily impacted by environmental factors (Reckling et al. 2021).

Estimation of G, E, and G × E components ensures valid recommendations of suitable varieties able to overcome the pressure due to variable occurring conditions. Modern breeding approaches have strongly backed the aim of conventional breeding programs, vigorously addressing the time taken and substantially minimizing the environmental influence (Bevan et al. 2017; Varshney 2016).

6.5.1 Expanding the Genetic Toolkit

Studies have shown that as genetic diversity increases, the variability in genotype performance across different environments also increases, and genetic diversity within populations can influence the magnitude of G × E (Dias et al. 2018). This implies that diverse genetic backgrounds may exhibit varying degrees of adaptability to environmental changes, affecting overall performance (Rogers et al. 2021). While elite cultivars offer high-yield potential, their narrow genetic base often limits novel stress-adaptive traits, which need a systematic exploration of a wider genetic diversity, including crop wild relatives (CWRs) and local landraces (Renzi et al. 2022; Zhang et al. 2017), to identify potential sources of germplasm for major stress(es) in TPE to integrate in the breeding pipeline aiming high-stable yields. The challenge associated with incorporating these resources into the main breeding pipeline has been the "linkage drag", where stress-tolerance genes are inherited along with undesirable traits (Hajjar and Hodgkin 2007), which is now possible to address using genomic tools, such as genomic-assisted introgression of desired alleles with a minimum transfer of undesirable once (Razzaq et al. 2021).

6.5.2 Genomic-Assisted Introgression of Desirable Alleles

Using novel genomic tools such as genome-wide association studies (GWAS), genomic selection (GS), SNP-mediated marker-assisted selection, and backcrossing for beneficial alleles within elite germplasm, rapid-cycle recurrent genomic selection (RC-GS), when combined with large-scale data analytics and DSS, enables breeding programs to effectively utilize diverse genetic resources for optimizing breeding strategies. This approach ultimately enhances crop improvement efforts by facilitating the development of varieties with stable and high-yielding performance under stress-prone conditions (Varshney et al. 2018). Genotypic data are widely incorporated in various phases of crop breeding for enhancing the selection accuracy, shortening the breeding cycle by balancing the genetic gain at the background for optimized resource utilization and operational efficiency (Hickey et al. 2019). As yield is dependent on the cumulative performance of several plant functions and traits, a simultaneous improvement of multiple traits is more efficient for interrelated low and high-heritable characteristics to achieve the high yield with stability (Hayashi and Iwata 2013; Das et al. 2020, 2021). Application of the multi-trait genotype-ideotype distance index (MGIDI) further enhances the efficiency of selection across environments, including optimal and multiple abiotic stresses (Singamsetti et al. 2023a).

Integrating genetic information, crop growth models (CGMs) can facilitate dissecting complex G × E and predict how specific genotypes might perform across a wide range of future climate scenarios (Chen et al. 2024). Platforms like CropGS-Hub have demonstrated how multi-omics data, combined with predictive algorithms, can be used to simulate genotype performance and support genomically designed breeding strategies. Recent advances in computing capacity offer a great potential in making highly precise and efficient pipelines for data analysis to enhance breeding performance and ultimately crop productivity (Singh et al. 2018; van Dijk et al. 2021).

6.5.3 Precision in Testing and Selection Environments

Relying solely on naturally occurring random stress events for selection is often non-repeatable. Modern strategies emphasize creating a managed stress environment (MSE) by mimicking TPE, enabling more relevant and effective selection (Bänziger 2000; Zaidi et al. 2020). Given the increasing frequency of compound stress scenarios, such as drought coupled with heat or excessive moisture (Zaidi et al. 2020; Prasanna et al. 2021), there is a critical need to develop phenotyping platforms that can simultaneously impose and evaluate multiple stress factors to better inform stress-resilient breeding efforts. Phenotyping under MSE not only helps in taking informed selection decisions but also underpins the development of robust genotype-phenotype associations and dissects complex G × E (Zaman-Allah et al. 2015; Zaidi et al. 2016a, b, 2019). Field-based precision phenotyping aids in identifying progeny with traits of interest, building robust statistical models for genotype–phenotype associations, and identifying genomic regions of influence for use in forward breeding or genomic selection (Desta and Ortiz 2014). Cairns et al. (2013), however, cautioned that G × E and large plot residuals might affect the repeatability of managed stress trials.

6.5.4 Selection Decisions Based on Multi-traits

Sole reliance on grain yield as a selection criterion can be misleading under variable climatic conditions, as it typically expresses low heritability and high environmental variance under stresses (Bolaños et al. 1993; Araus et al. 2018). Its complex inheritance and sensitivity to environmental fluctuations make it prone to false positives under stress. Furthermore, the relationship between grain yield under stress and non-stress conditions is often variable and lacks consistency (Amadu et al. 2025). Therefore, achieving breeding progress under abiotic stress conditions requires a strategic focus on both primary trait (grain yield) and key stress-adaptive secondary traits that are significantly associated with the performance of genotypes under stress conditions (Fig. 6.3). Several secondary traits have been identified as

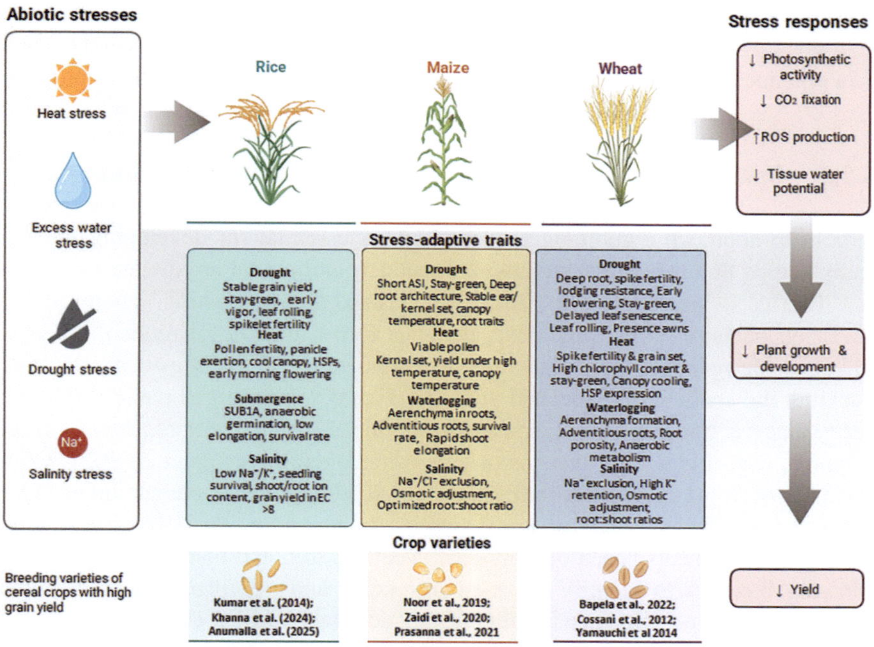

Fig. 6.3 Major effects of key abiotic stresses and stress-adaptive secondary traits identified in major cereal crops for use in selection index, along with grain yield under respective stresses

valuable indicators of stress adaptation in crops like maize (Alam et al. 2017; Cairns et al. 2013; Zaidi et al. 2020). However, for a secondary trait to be helpful in breeding programs, it must meet a few key critical criteria, such as being genetically variable and heritable, significantly correlated with yield under stress, not associated with yield penalty under optimal conditions, rapidly and reliably measurable in field conditions, non-destructive, and cost-effective (Banziger et al. 2000; Monneveux et al. 2011; Zaidi et al. 2020). To balance stress adaptation with agronomic performance, selection indices are commonly used to integrate multiple traits into breeding decisions. Such indices help reconcile trade-offs and improve selection efficiency across environments, for example, multi-trait selection for wheat under rainfed and irrigated conditions (Mohammadi and Geravandi 2024). These indices enable quantitative comparisons of genotype performance across stress and non-stress environments, facilitating the identification of genotypes that combine high yield potential with stability (Singamsetti et al. 2023a; Hallajian et al. 2024).

In recent years, the availability of high-throughput and precise phenotyping tools, based on remote sensing, multispectral, hyperspectral, fluorescence, and thermal sensors, has offered accuracy, efficiency, and cost-effectiveness in large-scale field-based phenotyping (Cerrudo et al. 2017; Makanza et al. 2018). Looking ahead, the integration of image processing algorithms and standardized analytical pipelines will democratize data analysis, allowing non-experts to process and interpret complex datasets. With rapid advancements in computational technologies, high-throughput field phenotyping tools will enhance trait heritability and accuracy, generate robust data pipelines, and facilitate the use of complex datasets in breeding decisions.

6.5.5 Integrated Approach in Breeding for Climate-Resilience

A holistic approach combining modern tools is essential for developing "future-ready" crops that can withstand environmental volatility and are the least prone to G × E effects. Varshney et al. (2021) emphasized that integrating genomics with advanced phenomics and predictive analytics forms the core of modern breeding strategies designed to accelerate the development of climate-resilient cultivars. Tackling the challenges associated with G × E under changing climatic conditions necessitates a transition from conventional linear breeding approaches to more dynamic, data-driven, and systems-oriented strategies (Reynolds et al. 2021; Cobb et al. 2019). At the core of this shift is the identification and characterization of TPEs, supported by climate typing and envirotyping frameworks (Xu 2016). These tools classify environments based on long-term weather, soil, and management patterns, allowing breeders to identify mega-environments and prioritize testing locations where genotypic performance is both relevant and stable (Millet et al. 2019). Models like APSIM and DSSAT simulate G × E, which, when integrated with genomic prediction frameworks, allow the development of climate-resilient ideotypes tailored to specific stress regimes such as drought, heat, or low-input systems (Chenu et al. 2011; Messina et al. 2018). Advanced breeding informatics systems manage vast multi-environment and multi-omic datasets, while AI and ML improve prediction accuracy under complex G × E landscapes (Montesinos-López et al. 2016; Crossa et al. 2017). Integrated DSS assists breeders in trait prioritization, trial planning, and varietal targeting, ensuring faster, more informed, and risk-aware breeding decisions (Runcie et al. 2021).

6.5.6 New Generation of High-Yielding Climate-Adaptive Crop Varieties

The swift replacement of high-yielding but stress-susceptible varieties with resilient, high-performing, and nutrient-rich alternatives can significantly contribute to food and nutritional security, uplifting the livelihoods of millions of smallholder families across the tropics (Prasanna et al. 2020). Developing climate-resilient crop varieties hinges on a deep understanding of traits vital for adaptation to evolving environmental stresses. Studies have demonstrated that the adoption of climate-resilient maize varieties can lead to yield gains of approximately 5–25% in many maize-growing regions across Eastern and Southern Africa (Shiferaw et al. 2011; Cooper et al. 2023).

Over the past four decades, CIMMYT has spearheaded the development and dissemination of climate-resilient maize germplasm, tailored to meet diverse client-preferred traits (Prasanna et al. 2021). To effectively address the challenges posed by substantial G × E, CIMMYT employs a decentralized breeding approach spanning sub-Saharan Africa (SSA), South Asia, and Latin America. This geographically distributed strategy facilitates broader testing and enhances the relevance of breeding outcomes. The approach led to the successful development and deployment of an array of elite stress-tolerant maize cultivars across sub-Saharan Africa, Asia, and Latin America (Zaidi et al. 2020; Prasanna et al. 2021). Unlike conventional hybrids, which suffer significant yield losses under abiotic stresses, the new stress-resilient hybrids maintain comparable yields under non-stressed conditions, while outperforming under drought and/or heat stress (Prasanna et al. 2021; Zaidi et al. 2023). With growing climate variability across lowland tropics, rice breeding has progressively shifted toward the development of varieties that are resilient to a spectrum of abiotic and biotic stresses (Anumalla et al., 2025; Bhosale et al. 2025). Marker-assisted selection (MAS) has enabled the successful introgression of key stress-tolerant loci, which helped in developing rice varieties demonstrating stability in both stress and non-stress conditions.

6.6 Mitigating G × E with Stress-Resilience Breeding—Evidence and Insights

6.6.1 Meta-Analysis of METs

Recent studies suggested that strategically harnessing G × E can facilitate the development of cultivars of food crops like rice, wheat, maize, chickpea, as well as in forestry species, that can perform across variable environmental conditions, demonstrating both yield stability and stress adaptation (Zaidi et al. 2020; Raffao and Jensen 2023; Crossa et al. 2017). Analysis of MET data from 2015 to 2018 for grain yield under heat stress showed that relative humidity (RH) and VPD were the primary environmental drivers underlying significant G × E effects on grain yield (Vinayan et al. 2020). Large-scale METs across the U.S. and Canada revealed that variation in solar radiation, temperature, rainfall, and humidity significantly contributed to G × E-driven yield differences, with solar radiation often being the primary environmental driver of genetic sensitivity (Sarzaeim and Muñoz-Arriola 2024).

In maize, large-scale breeding efforts by CIMMYT, integrating METs under managed stress and farmer-managed conditions across sub-Saharan Africa, Asia, and Latin America, have consistently shown that targeted selection under these environments yields climate-resilient hybrids with improved tolerance to drought, heat, and associated stresses (Prasanna et al. 2021; Cohen et al. 2021; Zaidi et al. 2023). Building on these meta-analytical insights, it becomes clear that understanding G × E patterns in stress-prone environments not only indicates the magnitude of variability but also guides the choice of analytical methods best suited to dissect it. Efforts are being made to leverage environmental stratification and multi-trait indices to enhance heat-stress MET interpretation (Yue et al. 2021 and 2022). Such methodologies not only enhanced the efficiency of selection by capturing both average performance and stability but also improved predictive accuracy for untested environments (Carvalho et al. 2024). A GWAS on a tropical heat-tolerance association mapping panel (543 inbred lines across nine South Asian locations) identified 269 environment-specific SNPs associated with grain yield under heat stress (Seetharam et al. 2021). From these, 26 haplotype blocks exhibited stability across up to five environments, enabling the development of molecular markers for breeding. These insights enable regionally targeted, genetically informed breeding strategies for climate-smart maize and enable synthesis across diverse stress-prone ecosystems, drought, heat, and waterlogging to quantify G × E and adaptive responses.

6.6.2 Prediction Accuracy of G × E-Inclusive Genomic Models

In a study on African maize, prediction accuracies under a CV2 framework where genotypes are evaluated in at least some correlated environments increased by ~240% under optimal and ~ 750% under drought conditions compared to models ignoring G × E effects, demonstrating the substantial benefit of environmental borrowing in prediction models (Edriss et al. 2017). Similarly, a study on maize hybrids in China using gBLUP integrated with climatic covariates (gBLUP-GE) achieved accuracies of 0.731 for grain moisture content and 0.331 for grain yield, significantly outperforming traditional gBLUP approaches (Wang et al. 2025).

Reaction-norm and factor-analytic models further enable genotype-specific reaction to environmental variables, effectively capturing high-dimensional marker-environment interactions, and showing superior accuracy across structured hybrid populations (Li et al. 2021; Westhues et al. 2021). Additionally, joint modeling of additive and dominance G × E effects in Brazilian multi-environment trials under well-watered and drought conditions improved predictive reliability, particularly when borrowing across trials (Dias et al. 2018). Models integrating environmental similarity matrices (G × E + G × W) consistently yield the highest predictive performance (Hu et al. 2025). Integration of G × E into genomic prediction models has demonstrated substantial improvements in prediction accuracy for maize traits across diverse environments through reaction-norm, gBLUP-GE, and mixed-effect frameworks (Edriss et al. 2017; Beyene et al. 2021). These enhancements are particularly notable under variable conditions such as drought stress and fluctuating soil fertility, where G × E play a pivotal role in phenotypic expression (Beyene et al. 2021).

6.6.3 Breeding for Specific Trait vs. Broad Adaptation

Breeding strategies often navigate the balance between stress-specific adaptation and broad adaptation to develop varieties that perform well across a range of environmental conditions. However, this approach may not always yield the highest performance under specific stress conditions, as the selection for general resilience may dilute the genetic potential for tolerance to individual stresses. Kusmec et al. (2023) conducted a landmark study, leveraging 81 years of public yield trial data from 4730 maize hybrids, alongside historical weather records, to model genetic responses to thermal stress. The study showed that continuous breeding has led to an incremental increase in genetic tolerance to moderate heat stress. Conversely, the study also identified an inadvertent reduction in tolerance to severe heat stress, a trend observed most notably from the mid-1970s onwards. This divergence suggests a genetic trade-off: improvements under moderate stress may compromise resilience under more extreme heat stress. Simultaneous testing for optimal conditions as well as key abiotic stresses, and selection decision based on across-stress and unstressed environments resulted in the development of high-yielding stress-resilient maize hybrids without significant yield penalty under optimal growing conditions (Vivek et al. 2017; Das et al. 2020, 2021, 2025; Singamsetti et al. 2023a). Genetic robustness, often manifesting as canalization and phenotypic plasticity, shapes G × E by delineating stable versus responsive phenotypic outcomes across environments. Schneider et al. (2020) demonstrated that maize genotypes that were highly plastic under water-deficit were genetically distinct from those that were plastic under phosphorus stress, highlighting trait-specific plasticity and the trade-offs in multi-stress environments and broad adaptation.

6.7 Breeding for High-Stable Yields—Challenges and Way Forward

Historically, crop breeding programs have focused on maximizing yield under optimal conditions, delivering substantial gains. However, in the era of increasing climate variability, this approach often falls short in capturing the dynamic nature of G × E. Climate change is not a static challenge, which necessitates dynamic and adaptive breeding strategies that evolve in response to shifting environmental baselines (Das et al.,2025; Zaidi et al. 2020; Anshori et al. 2024). Escalating climate variability, along with intensifying abiotic stresses and associated G × E, now poses a significant threat to the sustainability of global food systems. These realities necessitate a paradigm shift—from breeding for high yield to breeding for high-stable yield across diverse and often suboptimal environments. This transformation requires a strategic change in the crop improvement programs so that dynamic G × E is well captured, modelled, and leveraged for genetic gain. Central to the new paradigm are novel tools, including the definition of TPE-based mega-environment characterization, genomic-assisted selections, high-throughput precision phenotyping (HTPP) in a managed environment, trait network analysis, METs, and predictive modeling. Real-world implementation of this integrated approach is exemplified by IRRI's OneRice Framework (Bhosale et al. 2025) and CIMMYT's climate-resilient maize breeding strategy (Prasanna et al. 2021, 2022). However, to fully realize the benefits of G × E-informed breeding, the evaluation systems for varieties must be reframed. In many countries, varietal release protocols prioritize yield superiority under optimal conditions, often overlooking the importance of yield stability and climate resilience (Zaidi et al. 2023). This approach is increasingly misaligned with the realities of climate variability, which need to be corrected sooner instead of continuing business as usual.

Understanding G × E is not merely a statistical necessity; rather, it is a key to breeding the next generation of crop varieties for today and tomorrow's climatic conditions. Integrating G × E analysis with genomic selection, along with envirotyping and crop modeling, enables breeders to predict genotype performance under both current and future climates (Cooper et al. 2006; Xu 2016). This systems-based approach can support the development of climate-resilient cultivars with either broad or environment-specific adaptation, accelerating genetic gain under variable conditions (Ceccarelli and Grando 2020; Jiang et al. 2025). To operationalize a new breeding paradigm, modern breeding programs must integrate genotypic plasticity, resilience, and stability into selection strategies and decisions, supported by advanced tools such as genomic selection, trait networks, enviromics, and predictive modeling. The key aspects that need better appreciation and accounting for while designing a breeding pipeline aiming for improved cultivars with broad adaptation despite increasing climate variability with high G × E effects include:

6.7.1 G × E-Accounted Breeding Pipelines

Breeding pipelines often treat the environment as a static factor, limiting their ability to account for the spatial and temporal fluctuations and associated G × E, which is intrinsically dynamic, influenced by seasonal variations, agronomic practices, and localized microenvironments (Jiang et al. 2025). With intensified climate variability, increasing frequency of weather extremes, and episodes of compound stresses, breeding programs must transition from static, one-size-fits-all approaches to an agile, context-specific product development for this new normal for global agriculture. With improved understanding of G × E patterns, TPEs can be redefined using genetic correlations, the use of historical MET datasets, and environmental covariates to enhance model accuracy.

The emerging field of "enviromics" offers potential for enhancing trait prediction by providing suitable environmental predictors for G × E. Trevisan et al. (2025) introduced a novel framework for constructing enviromics matrices in mixed models to integrate genetic and environmental data to enhance phenotypic predictions in plant breeding. The framework is compatible with existing mixed model software, including rrBLUP and BGLR, and can be extended for more complex interactions. By combining genetic relationships and environmental influences, this approach offers a powerful tool for advancing G × E studies and accelerating the development of improved crop varieties.

6.7.2 Field-Based Scalable HTPP

Despite rapid technological progress in plant science, field-based precision phenotyping for major abiotic stresses, such as drought, heat, waterlogging, salinity, and their compound interactions, remains a critical bottleneck in crop improvement pipelines. Controlled environments offer precision in stress application but often fail to replicate the complexity of field conditions, limiting the data validity and scalability. Therefore, challenges in generating field-based precision phenotyping data using scalable tools need to be addressed for:

- *Capturing temporal dynamics of stress responses, such as canopy temperature fluctuations, diurnal transpiration patterns, and* in situ *root architecture.*
- *Non-invasive and high-resolution, high-throughput phenotyping of critical physiological responses, such as osmotic adjustment, membrane integrity, and metabolite fluxes, for understanding plant resilience under chronic or fluctuating stress conditions.*
- *Deployment of HTPP tools such as thermal imaging, hyperspectral sensors, and drone-based imaging in METs, not limited to a few on-station trials.*
- *Robust envirotyping platforms to quantify environmental variance contributing to G × E to link environmental descriptors with phenotypic and genomic data for predictive modeling.*

6.7.3 Rationalizing Genetic Gains Across Stressed and Unstressed Conditions

A persistent challenge in breeding for stress-resilience is the perceived trade-off between stress tolerance and yield potential. Recent research emphasizes the importance of trait plasticity as a key strategy to buffer performance across diverse conditions. For instance, Chapman et al. (2021) argue that selecting for plasticity, rather than fixed tolerance traits, can enhance both yield stability and adaptability across stress and non-stress environments. A compelling example comes from rapid-cycle genomic selection (RC-GS) in tropical maize, where Das et al. (2020, 2021) reported significant genetic gains for combined drought and waterlogging tolerance in tropical maize hybrids, without compromising on yields under optimal conditions. These findings underscore the importance of balanced selection strategies that integrate performance across stress and non-stress scenarios, thereby enhancing both broad adaptability and yield stability, key traits for navigating complex G × E landscapes. These findings are corroborated by a recent multi-country study by Tarekegne et al. (2024), which reported annual grain yield gains of 118 kg ha^{-1} under optimal conditions and 61 kg ha^{-1} under random stress, demonstrating that simultaneous improvement under both conditions is achievable. These studies and many others (Bänziger et al. 2006; Zaidi et al. 2020; Singamsetti et al. 2021; Prasanna et al. 2022) debunk the myth that selecting for stress tolerance or resilience inevitably results in a yield penalty under optimal conditions. These evidences demonstrate that new breeding strategies require a strategic shift from maximizing mean performance alone to considering minimizing downside risk as well, that is, ensuring yield stability under adverse conditions while maintaining high yield ceilings under optimal environments.

6.7.4 Institutional Appreciation for Climate-Adaptive Breeding

The successful implementation of G × E-informed, climate-smart breeding hinges not only on scientific innovation but also on robust institutional frameworks, sustained investment, and enabling policy environments. The development of MET networks has been instrumental in enabling reliable improvement of rice, wheat, and maize germplasm in the global South. These networks allow breeding programs to sample the TPE more comprehensively, improving the estimation of G × E effects and enhancing the predictive power of genomic selection models, and building climate-resilient food systems. However, despite the growing urgency of climate change, long-term funding for breeding programs targeting abiotic stress resilience remains insufficient, particularly in low- and middle-income countries. In the scenario of volatile investment support, the challenge is to maintain the regionally distributed shared phenotyping platforms and data pipelines developed after sizeable investment and efforts. To address these systemic gaps, strategic investments and a policy framework are required that facilitate establishing regionally distributed breeding hubs with shared access to high-throughput phenotyping and envirotyping platforms, which will enable more representative sampling of environmental variability and improve the scalability of breeding pipelines. Investments in data science, phenomics, and decision-support tools are critical for empowering local breeding programs.

References

Abebaw SE (2025) A global review of the impacts of climate change and variability on agricultural productivity and farmers' adaptation strategies. Food Sci Nutr 13:e70260. https://doi.org/10.1002/fsn3.70260

Abebe AT, Adewumi AS, Adebayo MA et al (2024) Genotype x environment interaction and yield stability of soybean (Glycine max l.) genotypes in multi-environment trials (METs) in Nigeria. Heliyon 10:e38097. https://doi.org/10.1016/j.heliyon.2024.e38097

Alam MA, Seetharam K, Zaidi PH et al (2017) Dissecting heat stress tolerance in tropical maize (Zea mays L.). Field Crop Res 204:110–119. https://doi.org/10.1016/j.fcr.2017.01.006

Alemu A, Åstrand J, Montesinos-López OA et al (2024) Genomic selection in plant breeding: key factors shaping two decades of progress. Mol Plant 17:552–578. https://doi.org/10.1016/j.molp.2024.03.007

Amadu MK, Beyene Y, Chaikam V et al (2025) Genome-wide association mapping and genomic prediction analyses reveal the genetic architecture of grain yield and agronomic traits under drought and optimum conditions in maize. BMC Plant Biol 25:135. https://doi.org/10.1186/s12870-025-06135-3

Amin A, Christopher J, Cooper M et al (2025) Envirotyping facilitates understanding of genotype × environment interactions and highlights the potential of stay-green traits in wheat. Field Crop Res 331:109940. https://doi.org/10.1016/j.fcr.2025.109940

Anshori MF, Musa Y, Farid M et al (2024) A comprehensive multivariate approach for GxE interaction analysis in early maturing rice varieties. Front Plant Sci 15:1462981. https://doi.org/10.3389/fpls.2024.1462981

Anumalla M, Khanna A, Catolos M et al (2025) Future flooding tolerant rice germplasm: resilience afforded beyond Sub1A gene. The Plant Genome 18:e70040. https://doi.org/10.1002/tpg2.70040

Araus JL, Kefauver SC, Zaman-Allah M et al (2018) Translating high-throughput phenotyping into genetic gain. Trends Plant Sci 23:451–466. https://doi.org/10.1016/j.tplants.2018.02.001

Argaw DM, Yoon S, Heilbron FC, Deilamsalehy H, Bui T, Wang Z, Dernoncourt F, Chung JS (2024) Scaling up video summarization pretraining with large language models. arXiv 2404.03398. https://doi.org/10.48550/arXiv.2404.03398

Argaw T, Fenta BA, Zegeye H et al (2025) Multi-environment trials data analysis: linear mixed model-based approaches using spatial and factor analytic models. Front Res Metr Anal 10: 1472282. https://doi.org/10.3389/frma.2025.1472282

Arshad M, Amjath-Babu TS, Krupnik TJ et al (2017) Climate variability and yield risk in South Asia's rice–wheat systems: emerging evidence from Pakistan. Paddy Water Environ 15:249–261. https://doi.org/10.1007/s10333-016-0544-0

Badu-Apraku B, Fakorede MAB, Menkir A, Sanogo D (2012) Conduct and management of maize field trials. IITA, Ibadan, p 59. https://hdl.handle.net/10568/87946

Badu-Apraku B, Abubakar AM, Adu GB et al (2023) Enhancing genetic gains in grain yield and efficiency of testing sites of early-maturing maize hybrids under contrasting environments. Genes 14:1900. https://doi.org/10.3390/genes14101900

Bänziger M (2000) In: Edmeades GO, Beck D, Bellon M (eds) Breeding for drought and nitrogen stress tolerance in maize: from theory to practice. CIMMYT, Mexico D.F

Bänziger M, Setimela PS, Hodson D, Vivek B (2006) Breeding for improved abiotic stress tolerance in maize adapted to southern Africa. Agric Water Manag 80:212–224. https://doi.org/10.1016/j.agwat.2005.07.014

Becker HC, Léon J (1988) Stability analysis in plant breeding. Plant Breed 101:1–23. https://doi.org/10.1111/j.1439-0523.1988.tb00261.x

Bevan M, Uauy C, Wulff B et al (2017) Genomic innovation for crop improvement. Nature 543: 346–354. https://doi.org/10.1038/nature22011

Beyene Y, Gowda M, Pérez-Rodríguez P et al (2021) Application of genomic selection at the early stage of breeding pipeline in tropical maize. Front Plant Sci 12:685488. https://doi.org/10.3389/fpls.2021.685488

Bhosale S, Thathapalli Prakash P, Kadaru SB et al (2025) Onerice breeding framework: an end-to-end system to develop better varieties faster. Crop Sci 65:e70087. https://doi.org/10.1002/csc2.70087

Blum A (2011) Drought resistance—is it really a complex trait? Funct Plant Biol 38:753–757. https://doi.org/10.1071/FP11101

Bolaños J, Edmeades GO, Martinez L (1993) Eight cycles of selection for drought tolerance in lowland tropical maize. III. Responses in drought-adaptive physiological and morphological traits. Field Crop Res 31:269–286. https://doi.org/10.1016/0378-4290(93)90066-V

Boote KJ, Jones JW, Hoogenboom G (2021) Incorporating realistic trait physiology into crop growth models to support genetic improvement. silico Plants 3:diab002. https://doi.org/10.1093/insilicoplants/diab002

Bustos-Korts D, Malosetti M, Chenu K et al (2019) From QTLs to adaptation landscapes: using genotype-to-phenotype models to characterize G×E over time. Front Plant Sci 10:1540. https://doi.org/10.3389/fpls.2019.01540

Cairns JE, Sonder K, Zaidi PH et al (2012) Chapter one—maize production in a changing climate: impacts, adaptation, and mitigation strategies. In: Sparks DL (ed) advances in agronomy. Academic Press, pp 1–58. https://doi.org/10.1016/b978-0-12-394275-3.00006-7
Cairns JE, Crossa J, Zaidi PH et al (2013) Identification of drought, heat, and combined drought and heat tolerant donors in maize. Crop Sci 53:1335–1346. https://doi.org/10.2135/cropsci2012.09.0545
Carvalho HF, Rio S, García-Abadillo J, Isidro y Sánchez J (2024) Revisiting superiority and stability metrics of cultivar performances using genomic data: derivations of new estimators. Plant Methods 20:85. https://doi.org/10.1186/s13007-024-01207-1
Casadebaig P, Zheng B, Chapman S et al (2016) Assessment of the potential impacts of wheat plant traits across environments by combining crop modeling and global sensitivity analysis. PLoS One 11:e0146385. https://doi.org/10.1371/journal.pone.0146385
Ceccarelli S (2015) Efficiency of plant breeding. Crop Sci 55:87–97. https://doi.org/10.2135/cropsci2014.02.0158
Ceccarelli S, Grando S (2020) Participatory plant breeding: who did it, who does it and where? Exp Agric 56:1–11. https://doi.org/10.1017/S0014479719000127
Cerrudo D, González Pérez L, Mendoza Lugo JA, Trachsel S (2017) Stay-green and associated vegetative indices to breed maize adapted to heat and combined heat-drought stresses. Remote Sens 9:235. https://doi.org/10.3390/rs9030235
Challinor AJ, Watson J, Lobell DB et al (2014) A meta-analysis of crop yield under climate change and adaptation. Nature Clim Change 4:287–291. https://doi.org/10.1038/nclimate2153
Chapman S, Birch CE, Galdos MV et al (2021) Assessing the impact of climate change on soil erosion in East Africa using a convection-permitting climate model. Environ Res Lett 16: 084006. https://doi.org/10.1088/1748-9326/ac10e1
Chen J, Tan C, Zhu M et al (2024) CropGS-hub: a comprehensive database of genotype and phenotype resources for genomic prediction in major crops. Nucleic Acids Res 52:D1519–D1529. https://doi.org/10.1093/nar/gkad1062
Chenu K, Cooper M, Hammer GL et al (2011) Environment characterization as an aid to wheat improvement: interpreting genotype-environment interactions by modelling water-deficit patterns in North-Eastern Australia. J Exp Bot 62:1743–1755. https://doi.org/10.1093/jxb/erq459
Cobb JN, Juma RU, Biswas PS et al (2019) Enhancing the rate of genetic gain in public-sector plant breeding programs: lessons from the breeder's equation. Theor Appl Genet 132:627–645. https://doi.org/10.1007/s00122-019-03317-0
Cohen I, Zandalinas SI, Huck C et al (2021) Meta-analysis of drought and heat stress combination impact on crop yield and yield components. Physiol Plant 171:66–76. https://doi.org/10.1111/ppl.13203
Colmer TD, Voesenek LACJ (2009) Flooding tolerance: suites of plant traits in variable environments. Functional Plant Biol 36:665–681. https://doi.org/10.1071/FP09144
Cooper M, Messina CD (2021) Can we harness "Enviromics" to accelerate crop improvement by integrating breeding and agronomy? Front Plant Sci 12. https://doi.org/10.3389/fpls.2021.735143
Cooper M, Messina CD (2023) Breeding crops for drought-affected environments and improved climate resilience. Plant Cell 35:162–186. https://doi.org/10.1093/plcell/koac321
Cooper M, van Eeuwijk F, Chapman SC et al (2006) Genotype-by-environment interactions under water-limited conditions. In: drought adaptation in cereals. CRC Press, p 46. https://doi.org/10.1201/9781003578338
Cooper M, Technow F, Messina C et al (2016) Use of crop growth models with whole-genome prediction: application to a maize multienvironment trial. Crop Sci 56:2141–2156. https://doi.org/10.2135/cropsci2015.08.0512
Cortés AJ (2024) Abiotic stress tolerance boosted by genetic diversity in plants. Int J Mol Sci 25: 5367. https://doi.org/10.3390/ijms25105367

Crossa J, Pérez-Rodríguez P, Cuevas J et al (2017) Genomic selection in plant breeding: methods, models, and perspectives. Trends Plant Sci 22:961–975. https://doi.org/10.1016/j.tplants.2017.08.011

Dar MH, Bano DA, Waza SA et al (2021) Abiotic stress tolerance-Progress and pathways of sustainable Rice production. Sustainability 13:2078. https://doi.org/10.3390/su13042078

Das RR, Vinayan MT, Patel MB et al (2020) Genetic gains with rapid-cycle genomic selection for combined drought and waterlogging tolerance in tropical maize (Zea mays L.). Plant Genome 13:e20035. https://doi.org/10.1002/tpg2.20035

Das RR, Vinayan MT, Seetharam K et al (2021) Genetic gains with genomic versus phenotypic selection for drought and waterlogging tolerance in tropical maize (Zea mays L.). Crop J 9: 1438–1448. https://doi.org/10.1016/j.cj.2021.03.012

Das RR, Vinayan MT, Seetharam K et al. (2025) Resilient yet productive: maize that can thrive under stress and in optimal conditions. Front Plant Sci 16:1690230. https://doi.org/10.3389/fpls.2025.1690230

Desta ZA, Ortiz R (2014) Genomic selection: genome-wide prediction in plant improvement. Trends Plant Sci 19:592–601. https://doi.org/10.1016/j.tplants.2014.05.006

Dias KODG, Gezan SA, Guimarães CT et al (2018) Improving accuracies of genomic predictions for drought tolerance in maize by joint modeling of additive and dominance effects in multi-environment trials. Heredity 121:24–37. https://doi.org/10.1038/s41437-018-0053-6

Edriss V, Gao Y, Zhang X et al (2017) Genomic prediction in a large African maize population. Crop Sci 57:2361–2371. https://doi.org/10.2135/cropsci2016.08.0715

Egea-Gilabert C, Pagnotta MA, Tripodi P (2021) Genotype × environment interactions in crop breeding. Agronomy 11:1644. https://doi.org/10.3390/agronomy11081644

El-Soda M, Malosetti M, Zwaan BJ et al (2014) Genotype × environment interaction QTL mapping in plants: lessons from Arabidopsis. Trends Plant Sci 19:390–398. https://doi.org/10.1016/j.tplants.2014.01.001

Fábián A, Sáfrán E, Szabó-Eitel G et al (2019) Stigma functionality and fertility are reduced by heat and drought co-stress in wheat. Front Plant Sci 10. https://doi.org/10.3389/fpls.2019.00244

Falconer DS, Mackay TFC (1996) Introduction to quantitative genetics, 4th edn. Longman

Filho CCF, Andrade MHML, Nunes JAR et al (2023) Genomic prediction for complex traits across multiples harvests in alfalfa (Medicago sativa L.) is enhanced by enviromics. Plant Genome 16: e20306. https://doi.org/10.1002/tpg2.20306

Gauch HG Jr, Zobel RW (1997) Identifying mega-environments and targeting genotypes. Crop Sci 37:311–326. https://doi.org/10.2135/cropsci1997.0011183X003700020002x

Gaur PM, Samineni S, Thudi M et al (2019) Integrated breeding approaches for improving drought and heat adaptation in chickpea (Cicer arietinum L.). Plant Breed 138:389–400. https://doi.org/10.1111/pbr.12641

Hajjar R, Hodgkin T (2007) The use of wild relatives in crop improvement: a survey of developments over the last 20 years. Euphytica 156:1–13. https://doi.org/10.1007/s10681-007-9363-0

Hallajian MT, Ebadi AA, Kordrostami M (2024) Advancing rice breeding for drought tolerance: a comprehensive study of traditional and mutant lines through agronomic performance and drought tolerance indices. BMC Plant Biol 24:1087. https://doi.org/10.1186/s12870-024-05771-5

Hatfield JL, Prueger JH (2015) Temperature extremes: effect on plant growth and development. Weather and Climate Extremes 10:4–10. https://doi.org/10.1016/j.wace.2015.08.001

Hayashi T, Iwata H (2013) A Bayesian method and its variational approximation for prediction of genomic breeding values in multiple traits. BMC Bioinformatics 14:34. https://doi.org/10.1186/1471-2105-14-34

Hayat K, Asad SA, Wahid N et al (2025) Drought resilience in oilseed crops: the role of phytohormones in stress mitigation. In: Abdel Latef AAH (ed) Oilseed crops under abiotic stress: mitigation strategies and future perspectives. Springer Nature, Singapore, pp 321–348. https://doi.org/10.1007/978-981-96-8346-8_10

Heslot N, Yang HP, Sorrells M, Jannink JL (2012) Genomic selection in plant breeding: a comparison of models. Crop Sci 52:146–160. https://doi.org/10.2135/cropsci2011.06.0297
Heslot N, Jannink JL, Sorrells M (2015) Perspectives for genomic selection applications and research in plants. Crop Sci 55:1–12. https://doi.org/10.2135/cropsci2014.03.0249
Hickey LT, N Hafeez A, Robinson H et al (2019) Breeding crops to feed 10 billion. Nat Biotechnol 37:744–754. https://doi.org/10.1038/s41587-019-0152-9
Hoegh-Guldberg O, Jacob D, Taylor M et al (2018) Impacts of 1.5°C global warming on natural and human systems. In: Global warming of 1.5°C. An IPCC special report. Cambridge University Press, Cambridge/New York, pp 175–312. https://doi.org/10.1017/9781009157940.005
Hu H, Rincent R, Runcie DE (2025) MegaLMM improves genomic predictions in new environments using environmental covariates. Genetics 229:iyae171. https://doi.org/10.1093/genetics/iyae171
Ismail AM, Singh US, Singh S et al (2013) The contribution of submergence-tolerant (Sub1) rice varieties to food security in flood-prone rainfed lowland areas in Asia. Field Crop Res 152:83–93. https://doi.org/10.1016/j.fcr.2013.01.007
Jagadish SVK, Murty MVR, Quick WP (2015) Rice responses to rising temperatures—challenges, perspectives and future directions. Plant Cell Environ 38:1686–1698. https://doi.org/10.1111/pce.12430
Jamil A, Riaz S, Ashraf M, Foolad MR (2011) Gene expression profiling of plants under salt stress. Crit Rev Plant Sci 30:435–458. https://doi.org/10.1080/07352689.2011.605739
Jarquín D, Crossa J, Lacaze X et al (2014) A reaction norm model for genomic selection using high-dimensional genomic and environmental data. Theor Appl Genet 127:595–607. https://doi.org/10.1007/s00122-013-2243-1
Jiang Z, van Zanten M, Sasidharan R (2025) Mechanisms of plant acclimation to multiple abiotic stresses. Commun Biol 8:655. https://doi.org/10.1038/s42003-025-08077-w
Kelly AM, Smith AB, Eccleston JA, Cullis BR (2007) The accuracy of varietal selection using factor analytic models for multi-environment plant breeding trials. Crop Sci 47(3):1063–1070. https://doi.org/10.2135/cropsci2006.08.0540
Kopeć P (2024) Climate change—the rise of climate-resilient crops. Plants 13:490. https://doi.org/10.3390/plants13040490
Kumar A, Dixit S, Ram T et al (2014) Breeding high-yielding drought-tolerant rice: genetic variations and conventional and molecular approaches. J Exp Bot 65:6265–6278. https://doi.org/10.1093/jxb/eru363
Kumar R, Kaur Y, Das AK et al (2023) Stability of maize hybrids under drought, rainfed and optimum field conditions revealed through GGE analysis. Ind J Genetic Plant Breed 83:499–507. https://doi.org/10.31742/ISGPB.83.4.6
Kusmec A, Attigala L, Dai X et al (2023) A genetic tradeoff for tolerance to moderate and severe heat stress in US hybrid maize. PLoS Genet 19:e1010799. https://doi.org/10.1371/journal.pgen.1010799
Lesk C, Rowhani P, Ramankutty N (2016) Influence of extreme weather disasters on global crop production. Nature 529:84–87. https://doi.org/10.1038/nature16467
Li Z, Howell SH (2021) Heat stress responses and thermotolerance in maize. Int J Mol Sci 22:948. https://doi.org/10.3390/ijms22020948
Li C, Camac J, Robinson A et al (2025) Predicting changes in agricultural yields under climate change scenarios and their implications for global food security. Sci Rep 15:2858. https://doi.org/10.1038/s41598-025-87047-y
Lobell DB, Sibley A, Ivan Ortiz-Monasterio J (2012) Extreme heat effects on wheat senescence in India. Nature Clim Change 2:186–189. https://doi.org/10.1038/nclimate1356
Lobell DB, Hammer GL, Chenu K et al (2015) The shifting influence of drought and heat stress for crops in Northeast Australia. Glob Chang Biol 21:4115–4127. https://doi.org/10.1111/gcb.13022

Makanza R, Zaman-Allah M, Cairns JE et al (2018) High-throughput method for ear phenotyping and kernel weight estimation in maize using ear digital imaging. Plant Methods 14:49. https://doi.org/10.1186/s13007-018-0317-4

Malosetti M, Ribaut J-M, van Eeuwijk FA (2013) The statistical analysis of multi-environment data: modeling genotype-by-environment interaction and its genetic basis. Front Physiol 4. https://doi.org/10.3389/fphys.2013.00044

Mekonnen A, Tadesse A, Melak W (2023) Effect of rhizobial inoculants and vermicompost application on growth and yield of faba bean (Vicia faba L.) in Arsi Zone, Southeastern Highlands of Ethiopia. Adv Crop Sci Tech 11:639. https://doi.org/10.4172/2329-8863.1000639

Mersmann K, Gray ET, Subba Rao M, Jägermeyr J (2021) Climate change could affect global agriculture within 10 years. NASA Scientific Visualization Studio. https://svs.gsfc.nasa.gov/13979

Messina CD, Technow F, Tang T et al (2018) Leveraging biological insight and environmental variation to improve phenotypic prediction: integrating crop growth models (CGM) with whole genome prediction (WGP). Eur J Agron 100:151–162. https://doi.org/10.1016/j.eja.2018.01.007

Meuwissen TH, Hayes BJ, Goddard ME (2001) Prediction of total genetic value using genome-wide dense marker maps. Genetics 157:1819–1829. https://doi.org/10.1093/genetics/157.4.1819

Mickelbart MV, Hasegawa PM, Bailey-Serres J (2015) Genetic mechanisms of abiotic stress tolerance that translate to crop yield stability. Nat Rev Genet 16:237–251. https://doi.org/10.1038/nrg3901

Millet EJ, Kruijer W, Coupel-Ledru A et al (2019) Genomic prediction of maize yield across European environmental conditions. Nat Genet 51:952–956. https://doi.org/10.1038/s41588-019-0414-y

Mittler R (2006) Abiotic stress, the field environment and stress combination. Trends Plant Sci 11:15–19. https://doi.org/10.1016/j.tplants.2005.11.002

Mohammadi R, Geravandi M (2024) Multi-trait selection for agronomic performance and drought tolerance among durum wheat genotypes evaluated under rainfed and irrigated environments. Exp Agric 60:e3. https://doi.org/10.1017/S0014479723000273

Monneveux P, Ribaut JM (eds) (2011) Drought phenotyping in crops: from theory to practice. CGIAR Generation Challenge Programme; International Maize and Wheat Improvement Center (CIMMYT), Texcoco. ISBN 978-970-648-178-8. 2 vols. https://hdl.handle.net/10568/67112

Monteleone B, Borzí I, Bonaccorso B, Martina M (2023) Quantifying crop vulnerability to weather-related extreme events and climate change through vulnerability curves. Nat Hazards 116:2761–2796. https://doi.org/10.1007/s11069-022-05791-0

Montesinos-López OA, Montesinos-López A, Crossa J et al (2016) A genomic Bayesian multi-trait and multi-environment model. G3 Genes|Genomes|Genetics 6:2725–2744. https://doi.org/10.1534/g3.116.032359

Neupane D, Adhikari P, Bhattarai D et al (2022) Does climate change affect the yield of the top three cereals and food security in the world? Earth 3:45–71. https://doi.org/10.3390/earth3010004

Noor JJ, Vinayan MT, Umar S et al (2019) Morpho-physiological traits associated with heat stress tolerance in tropical maize (Zea mays L.) at reproductive stage. Aust J Crop Sci 13:536–545. https://doi.org/10.21475/ajcs.19.13.04.p1448

Novick KA, Ficklin DL, Grossiord C et al (2024) The impacts of rising vapour pressure deficit in natural and managed ecosystems. Plant Cell Environ 47:3561–3589. https://doi.org/10.1111/pce.14846

Olivoto T, Lúcio ADC, da Silva JAG et al (2019) Mean performance and stability in multi-environment trials II: selection based on multiple traits. Agron J 111:2961–2969. https://doi.org/10.2134/agronj2019.03.0221

Onogi A, Watanabe M, Mochizuki T et al (2016) Toward integration of genomic selection with crop modelling: the development of an integrated approach to predicting rice heading dates. Theor Appl Genet 129:805–817. https://doi.org/10.1007/s00122-016-2667-5

Passioura JB (2012) Phenotyping for drought tolerance in grain crops: when is it useful to breeders? Funct Plant Biol 39:851–859. https://doi.org/10.1071/FP12079

Piepho H-P (2022) Prediction of and for new environments: what's your model? Mol Plant 15:581–582. https://doi.org/10.1016/j.molp.2022.01.018

Piepho HP, Möhring J (2005) Best linear unbiased prediction of cultivar effects for subdivided target regions. Crop Sci 45:1151–1159. https://doi.org/10.2135/cropsci2004.0398

Piepho HP, Möhring J, Melchinger AE, Büchse A (2008) BLUP for phenotypic selection in plant breeding and variety testing. Euphytica 161:209–228. https://doi.org/10.1007/s10681-007-9449-8

Pillai AJ, Walia P (2024) Heat stress in Indian mustard (Brassica juncea L.): a critical review of impacts and adaptation strategies. Plant Cell Biotechnol Mol Biol 25:1–11. https://doi.org/10.56557/pcbmb/2024/v25i5-68673

Poudel P, Naidenov B, Chen C et al (2023) Integrating genomic prediction and genotype specific parameter estimation in ecophysiological models: overview and perspectives. in silico Plants 5: diad007. https://doi.org/10.1093/insilicoplants/diad007

Prabhu KR, Kumar A, Yumkhaibam RS et al (2023) A review on conventional and modern breeding approaches for developing climate resilient crop varieties: NA. J Appl Nat Sci 15: 987–997. https://doi.org/10.31018/jans.v15i3.4653

Prasanna BM, Nair SK, Babu R et al (2020) Increasing genetic gains in maize in stress-prone environments of the tropics. In: Kole C (ed) Genomic designing of climate-smart cereal crops. Springer International Publishing, Cham, pp 97–132. https://doi.org/10.1007/978-3-319-93381-8_3

Prasanna BM, Cairns JE, Zaidi PH et al (2021) Beat the stress: breeding for climate resilience in maize for the tropical rainfed environments. Theor Appl Genet 134:1729–1752. https://doi.org/10.1007/s00122-021-03773-7

Prasanna BM, Burgueño J, Beyene Y et al (2022) Genetic trends in CIMMYT's tropical maize breeding pipelines. Sci Rep 12:20110. https://doi.org/10.1038/s41598-022-24536-4

Raffo MA, Jensen J (2023) Gene × gene and genotype × environment interactions in wheat. Crop Sci 63:1779–1793. https://doi.org/10.1002/csc2.20986

Rameshwaran P, Qadir M, Ragab R et al (2016) Tolerance of faba bean, chickpea and lentil to salinity: accessions' salinity response functions. Irrig Drain 65:49–60. https://doi.org/10.1002/ird.1922

Razzaq A, Saleem F, Wani SH et al (2021) De-novo domestication for improving salt tolerance in crops. Front Plant Sci 12:681367. https://doi.org/10.3389/fpls.2021.681367

Reckling M, Ahrends H, Chen T-W et al (2021) Methods of yield stability analysis in long-term field experiments. A review. Agron Sustain Dev 41:27. https://doi.org/10.1007/s13593-021-00681-4

Renzi JP, Coyne CJ, Berger J et al (2022) How could the use of crop wild relatives in breeding increase the adaptation of crops to marginal environments? Front Plant Sci 13:886162. https://doi.org/10.3389/fpls.2022.886162

Reynolds MP, Pierre CS, Saad ASI et al (2007) Evaluating potential genetic gains in wheat associated with stress-adaptive trait expression in elite genetic resources under drought and heat stress. Crop Sci 47:S-172–S-189. https://doi.org/10.2135/cropsci2007.10.0022IPBS

Reynolds M, Chapman S, Crespo-Herrera L et al (2020) Breeder friendly phenotyping. Plant Sci 295:110396. https://doi.org/10.1016/j.plantsci.2019.110396

Reynolds MP, Lewis JM, Ammar K et al (2021) Harnessing translational research in wheat for climate resilience. J Exp Bot 72:5134–5157. https://doi.org/10.1093/jxb/erab256

Rogers AR, Dunne JC, Romay C et al (2021) The importance of dominance and genotype-by-environment interactions on grain yield variation in a large-scale public cooperative maize experiment | G3 Genes|Genomes|Genetics | 11:jkaa050. https://doi.org/10.1093/g3journal/jkaa050

Runcie DE, Qu J, Cheng H et al (2021) MegaLMM: mega-scale linear mixed models for genomic predictions with thousands of traits. Genome Biol 22:213. https://doi.org/10.1186/s13059-021-02416-w

Sarzaeim P, Muñoz-Arriola F (2024) A method to estimate climate drivers of maize yield predictability leveraging genetic-by-environment interactions in the US and Canada. Agronomy 14: 733. https://doi.org/10.3390/agronomy14040733

Schneider HM, Klein SP, Hanlon MT et al (2020) Genetic control of root architectural plasticity in maize. J Exp Bot 71:3185–3197. https://doi.org/10.1093/jxb/eraa084

Seetharam K, Kuchanur PH, Koirala KB et al (2021) Genomic regions associated with heat stress tolerance in tropical maize (Zea mays L.). Sci Rep 11:13730. https://doi.org/10.1038/s41598-021-93061-7

Setter TL, Waters I (2003) Review of prospects for germplasm improvement for waterlogging tolerance in wheat, barley and oats. Plant Soil 253:1–34. https://doi.org/10.1023/A:1024573305997

Shahhosseini M, Hu G, Archontoulis SV (2020) Forecasting corn yield with machine learning ensembles. Front Plant Sci 11. https://doi.org/10.3389/fpls.2020.01120

Shahhosseini M, Hu G, Huber I, Archontoulis SV (2021) Coupling machine learning and crop modeling improves crop yield prediction in the US Corn Belt. Sci Rep 11:1606. https://doi.org/10.1038/s41598-020-80820-1

Shahi D, Guo J, Pradhan S et al (2022) Multi-trait genomic prediction using in-season physiological parameters increases prediction accuracy of complex traits in US wheat. BMC Genomics 23: 298. https://doi.org/10.1186/s12864-022-08487-8

Shiferaw B, Prasanna BM, Hellin J, Bänziger M (2011) Crops that feed the world 6. Past successes and future challenges to the role played by maize in global food security. Food Sec 3:307–327. https://doi.org/10.1007/s12571-011-0140-5

Singamsetti A, Shahi JP, Zaidi PH et al (2021) Genotype × environment interaction and selection of maize (Zea mays L.) hybrids across moisture regimes. Field Crop Res 270:108224. https://doi.org/10.1016/j.fcr.2021.108224

Singamsetti A, Zaidi PH, Seetharam K et al (2023a) Genetic gains in tropical maize hybrids across moisture regimes with multi-trait-based index selection. Front Plant Sci 14:1147424. https://doi.org/10.3389/fpls.2023.1147424

Singamsetti A, Shahi JP, Zaidi PH, Seetharam K, Mahawar M (2023b) Identification of waterlogging stress tolerant maize (Zea mays L.) hybrids using different selection indices. Agric Res J 60:126–134. https://doi.org/10.5958/2395-146X.2023.00020.0

Singh AK, Ganapathysubramanian B, Sarkar S, Singh A (2018) Deep learning for plant stress phenotyping: trends and future perspectives. Trends Plant Sci 23:883–898. https://doi.org/10.1016/j.tplants.2018.07.004

Singh VJ, Vinod KK, Krishnan SG, Singh AK (2021) Rice adaptation to climate change: opportunities and priorities in molecular breeding. In: Molecular breeding for rice abiotic stress tolerance and nutritional quality. John Wiley & Sons, Ltd, pp 1–25. https://doi.org/10.1002/9781119633174.ch1

Smith A, Cullis B, Thompson R (2001) Analyzing variety by environment data using multiplicative mixed models and adjustments for spatial field trend. Biometrics 57:1138–1147. https://doi.org/10.1111/j.0006-341X.2001.01138.x

Smith AB, Cullis BR, Thompson R (2005) The analysis of crop cultivar breeding and evaluation trials: an overview of current mixed model approaches. J Agric Sci 143:449–462. https://doi.org/10.1017/S0021859605005587

Steinke J, van Etten J (2017) Gamification of farmer-participatory priority setting in plant breeding: Design and validation of "AgroDuos". J Crop Improv 31:356–378. https://doi.org/10.1080/15427528.2017.1303801

Subba Rao M, Jägermeyr J (2022) Impact of climate change on global agricultural yields. NASA Scientific Visualization Studio. https://svs.gsfc.nasa.gov/4974

Tarekegne A, Wegary D, Cairns JE et al (2024) Genetic gains in early maturing maize hybrids developed by the International Maize and Wheat Improvement Center in Southern Africa during 2000–2018. Front Plant Sci 14:1321308. https://doi.org/10.3389/fpls.2023.1321308

Thomas K, Hardy RD, Lazrus H et al (2019) Explaining differential vulnerability to climate change: a social science review. WIREs Clim Change 10:e565. https://doi.org/10.1002/wcc.565

Thomson MJ, de Ocampo M, Egdane J et al (2010) Characterizing the Saltol quantitative trait locus for salinity tolerance in rice. Rice 3:148–160. https://doi.org/10.1007/s12284-010-9053-8

Trevisan BA, Junqueira VS, Florencio BM et al (2025) A framework for building enviromics matrices in mixed models. arXiv (Cornell University). https://doi.org/10.48550/arxiv.2501.04147

UN-Water (2013) Climate change (WWW document). https://www.unwater.org/sites/default/files/app/uploads/2019/10/UN_Water_PolicyBrief_ClimateChange_Water.pdf

Upadhyaya HD, Kashiwagi J, Varshney RK et al (2012) Phenotyping chickpeas and pigeonpeas for adaptation to drought. Front Physiol 3:179. https://doi.org/10.3389/fphys.2012.00179

van Dijk ADJ, Kootstra G, Kruijer W, de Ridder D (2021) Machine learning in plant science and plant breeding. iScience 24:101890. https://doi.org/10.1016/j.isci.2020.101890

van Eeuwijk FA, Bustos-Korts DV, Malosetti M (2016) What should students in plant breeding know about the statistical aspects of genotype × environment interactions? Crop Sci 56:2119–2140. https://doi.org/10.2135/cropsci2015.06.0375

van Etten J, de Sousa K, Aguilar A et al (2019) Crop variety management for climate adaptation supported by citizen science. Proc Natl Acad Sci USA 116:4194–4199. https://doi.org/10.1073/pnas.1813720116

Vargas M, Glaz B, Alvarado G et al (2015) Analysis and interpretation of interactions in agricultural research. Agron J 107:748–762. https://doi.org/10.2134/agronj13.0405

Varshney RK (2016) Exciting journey of 10 years from genomes to fields and markets: some success stories of genomics-assisted breeding in chickpea, pigeonpea and groundnut. Plant Sci 242:98–107. https://doi.org/10.1016/j.plantsci.2015.09.009

Varshney RK, Singh VK, Kumar A et al (2018) Can genomics deliver climate-change ready crops? Curr Opin Plant Biol 45:205–211. https://doi.org/10.1016/j.pbi.2018.03.007

Varshney RK, Bohra A, Roorkiwal M et al (2021) Fast-forward breeding for a food-secure world. Trends Genet 37:1124–1136. https://doi.org/10.1016/j.tig.2021.08.002

Vinayan MT, Zaidi PH, Seetharam K et al (2020) Genotype-by-environment interaction effects under heat stress in tropical maize. Agronomy 10:1998. https://doi.org/10.3390/agronomy10121998

Vivek BS, Krishna GK, Vengadessan V et al (2017) Use of genomic estimated breeding values results in rapid genetic gains for drought tolerance in maize. Plant Genome 10: eplantgenome2016.07.0070. https://doi.org/10.3835/plantgenome2016.07.0070

Vogel E, Donat MG, Alexander LV et al (2019) The effects of climate extremes on global agricultural yields. Environ Res Lett 14:054010. https://doi.org/10.1088/1748-9326/ab154b

Wang Q, Guan Y, Wu Y et al (2008) Overexpression of a rice OsDREB1F gene increases salt, drought, and low temperature tolerance in both Arabidopsis and rice. Plant Mol Biol 67:589–602. https://doi.org/10.1007/s11103-008-9340-6

Wang Y, Wang L, Zhou J et al (2019) Research Progress on heat stress of rice at flowering stage. Rice Sci 26:1–10. https://doi.org/10.1016/j.rsci.2018.06.009

Wang J, Liu L, He K et al (2025) Accurate genomic prediction for grain yield and grain moisture content of maize hybrids using multi-environment data. J Integr Plant Biol 67:1379–1394. https://doi.org/10.1111/jipb.13857

Westhues CC, Mahone GS, da Silva S et al (2021) Prediction of maize phenotypic traits with genomic and environmental predictors using gradient boosting frameworks. Front Plant Sci 12: 699589. https://doi.org/10.3389/fpls.2021.699589

Winans ND, Fonseca JMO, Perumal R et al (2024) Envirotyping can increase genomic prediction accuracy of new environments in grain sorghum trials depending on mega-environment. Crop Sci 64:2519–2533. https://doi.org/10.1002/csc2.21213

Wollmer A-C, Pitann B, Mühling K h (2018) Waterlogging events during stem elongation or flowering affect yield of oilseed rape (Brassica napus L.) but not seed quality. J Agron Crop Sci 204:165–174. https://doi.org/10.1111/jac.12244

Wu C, Cui K, Wang W et al (2017) Heat-induced cytokinin transportation and degradation are associated with reduced panicle cytokinin expression and fewer spikelets per panicle in Rice. Front Plant Sci 8:371. https://doi.org/10.3389/fpls.2017.00371

Xie J, Zhang D, Jin N et al (2025) Coupling crop growth models and machine learning for scalable winter wheat yield estimation across major wheat regions in China. Agric For Meteorol 372: 110687. https://doi.org/10.1016/j.agrformet.2025.110687

Xiong W, Reynolds M, Montes C, Crossa J, Snapp S, Akin B, Keser M, Ozdemir F, Li H, He Z, Wang D, Chen F (2024) New wheat breeding paradigms for a warming climate. Nat Clim Chang 14:869–875. https://doi.org/10.1038/s41558-024-02069-0

Xu Y (2016) Envirotyping for deciphering environmental impacts on crop plants. Theor Appl Genet 129:653–673. https://doi.org/10.1007/s00122-016-2691-5

Yan W, Kang MS (2002) GGE Biplot analysis: a graphical tool for breeders, geneticists, and agronomists. CRC Press, Boca Raton. https://doi.org/10.1201/9781420040371

Yan W, Nilsen KT, Beattie A (2023) Mega-environment analysis and breeding for specific adaptation. Crop Sci 63:480–494. https://doi.org/10.1002/csc2.20895

Yang Y, Yu J, Qian Q, Shang L (2022) Enhancement of heat and drought stress tolerance in rice by genetic manipulation: a systematic review. Rice 15:67. https://doi.org/10.1186/s12284-022-00614-z

Yang Q, Du T, Li N et al (2023) Bibliometric analysis on the impact of climate change on crop Pest and disease. Agronomy 13:920. https://doi.org/10.3390/agronomy13030920

Yuan X, Li S, Chen J et al (2024) Impacts of global climate change on agricultural production: a comprehensive review. Agronomy 14:1360. https://doi.org/10.3390/agronomy14071360

Yue H, Wei J, Xie J et al (2021) A study on genotype-by-environment interaction analysis for agronomic traits of maize genotypes across Huang-Huai-Hai region in China. Phyton 91:57–81. https://doi.org/10.32604/phyton.2022.017308

Yue H, Olivoto T, Bu J et al (2022) Multi-trait selection for mean performance and stability of maize hybrids in mega-environments delineated using envirotyping techniques. Front Plant Sci 13:1030521. https://doi.org/10.3389/fpls.2022.1030521

Zaidi PH (2019) Management of drought stress in field phenotyping. CIMMYT, Mexico. https://hdl.handle.net/10883/19998

Zaidi PH, Rafique S, Singh NN (2003) Response of maize (Zea mays L.) genotypes to excess soil moisture stress: morpho-physiological effects and basis of tolerance. Eur J Agron 19:383–399. https://doi.org/10.1016/S1161-0301(02)00090-4

Zaidi PH, Vinayan MT, Seetharam K (2016a) Phenotyping for abiotic stress tolerance in maize: waterlogging stress. In: A field manual. CIMMYT, Hyderabad

Zaidi PH, Zaman-Allah M, Trachsel S, Seetharam K, Cairns JE, Vinayan MT (2016b) Phenotyping for abiotic stress tolerance in maize—heat stress. In: A field manual. CIMMYT, Hyderabad

Zaidi PH, Nguyen T, Ha DN et al (2020) Stress-resilient maize for climate-vulnerable ecologies in the Asian tropics. Aust J Crop Sci:1264–1274. https://doi.org/10.21475/ajcs.20.14.08.p2405

Zaidi PH, Thayil Vinayan M, Nair SK et al (2023) Heat-tolerant maize for rainfed hot, dry environments in the lowland tropics: from breeding to improved seed delivery. The Crop Journal 11:986–1000. https://doi.org/10.1016/j.cj.2023.06.008

Zaman-Allah M, Vergara O, Araus JL et al (2015) Unmanned aerial platform-based multi-spectral imaging for field phenotyping of maize. Plant Methods 11:35. https://doi.org/10.1186/s13007-015-0078-2

Zhang X, Pérez-Rodríguez P, Semagn K et al (2015) Genomic prediction in biparental tropical maize populations in water-stressed and well-watered environments using low-density and GBS SNPs. Heredity 114:291–299. https://doi.org/10.1038/hdy.2014.99

Zhang H, Mittal N, Leamy LJ et al (2017) Back into the wild—apply untapped genetic diversity of wild relatives for crop improvement. Evol Appl 10:5–24. https://doi.org/10.1111/eva.12434

Zobel RW, Wright MJ, Gauch HG Jr (1988) Statistical analysis of a yield trial. Agron J 80:388–393. https://doi.org/10.2134/agronj1988.00021962008000030002x

Impact of Biotic Stresses Influence on Wheat G × E Interactions and Yield Performance

7

Pradeep Kumar Bhati, Velu Govindan, Manish Kumar Vishwakarma, and Arun Kumar Joshi

Abstract

Due to the increasing global food demand under climate change scenario, there is an urgent need for the development of climate-resilient, high-yielding, disease-resistant crop varieties that can perform well under diverse agroecological conditions. Genotype × environment (G × E) interactions significantly influence crop performance, making it essential to integrate G × E analysis into plant breeding programs. In this chapter, we explore the importance of incorporating G × E analysis, mainly in the context of biotic stresses, into multi-environment trials (METs) for crops like wheat, rice, and maize. Several case studies demonstrate how neglecting biotic stress in G × E assessments can lead to inaccurate conclusions about genotype adaptability and stability. We discuss strategies to integrate disease and pest pressures into trial designs, balance yield potential with resistance, and apply statistical tools like genotype and genotype × environment (GGE) biplot and additive main effects and multiplicative interaction (AMMI) models for precise selection. This chapter also highlights how advances in marker-assisted and genomic selection, combined with participatory breeding and pathogen monitoring, can be leveraged to enhance durability of resistance. By adopting evolution-informed breeding strategies and expanding METs to include hotspot environments, crop breeders can more effectively address the

P. K. Bhati (✉)
International Maize and Wheat Improvement Center (CIMMYT), New Delhi, India

Borlaug Institute for South Asia (BISA), New Delhi, India
e-mail: pk.bhati@cgiar.org

V. Govindan
International Maize and Wheat Improvement Center (CIMMYT), Texcoco, Mexico

M. K. Vishwakarma · A. K. Joshi
Borlaug Institute for South Asia (BISA), New Delhi, India

R. Mohammadi et al. (eds.), *Genotype x Environment Interactions and its Implications for Plant Breeding*, https://doi.org/10.1007/978-981-95-5664-9_7

challenges posed by dynamic biotic stress factors, ultimately facilitating the development of climate-resilient, farmers preferred cultivars and promoting sustainable crop production under diverse and evolving agroecological conditions.

7.1 Introduction

In plant breeding, genotype × environment (G × E) interaction is a fundamental challenge because it captures how different genotypes perform in different variable environments (Kang 1988; Dickerson 1962; Baker 1988; de Leon et al. 2016). It is crucial to differentiate G × E from phenotypic plasticity, which is the ability of a single genotype to exhibit various phenotypes in response to changes in the environment (Schlichting 1986; Grogan et al. 2016). G × E involves assessing at least two or more genotypes across a range of environments, while phenotypic plasticity is the ability of a plant to modify its growth, development, and metabolism in response to changes in its environment (Kang 1988). Moreover, genotype–environment correlation which occurs when genotypes are associated non-randomly with specific environmental conditions, beneficial or detrimental, is theoretically different from G × E (Crow 1986; Doolittle 1987). In most economically important quantitative characters, such as biomass, plant height, and grain yield, G × E interactions occur frequently. To estimate the performance, stability, and adaptability of varieties over different environments and seasons of growth, there is a need to understand these interactions and apply yield stability analyses. To facilitate selecting cultivars suitable for widespread adoption and mass production, these analyses attempt to find genotypes that perform consistently and better across a variety of conditions. Additive × environment, dominance × environment, and epistatic × environment components of the total G × E variance can be dissected. Successful hybridization schemes to be designed, superior combinations of parents to be discovered, and selection strategies optimized all rest upon knowledge about the relative contributions from each of these components. Though continued efforts to produce widely adapted cultivars are still a prime objective of breeding, G × E interactions tend to complicate early-generation and advanced selection phases and are commonly dealt with as nuisance effects to be reduced (Rogers et al. 2021). Breeding programs usually start by testing large numbers of genotypes in a few environments, and then multilocation testing of some selected entries is carried out to cover G × E effects.

To analyze G × E interactions and determine genotypes of broad or narrow adaptation, several statistical models have been utilized. Some of the most commonly used are the additive main effects and multiplicative interaction (AMMI) model (Gauch 2006) and the genotype and genotype × environment (GGE) biplot technique (Yan et al. 2007). The AMMI model combines analysis of variance with principal component analysis (PCA) to separate additive and multiplicative sources of variation, while the GGE biplot considers only genotype and G × E effects through PCA. Both approaches have been shown to be useful for projecting and explaining intricate MET data. Significantly, the GGE biplot provides a more intuitive graphical display of mega-environment relationships, particularly through its "which-won-where" functionality, enabling the selection of genotypes for target environments (Yan et al. 2007; Akter et al. 2015). Knowing G × E is critical for the discovery of genotypes that not only have high potential yield but also stable in performance across changing environmental circumstances. Such interactions emanate from differential expressions of genotypes in response to environmental stimuli, which comprise climatic fluctuation, soil fertility, water regime, and biotic stresses (Mungra et al. 2011). Due to the infeasibility of screening genotypes in every target environment of interest, METs are a strategic means of indirect selection. But G × E interaction might restrict the precision of genotype performance prediction for target environments using test site information (Kumar et al. 2013). Therefore, selection of proper test location becomes important to increase the reliability of selection and enhance genetic gains (Yan et al. 2011).

Current breeding paradigms emphasize the need for "envirotyping" as an essential complement to genotyping and phenotyping. Envirotyping deepens the knowledge on how the environment impacts crop performance and facilitates the construction of predictive models for genotype-specific reaction norms (Cooper et al. 2016; Van Eeuwijk 2016, Xu 2016). The incorporation of high-throughput genotyping, phenotyping, and envirotyping in multi-environment trials (METs) is increasingly acknowledged as the key to dealing with G × E and improving the accuracy of whole-genome prediction in contemporary plant breeding programs.

In general, this chapter emphasizes the importance of genotype × environment (G × E) interaction analysis in contemporary plant breeding, specifically in its utility for enhancing the important traits under varied agroecological conditions. Incorporating strong statistical methods and optimally choosing test locations, breeding programs can better produce resistant cultivars adapted to present and future climatic change.

7.2 Nature of Biotic Stress-Induced G × E Interactions

Biotic stresses such as those caused by pathogens (fungi, bacteria, viruses, and nematodes), insect pests, and parasitic weeds have a strong impact on plant growth and yield. In contrast to abiotic stresses, which tend to be relatively consistent over wide geographic regions, biotic stresses have greater spatial and temporal variability. This dynamic behavior presents a special challenge to plant breeders since it leads to very large genotype × environment (G × E) interactions when assessing genotypic resistance in various environments.

Fundamental to biotic stress-induced G × E interactions are the triadic interaction between the host-pathogen, pest, genotype, and the environment. Resistance or susceptibility expression is not just determined by the genetic composition of the plant but is regulated by environmental factors like temperature, humidity, soil water, nutritional status, and crop growth stage. For example, high humidity and moderate temperatures frequently establish favorable conditions for foliar fungal diseases such as rusts, blights, and mildews, augmenting disease severity and thereby altering genotype performance rankings across sites.

One of the primary contributors to these interactions is environmental sensitivity of resistance genes. Numerous resistance genes particularly those for race-specific (qualitative) resistance can be expressed conditionally, either in terms of their expression stage at plant development or environmental cues. In a few instances, the resistance genes might not express themselves under heavy pathogen pressure or under specific temperature levels, although they may be present in the genotype. For instance, adult plant resistance (APR) genes against rust diseases in wheat can prove to be ineffectual under infection during early seasons due to favorable weather, resulting in premature failure of resistance. Likewise, temperature-sensitive genes, like Yr36 in wheat for yellow rust resistance, can become ineffective under warmer temperature conditions and impact genotype stability over environments.

In addition, virulence diversity and pathogen evolution between regions are also likely to amplify G × E interaction. A genotype can exhibit high resistance in one region where the pathogen population does not have corresponding virulence but can perform poorly when grown in a second region with a diverse range of virulence. Differential expression of resistance results in non-crossover and crossover-type G × E interaction, whereby relative performance among genotypes varies with environments. Pressure from insects, also, is affected by the environment. Life cycles, migration, and intensity of infestation of insect pests like aphids, stem borers, or pod borers differ as a function of seasonal patterns of temperature, levels of precipitation, and availability of host crop. Consequently, the expression of resistance characters like antibiosis, antixenosis, or tolerance may differ significantly across test environments, particularly in multi-location tests.

Also, biotic stresses tend to induce indirect physiological responses that are not necessarily directly observable as disease symptoms or insect injury but may undermine photosynthesis, assimilate translocation, and reproductive development. Such subclinical manifestations tend to be responsible for yield loss and make genotype performance interpretation challenging, particularly in assessing multifaceted attributes such as yield and stress tolerance under natural infestation conditions. These interactions pose a problem in breeding for resistance to biotic stresses: the natural instability and environment-inducibility of the expression of resistance. Therefore, breeders have to assess genotypes in many environments and over several years to verify consistency and durability of resistance expressions. It also highlights the need to employ integrated disease and pest management practices such as utilization of multi-genic (quantitative) resistance, pyramiding resistance genes, and marker-assisted selection (MAS) that will protect against the effect of G × E interactions under biotic stress.

The biotic stress-induced G × E interactions are multifaceted with a complicated web of genetic, environmental, and biological contributors. Knowledge on these interactions will be necessary for developing successful breeding schemes, crop performance stability, and truly resilient genotypes appropriate for variable and high-pressure agro-ecological environments (Fig. 7.1).

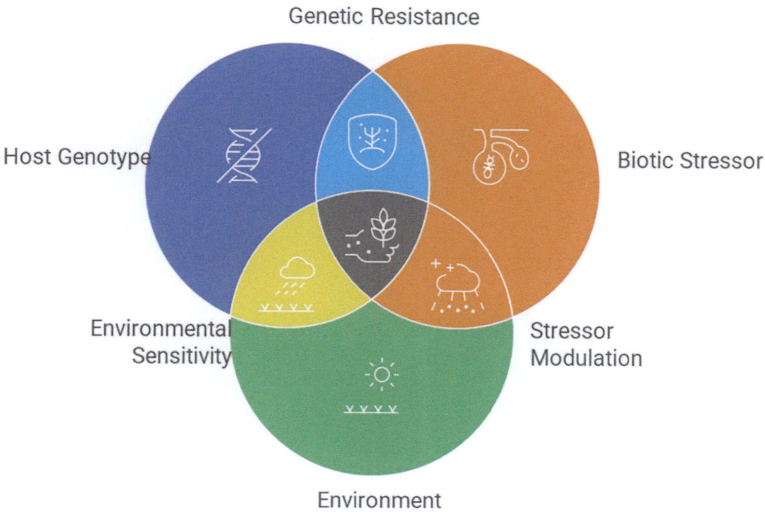

Fig. 7.1 The triadic nature of biotic stress-induced G × E interactions

7.2.1 How Biotic Stresses Modify G × E Outcomes

The presence or absence of biotic stress in any environment may drastically change a genotype's perceived performance. Under heavy disease or pest pressure, the genetic yield potential of susceptible genotypes might be drastically curtailed, while resistant genotypes have more stable productivity. On the other hand, in the absence of such stress, resistant genotypes, especially those with major resistance genes or polygenic resistance factors, might have no advantage and in certain situations even carry a yield penalty due to the physiological cost of expression of resistance.

For example, let us take two wheat genotypes: a high-yielding but stripe rust-susceptible one, and another moderately yielding but with APR (adult plant resistance) genes. The resistant genotype would outyield its susceptible counterpart in rust-endemic areas or epidemic years. Yet in disease-free environments, the susceptible high-yielding line would recover its superiority. These changes in relative performance between places or times are examples of crossover-type G × E interactions, whereby the rank of genotypes is altered as a function of the environmental biotic context (Fig. 7.2).

It highlights the masking phenomenon of biotic stress, whereby the true genetic capability of a genotype is masked unless subjected to appropriate conditions. Consequently, trials that do not incorporate biotic stress elements can underestimate the resilience or adaptability of resistant material and overestimate the stability of susceptible, high-yielding genotypes.

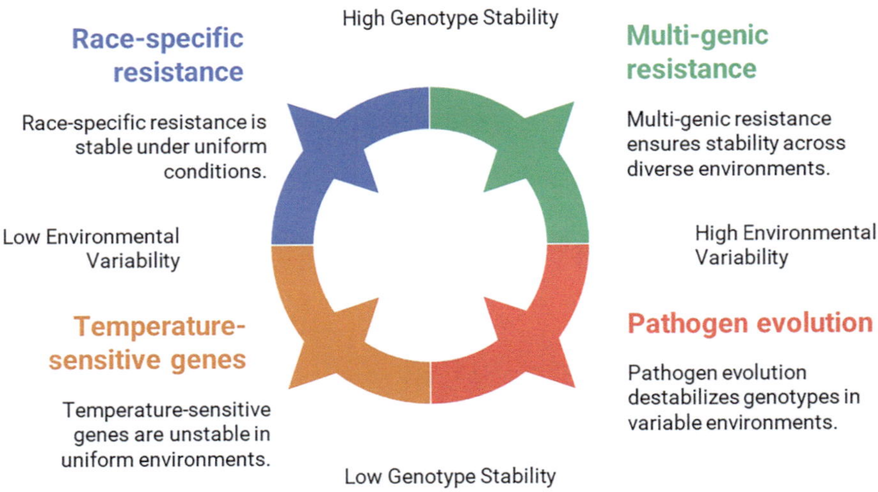

Fig. 7.2 Genotype × environment interactions in biotic stress

7.2.2 Evidence from Breeding Trials

Empirical data from multi-environment trials (METs) across various crops support the significant role of biotic stress in modulating G × E interactions. In wheat breeding, trials conducted across Indian hotspot locations (e.g., Ludhiana, Pantnagar, Wellington, Gurdaspur, Samastipur) consistently reveal that yield and stability are strongly influenced by the incidence of diseases such as yellow rust, leaf rust, and spot blotch. Genotypes with specific resistance genes often perform better in particular locations, while the same genotypes show no distinct advantage in rust-free or fungicide-protected trials location.

In maize, trials targeting regions prone to turcicum leaf blight, downy mildew, and stem borers demonstrate that hybrid performance rankings shift dramatically with pest and pathogen pressure. Similarly, rice varieties evaluated under natural or artificially induced bacterial blight, blast, or brown planthopper conditions show significant genotype re-ranking across locations and seasons. These outcomes clearly reveal that biotic stress is not an incidental variable but a critical environmental factor that must be integrated into MET designs.

Importantly, many breeding programs now incorporate artificial inoculation or conduct trials in known "hotspots" for specific diseases or pests to ensure meaningful evaluations. For example, in wheat, leaf rust, stem rust, and stripe rust nurseries in Kenya, Ethiopia, Mexico and India and Karnal bunt screening sites in Mexico and India provide controlled stress environments for precise phenotyping. Without such strategies, breeders risk recommending unstable cultivars whose performance collapses under real-world biotic stress scenarios.

7.2.3 G × E × Biotic Stress Interactions: A Conceptual Framework

Modern breeding increasingly recognizes that biotic stress cannot be fully separate from environmental variation. This has led to the evolution of a novel model: G × E × B interactions, where B explicitly represents biotic stresses. This expanded framework acknowledges that biotic stress is environment-dependent, that is, disease or pest prevalence varies across seasons, locations, and even microenvironments. The interaction among genotype, abiotic environment, and biotic stress creates complex performance patterns. Accounting for these three-way interactions improves the precision of selection and enhances prediction accuracy in genomic selection or whole-genome prediction models.

For instance, a rice variety carrying the *Xa21* gene for bacterial blight resistance may exhibit superior performance in endemic areas, especially under humid, warm conditions that promote disease development. In contrast, in upland or drier ecologies with low disease incidence, the performance difference between resistant and susceptible lines may be negligible or reversed. This variability in trait expression based on biological context is a classic case of G × E × B interaction.

Further, in crops such as chickpea and pigeon pea, major constraints like Fusarium wilt, pod borers, and root rot are strongly influenced by soil type, previous crop history, and prevailing weather conditions. Evaluations of genotypes in wilt-prone or pest-favorable zones consistently show substantial G × E × B interactions, which significantly affect the final varietal recommendations.

7.2.4 Implications for Breeding Strategy and Trial Design

G × E × B interactions have several practical implications for modern plant breeding. Some of the key strategies are as follows:

- *Targeted Environmental Design*: Selection of trial sites that reflect both stress-prone and non-stress environments to capture the entire range of genotype behavior.
- *Artificial Inoculation and Pest Pressure Simulation*: Standardization of biotic stress exposure to ensure fair evaluation and stable resistance identification.
- *Stability Analysis Tools*: Utilizing analytical models like AMMI, GGE biplots, and mixed-model methods with fixed and random biotic stress covariates.
- *Trait-Based Selection*: Assessing resistance as well as agronomic traits such as yield, maturity, and plant type to choose ideotypes that are resistant yet high-performing and resilient.
- *Integration of Molecular Tools*: Using marker-assisted or genomic selection to pyramid resistance genes and implement them in high-yielding backgrounds to buffer against environment- and biotic-driven instability (Fig. 7.3).

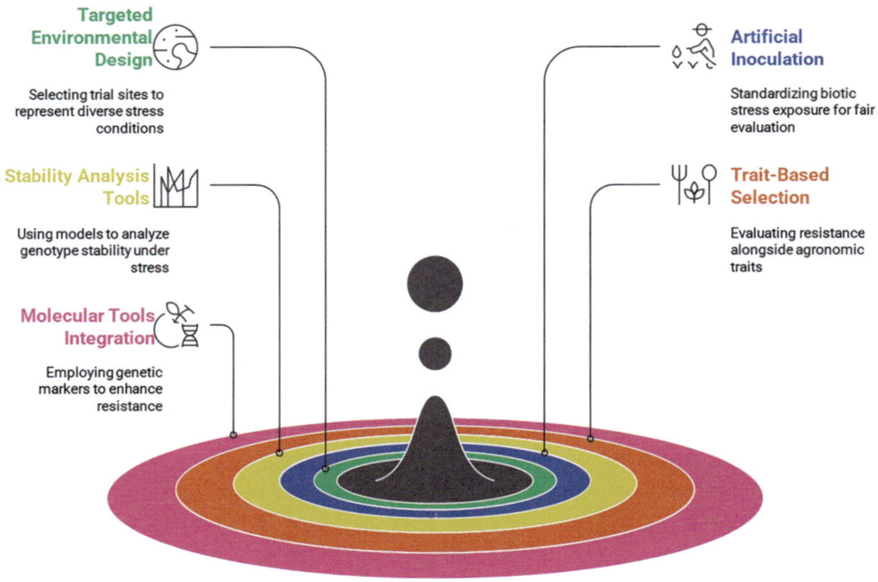

Fig. 7.3 Biotic stress interactions in plant breeding

7.3 Case Studies

In order to better understand the role of biotic stresses in G × E interactions, it is worth looking at actual case studies from leading crops where such interactions have been reported and managed by breeders and agronomists.

7.3.1 Wheat: Rust Diseases and G × E × B Interactions

Disease Overview and Environmental Dependency Stripe rust (*Puccinia striiformis f.* sp. *tritici*), leaf rust (*P. triticina*), and stem rust (*P. graminis f.* sp. *tritici*) are major biotic constraints in wheat production globally. Their expression is highly environment-dependent: stripe rust thrives under cool, moist conditions; stem rust under warmer, humid climates; and leaf rust under a broad range of warm, humid environments. This variability imposes inconsistent selective pressure, causing significant G × E interactions based on disease occurrence across locations and seasons (Liu et al. 2022).

7.3.1.1 Performance Shifts Due to Biotic Pressure

High-yielding but rust-susceptible genotypes frequently perform well in low rust pressure environments but collapse under epidemic conditions. Conversely, moderately yielding but resistant lines tend to have stable performance across variable conditions, documenting typical G × E interactions mediated by biotic stress (William et al. 2007; Singh et al. 2011a, b). Recent breakthrough breeding in International Maize and Wheat Improvement Center (CIMMYT) materials like "Mucuy" emphasizes the contributions of adult-plant resistance genes (i.e., Yr29/Lr46, YrMu) in ensuring uniformity across rust-susceptible environments (Liu et al. 2022).

7.3.1.2 CIMMYT's Strategy: Integrating Rust Screening in METs

CIMMYT integrates rust screening in their multi-environment trials (METs), with natural disease hotspots such as Toluca and Ludhiana (stripe rust) and Njoro (stem rust), along with artificial inoculation. Resistance loci such as Yr29, Lr34, Sr2, and others are integrated into elite backgrounds through MAS and genomic approaches to provide durable performance across G × E settings (Qureshi et al. 2023). This strategy enables CIMMYT to more effectively control G × E interactions because addition of biotic stress data enhances the accuracy of selection and minimizes the potential for encouraging genotypes that perform poorly under actual farmer conditions (Singh et al. 2008; He et al. 2019).

7.3.1.3 Implications for Breeding and Deployment

Incorporating biotic stress environments into METs helps to detect G × E × biotic stress (G × E × B) interactions, leading to more resilient cultivar recommendations for target regions. Furthermore, breeders are now increasingly applying marker-assisted selection and genomic selection approaches to integrate rust resistance genes (e.g., *Yr15*, *Lr34*, *Yr29*, *Yr36*, *Yr76, Sr2*) into high-yielding backgrounds, further enabling durable and predictable performance under variable biotic stress conditions. Inclusion of disease environments in METs enables uncovering G × E × B interactions and identifying lines with both high yield and durable rust resistance (Fig. 7.4). In recent past, newly released varieties in India, namely, DBW 187, PBW 826, PBW 872, DBW 303, DBW 327, WH 1270, DBW 370, DBW 371, and DBW 372 were chosen on both the basis of high yield and rust tolerance using data analysis and pathogen-informed targeting (Shahin et al. 2024).

7.3.2 Rice: Blast and Bacterial Blight in the Context of G × E Interactions

Rice blast (*Magnaporthe oryzae*) is favored by high humidity (>90%), temperature between 24–28 °C, and frequent leaf wetting commonly seen in upland and irrigated high-altitude zones. Bacterial blight (*Xanthomonas oryzae pv. oryzae*) flourishes in warm, rainy lowland ecosystems. Their spatial and temporal variability leads to strong G × E interactions through unpredictable genotype responses across locations and seasons (Salleh et al. 2022).

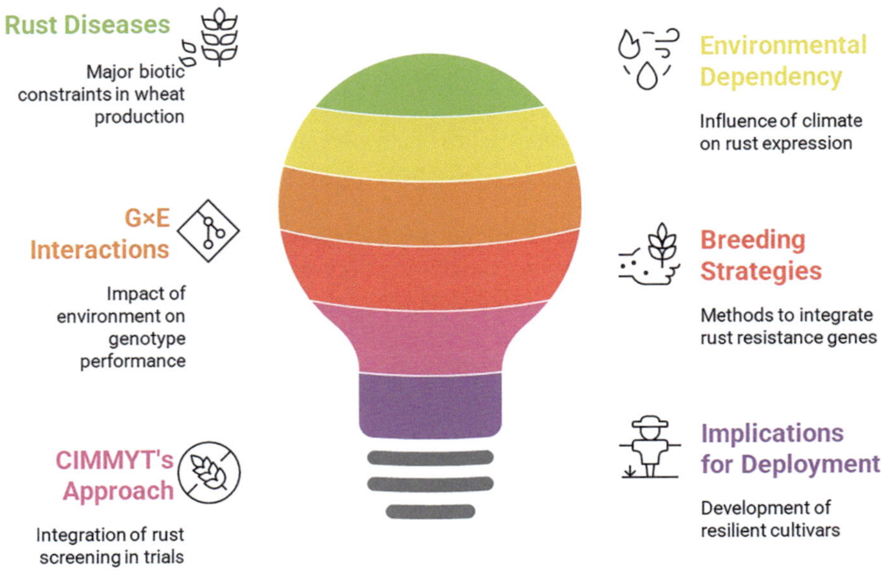

Fig. 7.4 Strategies for wheat rust management

7.3.2.1 Role of Resistance Genes and their Limitations

Blast resistance genes (e.g., *Pi1, Pi9, Pi54*) and blight genes (e.g., *Xa21, Xa4, Xa33*) provide high resistance under pathogen pressure, but can have yield penalties under low stress or become less effective as pathogen populations also adjust in environment. Past cultivars like IR64 illustrate this trade-off initially high-performing, later vulnerability emerged due to new races (He et al. 2025; Latif et al. 2024).

7.3.2.2 Disease Screening in METs

International Rice Research Institute (IRRI), an international organization, and Africa rice routinely use high-pressure disease nurseries Uniform Blast Nursery (UBN) and Bacterial Blight Screening Nursery (BBSN) within MET pipelines. This allows the detection of G × E × B interactions and lines such as DRR Dhan 45 and Improved Samba Mahsuri, which harmonize durable resistance (Pi54, Xa21/Xa13/Xa5) with stability of yield in stress and non-stress zones (Pradhan et al. 2015a, b).

7.3.2.3 Genomic and Haplotypic Innovations

GWAS studies with varied rice panels have pinpointed key marker trait associations for both blast and blight resistance. For example, eight better haplotypes for blast and five for blight were found among South Asian and Southeast Asian rice germplasm, giving breeders gene modules to include durable resistance into elite lines (Alam et al. 2024).

7.3.3 Maize: Maize Stalk Rots and Downy Mildew and Their Impact on G × E Interactions

Maize, one of the principal staple crops in the tropics and subtropics, is commonly subjected to biotic stresses that have considerable influence on its performance through environments. Stalk rots and downy mildew are two of the most prevalent diseases which exhibit strong interactions with genotype and environment, even producing large genotype × environment (G × E) interactions. Due to their dynamic nature, sensitivity to environmental factors, and unequal effect on genotypes, they require specific attention in breeding programs.

7.3.3.1 Overview of Diseases and Dependence on the Environment

Stalk Rots Stalk rots in maize are initiated by a complex of fungal pathogens, such as *Fusarium verticillioides*, *Macrophomina phaseolina*, *Diplodia maydis*, and *Colletotrichum graminicola*. These pathogens are most vicious under hot and drought-stressed conditions, particularly when combined with over-nitrogen application or close plantings. Infection tends to arise in the basal internodes, causing vascular occlusion, lodging, and substantial yield reduction (Mwatuni et al. 2022).

Downy Mildew Downy mildew of maize, *Peronosclerospora sorghi*, *P. philippinensis*, and *P. maydis*, grows well under cool, wet conditions usually aggravated in irrigated or continuously cropped rotations. It is most debilitating in the early stages of growth with leaf chlorosis, stunting, and eventual sterility. Outbreaks are frequent in monoculture regions without crop rotation (Gowda et al. 2021).

7.3.3.2 Influence on G × E Interactions

Stalk rots and downy mildew presence results in dramatic changes in genotype ranks between environments. A hybrid that is successful when disease pressure is low can be unsuccessful under epidemic levels of disease because of stalk lodging or vascular decline. Disease-tolerant hybrids, conversely, are sometimes middle-of-the-road producers but tend to exhibit higher yield stability over seasons and sites. This results in crossover interactions, increasing G × E variance and making selection more difficult when disease pressure is neglected (Beyene et al. 2023).

7.3.3.3 Responses and Innovations in Breeding Programs

Acknowledging the pivotal importance of these biotic stresses, a number of national and international maize breeding programs have adopted the following approaches: artificial inoculation methods-stalk rot: Toothpick or stem injection inoculation at flowering.

Downy Mildew: Whorl inoculation with spore suspension under early vegetative conditions in humid environments. These methods provide standardized screening and improve selection precision (Fisher et al. 2021).

7.3.3.4 Hotspot Testing Sites

There are several hotspot region for the disease screening: for example, in India—Tarai region for downy mildew and Andhra Pradesh, Gujarat, Karnataka for stalk rot; Africa—Western Kenya and Nigeria for integrated biotic stress; Southeast Asia—Vietnam and the Philippines for tropical downy mildew. These sites provide genotypes exposure to natural disease pressures that allow strong resistance validation (CIMMYT 2022).

7.3.3.5 G × E × B Modeling

Genomic prediction models are now including G × E × B frameworks, in which "B" stands for biotic stress, in the latest breeding programs. Hybrid performance and stability with fluctuating stress can now be predicted using QTLs for disease resistance like Rdm1 (downy mildew) and Rpr1 (stalk rot) in these models (Semagn et al. 2022).

7.3.3.6 Outstanding Examples and Success Stories

India: Bio-9637, a hybrid with high downy mildew tolerance, is commonly used in Andhra Pradesh and Karnataka (Gowda et al. 2021).

Vietnam & Indonesia: IRRI and national program-released hybrids tolerant to *P. philippinensis* had stable yields under variable rainfall and irrigation (IRRI 2023).

Nigeria & Kenya: IITA and CIMMYT-developed SAMMAZ 42 and WE1101 showed improved standability and resistance to stalk rot in trials (IITA 2022).

Philippines: NSIC Rc298, which is resistant to downy mildew and can be used across lowland irrigated systems, has had high adoption (Prasanna et al. 2020).

7.3.3.7 Breeding and Crop Recommendation Implications

This case eminently demonstrates the need to incorporate biotic stress screening throughout maize variety development. Failure to consider disease pressure in assessment can lead to the release of unstable hybrids with a tendency to collapse when stressed. Results in resistant genotypes are underestimated, especially for areas with cyclic epidemics. A disease-inclusive G × E strategy increases breeder ability to detect hybrids with both yield capacity and resilience, encouraging productivity and farmer risk minimization (Rahman et al. 2023).

7.3.4 Other Relevant Examples from Crops

Barley: Aphid-borne barley yellow dwarf virus exhibits high season and environmental dependence. Resistance genes (e.g., Bdv2) contribute to stabilizing performance across sites, though effectiveness is influenced by aphid dynamics (Oliver et al. 2022).

Sorghum: Outbreaks of anthracnose and grain mold result in huge losses during humid, late-season conditions. Resistant genotypes exhibit stable performance, but susceptible ones tend to fail in wet years (Menkir et al. 2021).

Pulses: Fusarium and Rhizoctonia root rots of chickpea and lentil induce inconsistent G × E effects based on temperature and soil moisture. Stress-tolerant lines perform better under stress but provide minimal benefit under low-pressure conditions (Gaur et al. 2023).

Vegetables: Bacterial wilt in tomato and potato late blight exhibit evident G × E × B interactions. Resistance genes bestow yield stability under hotspot areas but have limited applicability under non-endemic areas (Khan et al. 2021).

7.4 Implications for Plant Breeding

The case study evidence highlights that neglecting biotic stress in G × E analysis can lead to false assumptions regarding varietal stability and adaptation. For plant breeders, this has far-reaching implications:

7.4.1 Incorporating Biotic Stress in Trial Planning

Data from multi-crop breeding programs in wheat, rice, and maize unequivocally showed that the omission of biotic stress considerations in G × E analysis generates partial and frequently misleading information on genotype adaptability, particularly in tropical and subtropical environments where pest and disease stress is constant and variable (Singh et al. 2023). Biotic stresses like rusts, blast, stalk rots, and viral complexes are not episodic but are spatiotemporally dynamic and have a significant interaction with genotype and environment and hence are an integral part of G × E × B (genotype × environment × biotic Stress) analysis models (Zhao et al. 2021).

To allow for real-world application, trial design should mandatorily include biotic stress using the following methods: Artificial inoculation (e.g., wheat rusts, rice blast, maize downy mildew) should be normalized in multi-environment trials (METs), especially at early- to mid-stage selection (Zhang et al. 2022). Site selection for METs should emphasize identified disease hotspots or recurrent biotic stress areas to test genotype durability under endemic stress. Within-trial monitoring of disease (longitudinal pathogen data collection) must be deployed to capture seasonal biotic stress dynamics and relate them to genotype response.

Recent research emphasizes that such an approach that considers stress yields more accurate estimations of genotype stability and uncovers resistance features concealed under favorable conditions (He et al. 2023a, b; Velu et al. 2022).

7.4.2 Balancing Resistance with Yield in Selection Strategies

Plant breeders have the challenge of reconciling yield potential with resistance to disease. In traditional METs, high-yielding genotypes tend to prevail in environments with low disease pressure, yet their performance fails when conditions are stress-prone, hence leading to low adaptability and greater risk for farmers (Parlevliet 2021). Conversely, moderate-yielding but highly resilient genotypes tend to provide greater stability in yield and reduce the loss under disease epidemics (Joshi et al. 2023a, b).

Hence, selection strategies need to be revised to include biotic stress indices (e.g., disease severity index, canopy health ratings, stability score for resistance) in addition to traditional yield and phenology characteristics. Utilize GGE biplots and AMMI models under stress to select genotypes with both good average performance and poor genotype × environment interaction (Yadav et al. 2022). Apply stability metrics, that is, coefficient of variation (CV), Finlay–Wilkinson regression, and Shukla's stability variance, especially in trials that are carried out under disease-prone environments (Dawson et al. 2021).

This strategy places breeding success away from sole maximization of yield towards resilience-based selection for lines that maintain performance under pressure, playing a role in both productivity and risk protection in climate-susceptible agroecosystems (Sukumaran et al. 2023).

7.4.3 Marker-Assisted and Genomic Selection for Disease Resistance

Biotechnological progress has revolutionized breeding against diseases. Marker-assisted selection (MAS) allows the introgression and pyramiding of major resistance genes like Xa, Pi, Yr, Sr, Rlm, Rps, and Ht, considerably reducing the duration of breeding cycles (Singh et al. 2023). MAS has been extremely successful for major R genes that provide resistance to rusts in wheat or blast in rice, particularly when combined to create stacks of genes that provide durable, broad-spectrum resistance (Gupta et al. 2022; Sundaram et al. 2021). Genomic selection (GS) goes a step further by leveraging genome-wide markers to predict breeding values for resistance traits, even without phenotypic data. GS has improved selection accuracy for complex traits like quantitative resistance to Fusarium head blight, late blight, and rusts, particularly when trained on multi-environment datasets (Crossa et al. 2021; Wang et al. 2022).

To enhance precision and resistance durability, modern breeding integrates pathogen diagnostics, genotyping of pathogen populations, and host–pathogen interaction studies. These allow targeted deployment of resistance genes aligned with pathogen virulence profiles especially critical in rapidly evolving pathosystems like rust fungi or blast disease (Milus et al. 2023).

7.4.4 Participatory and Farmer-Centric Breeding Strategies

Biotic stress issues tend to be location-dependent and dependent on prevailing agronomic management at the local level. Participatory variety selection (PVS) guarantees that resistance breeding reflects the sophistication of on-farm disease pressures, such as mixed infections, nutrient-disease interactions, and heterogeneous stress intensities (Ceccarelli and Grando 2020).

Farmers contribute not only by the lines selected under conditions of actual stress but also by the detection of latent resistance-related traits, including seedling vigor at nematode load, stem strength at stalk rot, or seed-borne pathogen tolerance traits frequently neglected in station trials. Their input on disease resilience, grain quality, and market traits ensures adoptability and economic sustainability of new cultivars (Joshi et al. 2023a, b).

This strategy is progressively useful in marginal, stress-exposed ecologies such as rainfed, low-input, or tribal farming systems, where formal screening can overlook genotype × management interactions.

7.4.5 Anticipating Pathogen and Pest Evolution

One of the main challenges in resistance breeding is the development of pathogens and pests, which may develop rapidly to bypass solitary resistance genes particularly those that are qualitative R gene-based. Examples from history include Yr27, Sr31, and Xa21 breakdown in high pathogen pressure and monoculture areas (Singh et al. 2023; Sharma et al. 2022).

To enhance resistance durability, modern breeding strategies emphasize combining major R genes with polygenic/quantitative resistance to distribute selection pressure and slow pathogen adaptation (Parlevliet 2021); rotating resistance genes across time and space and diversifying resistance sources across landscapes and farming systems; monitoring pathogen virulence evolution through regional and international surveillance systems (e.g., the BGRI for wheat rust or IRRI's blast network); and ensuring breeding and deployment strategies are evolution informed (He et al. 2023a, b). This transition from static resistance deployment to dynamic, data-driven resistance management is necessary in the climate change era, where recombination, emergence, and migration of pathogens are increasing.

7.4.6 Integrating G × E × B Frameworks into Breeding Pipelines

In order to systematically enhance varietal performance in dynamic and variable biotic stress environments, current breeding programs need to incorporate a G × E × B (genotype × environment × biotic stress) conceptual framework. This holistic approach is particularly important under climate change, where increasing unpredictable pest and disease patterns are being caused by it.

Biotic stress factors such as controlled inoculations or natural hotspots are manipulated or observed in a series of agroecological environments in breeding trials conducted with the G × E × B paradigm. These experiments are being increasingly evaluated through spatial and temporal statistical models to enable partitioning of sophisticated interaction effects. In addition, machine learning programs and crop simulation models have started being utilized in order to predict genotype performance under projected pest/disease scenarios, taking into consideration weather, soil, and pathogen forecast information (Gill et al. 2023; Ramstein et al. 2020).

By integrating G × E × B into breeding pipelines, programs are more apt to identify robust, widely adapted cultivars with the ability to sustain yield under changing stress and produce niche-specific varieties for regions with endemic or emerging biotic stresses (e.g., stem rust pockets, downy mildew-belt maize). Future-proof breeding pipelines by planning ahead for epidemic threats and incorporating disease evolutionary trajectories into selection processes (Kozak et al. 2022).

7.5 Modeling and Statistical Strategies for G × E × B

Conventional G × E approaches like AMMI and GGE biplots work well in abiotic-dominated environments but can be unable to fully account for genotype performance under heterogeneous biotic stress loads. The development of models that focus on G × E × B is an important milestone in quantitative genetics and plant pathology since they enable breeders to account for interaction complexity caused by pests, diseases, and parasitic weeds.

This method utilizes mixed linear models, Bayesian hierarchical models, and reaction norm models to estimate the effect of biotic stress and its interaction with environment and genotype. For example, more recent models include disease pressure as a random or fixed covariate, enabling better estimations of yield resilience and resistance stability under stress (Crossa et al. 2021; Millet et al. 2019).

7.5.1 The G × E × B Interaction Framework

A growing number of breeding studies adopt the G × E × B framework, in which "B" explicitly captures biotic stress variability such as rust, blast, Striga, Fusarium, fall armyworm, and other constraints. The model partitions phenotypic variance into genotypic variance (G) due to genetic makeup; environmental variance (E) linked to location, season, soil, and climate; biotic variance (B) arising from disease or pest pressure (severity, incidence); and interaction terms (G × E, G × B, E × B, G × E × B) quantifying combined effects on performance.

This modeling structure enables breeders to precisely estimate resistance trait stability across biotic stress gradients, detect G × B interactions, critical for understanding resistance durability and potential pathogen race breakdown, assess yield resilience in the presence of high disease/pest loads, and make region-specific recommendations by revealing adaptive G × E × B response patterns (Gurung et al. 2022; Juliana et al. 2023).

Case Example: G × E × B in Wheat Rust Trials In wheat breeding programs across South Asia and East Africa, G × E × B analysis was applied to multi-location data from rust-prone environments by CIMMYT and national partners. Results revealed that several genotypes showed high G × E stability (good yield across climates), but low G × B resilience; that is, they collapsed under high stripe or stem rust pressure.

This insight proved critical: without G × B data, such lines might be mistakenly released as broadly adapted. Instead, programs could prioritize lines with balanced G × E and G × B stability for high-risk zones such as the Indo-Gangetic Plain and Kenyan highlands (Singh et al. 2023; Li et al. 2022).

7.5.2 Stability Analysis Under Biotic Stress Conditions

Stability analysis is the bedrock of multi-environment trials (METs), enabling scientists to measure consistency of genotype performance in different environments. Conventionally applied for yield stability evaluation across agroecological zones, statistical methods like AMMI (additive main effects and multiplicative interaction) and GGE (genotype + genotype × environment) biplot analyses are increasingly being converted to incorporate biotic stress indicators. This evolution allows for a more understanding of how genotypes respond to complex G × E × B (genotype × environment × biotic stress) interactions critical in the context of disease- and pest-prone production systems.

7.5.2.1 AMMI Analysis with Biotic Covariates

The AMMI model combines analysis of variance (ANOVA) to partition additive main effects (genotype and environment) with principal component analysis (PCA) to interpret non-additive interactions (G × E). When applied to incorporate biotic stress as a covariate like disease severity scores or pest infestation indices, the AMMI approach facilitates more profound understanding of genotype behavior under biotic stress: splitting performance across different levels of disease or pest pressure; allowing breeders to know which genotypes hold up when stressed; unveiling crossover interactions that arise from patchy pathogen outbreaks, which can reverse genotype ranks across sites; measuring the contribution of G × B and G × E × B terms to the overall performance variability, and allowing for more realistic genotype selection under conditions of actual stress. For example, Sharma et al. (2017) applied AMMI models to wheat multi-environment trials with stripe rust incidence data and demonstrated that biotic stress explained a significant portion of the genotype × environment interaction variance. This allowed the identification of genotypes with both high yield and rust stability across locations.

7.5.2.2 GGE Biplots Enhanced with Biotic Data

GGE biplot analysis displays both genotype performance and stability across environments by conjoining genotype main effects and G × E interaction effects in one graphical tool. When scores for biotic stress (e.g., percent disease incidence or pest density) are included in the analysis, GGE biplots provide some additional strategic benefits: Visualization of genotype-by-disease interactions, assisting in the detection of whether performance is associated with or independent of stress levels;

mega-environment identification driven by pathogen or pest pressures, which can be distinct from the mega-environments determined by abiotic factors only; and prioritization of genotypes with stable performance and wide resistance over biotic stress gradients. A good example is the application of GGE biplots in rice breeding in Eastern India, where Magnaporthe oryzae (the rice blast pathogen) is a constant threat. GGE analysis incorporating blast severity scores enabled breeders to identify genotypes that combined yield potential with durable blast resistance. These genotypes were recommended for release specifically in endemic zones, where both yield stability and resistance are vital (Singh et al. 2019).

7.5.2.3 Genomic and Digital Tools for Biotic G × E Analysis

Contemporary breeding is being revolutionized by a paradigm shift through the convergence of genomic tools, high-throughput phenotyping, and climate-based forecasting tools to comprehend and control genotype × environment × biotic (G × E × B) interactions. These tools improve dissection capability of complex characteristics such as disease and insect resistance in varied environments, allowing more accurate and proactive crop improvement approaches.

7.5.3 Genomic Prediction Models

Advanced genomics methodology enables the integration of resistance-associated markers into predictive models of breeding. These consist of marker-assisted selection (MAS) which introduces known, major-effect resistance loci into elite lines. Genomic Selection (GS) employs genome-wide marker data to predict biotic stress performance without previous knowledge of specific resistance genes. Genomic × environment interaction models allow for the detection of alleles with environment- or stress-specific effects, thereby enhancing prediction accuracy in multi-environment trials.

For example, in maize, genomic prediction models including QTLs associated with Fusarium stalk rot resistance and environmental covariates substantially enhanced hybrid performance prediction under disease-susceptible and variable agro-ecologies (Yu et al. 2022). This illustrates how combining pathogen-specific markers with environmental descriptors improve the biological utility of prediction models within breeding pipelines.

7.5.4 High-Throughput Phenotyping (HTP) and Remote Sensing

New digital technologies are transforming the way disease and pest stresses are monitored at scale and in real time: Unmanned aerial vehicles (UAVs) integrated multispectral and thermal cameras can identify early detection of disease, including chlorosis and rust pustules, before they can be scored visually. Spectral indices like the normalized difference vegetation index (NDVI) and photochemical reflectance index (PRI) act as non-destructive proxies for quantifying disease severity and stress-induced canopy responses. Automated phenotyping platforms allow rapid, standardized measurements of disease incidence and severity across large experimental plots.

These innovations improve both the spatial and temporal resolution of phenotyping, enabling deeper insights into the interaction between host genotype, environment, and biotic stress. Real-time phenotyping enhances the quality of G × E × B datasets and supports more precise selection decisions under actual field conditions (Araus and Cairns 2014).

7.5.5 Climate-Driven Pest and Disease Forecasting Models

Since most biotic stresses are strongly linked to environmental conditions, climate-driven modeling has become an important tool for anticipatory breeding and G × E analysis. They include: temperature and humidity thresholds to model the risk of pathogen infection, rainfall and wind trajectories to model dispersal patterns (e.g., for wheat rust spores), and host phenology to delimit windows of vulnerability and pathogen life-cycle synchronization.

Applications of such models are already proving transformative. The RustTracker system, for example, combines meteorological data with spore dispersal models to generate rust disease risk maps, aiding in strategic trial placement and selection of resistant wheat genotypes (Singh et al. 2011a, b). Similarly, Blitecast, a forecasting system for late blight in potato, supports optimal timing of disease assessments and fungicide applications (Kirk et al. 2013). Desert locust forecasting models through wind trajectory analysis have assisted in the identification of multi-environment testing (MET) zones for sorghum in East Africa, where desert locust outbreaks are a perennial threat (FAO 2020).

7.5.6 Towards Data-Rich, Disease-Aware Breeding Pipelines

The combination of genomic technologies, digital phenotyping, and climate-based pest and disease models produces a dynamic, data-intensive breeding environment. Together with genotype performance, these tools enable breeders to model future stress situations so that selection can be made for cultivars that are not just high-yielding and stable in their performance over varying environments but also biologically resilient to changing threats.

This integrated G × E × B-guided strategy enables to early selection of resistant genotypes with environment-specific adaptability, enhanced trial design and stress screening via predictive site selection, and enhanced breeding cycles via maximization of selection accuracy and resolution. This approach, finally, is a new frontier in plant breeding one that is biologically informed, digitally empowered, and climate ready.

7.6 Strategies for Managing Biotic Stresses in G × E Contexts

Biotic stress management under genotype × environment (G × E) interactions requires an integrated, system-level strategy that integrates genetics, agronomy, pest and disease management, and institutional infrastructure. Biotic stresses like disease and insect pests not only affect crop phenotypes but also change the environmental context itself and thereby feed back into G × E dynamics (Brachi et al. 2011). Hence, a proactive approach taking into account the intricacy of G × E × B (B for Biotic stress) is necessary for climatically stable and yield-promoting production systems under variability.

7.6.1 Integrated Pest and Disease Management (IPDM)

Integrated Pest and Disease Management is the basis for minimizing the risk and effect of biotic stress under field conditions. IPDM calls for a harmonious application of biological, cultural, physical, and chemical means to manage pests and pathogens in a sustainable manner (Koul et al. 2004). IPDM strategies can be made to stabilize genotype performance across environments by countering unpredictable pest and disease outbreaks when informed with G × E knowledge, avoid the selection pressure caused by excessive dependence on genetic resistance to prevent speedy breakdown of resistance, and supplement host resistance mechanisms through the lowering of pathogen or insect population pressure.

For instance, the intercropping of wheat with legumes has the ability to suppress pest build-up and microclimatic moderation, while crop rotation with non-host species reduces inoculum density (Zadoks and Schein 1979). In addition, early fungicide application utilizing forecasting models in combination with genotype susceptibility and local microclimates can refine protection against foliar pathogens like rusts or blights.

7.6.2 Agronomic Interventions

Agronomic management has a dual function in regulating abiotic and biotic stresses and is critical in reducing the manifestation of G × E interaction. Agronomic practices like conservation tillage, balanced nutrition, and precision irrigation have the ability to alter the microenvironment in such a manner that it decreases the pressure of biotic stress (Bakker et al. 2006). Adequate spacing of plants facilitates good air movement, thus decreasing the humidity conducive to fungal pathogens. A well-balanced use of nitrogen promotes healthy crop growth but avoids excessive vegetative growth that predisposes crops to lodging and disease (Fageria and Baligar 2005). Regulated irrigation avoids conditions conducive to waterborne pathogens like Pythium or Fusarium spp. Adapting these to genotypes and agro-ecologies increases the resilience of the cropping system while promoting the longevity of resistance genes.

7.6.3 Climate-Smart Approaches

With growing climate uncertainty, the occurrence and intensity of biotic stress incidences are becoming more and more unpredictable. Consequently, climate-smart approaches need to be integrated into breeding and management paradigms. These incorporate breeding for quantitative and broad-spectrum resistance that confers protection against a variety of pathogens or pests under changing conditions (Poland et al. 2009). Creation of decision support systems and disease/pest forecasting models based on real-time weather and crop data. Encouraging agroecological diversification like employing mixtures of varieties, polycultures, and diverse rotations to break the cycling of pathogens and pests. Climate-smart agriculture develops adaptive capacity by integrating G × E × B knowledge with projections of future climatic contexts and locally adapted management approaches (Lipper et al. 2014).

7.6.4 Policy and Institutional Support

Technical solutions are not sufficient to implement without the complement of policy frameworks and institutional capacity. Successful management of biotic stress in the G × E situation also relies on appropriate varietal release and regulatory approval procedures for resistant varieties: investment in plant protection services like diagnostics and early warning systems, extension programs for farmer training in IPDM and resistance stewardship, public–private partnerships for bulk seed multiplication and dissemination.

For example, rust monitoring networks and national resistance management tactics have been found to be essential in avoiding wheat rust epidemics (Singh et al. 2011a, b). Institutional support guarantees that farmers are reached efficiently and equitably by the benefits of breeding and agronomy.

7.7 Future Perspectives

The intersection of climate change, population pressure, and resource limitation is strengthening the imperative of future-proof mechanisms to address biotic stresses within G × E contexts. There is a need to switch towards predictive, proactive, and systems-oriented breeding and management in order to address future challenges.

7.7.1 Climate Change and New Challenges

Climate change is reconstructing the plant–pest–pathogen relationships. Climate change is likely to increase temperatures, modify patterns of rainfall, and intensify weather extremes, thereby widening the geographical distribution of most pests and diseases (Juroszek et al. 2020). Prediction tools that accommodate G × E × B data combined with climate change projections must be used to inform breeding and pest control. Durable resistance should be bred through pyramiding multiple genes and aiming for conserved pathogen effectors. Real-time monitoring systems (such as pest surveillance drones, climate stations) will be vital for predicting and reacting to developing threats. These strategies will make sure that multi-environment trial (MET) design and interpretation continue to be applicable under evolving conditions.

7.7.2 Genomic and Digital Innovations

Technological advances in genomics, data science, and digital farming provide significant capabilities to boost biotic stress resistance through G × E analysis: Genomic selection (GS) provides the opportunity for breeders to choose for resistance characteristics early in the breeding process, even prior to phenotyping (Crossa et al. 2017). Artificial intelligence (AI) and machine learning can simulate sophisticated G × E × B datasets to project genotype performance under different biotic pressures. UAVs (drones) integrated hyperspectral imaging enables early detection of disease symptoms through high-throughput precise phenotyping. These technologies reduce breeding cycles and enhance the possibility of varietal recommendations in heterogeneous environments.

7.7.3 Systems-Based Breeding

The future of plant breeding is in the use of a systems strategy that simultaneously handles multiple interacting constraints. Systems-based breeding involves combining G × E × B insight with farmer desire and socioeconomic environments, farmer participatory breeding and selection by farmers to ensure acceptability and adoption and cross-disciplinary collaboration among geneticists, agronomists, pathologists, and socio-economists. Such systems will deliver varieties not only well-suited in performance when tested under experimental conditions but also well-adapted to be productive in farmers' fields, under actual biotic and abiotic stresses (Ceccarelli and Grando 2007).

7.7.4 Research and Capacity Building

Capacity development will be key to effective implementation of these strategies. Investment in the future should include training scientists in multi-disciplinary G × E × B analysis and computer tools; the strengthening of infrastructure for METs, including facilities for controlled stress screening; education of farmers on integrated disease and pest management; and raising the level of public sector extension systems to enable rapid deployment of varieties. National and regional capacity-building activities can enable scaling up of innovations and readiness for emerging biotic challenges.

7.8 Conclusion

Biotic stresses inherently define the performance of genotypes over environments and need to be taken into account as an integral part of G × E interaction. Their neglect can yield incorrect conclusions regarding genotype adaptability and performance stability. With accelerating climate change, generating novel pathogen races and pest breakouts are becoming more common, and the frequency and severity of biotic stresses are likely to become even more variable and unpredictable. Overlooking biotic stresses in multi-environment tests threatens misclassifying better genotypes, understating susceptibility, and eventually disappointing farmers who have to deal with these stresses in the field. In contrast, the orderly inclusion of biotic stresses in trial planning, statistical modeling, and breeding programs can increase the chances of creating robust, high-yielding cultivars that provide consistent yields even under fluctuating pest and disease pressures. Integrated approaches, namely integrating breeding for resistance, good agronomic practice, climate-smart agriculture, and enabling environmental policy, are needed to cope with biotic stress in a changing climate. Genomics, digital farming, and predictive modeling will increasingly support the efforts, enabling breeders and farmers to deal with anticipated as well as unexpected threats. Altogether, biotic stresses significantly influence G × E interactions and hence should be at the center of crop improvement programs. Embracing a holistic, system-oriented approach that takes into account G × E × B interactions can make a meaningful advance toward sustainable, resilient, and productive agriculture of the future. Thus, strong breeding programs need to place high value on the planning of trials that capture actual field biotic stresses and employ analytical techniques that consider G × E × B interactions. It is only through such interconnected methods that breeders can generate cultivars that are at par high yielding but also stable and resistant under the varied, dynamic, and evolving biotic stress conditions encountered by growers globally.

References

Akter A, Hasan MJ, Kulsum U, Rahman MH, Khatun M, Islam MR (2015) GGE biplot analysis for yield stability in multi-environment trials of promising hybrid rice (Oryza sativa L.). Bangladesh Rice J 19(1):1–8

Alam S, Sundaram KT, Singh UM, Srinivas Prasad M, Laha GS, Sinha P, Singh VK (2024) Superior haplotypes towards the development of blast and bacterial blight-resistant rice. Front Plant Sci 15:1272326. https://doi.org/10.3389/fpls.2024.1272326

Araus JL, Cairns JE (2014) Field high-throughput phenotyping: the new crop breeding frontier. Trends Plant Sci 19(1):52–61. https://doi.org/10.1016/j.tplants.2013.09.008

Baker RJ (1988) Differential response to environmental stress. In: Weir BS, Eisen EJ, Goodman MM, Namkoong G (eds) Proceedings of the second international conference on quantitative genetics. Sinauer Associates, Sunderland, pp 492–504

Bakker MG, Schlatter DC, Otto-Hanson L (2006) Kinkel LL Diffuse symbioses: roles of plant-plant, plant-microbe and microbe-microbe interactions in structuring the soil microbiome. Molecular Ecology 23(6):1571–1583

Beyene Y et al (2023) Genotype by environment by biotic stress interactions in tropical maize breeding. Euphytica 219:134. https://doi.org/10.1007/s10681-023-03202-1

Brachi B, Faure N, Horton M, Flahauw E, Vazquez A, Nordborg M, Bergelson J, Cuguen J, Roux F (2011). Linkage and association mapping of Arabidopsis thaliana flowering time in nature. PLoS Genetics, 7(5): e1001318. https://doi.org/10.1371/journal.pgen.1001318

Ceccarelli S, Grando S (2007) Decentralized-participatory plant breeding: an example of demand driven research. Euphytica 155:349

Ceccarelli S, Grando S (2020) Participatory plant breeding: a tool for empowering farmers and achieving sustainable agriculture. Outlook Agric 49(1):36–43. https://doi.org/10.1177/0030727019898220

CIMMYT (2022) Hotspot-based screening for improved disease resistance in maize hybrids. CIMMYT Annual Report

Cooper M, Technow F, Messina C, Gho C, Totir LR (2016) Use of crop growth models with whole-genome prediction: application to a maize multi-environment trial. Crop Sci 56:2141–2156

Crossa J, Pérez-Rodríguez P, Cuevas J, Montesinos-López OA, Jarquín D, de Los Campos G, Burgueno J, González-Camacho JM, Pérez-Elizalde S, Beyene Y, Dreisigacker S, Gorjanc G, Hickey JM (2017) Genomic selection in plant breeding: Methods, models, and perspectives. Trends in Plant Science 22(11), 961–975. https://doi.org/10.1016/j.tplants.2017.08.011

Crossa J, Pérez-Rodríguez P, Cuevas J et al (2021) Genomic selection in plant breeding: methods, models, and perspectives. Trends Plant Sci 26(7):692–705. https://doi.org/10.1016/j.tplants.2021.03.012

Crow JF (1986) Basic concepts in population, quantitative and evolutionary genetics. WH Freeman, New York

Dawson IK, Endale A, Jones H (2021) Stability analysis in crop breeding under stress-prone conditions. Theor Appl Genet 134(8):2549–2564. https://doi.org/10.1007/s00122-021-03815-3

Dickerson GE (1962) Implications of genetic–environmental interaction in animal breeding. Animal Production 4:47–64

Doolittle DP (1987) Population genetics: basic principles. Springer, New York

Fageria NK, Baligar VC (2005) Enhancing nutrient use efficiency in crop plants. Advances in Agronomy, 88, 97–185. https://doi.org/10.1016/S0065-2113(05)88004-6

FAO (2020) Desert locust early warning and forecasting system. Food and Agriculture Organization of the United Nations, Rome

Fisher M et al (2021) Accelerated screening methods for biotic stress resistance in maize breeding. Plant Dis 105(5):1290–1297. https://doi.org/10.1094/PDIS-07-20-1693-RE

Gauch HG Jr (2006) Statistical analysis of yield trials by AMMI and GGE. Crop Sci 46(4): 1488–1500

Gaur PM et al (2023) Integrating biotic stress resistance into legume improvement programs. Legume Sci 5(1):e149

Gill MB, Velu G, Crespo-Herrera L, Joshi AK (2023) Breeding climate-resilient wheat with durable resistance to biotic stress: a multi-trait, multi-environment perspective. Theor Appl Genet 136: 134. https://doi.org/10.1007/s00122-023-04215-y

Gowda M et al (2021) Genetic dissection and molecular breeding for downy mildew resistance in maize. Theor Appl Genet 134(3):867–879. https://doi.org/10.1007/s00122-021-03742-3

Grogan SM, Anderson J, Baenziger PS, Frels K, Guttieri MJ et al (2016) Phenotypic plasticity of winter wheat heading date and grain yield across the US Great Plains. Crop Sci 56:2223–2236. https://doi.org/10.2135/cropsci2015.06.0357

Gupta PK, Balyan HS, Chhuneja P, Jaiswal JP, Tamhankar S, Mishra VK, Bains NS, Chand R, Joshi AK, Kaur S et al (2022) Pyramiding of genes for grain protein content, grain quality, and rust resistance in eleven Indian bread wheat cultivars: a multi-institutional effort. Mol Breed 42: 21. https://doi.org/10.1007/s11032-022-01277-w

Gurung S, Ravelombola WS, Buntjer D (2022) Modeling genotype × environment × disease interactions in peanut breeding. Crop Sci 62(2):647–662. https://doi.org/10.1002/csc2.20705

He X, Singh PK, Dreisigacker S, Singh RP, Ellis M (2019) Resistance to stripe rust in elite spring wheat germplasm derived from CIMMYT and ICARDA. Plant Dis 103(6):1168–1174. https://doi.org/10.1094/PDIS-05-18-0845-RE

He Z, Juliana P, Singh RP, Huerta-Espino J (2023a) Integrating disease monitoring with genetic gain strategies in wheat. Field Crop Res 297:108903. https://doi.org/10.1016/j.fcr.2023.108903

He Z, Singh RP, Huerta-Espino J et al (2023b) Global surveillance of wheat rusts: past, present, and future. Annu Rev Phytopathol 61:45–69. https://doi.org/10.1146/annurev-phyto-072720-095949

IITA (2022) Annual highlights: developing stress-resilient maize varieties. International Institute of Tropical Agriculture. https://www.iita.org

IRRI (2023) IRRI maize program highlights—Southeast Asia. International Rice Research Institute

Jan CZ, Richard DS (1979) Epidemiology and plant disease management. Oxford University Press. 427 pages. ISBN 9780195024517

Joshi AK, Velu G, Rane J (2023a) Breeding climate-resilient, disease-tolerant wheat cultivars for India's hotspots. J Genet Breed 77(1):12–27. https://doi.org/10.1016/j.jgb.2023.05.004

Joshi AK, Velu G, Singh RP et al (2023b) Delivering farmer-preferred, climate-resilient wheat cultivars using participatory breeding in South Asia. Field Crop Res 293:108867. https://doi.org/10.1016/j.fcr.2023.108867

Juliana P, Singh RP, Crespo-Herrera LA et al (2023) Genetic dissection of durable resistance to wheat stem rust using genome-wide association studies. Plant Dis 107(1):104–114. https://doi.org/10.1094/PDIS-05-22-1145-RE

Juroszek P, Racca P, Link S, Farhumand J, Kleinhenz B (2020) Overview of some recent advances in regional and field-scale modeling of plant diseases caused by airborne pathogens. European Journal of Plant Pathology 157(3):885–899. https://doi.org/10.1007/s10658-020-02138-4

Kang MS (1988) A rank-sum method for selecting high-yielding, stable corn genotypes. Cereal Res Commun 16:113–115

Khan MA et al (2021) Breeding for biotic stress resistance in vegetables under climate variability. Hortic Res 8:112

Kirk WW, Abu-El Samen FM, Muhinyuza JB, Hammerschmidt R (2013) Recent developments in potato late blight management. Am J Potato Res 90:35–45

Koul O, Cuperus GW, Elliott NC (Eds.) (2004) Integrated Pest Management: Potential, Constraints and Challenges. CABI Publishing, Wallingford, UK. https://doi.org/10.1079/9780851997170.0000

Kozak M, Piepho H-P, Smith A (2022) Modelling and interpretation of G×E×B interaction in plant breeding. Euphytica 218:155. https://doi.org/10.1007/s10681-022-03051-z

Kumar R, Dia M, Wehner TC (2013) Implications of mating behavior in watermelon breeding. HortScience 48:960–964

Latif MA, Nihad SAI, Hasan R, Rahman L, Anik T, Manidas AC, Rashid M, Khan M, Bhuiyan R (2024) Pyramiding of multiple resistant genes of blast and bacterial blight diseases in the background of rice (Oryza sativa) mega variety BRRI dhan29. Curr Plant Biol 40:100400. https://doi.org/10.1016/j.cpb.2024.100400

de Leon N, Jannink JL, Edwards JW, Kaeppler SM (2016) Introduction to a special issue on genotype by environment interaction. Crop Science 56(5):2081–2089. https://doi.org/10.2135/cropsci2016.07.0002i

Li F, Singh RP, Basnet BR et al (2022) Targeting both yield and disease resistance in South Asia: a case study in wheat breeding. Field Crop Res 279:108451. https://doi.org/10.1016/j.fcr.2021.108451

Lipper L, Thornton P, Campbell BM, Baedeker T, Braimoh A, Bwalya M, Caron P, Cattaneo A, Garrity D, Henry K, Hottle R, Jackson L, Jarvis A, Kossam F, Mann W, McCarthy N, Meybeck A, Neufeldt H, Remington T, Sen PT, Sessa R, Shula R, Tibu A, Torquebiau EF (2014) Climate-smart agriculture for food security. Nature Climate Change 4,1068–1072

Liu D, Yuan C, Singh RP, Randhawa MS, Bhavani S, Kumar U, Huerta-Espino J, Lagudah E, Lan C (2022) Stripe rust and leaf rust resistance in CIMMYT wheat line "Mucuy" is conferred by combinations of race-specific and adult-plant resistance loci. Front Plant Sci 13:880138. https://doi.org/10.3389/fpls.2022.880138

Menkir A et al (2021) Multi-trait selection for disease resistance and yield stability in sorghum. Euphytica 217:189

Millet EJ, Kruijer W, Coupel-Ledru A et al (2019) Genomic prediction of maize yield across European environmental conditions. Nat Genet 51(6):952–956. https://doi.org/10.1038/s41588-019-0414-y

Milus EA, Kolmer JA, Jin Y (2023) Pathogen variability and implications for resistance gene deployment in cereal rusts. Plant Dis 107(3):658–666. https://doi.org/10.1094/PDIS-08-22-1873-FE

Mungra KD, Jadhav BD, Khandelwal V (2011) Genetic analysis for yield and quality traits in forage sorghum [Sorghum bicolor (L.) Moench]. Indian J Genet Plant Breed 71(3):241

Mwatuni FM et al (2022) Stalk rot resistance in tropical maize: advances and screening methods. Front Plant Sci 13:1011331. https://doi.org/10.3389/fpls.2022.1011331

Oliver RE et al (2022) Genetics of barley yellow dwarf virus resistance in temperate climates. Theor Appl Genet 135(5):1417–1429

Parlevliet JE (2021) Durable resistance and polygenic control in host-pathogen interactions. Euphytica 217:176. https://doi.org/10.1007/s10681-021-02913-4

Poland J, Balint-Kurti P et al (2009) Shades of gray: the world of quantitative disease resistance. Trends Plant Sci 14:21

Pradhan SK, Nayak DK, Mohanty S, Behera L, Barik SR, Pandit E, Lenka S, Anandan A (2015a) Pyramiding of three bacterial blight resistance genes for broad-spectrum resistance in Deepwater rice variety, Jalmagna. Rice 8:1–14. https://doi.org/10.1186/s12284-015-0051-8

Pradhan SK, Nayak DK, Mohanty S, Behera L, Barik SR, Pandit E, Singh ON (2015b) Marker-assisted breeding for bacterial blight resistance and quality improvement in an elite rice cultivar, 'Swarna'. PLoS One 10(6):e0129611. https://doi.org/10.1371/journal.pone.0129611

Prasanna BM et al (2020) Developing resilient maize varieties through trait-based breeding. Plant Breed Rev 44:35–72

Qureshi N, Singh RP, Gonzalez BM, Velazquez-Miranda H, Bhavani S (2023) Genomic regions associated with resistance to three rusts in CIMMYT wheat line "Mokue#1". Int J Mol Sci 24: 12160. https://doi.org/10.3390/ijms241512160

Rahman MA et al (2023) Resilience-focused breeding: integrating disease resistance into G×E frameworks. Crop Sci 63(4):2157–2170. https://doi.org/10.1002/csc2.21087

Ramstein GP, Evans J, Kaeppler SM et al (2020) Accuracy of genomic prediction in tropical maize improved by modeling genotype × environment interaction. Plant Genome 13(1):e20001. https://doi.org/10.1002/tpg2.20001

Rogers AR, Dunne JC, Romay C, Bohn M, Buckler ES, Ciampitti IA, Edwards J, Ertl D, Flint-Garcia S, Gore MA, Graham C (2021) The importance of dominance and genotype-by-environment interactions on grain yield variation in a large-scale public cooperative maize experiment. G3 (Bethesda)

Salleh SB, Rafii MY, Ismail MR, Ramli A, Chukwu SC, Yusuff O, Hasan N'A (2022) Genotype-by-environment interaction effects on blast disease severity and genetic diversity of advanced blast-resistant rice lines based on quantitative traits. Front Agron 4:990397. https://doi.org/10.3389/fagro.2022.990397

Schlichting CD (1986) The evolution of phenotypic plasticity in plants. Annu Rev Ecol Syst 17: 667–693

Semagn K et al (2022) Incorporating disease resistance loci into genomic prediction models for tropical maize. Plant Genome 15(1):e20123. https://doi.org/10.1002/tpg2.20123
Shahin AA et al (2024) Evaluation of effectiveness of resistance genes in wheat genotypes using marker-assisted selection for stripe rust resistance breeding. BMC Plant Biology 24:1–24. https://doi.org/10.1186/s12870-024-05470-1
Sharma RC, Singh RP, Duveiller E (2017) Stability of wheat genotypes to spot blotch and terminal heat stress in South Asia. Euphytica 213:35. https://doi.org/10.1007/s10681-016-1803-2
Sharma TR, Rathour R, Singh B (2022) Breaking resistance: emerging threats of new pathogen races in South Asia. Front Plant Sci 13:840891. https://doi.org/10.3389/fpls.2022.840891
Singh RP, Hodson DP, Huerta-Espino J, Jin Y, Njau P, Wanyera R, Ward RW (2008) Will stem rust destroy the world's wheat crop? Adv Agron 98:271–309. https://doi.org/10.1016/S0065-2113(08)00205-8
Singh RP, Hodson DP, Huerta-Espino J et al (2011a) The emergence and spread of new races of wheat stem rust fungus: continued threat to food security and prospects of genetic control. Phytopathology 101(4):272–284
Singh RP, Huerta-Espino J, Rajaram S (2011b) Achieving near-immunity to leaf and stripe rusts in wheat by combining slow rusting resistance genes. Acta Phytopathol Entomol Hung 46(1):133–139
Singh D, Singh AK, Singh UD (2019) Integration of GGE biplot and disease resistance data to identify stable rice varieties in blast hot-spot regions. J Phytopathol 167(10):545–553. https://doi.org/10.1111/jph.12819
Singh RP, Huerta-Espino J, Juliana P et al (2023) Ten years of breeding for durable rust resistance in wheat: achievements and outlook. Annu Rev Phytopathol 61. https://doi.org/10.1146/annurev-phyto-030521-083938
Sukumaran S, Sansaloni C, Dreisigacker S (2023) Resilience breeding in wheat under pest and pathogen stress. Front Plant Sci 14:1224793. https://doi.org/10.3389/fpls.2023.1224793
Sundaram RM, Laha GS et al (2021) Marker-assisted pyramiding of multiple resistance genes for bacterial blight and blast in rice. Theor Appl Genet 134(3):719–733. https://doi.org/10.1007/s00122-020-03717-6
Van Eeuwijk FA, Bustos-Korts DV, Malosetti M (2016) What should students in plant breeding know about the statistical aspects of genotype × environment interactions? Crop Sci 56(5):2119–2140
Velu G, Ammar K, Singh RP (2022) Wheat breeding for yield and rust resistance under biotic and abiotic stress. Crop Sci 62(1):216–229. https://doi.org/10.1002/csc2.20653
Wang X, Bernal DL, Pérez-Rodríguez P (2022) Genomic prediction of rust resistance in wheat using multi-trait and multi-environment models. Plant Genome 15(1):e20147. https://doi.org/10.1002/tpg2.20147
William M, Singh RP, Huerta-Espino J, Islas SO, Hoisington D (2007) Molecular marker mapping of leaf rust resistance gene Lr46 and its association with stripe rust resistance gene Yr29 in wheat. Phytopathology 97(4):472–479
Xu Y (2016) Envirotyping for deciphering environmental impacts on crop plants. Theor Appl Genet 129:653–673
Yadav AK, Rai SK, Sharma RC (2022) GGE biplot analysis for identifying stable wheat genotypes under biotic stress. Indian J Genet Plant Breed 82(2):300–306. https://doi.org/10.5958/0975-6906.2022.00039.5
Yan W, Kang MS, Ma B, Woods S, Cornelius PL (2007) GGE biplot vs. AMMI analysis of genotype-by-environment data. Crop Sci 47(2):643–653
Yan W, Pageau D, Frégeau-Reid J, Durand J (2011) Assessing the representativeness and repeatability of test locations for genotype evaluation. Crop Sci 51:1603–1610. https://doi.org/10.2135/cropsci2011.01.0016

Yu X, Li X, Guo T et al (2022) Integrating environmental covariates and resistance QTLs improves genomic prediction for Fusarium stalk rot in maize. Theor Appl Genet 135:1123–1135. https://doi.org/10.1007/s00122-021-03964-z

Zhang P, Xie W, Li Y (2022) Artificial inoculation and hotspot trials for evaluating resistance in rice blast breeding. Plant Dis 106(6):1562–1572. https://doi.org/10.1094/PDIS-07-21-1423-RE

Zhao Y, Wang Y, Xu Y (2021) Dissecting G×E×B interactions in maize under biotic stress using mixed models. Theor Appl Genet 134(5):1325–1341. https://doi.org/10.1007/s00122-021-03777-6

Do Soil Properties Significantly Influence the G × E Interaction of Crop Varieties?

8

P. C. Ghasal, Jairam Choudhary, Kamlesh Kumar, M. Arif, Amit Kumar, D. Dutta, A. L. Meena, Ekta Narwal, Raghuveer Singh, and Y. S. Shivay

Abstract

Crop variety performance is determined by both genetic potential and the environment in which it is cultivated. Among environmental factors, including climate, management practices, and biotic and abiotic stresses, soil properties exert a particularly strong influence, shaping genotype × environment (G × E) interactions. Soil governs nutrient dynamics, water availability, root development, and microbial associations through its chemical (pH, nutrient-holding capacity, and fertility), physical (texture, structure, depth, and bulk density), and biological (organic carbon, microbial population and diversity, and enzymatic activity) attributes. These properties influence key plant traits such as root depth, length, density, surface area, volume, root hair density, and architecture, as well as nutrient uptake efficiency, water extraction capacity, and tolerance to edaphic stresses. They also interact with agronomic interventions such as tillage, irrigation scheduling, residue management, and organic inputs, creating a dynamic interface for genotype expression. Since varieties respond differently to edaphic variability, breeding programs must prioritize location- and soil-specific varietal development. Multi-location and multi-season trials are essential to identify genotypes that combine adaptability to specific soil constraints with stability across diverse conditions. Integrating soil-specific genotype selection with precision agriculture, targeted nutrient management, and specific

P. C. Ghasal · A. Kumar
ICAR-Vivekananda Parvatiya Krishi Anusandhan Sansthan, Almora, Uttarakhand, India

J. Choudhary · D. Dutta · E. Narwal · Y. S. Shivay (✉)
ICAR-Indian Agricultural Research Institute, New Delhi, India
e-mail: ysshivay@iari.res.in

K. Kumar · M. Arif · A. L. Meena · R. Singh
ICAR-Indian Institute of Farming Systems Research, Meerut, Uttar Pradesh, India

R. Mohammadi et al. (eds.), *Genotype x Environment Interactions and its Implications for Plant Breeding*, https://doi.org/10.1007/978-981-95-5664-9_8

amendments can improve yield stability, resource-use efficiency, and resilience to climatic variability. A deeper understanding of soil-mediated G × E interactions will enable the development of resilient cultivars and optimized management strategies, ultimately advancing climate-smart smart sustainable agriculture and ensuring productivity under variable edaphic and environmental conditions.

Keywords

Crop breeding · G × E interaction · Genotype selection · Soil properties

8.1 Introduction

The performance of crop varieties is influenced by both genetic variability and the specific traits embedded within their genotypes. However, genetics alone does not determine crop productivity or adaptability. A crucial factor in shaping varietal performance is the environment, with soil properties playing a central role (Mehdi Mohebodin et al. 2016). Soil characteristics spanning physical (texture, structure, porosity), chemical (pH, nutrient content, cation exchange capacity, etc.), and biological (microbial activity, microbial diversity, organic matter content) significantly contribute to the environment in which a genotype expresses its potential (Fig. 8.1). Even when grown on seemingly similar soil types, the same variety may perform differently due to variations in agro-climatic conditions and subtle differences in soil health indicators. Furthermore, soil conditions interact with crop management practices, leading to different responses among genotypes. For instance, the effectiveness of nutrient uptake, root proliferation, and resistance to soilborne pathogens can vary significantly across soil types, thereby influencing yield and quality. The complexity of genotype × environment (G × E) interactions is amplified when soil properties differ markedly across locations or within a given

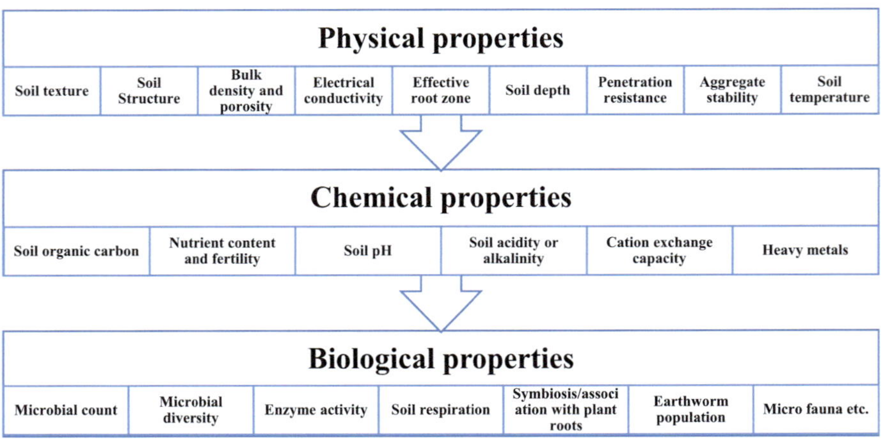

Fig. 8.1 Different soil properties affecting G × E interaction

region (Begna 2020). Such interactions result when genotypes respond inconsistently to changing soil environments, complicating varietal selection (Lal et al. 2010). As a result, two genetically similar varieties may show divergent performance due to contrasting soil moisture retention, salinity levels, or microbial diversity in the rhizosphere.

Agronomic interventions including tillage, irrigation scheduling, residue management, and organic amendments further modify the soil environment, creating a dynamic interface for genotype expression. Not all crop varieties respond uniformly to a particular soil management practice, emphasizing the importance of conducting context-specific evaluations of varietal performance under variable soil conditions. To better understand how soil factors shape varietal adaptability and stability, researchers conduct multi-location and multi-season trials (Natalia de Leon et al. 2016). These trials allow for the identification of genotypes with consistent performance under a range of soil conditions and help isolate genotypes that exhibit minimal G × E interaction. Such varieties are crucial in areas with high soil heterogeneity, where stability in performance becomes more desirable than high-yield potential alone. A major breakthrough in crop improvement lies in breeding genotypes that exhibit both adaptability to specific soil constraints and stability across varying soil environments (Dia et al. 2012). Stability refers to a genotype's ability to maintain steady performance across diverse soil types, while adaptability indicates its specific suitability to a given set of soil conditions (Becker and Leon 1988). Genotypes that consistently yield well across a range of soils with minimal fluctuation are considered both stable and adaptive, making them ideal for large-scale cultivation (Chandrakanth et al. 2016). Understanding the G × E interaction through the lens of soil properties provides valuable insights for developing soil-specific varietal recommendations. By aligning genotype selection with detailed knowledge of soil variability, it becomes possible to enhance crop performance, resource-use efficiency, and resilience in the face of environmental fluctuations.

8.2 Soil as a Component of the Environment

Soil is a fundamental component of the environment and plays a vital role in determining crop growth, productivity, and quality. It significantly influences the interaction between genotype and environment, as well as the effectiveness of crop management practices. Understanding the variability of soil within a field is crucial, as it affects how different genotypes respond to their growing conditions and how management strategies should be adapted (Duvick 2005). Soil properties influence crop performance in multiple ways. Chemically, soil determines the availability of nutrients through its pH, nutrient-holding capacity, and overall fertility status. For example, soil acidity or alkalinity directly affects the solubility and availability of micronutrients like zinc (Zn), copper (Cu), iron (Fe), and manganese (Mn) (Juhos et al. 2016). These nutrients are essential for plant physiological processes, and their availability varies with soil conditions (Katsenios et al. 2021). Physically, the texture, structure, and depth of the soil affect water retention, aeration, root penetration, and the overall effective root zone. Light-textured soils often require frequent irrigation due to their limited water-holding capacity, whereas deeper soils with better structure support more extensive root systems and water access. Crop varieties with deeper roots may perform better under drought conditions compared to those with shallow roots, particularly in soils with limited moisture availability. Biologically, the organic carbon content of soil plays a key role in sustaining microbial populations and enhancing nutrient cycling. Soils rich in organic carbon support beneficial microorganisms, such as plant growth-promoting rhizobacteria (PGPR), which contribute to nutrient availability, disease resistance, and water uptake. The release of root exudates from specific crop varieties influences microbial activity, including the production of siderophores that enhance the availability of micronutrients. Incorporation of biomass, leaf litter, and crop residues further improves soil chemical and biological properties, leading to better crop performance.

Management practices such as tillage, irrigation methods, residue management, and the use of machinery are also shaped by soil characteristics (Anderson 2008; Christensen and Jensen 2023). For instance, irrigation schedules must be tailored to the soil's water-holding capacity, and the choice of tillage methods depends on soil structure and compaction status. Additionally, nutrient-efficient crop varieties offer a significant advantage in soils with low fertility, including those managed under organic systems where nutrient release is gradual. Unlike nutrient-responsive cultivars that require high input levels, nutrient-efficient varieties maintain stable yields across diverse fertility gradients, making them suitable for both conventional and low-input systems (Ghasal et al. 2018). Overall, soil serves as a critical environmental factor in determining the expression of genotypic potential and the success of crop management strategies (Fu 2015). Its influence on nutrient dynamics, water availability, root development, microbial interactions, and physical support underscores its integral role in crop production systems. The subsequent discussion will focus on how different genotypes affect soil properties and how soil, in turn, impacts various physiological processes, yield attributes, and quality of crops.

8.3 Key Soil Properties Influencing G × E Interaction

Soil is considered as a dynamic living natural body, which has a vital role in the global ecosystems by a balance between living and dead (Doran and Zeiss 2000). The G × E interaction plays a vital role in determining crop performance across diverse agro-ecological conditions. Among the many environmental factors, soil properties are particularly influential, as they directly affect the availability of water, nutrients, and root development, thereby modulating genotypic expression. Key soil properties such as texture, structure, pH, organic matter content, cation exchange capacity (CEC), and nutrient availability significantly influence the degree and nature of G × E interactions.

Soil texture, which refers to the proportion of sand, silt, and clay, affects water holding capacity, aeration, and root penetration. Genotypes may perform differently on sandy soils, which drain quickly and have lower nutrient retention, compared to clayey soils, which retain more water but may restrict root growth (Ye et al. 2024). Soil structure, particularly the arrangement and stability of aggregates, governs root development and microbial activity, influencing nutrient cycling and availability. Soil pH determines the solubility of nutrients and the activity of soil microorganisms. Extremes in pH can lead to nutrient deficiencies or toxicities, affecting genotypic performance. Similarly, soils rich in organic matter and with high CEC support better nutrient retention and buffering capacity, enhancing the stability of genotypic performance across environments. Furthermore, site-specific variability in soil fertility and moisture availability due to differences in topography, erosion, or management history can lead to differential genotypic responses. Understanding the interaction between soil properties and genotype performance is thus crucial for breeding programs, precision agriculture, and site-specific management. Tailoring genotype selection and agronomic practices to the underlying soil conditions can help optimize crop productivity and stability across environments, ultimately enhancing agricultural sustainability.

8.3.1 Soil Texture

Soil texture refers to the relative proportion of sand, silt, and clay particles in the soil. Based on the composition of these particles, soils are commonly classified as light, medium, or heavy. The United States Department of Agriculture (USDA) recognizes 12 primary textural classes: sand, loamy sand, sandy loam, loam, silt loam, silt, sandy clay loam, clay loam, silty clay loam, sandy clay, silty clay, and clay. Soil texture profoundly influences several key soil physical properties such as water holding capacity, aeration, nutrient retention, and penetration resistance, as well as the soil microclimate, including temperature, which affects microbial activity essential for nutrient cycling (Dobarco et al. 2019; Ye et al. 2024). Clay particles, due to their high specific surface area, tend to absorb more soil organic matter (SOM; Six et al. 2002). Conversely, sandy soils have a lower water holding capacity due to the dominance of macropores, necessitating more frequent irrigation. Loamy soils, considered ideal for most crops, provide a balanced mix of drainage, aeration, and fertility.

Different genotypes respond uniquely to variations in soil texture. For instance, genotypes with deeper or more branched root systems often perform better in sandy soils, as they can explore deeper moisture zones. In contrast, genotypes with compact root systems may benefit more from nutrient-rich clayey soils (Gil et al. 2012). Moreover, short-duration crop varieties may perform better on light soils during the post-rainy season, using residual soil moisture more efficiently than longer-duration varieties suited for heavier soils. Varieties sensitive to waterlogging also tend to perform better in lighter soils due to better drainage. Water use efficiency, a critical trait under drought-prone conditions, is closely linked to a soil's capacity to retain and supply water. Some genotypes express their full potential only under specific soil textures where their root architecture or nutrient uptake mechanisms align with physical soil conditions. Soil texture also influences bulk density, which in turn affects root penetration and crop productivity (Chaudhari et al. 2013).

Research supports the importance of texture in soil fertility and organic matter dynamics. For example, Gentile et al. (2013) identified soil clay content as a key factor in maintaining the stability of soil organic matter. Similarly, Rakhsh and Golcchin (2018) reported that increased clay content is associated with higher mineral nitrogen, microbial nitrogen, and cation exchange capacity. Other studies, such as those by Wang et al. (2009), have shown that sand and silt content strongly influence the distribution of soil organic matter, total nitrogen, and total phosphorus in small river basins. However, some conflicting findings also exist. Several studies suggest that the correlation between soil texture and organic matter content may be weak or non-significant under certain conditions (Kaiser and Guggenberger 2010; Mao et al. 2024). A deep understanding of soil texture and its interaction with crop genotypes is critical for breeding programs, particularly in the selection of varieties suitable for marginal or stress-prone soils. It also informs the design of site-specific agronomic interventions to optimize crop productivity under varying soil conditions.

8.3.2 Soil Structure and Bulk Density

Soil structure refers to the arrangement of individual soil particles into aggregates or clumps, which governs the amount and connectivity of pore spaces within the soil matrix. Bulk density, defined as the mass of dry soil per unit volume (g/cm^3), indicates the degree of soil compaction and is inversely related to soil porosity. Together, soil structure and bulk density form critical components of the soil's physical environment, directly influencing plant growth by affecting root development, aeration, and water movement. A well-developed soil structure improves soil strength while maintaining sufficient porosity, allowing for efficient infiltration and retention of water and the free exchange of gases within the soil profile (Nichols et al. 2004). These conditions provide a more stable and accessible moisture supply to plant roots. Conversely, compacted or poorly structured soils restrict root penetration and limit the uptake of water and nutrients, negatively impacting the performance of genotypes not adapted to such conditions. In contrast, genotypes with deeper or more vigorous root systems are better equipped to exploit well-structured soils, accessing greater volumes of water and nutrients and thereby enhancing overall growth (Nichols et al. 2004).

Good soil structure also supports root expansion, enabling plants to explore a larger soil volume for resource acquisition. This significantly improves their ability to withstand environmental stresses such as drought or nutrient deficiency (Pardo et al. 2000). In addition, improved soil structure promotes microbial activity and organic matter decomposition, thereby enhancing soil fertility (Salehi and Maleki 2012). It also facilitates oxygen diffusion, which is essential for root respiration and plant metabolism (Franzluebbers 2002). The enhancement of water and oxygen movement due to favorable soil structure enables certain genotypes to more effectively express traits such as drought tolerance, efficient nutrient uptake, and greater biomass production. Thus, soil physical properties do not merely influence immediate plant growth but also shape the extent and expression of genotype-specific traits, determining their suitability, resilience, and productivity under varying soil conditions. Understanding the interaction between soil structure, bulk density, and genotype performance is essential for optimizing crop yields. Breeding programs focused on improving root traits such as root length, angle, branching pattern, and biomass can help develop genotypes better suited for compacted or poorly structured soils. Similarly, the adoption of conservation agriculture practices such as reduced tillage, cover cropping, and organic matter additions can maintain or improve soil structure. These practices create favorable conditions for the expression of desirable genotypic traits, thereby contributing to greater resilience and sustained productivity across diverse agro-ecological regions.

8.3.3 Soil pH and Buffering Capacity

Soil pH, a measure of the hydrogen ion concentration in the soil solution, is a fundamental chemical property that profoundly influences nutrient availability, microbial activity, and overall soil health. It ranges from acidic (<7.0) to alkaline (>7.0), with most crops thriving in a pH range of 6.0 to 7.5. Buffering capacity, on the other hand, refers to the soil's ability to resist changes in pH, primarily governed by clay content, organic matter, and types of minerals present. Soil pH directly affects the solubility of essential nutrients and toxic elements. In acidic soils (pH <5.5), the availability of macronutrients such as phosphorus, calcium, and magnesium decreases, while toxic metals like aluminum and manganese become more soluble, potentially harming plant roots. In alkaline soils (pH >7.5), micronutrients like iron, zinc, and boron become less available, leading to deficiencies that limit crop growth. So, they facilitate the decomposition of organic matter, which leads to an increase in the presence of phosphorus, manganese, and calcium in the soil (Li et al. 2013). The buffering capacity of a soil determines how rapidly these changes occur and how effectively amendments like lime or sulfur can adjust pH. The pH value is an indicator of nutrient cycling. Increasing pH will increase the adsorption of minerals (Bradl 2004).

Genotypic variation plays a critical role in crop response to different soil pH levels. Some genotypes exhibit greater tolerance to high aluminum concentrations in acidic soils through mechanisms like root exudation of organic acids, while others may perform better in calcareous soils with high pH by efficiently utilizing micronutrients under limited availability. These adaptive traits contribute to G × E interactions, as pH-related stress can either enhance or suppress the expression of genetic potential. Furthermore, soil pH strongly influences microbial communities and rhizosphere interactions. Neutral to slightly acidic soils support higher microbial diversity and activity, enhancing nutrient cycling and plant-microbe symbioses such as nitrogen fixation and mycorrhizal associations. Extreme pH levels, however, can suppress beneficial microbes and promote pathogens, negatively affecting plant health. Thus, managing soil pH and understanding genotypic responses to it are essential for optimizing crop performance, particularly under variable agro-climatic conditions.

8.3.4 Soil Organic Matter and Soil Carbon

Soil organic matter (SOM) and soil organic carbon (SOC) are critical determinants of soil health, fertility, and ecosystem productivity. Their roles extend far beyond simple nutrient reservoirs, profoundly influencing nutrient cycling, soil water dynamics, and the activity of soil biological communities. Understanding these processes is crucial for developing sustainable soil management practices that enhance agricultural productivity and environmental resilience (Zhang et al. 2023). Soil organic matter (SOM), a complex mixture of decomposing plant and animal residues, microbial biomass, and humic substances, plays a central role in maintaining soil functions. Soil carbon, a major component of SOM, is crucial for sustaining fertility, ecosystem productivity, and environmental quality (Lal 2004). One of the primary functions of SOM is its role in nutrient mineralization. Nutrient mineralization is the microbial conversion of organic forms of nutrients into inorganic forms that plants can readily uptake. Soil organic matter serves as the primary substrate for this biological process (Sahoo et al. 2023).

It contains essential nutrients, including nitrogen (N), phosphorus (P), and sulfur (S), bound in organic compounds. When soil microorganisms decompose SOM, they release these nutrients through mineralization. This process is tightly regulated by the C:N ratio of organic materials; lower C:N ratios generally promote faster mineralization, while higher ratios can result in immobilization (Brady and Weil 2016). The presence of stable organic carbon pools (e.g., humus) also helps buffer nutrient availability, ensuring a sustained release over time rather than short-term leaching losses. Thus, soils rich in SOM and SOC exhibit improved nutrient cycling efficiency, directly supporting plant growth. Water retention is another critical soil property improved by SOM. Organic matter acts like a sponge, holding substantial amounts of water often 2–20 times its weight. It improves soil structure by binding individual soil particles into stable aggregates. These aggregates increase soil porosity, enhancing infiltration and reducing surface runoff and erosion (Li et al. 2024).

Soils rich in organic matter maintain better moisture levels, which is particularly vital during dry periods. Additionally, SOM helps balance water availability: it holds water in micropores for plant use while maintaining air-filled macropores that ensure proper root respiration. In coarse-textured (sandy) soils, SOM dramatically improves water-holding capacity, while in fine-textured (clayey) soils, it promotes better drainage (Dey et al. 2024). Soil biological activity including microbial biomass, enzymatic activity, and faunal diversity is fundamentally linked to SOM and SOC. Organic matter serves as the primary energy source for soil microorganisms, including bacteria, fungi, actinomycetes, protozoa, and nematodes. Higher SOM levels generally correlate with greater microbial biomass and diversity. These soil organisms drive critical processes such as nutrient cycling, organic matter decomposition, disease suppression, and symbiotic relationships with plants (e.g., mycorrhizal associations) (Chatterjee et al. 2024). The biological activity resulting from abundant SOM contributes to forming humus, stabilizing soil structure, and promoting soil resilience against pests and environmental stresses.

Furthermore, a thriving soil biological community improves soil enzyme activities that facilitate biochemical transformations, enhancing soil fertility. As key source of soil biological activities, SOM provides energy and nutrients for soil organisms; SOC levels in soil influence the microbial diversity and enzyme activities and promote ecosystem resilience through enhanced biological activities (Li et al. 2023). SOM influences plant growth by enhancing soil structure, supplying essential nutrients through mineralization, and fostering beneficial soil microbial communities. Higher SOM generally improves soil health, but not all plant genotypes can utilize enhanced SOM equally. Some genotypes are specifically adapted to thrive under low SOM conditions, while others show significant improvements in growth and yield when SOM levels are high. Plant species and genotypes vary with respect to the degree of SOM level and SOM mineralization mediated by these genotypes because various genotypes of different crops differ in amount and composition of rhizodeposits that structure the rhizosphere microbial community and mineralization of SOM (rhizosphere priming effect) (Yin et al. 2019).

The SOM mineralization and nutrient availability in root–soil interaction can be manipulated through genotype breeding. Genotype traits, that is, root biomass, root diameter, specific root length and root density and root length define the nutrient absorption capacity of specific genotype roots. The genotype variation in these traits can impact the nutrient dynamics and SOM mineralization in soil which will be useful in improved fertilizer use efficiency (Kumar et al. 2018). Genotypic variation in response to SOM often involves traits related to root architecture, nutrient uptake efficiency (NUE), and microbial interactions. For instance, some genotypes may develop deeper or more branched root systems in response to higher SOM, enabling better exploitation of nutrient-rich zones. Others may exhibit more efficient nutrient transport mechanisms or form stronger symbiotic relationships with mycorrhizal fungi and nitrogen-fixing bacteria, enhancing their ability to capitalize on the organic matter present. Additionally, differences in carbon and nitrogen metabolism at the genetic level affect how plants respond to SOM (Guo et al. 2023).

Certain genotypes are better equipped to use organic nitrogen forms or exudate root compounds that stimulate microbial activity, which in turn can accelerate SOM decomposition and nutrient release. These interactions create positive feedback loops where plant genotype and soil organic processes mutually enhance each other. However, the relationship between genotype and SOM is not uniformly positive (Gomez-Becerra et al. 2010). In some cases, high SOM levels can lead to nutrient imbalances, particularly in systems where microbial immobilization of nitrogen limits plant availability. Some genotypes are more sensitive to these imbalances and may perform worse under high SOM compared to others that can maintain homeostasis. This variability highlights the importance of G × E interaction in determining plant performance. Breeding programs increasingly recognize the need to select for genotypic traits that optimize the use of SOM. This is particularly important in sustainable agriculture systems, where reliance on external chemical fertilizers is minimized.

By selecting genotypes that are highly responsive to organic nutrient sources and can perform well under varying SOM conditions, breeders can develop crops that are both high-yielding and environmentally resilient. Moreover, climate change is expected to alter SOM dynamics through its effects on temperature, moisture, and microbial activity. Genotypes that are adaptable to fluctuating SOM conditions will be critical for future food security. Research combining genomics, soil science, and agronomy is helping to identify key genes and traits associated with SOM responsiveness, providing new tools for crop improvement (Lal 2020). Soil organic matter (SOM) is a critical component of soil health, playing a multifaceted role in ecosystem stability and resilience. Comprising decomposed plant and animal residues, living microbial biomass, and complex organic compounds stabilized within the soil matrix, SOM serves as a dynamic buffer against environmental extremes such as drought, heavy rainfall, temperature fluctuations, and soil degradation.

One of the most vital functions of SOM is its impact on soil water dynamics. Organic matter improves soil structure by promoting the aggregation of soil particles, which enhances porosity and increases water infiltration rates. This structure allows soils to retain greater amounts of water in micropores, thus providing a reservoir of moisture accessible to plants during periods of drought. Conversely, during intense rainfall events, well-structured soils with higher organic matter content exhibit improved drainage and reduced surface runoff, thereby mitigating erosion and nutrient loss. Thermal buffering is another critical service provided by SOM (Oishy et al. 2025). Organic-rich soils have a higher specific heat capacity, meaning they warm up and cool down more slowly than soils low in organic matter. This moderation of soil temperature protects plant roots and soil organisms from rapid or extreme temperature swings, maintaining biological activity and nutrient cycling even under stressful climatic conditions. Soil organic matter also plays a crucial role in stabilizing soil fertility (Bashir et al. 2021).

It acts as a reservoir of essential nutrients such as nitrogen, phosphorus, and sulfur, releasing them slowly through microbial decomposition. This slow nutrient release ensures more consistent plant growth even when external inputs or environmental conditions fluctuate. Moreover, SOM enhances the soil's cation exchange capacity (CEC), improving the soil's ability to retain and supply positively charged nutrient ions to plants, particularly under conditions of leaching due to excessive rainfall. Beyond water and nutrient regulation, SOM fosters biological diversity and activity within the soil ecosystem (Wang et al. 2024). A diverse soil biota enhances ecosystem resilience by supporting various functional groups that can respond differently to environmental stresses. Soil organisms also contribute to the formation and stabilization of soil aggregates, reinforcing the positive feedback between organic matter and soil structure. The role of SOM in carbon sequestration is increasingly recognized in the context of climate change mitigation. By storing carbon in stable forms within the soil, SOM can help reduce atmospheric carbon dioxide concentrations, thereby indirectly buffering ecosystems against climate-induced environmental extremes (Geng et al. 2022).

8.3.5 Soil Nutrient Content and Fertility Status

Soil nutrients and fertility status are crucial factors influencing plant growth and agricultural productivity. Essential nutrients, including macronutrients, that is, nitrogen (N), phosphorus (P), potassium (K), sulfur (S), magnesium (Mg), and calcium (Ca), and micronutrients, that is, iron (Fe), zinc (Zn), copper (Cu), manganese (Mn), molybdenum (Mo), boron (B), and chlorine (Cl) must be present in balanced amounts. Soil fertility depends on nutrient availability, organic matter content, pH, and microbial activity. Depletion through intensive cropping, erosion, and leaching can reduce fertility, while sustainable practices like crop rotation, organic amendments, and balanced fertilization can restore it (Javed et al. 2022). Soil fertility, plant health, and agricultural productivity fundamentally depend on the availability of essential nutrients in the soil. These nutrients, categorized into macronutrients and micronutrients, are critical components that act as soil-driven variables, influencing not only plant growth but also the soil's chemical and biological dynamics. The primary macronutrients include nitrogen (N), phosphorus (P), and potassium (K), often abbreviated as N, P, K. Secondary macronutrients include calcium (Ca), magnesium (Mg), and sulfur (S).

In contrast, micronutrients are needed in much smaller amounts but are equally crucial. These include iron (Fe), manganese (Mn), zinc (Zn), copper (Cu), boron (B), molybdenum (Mo), chlorine (Cl), and nickel (Ni). The nutrients in soil plays a crucial role in various activities as soil driven variables. Efficient nutrient use is fundamental for sustainable agricultural production. Nutrient use efficiency (NUE) and phosphorus use efficiency (PUE) are critical components determining crop yield, quality, and environmental sustainability. Enhancing NUE and PUE is critical because the excessive use of fertilizers leads to environmental concerns like water eutrophication, greenhouse gas emissions, and soil degradation (Chowdhury and Zhang 2021). At the same time, resource limitations and rising fertilizer costs make it essential to produce more food using fewer inputs. Importantly, genotypic variability in NUE and PUE offers a promising avenue for breeding crops that perform better under low-input conditions or make better use of fertilizers. Understanding this genetic diversity is essential for developing nutrient-efficient varieties that can contribute to food security while minimizing environmental impacts.

Plants exhibit substantial genetic variability for nutrient acquisition, internal nutrient use, and response to nutrient deficiencies. This variability is a function of multiple traits, including root architecture, nutrient transport efficiency, nutrient assimilation pathways, and remobilization of stored nutrients. Among the genotype variability some of the important sources are:

Root Traits: Root length density, root hair length, root angle, and specific root architecture traits (e.g., deeper vs. shallow rooting) greatly influence nutrient acquisition, especially for immobile nutrients like phosphorus. Genotypes with more extensive or finer root systems often show higher nutrient uptake (Esfahani and Sonnewald 2024).

Physiological and Biochemical Traits: Variations in nutrient transporters, efficiency of nutrient assimilation enzymes (e.g., nitrate reductase), and internal recycling mechanisms contribute to differences in NUE and PUE. Some genotypes can maintain higher photosynthetic rates or better carbohydrate partitioning under nutrient-limited conditions.

Molecular and Genetic Regulation: Quantitative trait loci (QTL) mapping and genome-wide association studies (GWAS) have identified numerous genetic loci associated with nutrient efficiency traits.

Candidate genes such as *NRT1* (nitrate transporters) and *PHT1* (phosphate transporters) show significant polymorphisms across genotypes affecting their nutrient uptake capacity. *Phenological Traits*: Differences in growth duration and developmental timing influence nutrient uptake patterns. Early maturing varieties might differ significantly in their nutrient use dynamics compared to late-maturing ones. Different genotypes respond to NUE and PUE in significant ways. Significant genotypic differences are reported in crops like wheat, rice, maize, and legumes (Sharma et al. 2021). For example, some traditional landraces outperform modern cultivars in low-input systems due to their inherent adaptation to nutrient-poor soils. Conversely, modern breeding under high-input systems may have narrowed the genetic base for nutrient efficiency. Nitrogen use efficiency often comprises two components: Nitrogen uptake efficiency (NUpE) and nitrogen utilization efficiency (NUtE). Some genotypes excel in acquiring nitrogen under low availability (high NUpE), while others are more effective at converting absorbed nitrogen into biomass or grain (high NUtE).

Similarly, PUE includes phosphorus acquisition efficiency (PAE) and phosphorus utilization efficiency (PUTE). Genetic studies have identified cultivars capable of superior phosphorus uptake via increased root exploration or symbiotic associations (e.g., mycorrhizal fungi), and others that achieve better biomass production per unit phosphorus absorbed (Madan and Raghuram 2024). Among the key strategies to harness the genotype variability are:

- *Breeding and Selection*: Screening germplasm collections for NUE and PUE traits under low nutrient conditions can identify superior genotypes. Marker-assisted selection (MAS) and genomic selection (GS) offer tools to accelerate the breeding of nutrient-efficient varieties.
- *Biotechnological Approaches*: Genetic engineering, such as the overexpression of nutrient transporter genes or regulatory transcription factors, has shown promise in improving nutrient efficiency in several model and crop plants.
- *Root System Engineering*: Selection for desirable root traits, either via phenotypic selection or genomic approaches, can improve nutrient acquisition efficiency, particularly under nutrient-poor conditions.

- *Integrated Management*: Understanding genotype × environment × management (G × E × M) interactions is crucial. Some genotypes might perform well under certain fertilization regimes or soil types but not others, necessitating tailored agronomic practices.

The interaction between plant genotype and soil amendments—particularly organic and inorganic inputs—is a critical factor in sustainable crop production and agro-ecosystem resilience. This interaction, often described as G × E interaction, emphasizes that plant performance is not solely determined by genetics or soil management alone, but by their interplay (Yusuf et al. 2025). Soil amendments, whether organic (e.g., compost, manure, biochar) or inorganic (e.g., chemical fertilizers, lime), influence nutrient availability, microbial activity, and soil structure. Different genotypes respond uniquely to these inputs due to inherent physiological, morphological, and biochemical traits. Organic amendments contribute to soil fertility by enhancing soil organic matter, improving water retention, stimulating microbial activity, and gradually releasing nutrients. Genotypes vary in their ability to utilize these complex nutrient sources. For instance, some genotypes have a higher root surface area, increased root exudation, or a greater capacity for symbiosis with beneficial microbes like arbuscular mycorrhizal fungi or nitrogen-fixing bacteria (Laishram et al. 2024).

These traits allow them to more efficiently acquire nutrients from organic sources. Studies in rice and maize have demonstrated that certain varieties perform better with organic manures due to enhanced nutrient-use efficiency and better root–microbe interactions. Moreover, organic systems often favor genotypes with slower growth rates but better nutrient conservation and resilience to abiotic stress. Inorganic fertilizers offer immediate nutrient availability but can lead to leaching and long-term soil degradation if not managed properly (Liu et al. 2024). Plant genotypes differ in their response to synthetic inputs depending on traits like nutrient uptake efficiency, root architecture, and transport protein expression. For example, wheat and maize varieties have shown variable responses to nitrogen or phosphorus fertilization due to genetic differences in root length density or nitrate transporter activity. These differences are often exploited in breeding programs targeting fertilizer-use efficiency. Some genotypes are "responsive" (high yield under high inputs), while others are "efficient" (maintain yield under low inputs) (Hu et al. 2023).

Understanding genotype interactions with soil amendments supports the development of nutrient-efficient varieties tailored to specific management systems. In low-input systems, such as organic farming, breeding efforts focus on genotypes that can thrive under nutrient limitations. In contrast, conventional systems may require varieties optimized for high responsiveness to chemical fertilizers (Van Bueren et al. 2011). Integrating genotype selection with site-specific soil amendment practices can enhance yield stability and reduce environmental impacts. Quantitative trait loci (QTL) mapping and genome-wide association studies (GWAS) have identified genes linked to nutrient acquisition and utilization that are influenced by soil amendments. Incorporating this knowledge into breeding programs enables the development of genotypes optimized for particular soil amendment strategies. Furthermore, long-term field trials and participatory breeding approaches involving farmers are crucial to understand real-world G × E interactions and promote the adoption of suitable genotype-input combinations (Zhu et al. 2021).

8.3.6 Soil Moisture Regimes and Water Holding Capacity

Soil moisture regimes and water holding capacity play a crucial role in determining how long soil can supply moisture to plants, influencing their tolerance to both drought and waterlogging. Plant genotypes exhibit plasticity by adjusting traits like root structure and leaf behavior to cope with varying soil moisture levels. These responses vary across environments, as water stress triggers G × E interaction that modify root and shoot traits to optimize water use. Understanding and managing these dynamics are key to improving crop performance under fluctuating moisture conditions.

8.3.6.1 Role of Available Water Capacity (AWC) in Drought and Waterlogging Responses

Available water capacity (AWC), defined as the range of water held between field capacity and the permanent wilting point, is a vital soil property influencing plant responses to both drought and waterlogging (Hillel 2004). AWC determines how much usable water is stored in the soil and, consequently, how long plants can access moisture without experiencing stress. In drought-prone conditions, soils with higher AWC typically clay loams or loamy soils act as a moisture reservoir, sustaining water availability during dry periods. This extended water retention helps maintain physiological functions such as stomatal conductance, photosynthesis, and transpiration for a longer duration, thereby supporting plant growth and yield stability (Blum 2011). Crops like cassava (*Manihot esculenta*), which are frequently grown in marginal soils, are particularly influenced by AWC during sensitive growth stages like tuber initiation and bulking (30–150 days after planting) (Pereira et al. 2018; Silva et al. 2021). In contrast, sandy or degraded soils with low AWC lose water quickly through deep percolation or surface evaporation, intensifying drought effects. These conditions result in early stomatal closure, reduced leaf water potential, and inhibited cell growth, ultimately limiting biomass accumulation and productivity (Chaves et al. 2003; Farooq et al. 2009). Conversely, in regions with high rainfall or poor irrigation management, soils with high AWC may become prone to waterlogging if drainage is inadequate. In such cases, excess water fills pore spaces, displacing oxygen and creating anaerobic conditions that impair root respiration and nutrient uptake (Kirk et al. 2014). Even drought-tolerant crops like cassava can suffer root damage and reduced metabolic activity under prolonged saturation. Thus, optimizing AWC through soil amendments, mulching, or drainage improvement is essential for mitigating both drought and waterlogging stress. Understanding the dual role of AWC in supporting plant water availability while preventing saturation stress is critical for designing climate-resilient cropping systems.

8.3.6.2 Genotypic Plasticity Under Different Soil Moisture Scenarios

Genotypic plasticity describes a plant genotype's capacity to adjust its physiological and structural traits in response to fluctuating soil moisture conditions. This adaptability is essential for maintaining productivity and survival in environments characterized by drought or excess water. Different genotypes respond uniquely to moisture extremes through various adaptive traits that enhance resilience. Under drought conditions, genotypes with high plasticity often develop deeper or more extensively branched root systems that enable access to deeper water reserves. These genotypes may also exhibit traits such as reduced leaf area or stomatal regulation to conserve water. In contrast, genotypes exposed to waterlogged conditions typically produce adventitious roots and form aerenchyma tissues, which facilitate oxygen movement within the root zone, mitigating the effects of hypoxia (Colmer and Voesenek 2009; Tardieu 2012). Research on Brachypodium species has demonstrated wide variation in root development and biomass allocation among genotypes across a continuous moisture spectrum, emphasizing the diversity of plant responses to soil moisture gradients (Des Marais et al. 2017). Similarly, cassava (*Manihot esculenta*) exhibits notable genotypic differences under moisture stress. Certain genotypes modify their root architecture, leaf anatomy, and stomatal behavior to optimize water uptake or limit transpiration losses, thus maintaining growth under both drought and water-saturated conditions (Setter and Fregene 2007; Miranda et al. 2020). Additionally, many plant genotypes activate osmotic adjustment mechanisms under water stress by synthesizing osmoprotectants like proline and soluble sugars. These compounds help preserve cellular function and structure during dehydration (Bartels and Sunkar 2005). Therefore, genotypic plasticity, expressed through an integrated suite of morphological and biochemical traits, plays a central role in plant adaptation to variable soil moisture environments. Harnessing this variability is key to breeding climate-resilient crops suited for diverse agro-ecological zones.

8.3.6.3 Water Stress-Induced G × E Variation in Root and Shoot Traits

Water stress induces significant G × E interactions, particularly in root and shoot traits, which are key to plant adaptation in variable soil moisture conditions. Traits such as root depth, root length density, biomass allocation, leaf water potential, and shoot elongation are directly influenced by water availability and exhibit marked variation across genotypes. In drought-prone environments, plants often adjust by increasing root-to-shoot ratios to optimize water uptake (Comas et al. 2013). However, these morphological adjustments vary across genotypes, indicating that G × E interactions play a critical role in plant responses to water stress. In crops like maize and wheat, deep-rooted genotypes tend to perform better under water deficit by maintaining stable yields, whereas shallower-rooted genotypes are more suited to environments with adequate moisture (Cooper et al. 2014). This variation emphasizes the importance of selecting specific traits based on the environmental conditions in crop breeding programs. Cassava (*Manihot esculenta*), known for its drought tolerance, exhibits genotype-specific changes in root depth, stomatal regulation, and shoot biomass under water stress (El-Sharkawy 2004). These modifications are coupled with the production of osmoprotectants and antioxidant enzymes, which help mitigate oxidative stress and enhance plant resilience (Hasanuzzaman et al. 2018). Furthermore, G × E interactions under water stress also influence reproductive traits, such as seed dormancy and germination, which are crucial for plant establishment and survival in marginal environments (Tuberosa 2012; Blum 2011). Understanding these complex G × E dynamics is essential for developing drought-resistant genotypes that are tailored to different agro-ecological zones, ensuring stable yields under varying moisture conditions.

8.3.7 Soil Salinity and Sodicity

Soil salinity and sodicity severely impair plant growth by increasing osmotic pressure and disrupting nutrient uptake. Elevated concentrations of sodium and chloride limit water availability and compete with essential ions like potassium and calcium, leading to physiological drought and metabolic dysfunction (Munns and Tester 2008; Rengasamy 2010). Crop tolerance to salinity is highly influenced by G × E interactions, as genotypes respond differently under varying soil textures, pH, and salt concentrations (Roy et al. 2014; Ismail and Horie 2017). Sustainable solutions require integrating soil amendments such as gypsum and organic matter with salt-tolerant cultivars to rehabilitate degraded soils and maintain yields (Qadir et al. 2007). This combined strategy is particularly relevant in salt-prone regions.

8.3.7.1 Accumulation of Salts Affecting Osmotic and Ionic Balance

Soil salinity and sodicity significantly impair plant growth by altering the osmotic and ionic balance in the root environment. Salinity arises from the accumulation of soluble salts such as sodium (Na^+), chloride (Cl^-), and sulfate (SO_4^{2-}), while sodicity is characterized by high exchangeable sodium percentages on the soil's cation exchange complex (Munns and Tester 2008). Both conditions reduce the soil's osmotic potential, limiting water uptake even when soil moisture is present, this phenomenon is commonly termed physiological drought (Greenway and Munns 1980). Under saline conditions, sodium ions compete with essential nutrients like potassium (K^+), magnesium (Mg^{2+}) and calcium (Ca^{2+}), disrupting ionic homeostasis within plant tissues. This imbalance leads to membrane instability, enzyme inhibition, and nutrient deficiencies, which manifest as chlorosis, leaf burn, and overall stunted growth (Rengasamy 2010; Reid and Hayes 2003). In salt-affected soils, particularly those in arid and semi-arid zones, these effects are aggravated due to low rainfall and high evapotranspiration rates, which favor salt accumulation near the soil surface (Rhoades 1997; Qadir et al. 2001). The use of poor-quality irrigation water, inadequate drainage, and rising water tables further contribute to salt buildup. The excessive accumulation of salts such as sodium bicarbonate and chloride deteriorates soil physical properties, leading to crusting, poor infiltration, and limited aeration. Within the plant, toxic levels of Na^+ and Cl^- disrupt metabolic activities, cause oxidative stress, and reduce photosynthetic efficiency (Litalien and Zeeb 2020). To mitigate salt stress, plants employ mechanisms like ion exclusion, vacuolar sequestration, and osmoprotectant synthesis, but these adaptations demand high energy and often reduce overall productivity (Munns et al. 2019). Thus, salinity-induced osmotic and ionic stress poses a serious challenge to crop growth and food security in vulnerable agro-ecosystems.

8.3.7.2 Soil-Specific G × E for Salt-Tolerant Genotypes

The response of crops to salinity and sodicity is significantly influenced by the interaction between genotype and environment, with soil factors such as texture, pH, electrical conductivity (ECe), and sodium adsorption ratio (SAR) playing a crucial role. Different genotypes exhibit varied mechanisms for coping with salt stress, such as limiting sodium uptake, compartmentalizing salts in vacuoles, or synthesizing compatible solutes like proline to manage osmotic stress (Roy et al. 2014). These adaptive strategies are not uniform across all genotypes, indicating that salinity tolerance varies based on both the genotype and the soil characteristics. According to Ismail and Horie (2017), wheat and rice genotypes show considerable variation in their tolerance to salinity under different environmental conditions, necessitating that genotype evaluations for salt tolerance be conducted in the context of specific local soil types. Salt tolerance is therefore not a fixed trait for any given genotype but is highly dependent on the local soil environment, emphasizing the need for tailored breeding programs that address these specific environmental challenges.

The G × E interaction is particularly important when considering the complex mechanisms through which plants respond to salt stress. These include ion exclusion, osmotic adjustment, and the production of protective molecules (Munns and Tester 2008). Environmental factors such as soil salinity levels, pH, and moisture availability heavily influence how well a genotype performs in saline conditions (Yamaguchi and Blumwald 2005). Studies in Ethiopia have found that quinoa, a salt-tolerant crop, shows differing levels of salt tolerance depending on the specific environmental conditions, highlighting the importance of selecting genotypes suited to local conditions (Fikadu et al. 2021). Selecting salt-tolerant genotypes that are specifically adapted to the unique environmental and soil conditions of these areas is critical for enhancing agricultural productivity in salt-affected regions (Qadir et al. 2007). Crops like barley, quinoa, and various legumes have shown strong salt tolerance and are actively used in reclamation projects, where their ability to regulate water and ions plays a key role in improving soil health and crop productivity (Ben-Gal et al. 2010). Thus, integrating G × E interactions into breeding programs is essential to develop varieties that thrive in specific environmental conditions. Region-specific breeding strategies will allow for the identification of salt-tolerant genotypes that can withstand the challenges posed by saline and sodic soils, ensuring sustainable agricultural productivity.

8.3.7.3 Importance of Soil Reclamation and Genotype Adaptation

Addressing the challenges of salinity and sodicity in agriculture requires both agronomic and genetic strategies. Soil reclamation methods, such as the application of gypsum to address sodicity, the addition of organic matter, and the use of good-quality water for leaching, help restore soil productivity (Qadir et al. 2007). When combined with the use of salt-tolerant crop varieties, these reclamation practices can help sustain crop yields in areas with moderate salinity, providing an effective and sustainable solution for salt-affected soils. The integration of soil rehabilitation and the selection of appropriate genotypes has been emphasized as an effective strategy for managing saline and sodic soils (Dong et al. 2023). This combined approach is crucial for ensuring long-term soil fertility and productivity, especially in areas where salinity and sodicity are significant challenges to agriculture.

Soil reclamation strategies aim to reduce the salt content of soils and restore their fertility through a variety of chemical, biological, and physical methods (Qadir et al. 2014). Among these, chemical treatments like gypsum and sulfuric acid help displace sodium ions with calcium ions, improving soil structure and enhancing its capacity to retain water (Rengasamy 2006). Moreover, effective irrigation practices, including leaching and maintaining proper groundwater levels, are essential in preventing further salinization and promoting sustainable soil productivity (Qadir et al. 2001). In addition to chemical amendments, other techniques such as precision irrigation and fertigation help manage salinity by improving water use efficiency and nutrient delivery (Tal 2016; Azad et al. 2020). Crop-based strategies, such as intercropping with salt-tolerant legumes and green manuring, also improve soil structure and microbial activity, which reduces salt stress (Cuevas et al. 2019). Furthermore, agroforestry systems, using salt-tolerant trees like mesquite (*Prosopis juliflora*), help reduce salt accumulation and improve soil stability (Dagar et al. 2001).

On the other hand, genotype adaptation focuses on selecting and breeding crop varieties that can tolerate high salinity while still producing viable yields. Salt-tolerant genotypes play a critical role in overcoming the limitations posed by salinity and sodicity (Zhu 2001). In Ethiopia, the development of salt-tolerant varieties such as quinoa and barley has significantly improved crop yields under saline conditions (Fikadu et al. 2021). These varieties possess traits like enhanced ion exclusion, osmotic adjustment, and improved root growth, which enable them to perform well in saline environments (Munns and Tester 2008). Integrating these tolerant genotypes with soil reclamation practices provides a comprehensive approach to improving agricultural output in salt-affected regions (Shannon et al. 2017). Therefore, integrating soil reclamation techniques with the use of salt-tolerant genotypes is essential for sustainable agriculture in saline-prone regions. This combination helps mitigate the negative impacts of salinity, restores soil health, and ensures long-term agricultural productivity (Hassan and Tsegaye 2019).

8.3.8 Soil Biological Properties

Soil microbiological properties play a crucial role in the G × E interaction of crop varieties, influencing plant growth, nutrient uptake, and stress tolerance. The composition and activity of soil microbes directly impact the availability of nutrients and water to plants, as well as their ability to withstand biotic and abiotic stresses. Different crop varieties exhibit varying degrees of interaction with the soil microbiome, leading to differences in performance under the same environmental conditions due to three-way interaction between plant genotype (G), environment (E), and microbiomes (M). Both plant varieties (genotypes) and microbial communities can shape and be shaped by each other, and the environment (like soil) mediates this relationship. Some genotypes are more efficient in nutrient uptake or more tolerant to abiotic or biotic stresses. Root architecture and microbial associations (like mycorrhizae) vary by genotype and can be influenced by soil. Soil-driven stress (like drought or nutrient deficiency) may trigger different gene expression patterns in different genotypes.

8.3.8.1 Role of Plant Genotypes/Varieties in Influencing G × E for Microbes

Crop varieties influence G × E interaction with microbes by affecting the plant's phenotype, which in turn affects microbial communities. Different variants/strains/varieties of a host can shape a unique microbiome (Berlow et al. 2024). Different plant genotypes can alter microbial communities through various mechanisms, including root and leaf morphology, exudate production, and plant defense compounds. These changes in the plant's physical and chemical characteristics can influence the types and abundance of microbes present in the soil and on plant surfaces, which can directly affect the type of plant growth-promoting activities in the plant rhizosphere.

Root and Leaf Morphology

The structure of roots and leaves can influence the types of microbes that colonize these tissues. For instance, certain root architectures may be more conducive to colonization by specific types of bacteria or fungi. Plant genotypes influence microbial communities through variations in their phenotypic traits, such as root and leaf morphology (King et al. 2021), exudate production (Saunders and Kohn 2009; Bailey et al. 2005), and G × E interactions affecting traits like root architecture (Lareen et al. 2016). Genotypes also differ in the type and quantity of compounds they produce either as direct defenses against microbial attack (Van Der Ent et al. 2008) or as secondary metabolites that, while not primarily antimicrobial, still influence the structure of associated microbial communities (Voges et al. 2019). These genetically determined phenotypic differences have been shown to affect both specific microbial taxa and the overall composition of plant-associated microbiomes (Ahlholm et al. 2002). The strong existing evidence that plant genotype plays a critical role in shaping microbial associations suggests that evolutionary changes in the genotype of invasive species could drive shifts in their microbial partners-potentially influencing their success and spread in novel environments.

The growth rates and gravitropic responses of different root types within a root system, such as the primary axis, seminal axes, adventitious roots, and lateral (branch) roots, determine the timing and spatial distribution of rhizosphere development within the soil profile. Gaining a detailed understanding of when and where each root type develops in the soil can help predict which microbial communities are likely to colonize the rhizospheres of specific root types at various depths. Different root types are often colonized by distinct bacterial and fungal communities. For instance, in maize, seminal and adventitious roots harbor similar bacterial populations, yet adventitious roots support fewer fungi (Watt et al. 2006). Likewise, vesicular-arbuscular mycorrhizal (VAM) fungi infect first-order lateral roots of subclover at twice the rate, per unit length, compared to primary roots (Watt et al. 2006). In cluster roots, rootlets exhibit stage-specific bacterial communities that correlate with their developmental phase and patterns of anion release. These differ from the microbial populations associated with non-proteoid roots (Marschner et al. 2002). As research continues to identify the genes controlling the formation of distinct root types (e.g., in maize; Hochholdinger et al. 2004), there is growing potential to breed plants with root architectures that selectively foster beneficial rhizosphere microorganisms through targeted root type development.

Root Exudates Vary by Genotype

Different plant varieties secrete different sugars, amino acids, and secondary metabolites into the rhizosphere. These compounds act as signals or nutrients for specific microbes and play a crucial role in deciding microbial communities (bacteria, fungi, etc.) in rhizosphere region. Root exudates can suppress or promote pathogens or beneficial microbes. The composition of root exudates is highly intricate, encompassing a wide range of substances such as sugars, sugar alcohols, sugar phosphates, amino acids, organic acids, nucleotides, peptides, fatty acids, inorganic compounds, secondary metabolites, and hormones (Sasse et al. 2018; Preece and Peñuelas 2020). Moreover, both the type and quantity of these exudates are greatly affected by factors like plant genotype or variety and the stage of plant development (Yang et al. 2015; Sasse et al. 2018; Liu et al. 2023). For example, wild-type tomatoes produce higher amounts of specific root exudates, including tartaric acid, esculetin, 3-hydroxyflavone, riboflavin, and salicylic acid compared to rin mutant tomatoes (Yang et al. 2023). Additionally, the processes involved in the biosynthesis, secretion, and transport of root exudates are likely governed by numerous QTLs or genes. It is also important to recognize that the recruitment of microbes associated with disease resistance often depends on a complex mixture of various exudates, rather than a single compound. In essence, variations in the composition and concentration of root exudates can shape distinct rhizobiome profiles. Therefore, root exudates are considered as a bridge connecting rhizosphere microorganisms and plant roots.

Plant Immunity and Genotype

Plant immune responses can be triggered through interactions with various microbes. While pathogenic microbes typically induce systemic acquired resistance (SAR), beneficial microbes are known to activate induced systemic resistance (ISR) in plants (Vlot et al. 2021). Recently, there has been a growing body of evidence showing that many rhizosphere-associated microbes can stimulate plant immunity. For instance, four bacterial strains *Bacillus niacini* HRS2, *Bacillus luciferensis* HRS4, *Brevibacterium frigoritolerans* HRS1, and *Solibacillus silvestris* HRS3 are specifically enriched in the rhizosphere of healthy tomato plants. Although these strains do not directly inhibit *Ralstonia solanacearum*, a synthetic community composed of all four strains significantly enhances the expression of jasmonic acid (JA) and salicylic acid (SA) pathway marker genes, leading to a stronger immune response compared to the response elicited by individual strains (Lee et al. 2021). In a similar case, three *Bacillus* strains (B56, B18, and B76) have been reported to not only suppress the in-vitro growth of *Verticillium dahliae* but also to induce ISR in tomato plants using a split-root experimental setup (Zhou et al. 2023).

8.3.8.2 How Microbes Influence G × E for Plants

Modifying Genotype Performance

Microbes significantly influence G × E interactions in plants by affecting various aspects of plant growth and development, including nutrient uptake, stress tolerance, and disease resistance. This influence arises from the diverse ways microbes interact with plants, impacting their physiology, gene expression, and overall response to environmental conditions. Microbes can enhance nutrient uptake (e.g., N-fixing bacteria, P, K, Zn, Fe, etc.-solubilizing bacteria), boost stress tolerance (e.g., drought, salinity), suppress diseases or trigger systemic resistance. These benefits can vary depending on genotype; some genotypes respond strongly, others weakly or not at all (Table 8.1).

Microbes play significant role in building resistance or tolerance against various kind of stresses. Microbes support plants under drought conditions through different mechanisms such as increased root colonization by mycorrhiza, increased leaf chlorophyll activity, improved root, shoot biomass by producing growth hormones, increased Abscisic acid level and decreased stomatal conductance thus retarded water loss and help the plant to survive. In the same way, microbes perform different defense/survival mechanisms in interaction with plant under salinity, heat, and heavy metal stress (Table 8.2).

8.4 Spatial and Temporal Variability of Soil Properties

Soil is a complex dynamic medium and is critical to terrestrial plant growth and productivity. It provides essential nutrients, water, and a medium for root development while hosting a diverse array of biological organisms. Soil properties are inherently variable across space and time, influencing plant performance and agricultural outcomes. The uneven distribution of soil properties within a specific location, known as spatial variability, is a common characteristic of most terrestrial ecosystems. Spatial variability describes differences in soil attributes across a given area, whereas temporal variability pertains to changes in these attributes over time. A thorough understanding of both types of variability is essential for enhancing crop productivity and refining management strategies. Spatial heterogeneity occurs at multiple scales—from micro-variations within a field to macro-scale differences between fields or regions. Similarly, temporal fluctuations in soil properties due to

Table 8.1 Microbial functions/activities affecting the plant growth and development

Plant growth promotion activity	Microbes involved	Mechanisms
Nitrogen fixation	*Rhizobium, Azotobacter, Azospirillum,* Frankia, Blue green algae, etc.	Symbiotic, free living and associative symbiotic N fixation
Phosphate, solubilization	*Bacillus, Pseudomonas, Aspergillus,* etc.	Organic acids production, Phytase enzymes, Chelation of iron by siderophores, H^+ ion production
Phosphate mobilization	Ecto and Endomycorrhiza	Movement from deeper zone to root zone through fungal hyphae
Organic matter decomposition	Actinomycetes, *Trichoderma, Aspergillus, Phanerochaete*, etc.	Production of cellulolytic, lhemicellulolytic and ignolytic enzymes
Root and shoot development	*Bacillus, Pseudomonas, Azotobacter, Azospirillum*, etc.	Production of plant growth hormones
Disease suppression	*Trichoderma harzianum, Pseudomonas florescence*, *Bacillus,* Actinobacteria, *Stenotrophomonas*	Modification in root exudate composition with higher amounts of small lipophilic compounds, abundance of esters and phenolic compounds, antibiotic production, siderophore production, Hydrogen Cyanide (HCN) production, hydrolytic enzymes, growth hormones production, increased expression of defense-associated enzymes such as peroxidase and polyphenol oxidase, high levels of antioxidants, 1-aminocyclopropane-1-carboxylate (ACC) deaminase activity, callose deposition, etc.
Insect biocontrol	*Bacillus, Metarhizium, Beauveria bassiana*, Nuclear Polyhedrosis Virus (NPV), etc.	Predation, parasitism, herbivory

seasonal changes in moisture, temperature, and biological activity further complicate the soil-plant relationship (Cambardella and Karlen 1999).

8.4.1 Intra-Field and Inter-Field Soil Variability

Intra-field variability refers to the differences in soil properties observed within a single field. These variations are driven by factors such as micro-topography, organic matter distribution, water flow patterns, historical management and land use patterns, and biological activity (Robertson et al. 1997). Soil properties vary spatially due to combined impact of physical, chemical, and biological processes operating in soil along with human/animal activities (Goovaerts 1998). Within-

Table 8.2 Role of microbes in imparting tolerance through various mechanisms in the plants against different kind of stresses

Type of stress tolerance	Microbes involved	Mechanisms	References
Drought tolerance	*Pseudomonas* sp.	Accelerated levels of catalase, peroxidase, and superoxide dismutase	Sarma and Saikia (2014)
	Bacillus thuringiensis	Production of IAA and ACC deaminase, increase in lateral root density and length, alginate production	Timmusk et al. (2014)
	Azospirillum brasilense	Elevated Abscisic acid (ABA) level, decreased stomatal conductance thus retarded water loss	Cohen et al. (2015)
	Azospirillum and Pseudomonas	Biosynthesis of soluble carbohydrates and proline	Heidari et al. (2011)
Salinity tolerance	*Enterobacter cloacae*	Decreased accumulation of Na^+, increased uptake of K^+ in roots and shoots, improved water use efficiency	Singh et al. (2017)
	Bacillus subtilis	Downregulation of the high-affinity K+ transporter (HKT1) in the roots, hence reducing Na^+ uptake	Zhang et al. (2008)
	Bacillus subtilis	Decreased shoot and root Na^+ accumulation, improved K^+/Na^+ ratio, enhanced chlorophyll content, osmotic potential and cell membrane integrity	Han et al. (2014)
	Bradyrhizobium diazoefficiens, Bacillus velezensis	Enhanced plant growth due to increased degree of nodulation by producing larger nodules and enhanced N_2-fixing efficiency	Sibponkrung et al. (2020)
	Hartmannibacter diazotrophicus	Growth enhancement due to increased ACC deaminase activity, increased root dry weight and shoot dry weight	Suarez et al. (2015)
	Pseudomonas aeruginosa	ACC deaminase activity, increased plant root length, height, relative water content, dry matter, leaf area, chlorophyll content and lower injury to cell membrane	Meena et al. (2015)
Heat tolerance	*Brevibacillus brevis*	IAA, biological nitrogen fixation, ammonia production, and antifungal activity at high temperature	Nehra et al. (2016)
	Pseudomonas sp.	Induction of heat shock proteins (HSPs) in leaves, reduced membrane injury, increase in proline, glycine betaine and amino acid content	Ali et al. (2009)
	Pseudomonas aeruginosa	ACC deaminase activity, increased plant root length, height, relative water content, dry matter, leaf area, chlorophyll content and lower injury to cell membrane	Meena et al. (2015)

(continued)

Table 8.2 (continued)

Type of stress tolerance	Microbes involved	Mechanisms	References
Heavy metal stress tolerance	*Pseudomonas* sp. and *Bacillus* sp.	Production of ACC deaminase, siderophores, IAA and P solubilization	Shilev et al. (2020)
	Pseudomonas sp.	Enhanced growth of wheat due to IAA and ammonia production, P solubilization	Verma et al. (2015)
	Enterobacter aerogenes	Nitrogen fixation, P solubilization, IAA production, ACC deaminase activity	Pramanik et al. (2018)

ecosystem spatial variability of soil properties and functions is largely controlled by the interaction of multiple biological, chemical, and physical attributes. The spatial variability of soil properties is known to play essential roles in controlling ecosystem functioning (Franklin and Mills 2009; Fierer 2017), plant performance and competitive ability, ecosystem productivity (Maestre and Reynolds 2007), trophic interactions, and soil nutrient cycling. On the spatial scale, heterogeneous growing conditions are usually affected by spatial variability of farmland management, soil characteristics, and the interaction of these factors. Attributes such as soil texture, organic carbon content, nutrient availability, pH, and bulk density can vary significantly even within a few meters, affecting plant growth heterogeneously. Technologies like electromagnetic induction sensors, grid-based soil sampling, and remote sensing are now routinely used to map intra-field variability (Corwin and Lesch 2005). Inter-field variability denotes differences between distinct fields or sites. This variability arises from variations in parent material, climate, land use history, and geomorphological processes (Jenny 1941). Differences in soil mineralogy, depth, fertility status, and microbial communities contribute to diverse crop growth environments across fields.

Soil exhibits both horizontal and vertical variability due to processes such as additions, removals, movements, and transformations, influenced by various soil-forming factors including climate, organisms, topography, parent material, and time (Simonson 1959). Soil variability depends on the scale of observation, with different factors influencing it at varying spatial and temporal scales (Corstanje et al. 2007). Wiesmeier et al. (2019) highlighted key drivers influencing soil organic carbon (SOC) storage across different spatial scales, ranging from soil particles and aggregates at the micro-scale to the global level. They found that climate and vegetation predominantly governed SOC storage at broader scales; soil physical and chemical properties were more influential at finer spatial scales. At the landscape or catchment scale (5–250 meters resolution), soil variability is largely influenced by topography (Minasny et al. 2013); however, at the finer spatial scales (within a few meters), soil differences are primarily driven by micro-topography and microclimatic conditions (Zhang and Hartemink 2017). At soil profile level, vertical variability is common due to differences in parent material and pedogenic processes

(McBratney 1993). Although soil horizons are typically considered homogeneous, significant variations in their physical and chemical properties have been documented (Hartemink et al. 2020).

Inherent properties like soil texture and mineralogy play a significant role in determining the amount of storable carbon and nutrients, soil pH, soil structure and aggregation, water holding capacity, and other key factors (Libohova et al. 2018). In rainfed areas, sandy soils typically have low organic matter levels, limited water retention capacity, and high permeability. These characteristics often lead to water shortages during the growing season, which significantly impacts their overall productivity. Additionally, such soils are usually acidic (pH less than 5.5), have low cation exchange capacity due to their minimal clay and organic carbon contents, and thus possess a low specific surface area. As a result, they are prone to the leaching of basic exchangeable ions (Behera and Shukla 2015). However, sandy soils also offer advantages such as requiring less energy for tillage (Novak et al. 2014) and warming up quickly in early spring, helping achieve the minimum temperatures needed for plant growth. On the other hand, clay soils with fine particles affect crop performance by influencing water availability and drainage, nutrient retention, and soil structure. The high water retention character can be beneficial during dry periods, though excessive retention can also result in poor drainage and root issues, especially in wetter years. Furthermore, clay soils can develop surface crusts, which can obstruct seedling emergence and hinder root growth. Usowicz and Lipiec (2017) found that cereal grain yields showed a significant positive correlation with the cation exchange capacity of both topsoil and subsoil, topsoil water content, clay content, and subsoil organic carbon content, while a significant negative correlation was observed with topsoil bulk density.

8.4.2 Seasonal Changes in Soil Moisture, Temperature, and Biology

In the terrestrial ecosystems, soil moisture plays a vital role in the hydrological cycle and is very crucial link between soil properties and plant functions. It is critical for vegetation restoration and affects key hydrological processes such as infiltration, evaporation, and absorption of groundwater and root-zone water (Puri et al. 2011). Adequate soil moisture is vital for the healthy crop growth, survival, and crop quality (Yang et al. 2015) and plays an important role in shaping the microenvironment (Song et al. 2009). The soil moisture tends to be highly variable across both spatial and temporal scales, even within small areas (Gómez-Plaza et al. 2000). This variability is influenced by factors such as vegetation (Vivoni et al. 2010), topography, soil type, and land use practices. The variability, spatial patterns, and distribution of surface soil moisture have gained attention across local, regional, and continental scales (Parent et al. 2006). On a global scale, seasonal shifts in precipitation drive corresponding changes in soil moisture, resulting in distinct wet and dry seasons. Soil moisture status also changes due to evapo-transpiration demand and agricultural management practices. During rainy seasons, soils may experience high moisture levels conducive to microbial activity and nutrient mineralization, whereas dry periods result in moisture stress and reduced biological functioning (Lal and Shukla 2004). In rainfed systems, these moisture fluctuations are particularly critical, influencing root growth, nutrient uptake, and ultimately, yield potential. Seasonal precipitation serves as a major limiting factor controlling primary productivity, soil microbial activity, and overall ecosystem dynamics in arid and semiarid regions. Therefore, characterizing spatial and temporal patterns of soil moisture is essential for effective water resource management and for studying hydrological processes.

Soil temperature is determined by the heat flux within the soil and the heat exchange occurring between the soil and the atmosphere (Elias et al. 2004). It fluctuates both seasonally and daily, influencing soil physico-chemical and biological activities (Buchan 2001), as well as nutrient availability and plant development. Soil temperature showed a strong relationship with the air temperature, which indicates that if air temperature is going to change over the periods, it will have also an impact on soil temperature. Soil temperature variations do not affect soil pH or $CaCO_3$ content but influence electrical conductivity (EC) and micronutrient availability. Higher temperatures increase DTPA-extractable Fe, Zn, and Mn, while Cu availability decreases. Micronutrient concentrations are highest at the surface (0–5 cm) and decline with depth. Availability of Fe, Zn, and Mn is greatest during the monsoon, whereas Cu remains stable across seasons (Yadav and Gupta 2019). Environmental factors impact soil temperature by regulating the amount of heat reaching the soil surface and the extent of heat loss from the surface into deeper layers. Soil temperature impacts enzymatic activities, microbial processes, and seed germination rates (Campbell and Norman 1998). Optimal soil temperature ranges differ among crops; deviations can lead to suboptimal growth, delayed phenology, and yield penalties.

Soil biological activity encompassing enzymatic activity, basal respiration, and microbial biomass is closely linked to soil properties like pH, organic matter, and texture, as well as environmental factors such as climate, moisture, and temperature (Bastida et al. 2008). The dynamics of soil microbial biomass across various agricultural systems are strongly influenced by climate, with only minor fluctuations observed in temperate regions (Murphy et al. 2007; Bastida et al. 2008). Although organic carbon and total nitrogen levels remain stable and are not affected by the rotation phase throughout the season, other more labile soil biochemical properties exhibit variability. This temporal variation is largely associated with changes in soil moisture, temperature, and precipitation (Campbell et al. 1999; Feng and Simpson 2009). Similarly, Chen et al. (2003) reported that soil respiration rates closely align with soil temperature and moisture, showing higher activity during the summer and lower rates during other seasons. Warm and moist conditions generally promote microbial activity, enhancing nutrient mineralization and soil organic matter turnover. Seasonal stresses such as drought or freezing temperatures reduce microbial populations and enzymatic activities, thereby affecting nutrient availability. Incorporating seasonal biological monitoring into soil assessments provides a deeper understanding of soil fertility dynamics influencing crop responses. Soil organisms, such as earthworms and arthropods, also exhibit seasonal changes in activity. These organisms are sensitive to temperature and moisture levels, with their activity levels peaking during warmer months. For instance, earthworms tend to be more active during spring and summer, when temperatures are optimal for their growth and reproduction. In the winter, they retreat to deeper soil layers to avoid freezing conditions. Temperature and moisture are critical factors influencing soil respiration rates (Feng and Simpson 2009). Studies have shown that changes in soil respiration rates correlate with variations in soil temperature and moisture, with seasonal trends indicating higher respiration rates in summer and lower rates in other seasons (Chen et al. 2003). Soil respiration and microbial efficiency are notably higher in summer (June) compared to spring (April) and autumn (September). This seasonal variation is primarily due to the enhanced microbial activity driven by elevated soil temperatures and higher moisture content in the summer. In contrast, the lower temperatures and reduced moisture content during the other seasons suppress microbial activity, leading to a decrease in soil respiration (Ge et al. 2010). Thus, soil microbial communities, as key decomposers, drive essential ecosystem processes such as nutrient cycling and organic matter decomposition, thereby influencing ecosystem stability and functioning across all trophic levels. Their activity, crucial for maintaining soil health and ecosystem balance, is largely governed by moisture availability, which plays a pivotal role in regulating their effectiveness in these processes.

8.4.3 Consequences for G × E Studies and Trial Interpretations

Recent advancements in molecular biology have greatly expanded our knowledge of the genetic factors that affect crop performance. The effects of genotype, environment, and G × E interaction influence phenotypic performance, as well as general and specific adaptation across different environments. Studying gene environment interactions is essential for enhancing the precision and accuracy of evaluating both genetic and environmental influences. The G × E refers to the varying effects of environmental factors on crops with different genotypes, or the differing responses of genotypes under varying environmental conditions. In the era of global climate change understanding, these interactions are crucial for breeding programs, agronomic practices, and ecological studies. To develop cultivars suitable for multiple environments, understanding grain yield (GY)—a complex, quantitative trait highly influenced by environmental interactions—is crucial. Selecting genotypes based solely on performance in a single environment is ineffective when the goal is to identify new stable and adaptable cultivars for diverse locations and broader ecologies (Shrestha et al. 2021). For this evaluation of large number of genotypes in multi-location trials is need of hour. It has been found that due to restricted genetic diversity in multi-location trials and various environmental factors, the average annual increase in cereals yield has fallen from 3.2% per year in 1960 to 1.5% in 2000 (Powell et al. 2012). One proven method to overcoming climate stresses is to enhance genetic gains by exploring a wide range of genotypes and selecting the best-suited ones for the specific target environment (Ceccarelli 2015). Therefore, multi-environment trials (MET) are crucial for selecting genotypes based on yield stability rather than just average performance across diverse environmental conditions (Islam et al. 2016). Statistical models like additive main effects and multiplicative interaction (AMMI) and genotype + genotype × environment interaction (GGE) biplots can be utilized to analyze G × E effects, aiding in the selection of stable genotypes. This knowledge aids in developing varieties that are resilient to environmental stresses, thereby enhancing food security and sustainable agricultural practices. Incorporating G × E interactions into trial designs and interpretations is essential for the accurate assessment of genotype performance.

References

Ahlholm JU, Helander M, Henriksson J, Metzler M, Saikkonen K (2002) Environmental conditions and host genotype direct genetic diversity of *Venturia ditricha*, a fungal endophyte of birch trees. Evolution 56(8):1566–1573

Ali SZ, Sandhya V, Grover M, Kishore N, Rao LV, Venkateswarlu B (2009) Pseudomonas sp. strain AKM-P6 enhances tolerance of sorghum seedlings to elevated temperatures. Biol Fertil Soils 46:45–55

Anderson RL (2008) Growth and yield of winter wheat as affected by preceding crop and crop management. Agron J 100:977–980

Azad MAK, Salam MA, Rahman MM, Islam MS (2020) Enhancing fertilizer use efficiency through fertigation in saline soil. Agric Water Manag 231:105987

Bailey JK, Deckert R, Schweitzer JA, Rehill BJ, Lindroth RL, Gehring C, Whitham TG (2005) Host plant genetics affect hidden ecological players: links among Populus, condensed tannins, and fungal endophyte infection. Can J Bot 83(4):356–361

Bartels D, Sunkar R (2005) Drought and salt tolerance in plants. Crit Rev Plant Sci 24(1):23–58. https://doi.org/10.1080/07352680590910410

Bashir O, Ali T, Baba ZA, Rather GH, Bangroo SA, Mukhtar SD, Naik N, Mohiuddin R, Bharati V, Bhat RA (2021) Soil organic matter and its impact on soil properties and nutrient status. In: Springer eBooks. Springer, Cham, pp 129–159. https://doi.org/10.1007/978-3-030-61010-4_7

Bastida F, Zsolnay A, Hernández T, García C (2008) Past, present and future of soil quality indices: a biological perspective. Geoderma 147:159–171

Becker HC, Leon J (1988) Stability analysis in plant breeding. Plant Breed 101:1–23. https://doi.org/10.1111/j.1439-0523.1988.tb00261.x

Begna T (2020) The role of genotype by environmental interaction in plant breeding. Int J Agric Biosci 9(5):209–215

Behera SK, Shukla AK (2015) Spatial distribution of surface soil acidity, electrical conductivity, soil organic carbon content and exchangeable potassium, calcium and magnesium in some cropped acid soils of India. Land Degrad Dev 26(1):71–79

Ben-Gal A, Shani U, Zaslavsky D (2010) Effect of salinity on wheat growth, water use, and yield. Field Crop Res 115(2–3):63–70

Berlow M, Mesa M, Creek M, Duarte JG, Carpenter E, Phinizy B, Andonian K, Dlugosch KM (2024) Plant G× Microbial E: plant genotype interaction with soil bacterial community shapes rhizosphere composition during invasion. Microb Ecol 87(1):113

Blum A (2011) Plant breeding for water-limited environments. Springer, New York. https://doi.org/10.1007/978-1-4419-7491-4

Bradl HB (2004) Adsorption of heavy metal ions on soils and soils constituents. J Colloid Interface Sci 277:1–18

Brady NC, Weil RR (2016) The nature and properties of soils, 15th edn. Pearson Education

Buchan GD (2001) Soil temperature regime. In: Smith KA et al (eds) Soil and environmental analysis: physical methods. M. Dekker, New York

Cambardella CA, Karlen DL (1999) Spatial analysis of soil fertility parameters. Precis Agric 1(1): 5–14

Campbell GS, Norman JM (1998) The light environment of plant canopies. In An introduction to environmental biophysics (pp. 247-278). New York, NY: Springer New York

Campbell CA, Lafond GP, Biederbeck VO, Wen G, Schoenau J, Hahn D (1999) Seasonal trends in soil biochemical attributes: effects of crop management on a Black Chernozem. Can J Soil Sci 79(1):85–97

Ceccarelli S (2015) Efficiency of plant breeding. Crop Sci 55(1):87–97

Chandrakanth N, Moorthy SM, Rekha M, Sivaprasad V (2016) Stability and path analysis for yield and related traits in silkworm (Bombyx mori L.) reared under stress conditions. Genetika 48(1): 271–284

Chatterjee D, Das SR, Saha S, Sarkar A, Pathak H (2024) Impacts of Climate Change on Soil Processes. In Climate Change Impacts on Soil-Plant-Atmosphere Continuum (pp. 3–36). Singapore: Springer Nature Singapore

Chaudhari PR, Ahire DV, Ahire VD, Chkravarty M, Maity S (2013) Soil bulk density as related to soil texture, organic matter content and available total nutrients of Coimbatore soil. Int J Sci Res Publ 3(2):1–8

Chaves MM, Maroco JP, Pereira JS (2003) Understanding plant responses to drought—from genes to the whole plant. Funct Plant Biol 30(3):239–264

Chen QS, Li LH, Han XG, Yan ZD, Wang YF, Yuan ZY (2003) Influence of temperature and soil moisture on soil respiration of a degraded steppe community in the Xilin river basin of Inner Mongolia. Chin J Plant Ecol 27(2):202–209

Chowdhury RB, Zhang X (2021) Phosphorus use efficiency in agricultural systems: a comprehensive assessment through the review of national scale substance flow analyses. Ecol Indic 121: 107172

Christensen S, Jensen SM (2023) Unravelling the complexities of genotype-soil-management interaction for precision agriculture. Agronomy 13:2727. https://doi.org/10.3390/agronomy13112727

Cohen AC, Bottini R, Pontin M, Berli FJ, Moreno D, Boccanlandro H, Travaglia CN, Piccoli PN (2015) Azospirillum brasilense ameliorates the response of Arabidopsis thaliana to drought mainly via enhancement of ABA levels. Physiol Plant 153:79–90

Colmer TD, Voesenek LACJ (2009) Flooding tolerance: suites of plant traits in variable environments. Funct Plant Biol 36(8):665–681. https://doi.org/10.1071/FP09144

Comas LH, Becker SR, Von Mark VC, Byrne PF, Dierig DA (2013) Root traits contributing to plant productivity under drought. Front Plant Sci 4:442. https://doi.org/10.3389/fpls.2013.00442

Cooper M, Gho C, Leafgren R, Tang T, Messina C (2014) Breeding drought-tolerant maize hybrids for the US corn-belt: discovery to product. J Exp Bot 65(21):6191–6204. https://doi.org/10.1093/jxb/eru064

Corstanje R, Schulin R, Lark RM (2007) Scale-dependent relationships between soil organic carbon and urease activity. Eur J Soil Sci 58:1087–1095

Corwin DL, Lesch SM (2005) Characterizing soil spatial variability with apparent soil electrical conductivity. Comput Electron Agric 46(1–3):103–133

Cuevas J, Pinillos V, Policarpo M, Rallo G (2019) Intercropping to reduce salinity and improve soil health in degraded agroecosystems. Agric Water Manag 217:1–10

Dagar JC, Singh G, Singh NT (2001) Agroforestry for saline soils and water. Int Tree Crops J 11(4): 281–309

de Leon N, Jannink JL, Edwards JW, Kaeppler SM (2016) Introduction to a special issue on genotype by environment interaction. Crop Sci 56:2081–2089

Des Marais DL, Auchincloss LC, Sukamtoh E, Juenger TE (2017) The genetic architecture of variation in the physiological response of Brachypodium distachyon to soil water availability. Plant Cell Environ 40(9):2029–2046. https://doi.org/10.1111/pce.13002

Dey A, Dwivedi BS, Bhattacharyya R, Datta SP, Meena MC, Jat RK, Kumar R (2024) Functional groups and mineralization kinetics of soil organic matter under contrasting hydro-thermal regimes under conservation agriculture-based rice–wheat system in eastern Indo-Gangetic Plains. Soil Use Manag 40(1):e12962

Dia M, Wehner TC, Hassell R, Price DS, Boyhan GE, Olson S, King S, Davis AR, Tolla GE, Bernier J, Juarez B, Sari N, Solmaz I, Aras V (2012) Stability of fruit yield in watermelon genotypes tested in multiple US environments. In: Proceedings of the Xth EUCARPIA Meeting on Genetics and Breeding of Cucurbitaceous, Antalya, Turkey, 15–18 October 2012, pp 84–88

Dobarco MR, Bourennane H, Arrouays D, Saby NPA, Cousin I, Martin MP (2019) Uncertainty assessment of Global Soil Map soil available water capacity products: a French case study. Geoderma 344:14–30

Dong Y, Liu Y, Hu J, Wang L, Zhang X (2023) Soil salinization in agriculture: mitigation and adaptation strategies. Sci Total Environ 904:167497. https://doi.org/10.1016/j.scitotenv.2023.167497

Doran JW, Zeiss MR (2000) Soil health and sustainability: managing the biotic component of soil quality. Appl Soil Ecol 15:3–11

Duvick DN (2005) The contribution of breeding to yield advances in maize (Zea mays L.). Adv Agron 86:83–145

Elias EA, Cichota R, Torraiani HH (2004) Analytical soil temperature model: correction for temporal variation of daily amplitude. Soil Sci Soc Am J 68:784–788

El-Sharkawy MA (2004) Cassava biology and physiology. Plant Mol Biol 56(4):481–501. https://doi.org/10.1007/s11103-005-2270-7

Esfahani MN, Sonnewald U (2024) Unlocking dynamic root phenotypes for simultaneous enhancement of water and phosphorus uptake. Plant Physiol Biochem 207:108386

Farooq M, Wahid A, Kobayashi N, Fujita D, Basra SMA (2009) Plant drought stress: effects, mechanisms and management. Agron Sustain Dev 29:185–212

Feng X, Simpson MJ (2009) Temperature and substrate controls on microbial phospholipid fatty acid composition during incubation of grassland soils contrasting in organic matter quality. Soil Biol Biochem 41(4):804–812

Fierer N (2017) Embracing the unknown: disentangling the complexities of the soil microbiome. Nat Rev Microbiol 15:579–590

Fikadu D, Mulugeta S, Gedefaw M (2021) The effect of salinity on growth and yield of quinoa in salt-affected soils of Ethiopia. Int J Agron 2021:101–110

Franklin RB, Mills AL (2009) Importance of spatially structured environmental heterogeneity in controlling microbial community composition at small spatial scales in an agricultural field. Soil Biol Biochem 41:1833–1840

Franzluebbers AJ (2002) Water infiltration and soil structure related to organic matter and its stratification with depth. Soil Tillage Res 66:197–205

Fu YB (2015) Understanding crop genetic diversity under modern plant breeding. Theor Appl Genet 128:2131–2142

Ge G, Li Z, Fan F, Chu G, Hou Z, Liang Y (2010) Soil biological activity and their seasonal variations in response to long-term application of organic and inorganic fertilizers. Plant Soil 326:31–44

Geng N, Kang X, Yan X, Yin N, Wang H, Pan H, Zhuge Y (2022) Biochar mitigation of soil acidification and carbon sequestration is influenced by materials and temperature. Ecotoxicol Environ Saf 232:113241

Gentile RM, Vanlauwe B, Six J (2013) Integrated Soil Fertility Management: Aggregate carbon and nitrogen stabilization in differently textured tropical soils. Soil Biology and Biochemistry 67: 124–132

Ghasal PC, Shivay YS, Pooniya V, Choudhary M, Verma RK (2018) Zinc partitioning in basmati rice varieties as influenced by Zn fertilization. Crop J 6(2):136–147

Gil PM, Bonomelli C, Schaffer B, Ferreyra R, Gentina C (2012) Effect of soil water-to air ratio on biomass and mineral nutrition of avocado trees. J Soil Sci Plant Nutr 12:609–630

Gomez-Becerra HF, Yazici A, Ozturk L, Budak H, Peleg Z, Morgounov A, Fahima T, Saranga Y, Cakmak I (2010) Genetic variation and environmental stability of grain mineral nutrient concentrations in Triticum dicoccoides under five environments. Euphytica 171:39–45

Gómez-Plaza A, Alvarez-Rogel J, Albaladejo J, Castillo V (2000) Spatial patterns and temporal stability of soil moisture across a range of scales in a semi-arid environment. Hydrol Process 14: 1261–1277

Goovaerts P (1998) Geostatistical tools for characterizing the spatial variability of microbiological and physico-chemical soil properties. Biol Fertil Soils 27:315–334

Greenway H, Munns R (1980) Mechanisms of salt tolerance in non-halophytes. Annu Rev Plant Physiol 31(1):149–190

Guo J, Xiong W, Qiu J, Wang G (2023) Linking soil organic carbon mineralization to soil physicochemical properties and bacterial alpha diversity at different depths following land use changes. Ecol Process 12(1):39

Han QQ, Lü XP, Bai JP, Qiao Y, Paré PW, Wang SM, Zhang JL, Wu YN, Pang XP, Xu WB (2014) Beneficial soil bacterium Bacillus subtilis (GB03) augments salt tolerance of white clover. Front Plant Sci 5:525

Hartemink AE, Zhang Y, Bockheim JG, Curi N, Silva SHG, Grauer-Gray J, Lowe DJ, Krasilnikov P (2020) Soil horizon variation: a review. Adv Agron 160:125–185

Hasanuzzaman M, Nahar K, Alam MM, Roychowdhury R, Fujita M (2018) Physiological, biochemical, and molecular mechanisms of heat stress tolerance in plants. Int J Mol Sci 14(5): 9643–9684. https://doi.org/10.3390/ijms14059643

Hassan NM, Tsegaye A (2019) Effect of irrigation management on salinity in Amibara irrigation scheme of Ethiopia. Agric Ecosyst Environ 282:69–78

Heidari M, Mousavinik SM, Golpayegani A (2011) Plant growth promoting rhizobacteria (PGPR) effect on physiological parameters and mineral uptake in basil (Ocimum basilicum L.) under water stress. J Agric Biol Sci 6:6–11

Hillel D (2004) Introduction to environmental soil physics. Elsevier. https://www.elsevier.com/books/introduction-to-environmental-soil-physics/hillel/978-0-12-348655-4

Hochholdinger F, Park WJ, Sauer M, Woll K (2004) From weeds to crops: genetic analysis of root development in cereals. Trends Plant Sci 9(1):42–48

Hu B, Wang W, Chen J, Liu Y, Chu C (2023) Genetic improvement toward nitrogen-use efficiency in rice: lessons and perspectives. Mol Plant 16(1):64–74

Islam MR, Sarker MRA, Sharma N, Rahman MA, Collard BCY, Gregorio GB, Ismail AM (2016) Assessment of adaptability of recently released salt tolerant rice varieties in coastal regions of South Bangladesh. Field Crop Res 190:34–43

Ismail AM, Horie T (2017) Genomics, physiology, and molecular breeding approaches for improving salt tolerance. Annu Rev Plant Biol 68:405–434. https://doi.org/10.1146/annurev-arplant-042916-040936

Javed A, Ali E, Afzal KB, Osman A, Riaz S (2022) Soil fertility: factors affecting soil fertility, and biodiversity responsible for soil fertility. Int J Plant Anim Environ Sci 12(1):21–33

Jenny H (1941) Factors of soil formation: a system of quantitative pedology. McGraw-Hill

Juhos K, Szabó S, Ladányi M (2016) Explore the influence of soil quality on crop yield using statistically-derived pedological indicators. Ecol Indic 63:366–373

Kaiser K, Guggenberger G (2010) Mineral surfaces and soil organic matter. Eur J Soil Sci 54:219–236

Katsenios N, Sparangis P, Chanioti S, Giannoglou M, Leonidakis D, Christopoulos MV, Katsaros G, Efthimiadou A (2021) Genotype environment interaction of yield and grain quality traits of maize hybrids in Greece. Agronomy 11:357. https://doi.org/10.3390/agronomy11020357

King WL, Yates CF, Guo J, Fleishman SM, Trexler RV, Centinari M, Bell TH, Eissenstat DM (2021) The hierarchy of root branching order determines bacterial composition, microbial carrying capacity and microbial filtering. Commun Biol 4(1):483

Kirk GJD, Santos EE, Findenegg GR (2014) Phosphorus dynamics in upland soils with contrasting phosphorus status. Plant Soil 206:89–97

Kumar J, Thavarajah D, Kumar S, Sarker A, Singh NP (2018) Analysis of genetic variability and genotype × environment interactions for iron and zinc content among diverse genotypes of lentil. J Food Sci Technol 55:3592–3605

Laishram B, Devi OR, Dutta R, Senthilkumar T, Goyal G, Paliwal DK, Rasool A (2024) Plant-microbe interactions: PGPM as microbial inoculants/biofertilizers for sustaining crop productivity and soil fertility. Curr Res Microb Sci 8:100333

Lal R (2004) Soil carbon sequestration to mitigate climate change. Geoderma 123(1–2):1–22

Lal R (2020) Soil organic matter and water retention. Agron J 112(5):3265–3277

Lal R, Shukla MK (2004) Principles of soil physics. CRC Press

Lal HA, Muhammad K, Muhammad A, Tariq A (2010) Stability analysis for grain yield in Mungbean (Vigna radiata L. Wilczyek) grown in different agro-climatic regions. Emir J Food Agric 22(6):490–497

Lareen A, Burton F, Schäfer P (2016) Plant root-microbe communication in shaping root microbiomes. Plant Mol Biol 90:575–587

Lee SM, Kong HG, Song GC, Ryu CM (2021) Disruption of Firmicutes and Actinobacteria abundance in tomato rhizosphere causes the incidence of bacterial wilt disease. ISME J 15(1): 330–347

Li SX, Wang ZH, Stewart BA (2013) Chapter five—responses of crop plants to ammonium and nitrate N. In: Donald LS (ed) Advances in agronomy. Academic, pp 205–397

Li D, Meng M, Ren B, Ma X, Bai L, Li J, Bai G, Yao F, Tan C (2023) Different responses of soil fungal and bacterial communities to nitrogen addition in a forest grassland ecotone. Frontiers in Microbiology 14:1211768

Li X, Feng G, Tewolde H, Adeli A, Jenkins JN (2024) Critical soil organic carbon for improving available water content of silt loam soils. Soil Use Manag 40(4):e13144

Libohova Z, Seybold C, Wysocki D, Wills S, Schoeneberger P, Williams C, Owens PR (2018) Reevaluating the effects of soil organic matter and other properties on available water-holding capacity using the National Cooperative Soil Survey Characterization Database. J Soil Water Conserv 73(4):411–421

Litalien A, Zeeb B (2020) Curbing salinization with halophytes: a sustainable solution? Sci Total Environ 740:139760

Liu C, Geng HY, Li WX, Li YY, Lu YS, Xie KZ, Sun LL, Zhang JX, Peng HL, Shi CH, Li WL (2023) Innate root exudates contributed to contrasting coping strategies in response to Ralstonia solanacearum in resistant and susceptible tomato cultivars. J Agric Food Chem 71(50): 20092–20104

Liu Y, Lan X, Hou H, Ji J, Liu X, Lv Z (2024) Multifaceted ability of organic fertilizers to improve crop productivity and abiotic stress tolerance: review and perspectives. Agronomy 14(6):1141

Madan B, Raghuram N (2024) Phenotypic, physiological, and gene expression analysis for nitrogen and phosphorus use efficiencies in three popular genotypes of Rice (Oryza sativa Indica). Plants 13(18):2567

Maestre FT, Reynolds JF (2007) Biomass responses to elevated CO_2, soil heterogeneity and diversity: an experimental assessment with grassland assemblages. Oecologia 151:512–520

Mao HR, Cotrufo MF, Hart SC, Sullivan BW, Zhu XF, Zhang JC, Liang C, Zhu MQ (2024) Dual role of silt and clay in the formation and accrual of stabilized soil organic carbon. Soil Biol Biochem 192:109390

Marschner P, Neumann G, Kania A, Weiskopf L, Lieberei R (2002) Spatial and temporal dynamics of the microbial community structure in the rhizosphere of cluster roots of white lupin (Lupinus albus L.). Plant Soil 246:167–174

McBratney AB (1993) Some remarks on soil horizon classes. Catena 20:427–430

Meena RK, Singh RK, Singh NP, Meena SK, Meena VS (2015) Isolation of low temperature surviving plant growth–promoting rhizobacteria (PGPR) from pea (Pisum sativum L.) and documentation of their plant growth promoting traits. Biocatal Agric Biotechnol 4:806–811

Minasny B, McBratney AB, Malone BP, Wheeler I (2013) Digital mapping of soil carbon. Adv Agron 118:1–47

Miranda LA, Spinosa WA, Destro TM, Souza-Junior H, Nascimento V (2020) Sweet cassava cooking time. Agron Sci Biotechnol 6:1–16. https://doi.org/10.33158/asb.r109.v6.2020

Mohebodin M et al (2016) The influence of genotype × environment interaction on the grain yields of lentil (Lens culinaris Medik) genotypes. Department of Horticulture Science, University of Mohaghegh Ardabili, Iran

Munns R, Tester M (2008) Mechanisms of salinity tolerance. Annu Rev Plant Biol 59:651–681

Munns R, James RA, Gilliham M (2019) Plant adaptations to salt and water stress: differences and commonalities. Adv Bot Res 91:1–32

Murphy DV, Stockdale EA, Poulton PR, Willison TW, Goulding KWT (2007) Seasonal dynamics of carbon and nitrogen pools and fluxes under continuous arable and ley-arable rotations in a temperate environment. Eur J Soil Sci 58(6):1410–1424

Nehra V, Saharan BS, Choudhary M (2016) Evaluation of Brevibacillus brevis as a potential plant growth promoting rhizobacteria for cotton (Gossypium hirsutum) crop. Springerplus 5:1–10

Nichols KA, Wright SE, Liebig MA, Pikul JL Jr (2004) Functional significance of glomalin to soil fertility. In: Aj S (ed) Great Plains Soil Fertility. Kansas State University, pp 219–224

Novak P, Chyba J, Kumhala F, Prochazka P (2014) Measurement of stubble cultivator draught force under different soil conditions. Agron Res 12(1):135–142

Oishy MN, Shemonty NA, Fatema SI, Mahbub S, Mim EL, MBH R, Anik AH (2025) Unravelling the effects of climate change on the soil-plant-atmosphere interactions: a critical review. Soil Environ Health 3:100130

Pardo A, Amato M, Chiarandà FQ (2000) Relationships between soil structure, root distribution and water uptake of chickpea (Cicer arietinum L.). Plant growth and water distribution. Eur J Agron 13:39–45

Parent AC, Anctil F, Parent LE (2006) Characterization of temporal variability in near-surface soil moisture at scales from 1 h to 2 weeks. J Hydrol 325(1–4):56–66

Pereira LFM, Zanetti S, Silva MA (2018) Water relations of cassava cultivated under water-deficit levels. Acta Physiol Plant 40(1):1–13. https://doi.org/10.1007/s11738-017-2590-7

Powell N, Ji X, Ravash R, Edlington J, Dolferus R (2012) Yield stability for cereals in a changing climate. Funct Plant Biol 39(7):539–552

Pramanik K, Mitra S, Sarkar A, Maiti TK (2018) Alleviation of phytotoxic effects of cadmium on rice seedlings by cadmium resistant PGPR strain Enterobacter aerogenes MCC 3092. J Hazard Mater 351:317–329

Preece C, Peñuelas J (2020) A return to the wild: root exudates and food security. Trends Plant Sci 25(1):14–21

Puri S, Stephen H, Ahmad S (2011) Relating TRMM precipitation radar land surface backscatter response to soil moisture in the Southern United States. J Hydrol 402:115–125

Qadir M, Ghafoor A, Murtaza G (2001) Use of saline-sodic waters through phytoremediation of calcareous saline-sodic soils. Agric Water Manag 50(3):197–210

Qadir M, Ghafoor A, Murtaza G, Saqib M, Hussain M, Schubert S (2007) Productivity enhancement of salt-affected lands through integrated management. Land Degrad Dev 18(1):49–66. https://doi.org/10.1002/ldr.770

Qadir M, Farooq M, Murtaza G (2014) Reclamation of salt-affected soils for improving crop production: a review. Agric Ecosyst Environ 163:3–12

Rakhsh F, Golcchin A (2018) Investigation of mineralization of organic nitrogen under the influence of type, content of clay and exchangeable cations. Water and Soil 31(6):1691–1711

Robertson GP, Klingensmith KM, Klug MJ, Paul EA, Crum JR, Ellis BG (1997) Soil resources, microbial activity, and primary production across an agricultural ecosystem. Ecological Applications 7(1):158–170

Reid RJ, Hayes JE (2003) Mechanisms and genes involved in the uptake and transport of chloride in plants. Soil Sci Plant Nutr 49(3):619–632

Rengasamy P (2006) World salinization with emphasis on Australia. J Exp Bot 57(5):1017–1023

Rengasamy P (2010) Soil processes affecting crop production in salt-affected soils. Funct Plant Biol 37(7):613–620. https://doi.org/10.1071/FP09249
Rhoades JD (1997) Diagnosis and improvement of saline and sodic soils. Soil Sci Soc Am J 61(2): 405–412
Roy SJ, Negrão S, Tester M (2014) Salt resistant crop plants. Curr Opin Biotechnol 26:115–124. https://doi.org/10.1016/j.copbio.2013.12.004
Sahoo SK, Mishra KN, Panda N, Sethi D, Dey P (2023) Carbon and nitrogen mineralization kinetics in soil under differential nutrient management regimes in Inceptisols of coastal India. Commun Soil Sci Plant Anal 54(20):2782–2799
Salehi A, Maleki M (2012) Evaluation of soil physical and chemical properties in poplar plantations in North of Iran. Ecología 4:69–76
Sarma RK, Saikia R (2014) Alleviation of drought stress in mung bean by strain Pseudomonas aeruginosa GGRJ21. Plant Soil 377:111–126
Sasse J, Martinoia E, Northen T (2018) Feed your friends: do plant exudates shape the root microbiome? Trends Plant Sci 23(1):25–41
Saunders M, Kohn LM (2009) Evidence for alteration of fungal endophyte community assembly by host defense compounds. New Phytol 182(1):229–238
Setter TL, Fregene MA (2007) Recent advances in breeding cassava for improved drought tolerance. In: Jenks MA et al (eds) Advances in molecular breeding toward drought and salt tolerant crops. Springer, pp 701–711. https://doi.org/10.1007/978-1-4020-5578-2_27
Shannon MC, Grieve CM, Munns R (2017) Salinity and plant growth. In: Läuchli A, Lüttge U (eds) Salinity: Environment-Plants-Molecules. Springer, pp 46–60
Sharma N, Sinha VB, Prem Kumar NA, Subrahmanyam D, Neeraja CN, Kuchi S, Raghuram N (2021) Nitrogen use efficiency phenotype and associated genes: roles of germination, flowering, root/shoot length and biomass. Front Plant Sci 11:587464
Shilev S, Babrikova I, Babrikov T (2020) Consortium of plant growth-promoting bacteria improves spinach (Spinacea oleracea L.) growth under heavy metal stress conditions. J Chem Technol Biotechnol 95:932–939
Shrestha J, Subedi S, Kushwaha UKS, Maharjan B (2021) Evaluation of growth and yield traits in rice genotypes using multivariate analysis. Heliyon 7(9):e07940
Sibponkrung S, Kondo T, Tanaka K, Tittabutr P, Boonkerd N, Yoshida K, Teaumroong N (2020) Co-inoculation of bacillus velezensis strain S141 and Bradyrhizobium strains promotes nodule growth and nitrogen fixation. Microorganisms 8:678
Silva MA, Zanetti S, Santos HL, Pereira LFM (2021) Nutritional status, yield components, and yield of cassava as influenced by phenological stages and water regimes. J Plant Nutr 44: 1927084. https://doi.org/10.1080/01904167.2021.1927084
Simonson RW (1959) Outline of a generalized theory of soil genesis. Soil Sci Soc Am J 23:152–156
Singh RP, Jha P, Jha PN (2017) Bio-inoculation of plant growth-promoting rhizobacterium Enterobacter cloacae ZNP-3 increased resistance against salt and temperature stresses in wheat plant (Triticum aestivum L.). J Plant Growth Regul 36:783–798
Six J, Conant RT, Paul EA, Paustian K (2002) Stabilization mechanisms of soil organic matter: implications for C-saturation of soils. Plant Soil 241:155–176
Song TQ, Peng WX, Zeng FP, Ouyang ZW, Wu HY (2009) A spatial heterogeneity of surface soil moisture content in dry season in Mulun National Natural Reserve in Karst area. J Appl Ecol 20(1):98–104
Suarez C, Cardinale M, Ratering S, Steffens D, Jung S, Montoya AMZ, Geissler-Plaum R, Schnell S (2015) Plant growth-promoting effects of Hartmannibacter diazotrophicus on summer barley (Hordeum vulgare L.) under salt stress. Appl Soil Ecol 95:23–30
Tal A (2016) Rethinking the sustainability of Israel's irrigation practices in the drylands. Water Res 90:387–394
Tardieu F (2012) Any trait or trait-related allele can confer drought tolerance: just design the right drought scenario. J Exp Bot 63(1):25–31. https://doi.org/10.1093/jxb/err307

Timmusk S, Abd El-Daim IA, Copolovici L, Tanilas T, Kännaste A, Behers L, Nevo E, Seisenbaeva G, Stenström E, Niinemets Ü (2014) Drought-tolerance of wheat improved by rhizosphere bacteria from harsh environments: enhanced biomass production and reduced emissions of stress volatiles. PLoS One 9:e96086

Tuberosa R (2012) Phenotyping for drought tolerance of crops in the genomics era. Front Physiol 3: 347. https://doi.org/10.3389/fphys.2012.00347

Usowicz B, Lipiec J (2017) Spatial variability of soil properties and cereal yield in a cultivated field on sandy soil. Soil Tillage Res 174:241–250

Van Bueren EL, Jones SS, Tamm L, Murphy KM, Myers JR, Leifert C, Messmer MM (2011) The need to breed crop varieties suitable for organic farming, using wheat, tomato and broccoli as examples: a review. NJAS Wageningen J Life Sci 58(3–4):193–205

Van der Ent S, Verhagen BW, Van Doorn R, Bakker D, Verlaan MG, Pel MJ, Joosten RG, Proveniers MC, Van Loon LC, Ton J, Pieterse CM (2008) MYB72 is required in early signaling steps of rhizobacteria-induced systemic resistance in Arabidopsis. Plant Physiol 146(3): 1293–1304

Verma C, Singh P, Kumar R (2015) Isolation and characterization of heavy metal resistant PGPR and their role in enhancement of growth of wheat plant under metal (cadmium) stress condition. Arch Appl Sci Res 7:37–43

Vivoni ER, Rinehart AJ, Mendez-Barroso LA, Aragon CA, Bisht G, Cardenas MB (2010) Vegetation controls on soil moisture distribution in the Valles Caldera, New Mexico, during the North American monsoon. Ecohydrology 1:225–238

Voges MJ, Bai Y, Schulze-Lefert P, Sattely ES (2019) Plant-derived coumarins shape the composition of an Arabidopsis synthetic root microbiome. Proc Natl Acad Sci 116(25):12558–12565

Vlot AC, Sales JH, Lenk M, Bauer K, Brambilla A, Sommer A, Chen Y, Wenig M, Nayem S (2021) Systemic propagation of immunity in plants. New Phytologist 229(3):1234–1250

Wang HJ, Shi XZ, Yu DS, Weindorf DC, Huang B, Sun WX, Ritsema CJ, Milne E (2009) Factors determining soil nutrient distribution in a small-scaled watershed in the purple soil region of Sichuan Province, China. Soil and Tillage Research 105(2):300–306

Wang E, Yu B, Zhang J, Gu S, Yang Y, Deng Y, Fan F (2024) Low carbon loss from long-term manure-applied soil during abrupt warming is realized through soil and microbiome interplay. Environ Sci Technol 58(22):9658–9668

Watt M, Silk WK, Passioura JB (2006) Rates of root and organism growth, soil conditions, and temporal and spatial development of the rhizosphere. Ann Bot 97(5):839–855

Wiesmeier M, Urbanski L, Hobley E, Lang B, von Lützow M, Marin-Spiotta E, van Wesemael B, Rabot E, Ließ M, Garcia-Franco N, Wollschläger U (2019) Soil organic carbon storage as a key function of soils – a review of drivers and indicators at various scales. Geoderma 333:149–162

Yadav BR, Gupta N (2019) Seasonal variation in soil temperature, physico-chemical properties and micronutrient availability in soils of pear (Pyrus communis L.) orchard under semi-arid environment. J Agric Phys 19(2):236–238

Yang K, Fu R, Feng H, Jiang G, Finkel O, Sun T, Liu M, Huang B, Li S, Wang X, Yang T (2023) RIN enhances plant disease resistance via root exudate-mediated assembly of disease-suppressive rhizosphere microbiota. Molecular plant 16(9):1379–1395

Yamaguchi T, Blumwald E (2005) Developing salt-tolerant crop plants: challenges and opportunities. Trends Plant Sci 10(12):607–614

Yang L, Chen L, Wei W (2015) Effects of vegetation restoration on the spatial distribution of soil moisture at the hillslope scale in semi-arid regions. Catena 124:138–146

Ye C, Zheng G, Tao Y, Xu Y, Chu G, Xu C, Chen S, Liu Y, Zhang X, Wang D (2024) Effect of soil texture on soil nutrient status and rice nutrient absorption in paddy soils. Agronomy 14:1339. https://doi.org/10.3390/agronomy14061339

Yin L, Corneo PE, Richter A, Wang P, Cheng W, Dijkstra FA (2019) Variation in rhizosphere priming and microbial growth and carbon use efficiency caused by wheat genotypes and temperatures. Soil Biol Biochem 134:54–61

Yusuf A, Li M, Zhang SY, Odedishemi-Ajibade F, Luo RF, Wu YX, Duan S (2025) Harnessing plant–microbe interactions: strategies for enhancing resilience and nutrient acquisition for sustainable agriculture. Front Plant Sci 16:1503730

Zhang Y, Hartemink AE (2017) Sampling designs for soil organic carbon stock assessment of soil profiles. Geoderma 307:220–230

Zhang H, Kim MS, Sun Y, Dowd SE, Shi H, Paré PW (2008) Soil bacteria confer plant salt tolerance by tissue-specific regulation of the sodium transporter HKT1. Mol Plant-Microbe Interact 21:737–744

Zhang S, Lin J, Wang P, Zhu B (2023) The direct and legacy effects of drying rewetting cycles on active and relatively resistant soil carbon decomposition. Land Degrad Dev 34:2124–2135

Zhou X, Zhang J, u Rahman MK, Gao D, Wei Z, Wu F, Dini-Andreote F (2023) Interspecific plant interaction via root exudates structures the disease suppressiveness of rhizosphere microbiomes. Mol Plant 16(5):849–864

Zhu JK (2001) Plant salt tolerance. Trends Plant Sci 6(2):66–71

Zhu G, Hou S, Song X, Wang X, Wang W, Chen Q, Guo W (2021) Genome-wide association analysis reveals quantitative trait loci and candidate genes involved in yield components under multiple field environments in cotton (Gossypium hirsutum). BMC Plant Biol 21(1):250

Efficiency of Multivariate Statistical Models in Analysis of G × E Interactions

9

Praveen Kona, Vinayaka, Amaresh, T. Lakshmi Pathy, K. Gopalareddy, R. T. Maruthi, H. K. Mahadeva Swamy, K. Mohanraj, A. Anna Durai, R. M. Shanthi, and P. Govindaraj

Abstract

Understanding genotype × environment (G × E) interactions is a critical component of plant breeding programs aimed at developing high-yielding, stable, and climate-resilient crop varieties. Multivariate statistical models have become essential instruments for unravelling the intricacies of G × E interactions. They enable the recognition of genotypes that demonstrate both extensive adaptability and particular stability across various agro-environmental conditions. By capturing the interactive effects of genotypic and environmental variables, multivariate approaches provide a robust framework for evaluating genotype performance, ranking stability, and identifying mega-environments. Additionally, they allow breeders to assess the influence of specific environmental covariates, thereby informing region-specific selection strategies. Integration of multivariate models with genomic tools such as genome-wide association studies (GWAS), high-throughput phenotyping, and genomic selection (GS) further enhances their predictive power. These approaches help uncover the genetic architecture underlying yield stability and adaptability by linking genomic regions to phenotypic plasticity. Moreover, incorporating omics technologies (transcriptomics, proteomics, metabolomics) with multivariate analysis enables a systems-level understanding of stress-responsive pathways that drive stable performance under adverse conditions. The synergy of multivariate modelling and next-generation

P. Kona (✉) · Vinayaka · Amaresh · T. Lakshmi Pathy · R. T. Maruthi · H. K. Mahadeva Swamy · K. Mohanraj · A. Anna Durai · R. M. Shanthi · P. Govindaraj
ICAR-Sugarcane Breeding Institute, Coimbatore, Tamil Nadu, India
e-mail: kona.praveen@icar.gov.in

K. Gopalareddy
ICAR-Sugarcane Breeding Institute, Coimbatore, Tamil Nadu, India

National Institute of Seed Science and Technology, Regional Station, Bengaluru, India

R. Mohammadi et al. (eds.), *Genotype x Environment Interactions and its Implications for Plant Breeding*, https://doi.org/10.1007/978-981-95-5664-9_9

sequencing accelerates the finding of molecular markers linked with key traits, improving the efficiency of marker-assisted selection (MAS). As climate variability intensifies, multivariate models will play a crucial role in developing resilient cultivars by enabling more accurate predictions of genotype performance across environments. Their integration into molecular breeding platforms offers a transformative approach to sustainable crop improvement in the context of global environmental challenges.

Keywords

Genotype × environment Interaction (G × E) · Multivariate statistical models · Yield stability · AMMI · GGE · Machine learning

9.1 Introduction

The effectiveness of any plant breeding initiative depends on its capacity to identify and cultivate genotypes that consistently demonstrate exceptional phenotypic performance, particularly regarding yield and quality, across diverse environmental conditions. Accomplishing this necessitates a comprehensive understanding of the fundamental factors that affect phenotype expression. Although phenotypic characteristics such as grain yield or thousand-grain weight (TGW) are generally assessed at the conclusion of the growing season, they arise from a complex interaction between the genotype (genetic makeup of plant) and the environment in which it was grown. Plants differ significantly in how efficiently they utilize environmental inputs, such as water, nutrients, and light, and convert them into biomass. These differences are governed by specific genetic traits. Simultaneously, environments vary in the quantity, timing, and quality of the inputs they provide. Therefore, one of the core objectives in plant breeding is to identify suitable genotypes with specific environments to maximize performance. For example, under water-limited or low nitrogen conditions, breeders aim to select genotypes that maintain high productivity despite the stress.

Some genotypes exhibit broad adaptability, performing consistently across diverse environments, while others display specific adaptation, thriving only under particular conditions. This variation in performance is explained by genotype-by-environment interaction (G × E)—a phenomenon where the relative performance of genotypes changes depending on environmental conditions. The G × E complicates the process of selection because it reduces the predictability of genotype performance across locations or years. It is manifested as changes in genetic variance, covariance, and correlations across environments. For instance, greater genetic variance is often observed under optimal growing conditions compared to stress environments, although exceptions do occur (Przystalski et al. 2008). In environments with significant G × E, the ranking of genotypes becomes unstable, making selection decisions more difficult. This underscores the importance of selecting analytical tools that can quantify, interpret, and exploit G × E patterns.

Such tools are vital not only for identifying stable and high-performing cultivars but also for understanding the nature of environmental responsiveness.

Multivariate statistical models have become essential tools for analyzing genotype-by-environment interactions in modern plant breeding. These models simultaneously analyze multiple traits or responses across environments, enabling a more holistic assessment of genotype performance. They facilitate the detection of genetic relationships, classification of genotypes, understanding of trait interdependencies, and identification of mega-environments (Krishnappa et al. 2025). By reducing dimensionality (e.g., through principal component analysis [PCA] or GGE biplot), these models reveal patterns that univariate analyses often fail to predict, providing breeders with actionable insights to guide selection for stability, adaptability, and efficiency.

The G × E is typically defined as the non-additive interaction between the genotype and the environment (Freeman 1973), where the combined effect deviates from the simple sum of individual effects. Addressing this interaction through multivariate approaches is now considered essential for improving prediction accuracy, selecting genotypes with superior and stable performance, and ultimately developing varieties that meet the challenges of diverse agro-ecological zones.

9.2 Multivariate Statistical Techniques: Theory

Plant breeders and agronomists face a significant problem with G × E interaction. For a quantitative trait like grain yield, a significant G × E interaction can significantly impede selection progress. It is standard procedure to test selected materials across locations and years to guarantee that future cultivars will perform consistently across a variety of conditions. Various statistical models exist for G × E interaction analysis. For a comprehensive overview, we recommend consulting recent reviews, including those by Bustos-Korts et al. (2018), van Eeuwijk et al. (2016), and Malosetti et al. (2013). In this context, multivariate tools such as principal component analysis (PCA), cluster analysis (CA), canonical correspondence analysis (CCA), discriminant analysis (DA), AMMI-GGE models, and factorial regression and their application are pivotal in modern plant breeding and crop research. These techniques enable the simultaneous analysis of multiple traits, thereby revealing complex interrelationships that aid in optimizing selection strategies and enhancing the accuracy of genetic evaluations (Fig. 9.1). By integrating these methods, breeders can make more informed decisions for selecting superior genotypes. For more simplicity, these statistical methods and their applications are clearly detailed in the sequel.

9.2.1 PCA (Principal Component Analysis)

Karl Pearson (1901) introduced the principal component analysis (PCA) formally and further Harold Hotelling (1933) developed it. This is a statistical method that

Fig. 9.1 Applications of distinct multivariate tools in plant breeding

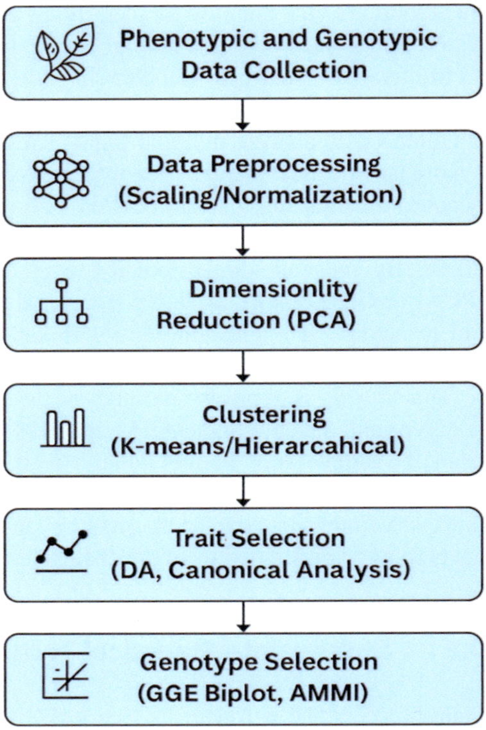

employs an orthogonal transformation to transform a collection of observations of potentially correlated variables into a set of values representing linearly uncorrelated variables, referred to as principal components (Jolliffe and Cadima 2016). It's one among the unsupervised statistical techniques used mainly for (i) the reduction in data dimension, (ii) AMMI-GGE biplot analysis (see Fig. 9.2, Sec. 2–3), (iii) eliminating the multicollinearity, (iv) visualization of multidimensional datasets, and (v) selection of useful features. Further, principal components are beneficial as input for many statistical treatments, such as multiple regression, cluster analysis, and factor analysis. Mathematically, let $x_1, x_2, \ldots, x_k$ are set of correlated variables then $y_1, y_2, \ldots, y_k$ are set of principal components, where $y_i = \sum_{j=1}^{k} l_j x_j$. Consider X be a matrix of k- dimension (i.e., includes k variables) with variance–covariance matrix Σ, then we can write $\Sigma = P\Delta P'$ using spectral decomposing, where $P'P = I$, $\Delta = \text{diag}\ (\lambda_1, \lambda_2, \ldots, \lambda_k)$, here $P = [\xi_1, \xi_2, \ldots, \xi_k]$ is an orthogonal matrix, ξ_i represents eigen vectors corresponding to the eigenvalues λ_i such that $\lambda_1 \geq \lambda_2 \geq \ldots \geq \lambda_k$. Variance structure of principal components is given by $V(y_i) = \xi_i' \Sigma \xi_i = \lambda_i$. Thus, the proportion of total variation accounted for by the ith principal component is expressed as $\lambda_i/(\lambda_1 + \lambda_2 + \ldots + \lambda_k)\ \forall\ i = 1, 2, \ldots, k$. In cases where k is large, the overall population variance can be largely explained by the initial few principal components, which can effectively substitute the original k variables with minimal

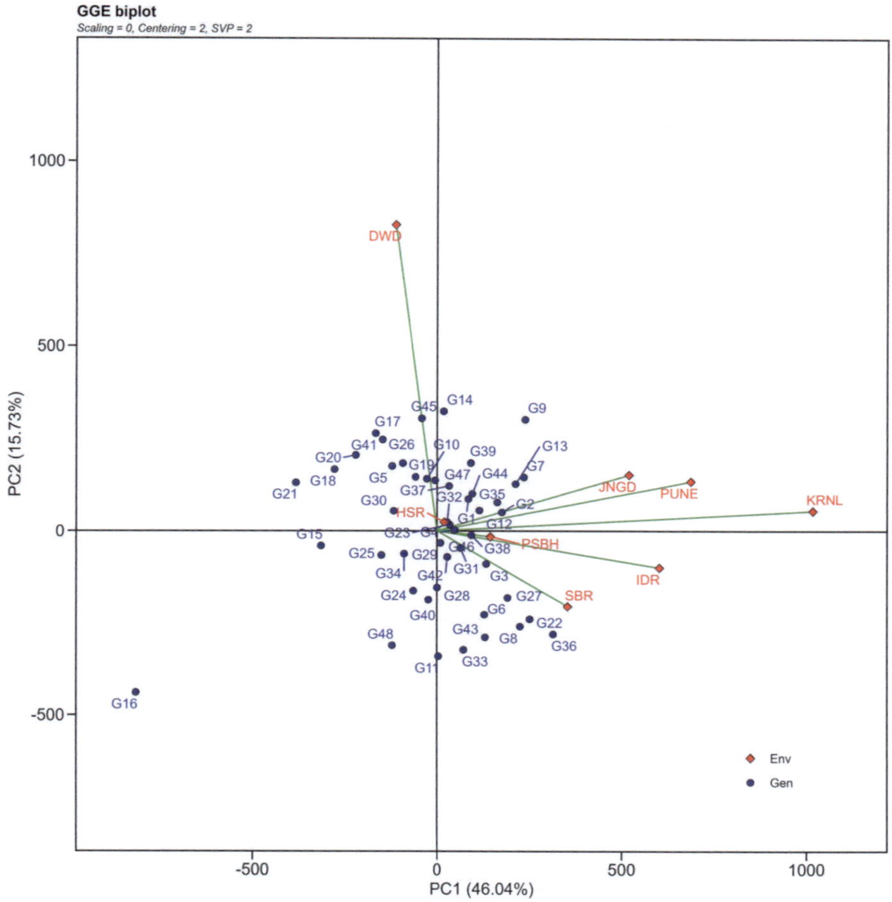

Fig. 9.2 GGE biplot denotes genotypes and environments interactions

information loss. A valuable method for identifying the suitable number of principal components is the Scree plot. The eigenvalues are arranged in descending order, and the Scree plot is created such that λ_i on the Y-axis and the index i (representing the magnitude of eigenvalues against their respective numbers) on the X-axis. To ascertain the correct number of principal components, one should look for an elbow (bend) in the Scree plot (see Fig. 9.3).

9.2.2 Additive Main Effects and Multiplicative Interaction (AMMI) Model

Gollob (1968) and Mandel (1969, 1971) first proposed the AMMI model within a fixed effects framework. Its application in agricultural research was later suggested

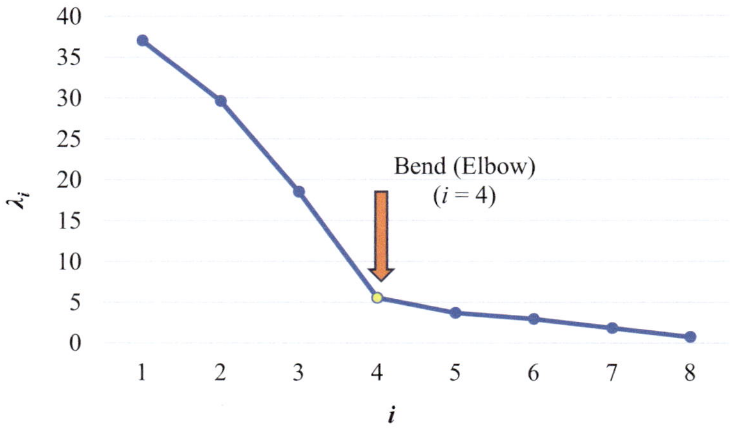

Fig. 9.3 Scree plot indicating appropriate selection of eigenvalues (principal components)

by Kempton (1984) and Zobel et al. (1988). This model combines analysis of variance (ANOVA) for main effects with principal component analysis (PCA) to assess interaction effects. According to Gauch and Zobel (1988), the AMMI model is proficient in identifying G × E interaction and in recognizing stable genotypes. The additive main effects of genotypes and environments/seasons are evaluated through univariate ANOVA, followed by the application of PCA to model G × E interactions as per the AMMI framework. The corresponding mathematical representation is as follows:

$$y_{ij} = \mu + g_i + s_j + \sum_{k=1}^{n} \lambda_k \alpha_{ik} \gamma_{ij} + \varepsilon_{ij}$$

where, y_{ij} = Trait mean of ith genotype in the jth season; μ = experimental mean trait value; g_i and s_j are the ith genotype and jth environmental mean deviation from the experimental mean trait value, respectively. λ_k = square root of the eigenvalue corresponding to the kth interaction principal component (IPC) axis. The interaction IPC scores for the kth IPC of the ith genotype and the jth environment are represented by α_{ik} and γ_{ij}, respectively. ε_{ij} is the residual effect. Least square principle was used for estimating the parameters of the AMMI model. The replication-wise quantitative trait means of genotypes are used for all statistical analyses as described in the following sections. The structure of ANOVA as per AMMI model is as follows:

Source of variation	df	Mean sum of squares (MSS)	Percentage of variance
Genotype (g)	g − 1	MSS_g	$(SS_g/TSS) \times 100$
Season (s)	s − 1	MSS_s	$(SS_s/TSS) \times 100$

(continued)

Source of variation	df	Mean sum of squares (MSS)	Percentage of variance
Genotype (g) × season (s)	(g − 1) (s − 1)	MSS_{gs}	$(SS_{gs}/TSS) \times 100$
IPCA 1	$g + s - 1 - 2n$	MSS_{IPCA1}	$(SS_{IPCA1}/SS_{gs}) \times 100$
IPCA 2	$g + s - 1 - 2n$	MSS_{IPCA2}	$(SS_{IPCA2}/SS_{gs}) \times 100$
Residuals	((g − 1) (s − 1) − {DF PCA1 + DF PCA2 + ... + DF PCAn})	MSS_r	
Error	(gs − 1)	MSS_e	

A commonly employed strategy within the AMMI framework involves the creation of biplots that represent averages multiplied by IPCA1, known as AMMI1 (Oliveira et al. 2003; Carbonell et al. 2004) and between IPCA1 and IPCA2 as AMMI 2. These methods are crucial for recommending new cultivars, as it complements the traits of adaptability and phenotypic stability, particularly through the average yields of genotypes. In the AMMI1 context, genotypes positioned near the center of the IPCA1 axis are characterized by enhanced stability and minimal interaction effects, indicating their broad adaptability to varying environmental conditions. Conversely, a positive interaction is identified when there is aligned polarity between environment and the genotype along IPCA axis, while a negative interaction occurs when there is a misalignment in polarity. The AMMI2 biplot utilizes the distances from the origin to elucidate the extent of interaction displayed by genotypes across different environments. Environments having low scores for IPCA1 and IPCA2 situated near to the origin significantly affect genotype stability and exhibit a limited impact on genotype-by-environment interaction, suggesting a wide adaptability to various developmental scenarios.

9.2.2.1 AMMI-Based Parameters for Identifying High-Yielding Stable Genotypes

The stability index (SI) and AMMI stability value (ASV) can be used to quantitatively measure the relative stability of genotypes. The next sections show how to estimate ASV and SI and how to evaluate genotypes stability performance based on ASV and SI.

AMMI Stability Value (ASV) The ASV is calculated as follows to enable an objective approach to finding genotypes with consistent performance across many sowing seasons (Purchase et al. 2000):

$$\mathrm{ASV} = \sqrt{\left[\frac{SS_{IPC1}}{SS_{IPC2}}(\mathrm{IPC1\ score})\right]^2 + (IPC2\ score)^2}$$

where the sum of squares (SS) attributed to the first two IPCs is represented by SS_{IPC1} and SS_{IPC2}. In a two-dimensional scatter map of IPC 1 against IPC 2 scores,

ASV is conceptually defined as the distance from zero. To account for the individual contributions of IPC 1 and IPC 2 scores to the total sum of squares of G × E interaction, the IPC 1 score is adjusted by the relative difference between the two scores because it typically contributes proportionately more to the sum of squares of GE. Greater stability of genotypes is indicated by lower ASV estimates, whereas lesser stability is indicated by higher ASV estimates.

Stability Index (SI) Since ASV just takes stability into account, irrespective of genotypes' yield potential, SI was calculated to make it easier to choose genotypes for stability and desired performance for several quantitative parameters at the same time. SI = RASV + RY is how it is expressed, with RASV standing for genotype rank based on RY and ASV for genotype rank based on mean over seasons of the quantitative trait (Farshadfar 2011). High trait expression and greater stability are considered to be characteristics of genotypes with low SI.

9.2.3 GGE Biplot Analysis

The genotype + genotype × environment (season) (GGE) biplot, which combines GGE principles with AMMI biplot, is a subjective method of defining genotypes × season interaction patterns and evaluating stability (Yan et al. 2000). The G × E interaction patterns were visually interpreted using a GGE biplot. The following model serves as the foundation for the GGE biplot.

$$Y_{ij} - Y_i = \lambda_1 \alpha_{i1} \gamma_{j1} + \lambda_2 \alpha_{i2} \gamma_{j2} + \varepsilon_{ij}$$

where, Y_{ij} is the ith genotype trait mean in jth season, Y_i is all the genotypes trait mean in the jth season, $_1$ and $_2$ represent square root of eigenvalues of first and second IPC axes, λ_1 and λ_2, α_{i1} and α_{i2} represent scores of the first and second IPC, respectively, for the ith genotype and γ_{j1} and γ_{j2} are first and second IPCs, respectively, for jth season. Use and interpretation of the GGE biplot can be done in a variety of ways. Instead, the most pertinent GGE biplot views are four. These include (i) the average-seasonal environment coordination (AEC) view of the GGE biplot based on genotype-focused scaling for ranking the test genotypes in relation to the ideal genotype, which is the genotype whose point is in the center of the concentric circles in the GGE biplot; (ii) the AEC perspective of the biplot utilizing seasonal environment-oriented scaling to assess the average performance of genotypes in relation to their stability patterns; (iii) the polygon representation of the GGE biplot employing symmetrical scaling to determine the "which-won-where" pattern of genotypes within the seasonal testing environment; and (iv) the evaluation of the discriminative and representative nature of the seasonal test environments as depicted in the GGE biplot.

9.2.4 Factorial Regression

Factorial regression is a powerful statistical tool widely applied in plant breeding to dissect G × E interaction and improve genetic gain through informed selection. It extends traditional analysis of variance (ANOVA) by incorporating genetic, physiological, and environmental covariates directly into the model. This approach allows for a more nuanced understanding of how genotypes respond to different environmental conditions, making it a powerful tool for plant breeders (van Eeuwijk et al., 2005). Mathematically, the factorial regression model can be expressed as:

$$Y = \mu + G + E + GE +$$

where Y is the phenotypic response, μ is the overall mean, G is the genotypic effect, E is the environmental effect, GE represents G × E interaction, and is the random error.

This model can be extended to include covariates:

$$Y = \mu + G + E + \beta_g X_g + \beta_e X_e +$$

where X_g and X_e are covariates for genotypes and environments, respectively. β_g and β_e are regression coefficients, respectively (Crossa et al. 2010).

It facilitates the identification of how genotypes respond across environments, supports quantitative trait loci (QTL) mapping by linking genetic markers to environmental responses, and enhances genomic selection (GS) accuracy using genome-wide data (Hébert et al. 1995). Several case studies illustrate its versatility: in groundnut, factor analytic models were employed to interpret G × E interaction and assess stability across environments (Coulibaly et al. 2024); in winter wheat, biadditive factorial regression identified key environmental covariates influencing yield (Brancourt-Hulmel et al. 2000); and in potato and ryegrass, comparisons with alternative models demonstrated its efficacy in yield stability and G × E interaction analysis (Baril et al. 1995; Balfourier et al. 1997). These applications are often embedded within multi-environment trials (METs) and factorial treatment structures, incorporating genetic and environmental covariates—such as climatic, soil, and molecular marker data—to improve model precision and interpretability (Smith et al. 2019; Piepho and Blancon 2023).

9.2.5 Cluster Analysis

Cluster analysis (CA) is a fundamental technique that operates without any assumptions regarding the number of groups or their structure. It categorizes data based on the similarities or differences observed. Different measures based on which clustering is done are (i) Euclidean distance: $\mathrm{d}(x, y) = \sqrt{(x-y)'(x-y)}$, where, $x' = (x_1, x_2, \ldots, x_k)$, $y' = (y_1, y_2, \ldots, y_k)$ are $k-$dimensional observations;

(ii) Minkowski measure: $d(x, y) = \left[\sum_{i=1}^{k} |x_i - y_i|^m\right]^{1/m}$, in general case, varying m changes the weight given to larger and smaller differences, for $m = 1$, it is reduced to city-block distance between two points in k dimensions, but for $m = 2$, it coincides Euclidean distance; (iii) Canberra metric: $d(x, y) = \sum_{i=1}^{k} \frac{|x_i - y_i|}{(x_i + y_i)}$; and (iv) Czekanowski coefficient: $d(x, y) = 1 - \frac{2\sum_{i=1}^{k} \min(x_i, y_i)}{\sum_{i=1}^{k} (x_i + y_i)}$. Nonetheless, in the absence of prior understanding of the grouping structure, it is not possible to compute these sample variance–covariance. Consequently, the Euclidean distance measure is frequently favored for cluster analysis.

CA is an essential tool in plant breeding for exploring genetic diversity, classifying germplasm, and selecting optimal parental lines. Among its methodologies, hierarchical clustering and K-means clustering are widely adopted due to their effectiveness in handling complex phenotypic and genotypic datasets. Hierarchical clustering constructs a dendrogram to visualize genetic relationships, using similarity measures such as Euclidean and Gower distances and linkage methods like Ward's and UPGMA (Houmanat et al. 2021). Its flexibility and visual interpretability make it suitable for identifying core sets and assessing diversity without predefined cluster numbers. In contrast (See Table 9.1), K-means clustering divides data into a definite number of clusters through iterative centroid optimization, offering computational efficiency for large-scale datasets (Raman et al. 2024; Acito 2023). Although sensitive to initial centroid selection, its performance can be enhanced using hybrid approaches that integrate hierarchical clustering. Both methods have been successfully employed in various crops, including rice, maize,

Table 9.1 Comparison of hierarchical clustering and K-means clustering

Feature	Hierarchical clustering	K-Means clustering
Cluster formation	Builds a tree-like structure (dendrogram) by merging or splitting clusters	Partitions data into K predefined clusters based on centroids
Number of clusters	Does not require predefining the number of clusters	Requires the number of clusters (K) to be specified beforehand
Handling mixed data	Effective for combining qualitative and quantitative data	Primarily designed for quantitative data
Computational efficiency	Computationally intensive due to the construction of dendrograms	Computationally efficient and suitable for large datasets
Visual representation	Provides a dendrogram for visualizing cluster relationships	Lacks a visual representation of cluster relationships
Sensitivity to initial conditions	Less sensitive to initial conditions as it does not rely on centroids	Highly sensitive to initial centroid selection

and safflower, to support strategic breeding decisions. Future advancements lie in integrating these techniques with machine learning (ML) and improving their scalability for high-dimensional genomic data.

9.2.6 Canonical Correspondence Analysis (CCA)

CCA is a multivariate statistical method that analyzes the relationships among two sets of variables: species data (response variables) and environmental data (explanatory variables). It is an extension of correspondence analysis (CA) and redundancy analysis (RDA), but it is particularly suited for ecological data where the response variables are species abundances (Makarenkov and Legendre 2002). CCA assumes a unimodal response of species to environmental gradients, which makes it more flexible than linear methods like RDA in modeling nonlinear relationships (Makarenkov and Legendre 2002). CCA is widely used in plant ecology to:

Identify environmental factors influencing plant distribution: By analyzing the relationships between plant species and environmental variables, CCA helps in understanding how factors like soil type, topography, and climate influence plant communities (Lee et al. 2020).

Classify plant communities: CCA can group plant species into communities based on their environmental preferences, which are useful for conservation and management strategies (Keymeulen and Beeckman 1990).

Assess ecological gradients: CCA is effective in identifying and visualizing ecological gradients, such as moisture, light, and nutrient gradients, which are critical for plant breeding and habitat restoration (Destan and Yurtseven 2012).

Evaluate species-environment relationships: CCA provides insights into the niche differentiation of plant species, which is essential for understanding how species adapt to specific environmental conditions (Fängström and Willén 1987).

9.2.7 Discriminant Analysis (DA)

Fisher (1938) introduced the concept of discrimination to characterize the distinguishing attributes of objects (observations) from various known collections (populations), either through algebraic or graphical means (in three dimensions or fewer). The objective is to identify "discriminants" with numerical values that maximize the separation between the collections. To achieve the greatest distinction between the y-values derived from populations π_1 and π_2, Fisher (1938) proposed transforming the multivariate observations x into univariate observations y. Fisher suggested employing linear combinations of **x** to produce ***y*'s**, as they are manageable functions of x. His approach does not require the assumption of normality in the populations. However, it does implicitly assume that the covariance matrices of the populations are identical, since a pooled estimate of the common covariance matrix is utilized. A fixed linear combination of $\boldsymbol{x}$'s yields the values $y_{11}, y_{12}, \ldots, y_{1n_1}$ for the first population's observations and $y_{21}, y_{22}, \ldots, y_{2n_2}$ for the second population's observations. The distinction between these two sets of univariate **y**'s is evaluated based on the difference between $\overline{y}_1$ and $\overline{y}_2$, measured in standard deviation units. That is, separation $= \frac{|\overline{y}_1 - \overline{y}_2|}{s_y}$, where $s_y^2 = \sum_{j=1}^{n_1} \left(y_{1j} - \overline{y}_1\right)^2 + \sum_{j=1}^{n_2} \left(y_{2j} - \overline{y}_2\right)^2 / n_1 + n_2 - 2$ is the pooled estimate of the variance. The expression for Fisher linear discriminant function which maximally separates the two populations and with maximum separation in the samples is $D^2 = (\overline{x}_1 - \overline{x}_2)' S_{\text{pooled}}^{-1} (\overline{x}_1 - \overline{x}_2)$. In plant breeding, discriminant analysis (DA), a potent statistical technique is frequently used to categorize and identify plant traits that are otherwise challenging to observe directly. Its strength is its capacity to support the selection of desirable genotypes by discriminating among predefined groups based on multivariate data. For example, by examining physiological characteristics like photosynthetic rate, transpiration, and stomatal conductance, DA has been successfully used to differentiate photosynthetic types (C3 vs. C4) with 98.48% classification accuracy (Hai-Ping et al. 1999). By employing shoot-related traits, DA enabled indirect selection for deep-rooted genotypes, which are crucial for drought tolerance, in rice breeding, attaining 78.6% classification accuracy (Kanbar et al. 2010). Further aiding in genetic mapping and trait improvement, DA has been crucial in locating DNA markers linked to important agronomic traits (Zhang et al. 2005). By determining the best linear combinations of predictor variables, a variation called canonical discriminant analysis (CDA) has also been used in horticulture to distinguish fruit groups according to storage disorders (Cruz-Castillo et al. 1994). These illustrations highlight DA's adaptability and usefulness in contemporary plant breeding and genetics.

9.2.8 Machine Learning (ML)-Based Approaches

In plant breeding, ML techniques have become revolutionary instruments that improve the efficacy and efficiency of breeding initiatives. These techniques tackle the enormous datasets produced by contemporary omics technologies as well as the complexity of G × E interaction. The theoretical reinforcements and real-world uses of ML in plant breeding are described in the sections that follow.

9.2.8.1 Theoretical Foundations of ML in Plant Breeding

Nonlinear Modeling: The nonlinear relationships in plant traits are frequently missed by conventional statistical techniques. Better predictions of plant characteristics are made possible by the superior modeling of these complexities by ML techniques like neural networks and support vector machines (Niazian and Niedbała 2020; Gabur et al. 2022).

Data Integration: ML makes it easier to combine various datasets (genomic, phenomic, and environmental), which improves selection procedures and advances our knowledge of plant genetics (Yoosefzadeh Najafabadi et al. 2023; Montesinos-López et al. 2024).

9.2.8.2 Applications of ML in Plant Breeding

Genotype Classification: To help identify desirable traits, ML algorithms categorize plant genotypes according to morphological and molecular markers (Niazian and Niedbała 2020).

Yield Prediction: The ability of ML models to predict yield results through the analysis of sizable datasets is essential for creating crop varieties that are resilient to the changing climate (Yoosefzadeh Najafabadi et al. 2023; Sharma et al. 2025).

Precision Phenotyping: ML and high-throughput phenotyping allow for accurate measurements and analyses of plant characteristics, which results in better breeding choices (Niazian and Niedbała 2020).

Even though ML has greatly advanced plant breeding, problems like the requirement for large amounts of computational power and the difficulty of implementing models still exist. However, ML has the potential to significantly increase food security and transform breeding methods. Furthermore, Table 9.2 comprises several R packages, Python libraries, and specialized software that are most regularly used in plant breeding, crop research, and agricultural data analysis.

Table 9.2 Applications of software and R-Python-packages in crop research

Tool(s)	Use	Importance	Example
metan, agricolae (R)	Analyze stability and adaptability in plant breeding trials	Helps in selecting consistent and high-performing genotypes across environments (Olivoto and Lúcio 2020; de Mendiburu 2023)	metan:: ge_means(), agricolae:: AMMI()
Factoextra, FactoMineR (R)	Perform PCA and clustering on multivariate data	Reduces dimensionality and identifies patterns in trait data (Kassambara and Mundt 2020; Lê et al. 2008)	FactoMineR:: PCA() for trait analysis
GGEBiplot (R)	Generate GGE biplots for G × E interaction	Visualizes genotype and environment interactions for selecting ideal genotypes (Olivoto and Lúcio 2020)	GGEBiplot:: GGEPlot() for environment grouping
PhysioIndexR and CANE (R)	Pooled ANOVA for multi-location or season data and Stress indices analysis	Helps in selecting consistent and high-performing genotypes across locations/ seasons and best at multi-stress-indices (Vinayaka et al. 2025a, b)	CANE:: PooledLSD() CANE:: PooledRBD() CANE:: PooledSPD() CANE:: PooledStPD(); WASI(); WMFVST()

(continued)

Table 9.2 (continued)

Tool(s)	Use	Importance	Example
sklearn (Python)	PCA, clustering, classification in ML pipelines	Automates pattern recognition in large datasets (Pedregosa et al. 2011)	sklearn.decomposition.PCA(), sklearn.cluster.KMeans()
statsmodels and matplotlib (Python)	Statistical modeling and data visualization	Presents and interprets results effectively for researchers (Seabold and Perktold 2010; Hunter 2007)	statsmodels.ols(), matplotlib.pyplot.plot()
GenStat	Statistical analysis for agricultural experiments	Supports complex analyses like ANOVA and BLUPs in breeding trials	Estimate heritability, run multi-environment analysis
CropStat	Analyze RCBD, LSD, and multi-location trials	Tailored for crop improvement research with built-in trial designs	Analyze genotype performance across multiple sites
PBTools	Analyze plant breeding trials and p-rep designs	Provides reliable statistical summaries and helps in selection decisions	Compute LSD, variety effects, trait correlations

9.3 Univariate Models Versus Multivariate LModels

Univariate methods are mathematically straightforward and easily understood for agronomists, plant breeders, and geneticists. However, these approaches have a significant drawback: they often oversimplify response patterns by representing interaction variation in a one dimension, despite the fact that responses may be multidimensional. This approach risks omitting crucial information in favor of a simpler statistical or biological interpretation. While these easy methods can be beneficial for breeders assessing a large number of lines, multivariate methods should be employed when a more precise analysis is required, such as in the study of "mega-environments" utilizing data from multilocation yield trials (Table 9.3).

9.4 Applications of Multivariate Analysis in Plant Breeding

Multivariate analysis techniques like D^2 analysis and principal component analysis (PCA) help to quantify genetic variation within plant populations (genetic diversity). Most of the techniques used to identify superior genotypes or diverge parents by analyzing multiple traits that are likely to produce offspring with high heterosis

Table 9.3 Key differences between univariate and multivariate models

Particulars	Univariate statistical tools	Multivariate statistical tools
Number of variables analyzed	Single variable at a time	Two or more variables simultaneously
Purpose	Describe and analyze patterns in one variable (e.g., mean, variance)	Analyze relationships among multiple variables and detect interaction patterns
Data complexity	Simpler datasets	Complex, multi-dimensional datasets
Examples of techniques	ANOVA, t-test, regression (with one dependent variable), mean, SD	ANCOVA, MANOVA, PCA, cluster analysis, canonical correlation, multiple regression, etc.
G × E interaction analysis	Limited; typically evaluates one trait across environments	Captures interactions across multiple traits and environments simultaneously
Application in molecular studies	Used to analyze expression of a single gene or trait	Used for integrative omics (e.g., gene expression + phenotype + environment)
Outcome	Individual trait performance or comparison	Trait interdependence, trait–environment interaction, and genotype classification
Visualization	Histograms, box plots, univariate line plots	Biplots, heatmaps, dendrograms, 3D scatter plots

(hybrid vigor). Several selection indices can be used to identify genotypes with desirable combinations of traits, potentially leading to improved crop varieties. Techniques like path analysis can reveal the direct and indirect effects of different traits on each other, allowing breeders to make more informed decisions about selection. But most of multivariate methods can be used to assess how different plant genotypes perform to changing environmental conditions. Multivariate analysis is key in plant breeding for understanding and managing G × E interaction. It allows plant breeders to analyze complex data, identify patterns, and make more informed decisions about which genotypes are best suited for different environments. By considering multiple traits and environments simultaneously, multivariate methods can provide a more complete picture of G × E interaction, leading to more effective selection and breeding strategies which are given below.

9.4.1 Understanding G × E Interactions

AMMI and GGE biplots are used to visualize and analyze G × E interaction. The AMMI (additive main effects and multiplicative interactions) and GGE (genotype plus G × E interaction) biplots facilitate the identification of genotypes that exhibit both high performance and stability across various environmental conditions. Some examples where the multivariate models help in understating the G × E interaction are given in Table 9.4.

Table 9.4 Examples where multivariate models help in understating the G × E interaction

Crop	Reference	Details
Groundnut	Dabessa et al. (2016)	A total of nine groundnut varieties were assessed in six different environments in western Oromia through AMMI and GGE biplot analyses. The findings indicated highly significant effects ($p \leq 0.01$) of genotype, environment, and their interaction on kernel yield. In the AMMI analysis, the first and second Interaction Principal Component Axes (IPCA1 and IPCA2) contributed to 41.32 and 7.2% of the total sum of squares for genotype-by-environment interaction, respectively.
	Lal et al. (2021)	Fourteen groundnut genotypes were evaluated under mid-season water stress using AMMI and GGE biplot techniques to study the impact of environmental factors on nodulation traits. GGE analysis highlighted the environment as the most influential factor, with PBS-14060 identified as stable and highly resilient, making it suitable for breeding drought-tolerant varieties.
Wheat	Sharma et al. (2012)	The winter wheat genotypes performance was assessed in semi-arid environments using 133 advanced breeding lines and four check varieties over six years (2005–2010). The study revealed that 11 out of the top 30 high-yielding genotypes performed consistently well across both low and high productivity environments. The findings indicated that different genetic materials are generally needed for semi-arid and irrigated conditions. However, certain genotypes bred for stress-prone environments demonstrated yield plasticity, allowing them to perform effectively under both dryland and irrigated conditions.
	Verma and Singh (2020)	Significant effects of genotypes, environment, and genotype × environment interaction were observed in wheat during 2018–19 and 2019–20 seasons. The superiority index identified HD3345, PBW822, and NIDW1158 as stable, high-yielding genotypes, while PRVG and MHPRVG identified HI8811, GW322, and HI8737 as superior, emphasizing the value of integrating AMMI analysis with yield.
Soybean	Bhartiya et al. (2017)	G × E was analyzed using GGE biplot and AMMI models across 04 environments with 32 diverse genotypes evaluated for four agronomic traits. The GGE biplot revealed that the first two principal components (PC1 and PC2) accounted for 84.68% of the variation in 100-seed weight, 74.40% in grain yield, 83.27% in days to maturity, and 91.98% in days to 50% flowering. Based on grain yield performance, genotype "C 17" ("PS 1556") was identified as the top yielder. However, considering both high yield and stability, "C 11" ("VLS 89") was recognized as the most ideal and stable genotype across environments.
Rice	Sharifi et al. (2017)	An experiment was conducted to identify high-yielding and stable rice genotypes and to determine their suitable cultivation areas using the AMMI model. The analysis revealed significant effects of genotype, environment, and G × E interaction on grain yield. Genotype G6 was recognized as both high-yielding and highly stable across different environments. The study

(continued)

Table 9.4 (continued)

Crop	Reference	Details
		highlighted that GE interactions were a major contributor to yield variability, and the AMMI biplots proved to be effective tools for visualizing genotype responses across environments, facilitating better selection decisions.
Linseed	Chobe and Ararsa (2018)	A study was conducted on 12 linseed genotypes evaluated across 13 environments in Ethiopia to assess the extent and pattern of G × E interaction and yield stability using AMMI and GGE biplot analyses. It revealed significant effects of the environment, genotype, and their interactions on seed yield, emphasizing the importance of considering G × E in breeding programs aimed at improving linseed productivity and stability across diverse environments.
Rapeseed	Agahi et al. (2020)	A study evaluating 22 spring oilseed rape genotypes across five sites during 2015–17 using AMMI analysis found that the first three IPCAs closely matched the G × E interaction signal, making AMMI3 the most reliable model. Yield gains were projected at 55.80% through broad adaptations, 26.73% through narrow adaptations across four mega-environments, and 17.47% with five mega-environments.
Sunflower	Ansarifard et al. (2020)	An experiment was conducted to assess the stability and adaptability of grain yield in 12 sunflower genotypes across five regions over two agricultural years. ANOVA showed that genotype × location, year × location, genotype × year, and genotype × year × location contributed 10.8, 17, 1.04, and 7.48% of the total variance, respectively. Biplot analysis indicated that the first and second principal components combine explained 73.4% of the total grain yield variance, accounting for a cumulative 73.4%.
Pigeon pea	Rao et al. (2020)	An experiment using AMMI and GGE biplot methods was conducted across 05 environments to identify stable pigeon pea genotypes and genotype-by-environment interactions. The genotype WRG-330 outperformed all other genotypes and checks across locations, while WRG-327 showed minimal interaction with environments, ensuring consistent performance. Environments at Adilabad and Tandur were found to be highly representative with strong discriminating ability.
	Kumar et al. (2021)	A study of 28 pigeon pea genotypes across 10 rainfed locations in Telangana and Karnataka, India, using AMMI and GGE biplot analysis revealed that grain yields were most influenced by the environment (56.8%), followed by genotype × environment interaction (27.6%) and genotype (18.6%) variances. Genotypes ICPH 2740 and PRG 176 were identified as suitable for large-scale recommendation to farmers with small holdings to boost pigeon pea productivity and enhance food security.

(continued)

Table 9.4 (continued)

Crop	Reference	Details
Maize	Katsenios et al. (2021)	In a two-year field experiment at six locations, five maize genotypes were evaluated, revealing that the G × E interaction explained 80.36% of the variation, while genotype (G) accounted for 12.79%, and environment (E) explained 6.85%. The AMMI model identified GEN2, GEN4, and GEN5 as the most suitable genotypes for three delineated mega-environments. GEN4 showed a strong positive interaction with environments characterized by high rainfall during the cultivation period.

9.4.2 Identifying Stable and Widely Adapted Genotypes and Best Environments

Multivariate methods have become essential tools in identifying genotype stability and adaptability, especially in METs. These methods, such as AMMI (additive main effects and multiplicative interaction) and GGE (genotype and G × E interaction) biplot analysis, allow for the calculation of stability parameters, including regression slope, Shukla's stability variance, and various indices like the yield stability index (YSI), which quantify a genotype's ability to maintain high performance across different environmental conditions. They also facilitate the prioritization of genotypes based on their performance across multiple traits and environments. These tools help breeders select the most stable and high-yielding genotypes, ensuring better productivity across varying environmental conditions.

In the study, Farshadfar et al. (2011) assessed 14 bread wheat genotypes over three consecutive years (2008–2011) under both irrigated and rainfed conditions using a complete randomized block design. The results from AMMI analysis indicated that the AMMI stability value was effective in discriminating stable genotypes, with genotypes 10 and 6 showing stable performance across different environments. Moreover, the study found that the yield stability index (YSI) and the new rank-sum (RS) method were better at identifying the most stable genotypes with high grain yield than traditional stability indices like the sustainability index (SI) and AMMI stability index (ASI). This highlights the importance of using more comprehensive stability indices that consider environmental interactions for more accurate genotype ranking and selection.

Zali et al. (2012) conducted a stability analysis on 17 chickpea genotypes in Iran, utilizing AMMI and Spearman's rank correlation to assess the relationship between different stability measures. They identified three distinct groups of stability indices, revealing that some indices, such as Sums of the Absolute Value of the Interaction Principal Component Scores (SIPC4), Averages of the Squared Eigenvector Values (EV4), and modified AMMI stability value (MASV), were correlated with mean yield, while others like stability statistic based on the first IPC axes (FPi) were not associated with stability and mean yield. Based on the comprehensive stability analysis, they recommended genotype G13 (FLIP 97–114) for national release due to its wide adaptation across different environments, showing both high yield and stability.

Jambhulkar et al. (2014) introduced the modified AMMI stability index (MASI), a modification of the traditional AMMI stability index (ASI). Unlike ASI, which only considers the first few interaction principal components (IPCs), MASI takes into account all significant IPCs, providing a more comprehensive measure of genotype stability. They also noted that the simultaneous selection index for yield and stability (SSI) could be computed based on MASI, allowing for better selection of genotypes that are both stable and high-yielding.

Further advancements were made by Ajay et al. (2018), who examined several stability parameters derived from the AMMI model, including the modified AMMI stability value (MASV), stability measure based on fitted AMMI model (FA), and the Annicchiarico's D Parameter (DA). They calculated the simultaneous selection index for yield and stability (SSI) from these stability parameters and highlighted the importance of using all significant IPCs in the AMMI model for accurate genotype ranking. They also noted that using the correct formula for MASV, such as MASV2, provided more reliable estimates of genotype stability compared to previous versions of the index.

In their 2020 study, Ajay et al. (2020) further extended their research to peanut genotypes, where they used the AMMI model and analyzed the first three significant interaction principal components (IPC1–IPC3). By applying the AMMI stability parameters and SSI, they identified genotypes PBS-22083, PBS-22088, and PBS-29098 as the most stable and high-yielding. They emphasized the importance of using all significant IPCs for genotype ranking, as MASI and MASV provided more accurate results compared to AMMI stability indices like ASI and ASV, which only considered the first two IPCs.

Cheloei et al. (2020) conducted a study on 13 M8 mutant genotypes derived from Tarrom-Mahalli rice land races in Iran, evaluating their grain yield stability and adaptability. They used AMMI analysis and identified several stability indices such as AMGE, ASI, MASI, and MASV, which helped in identifying genotypes 33, 26, and 30 as the most stable and high-yielding. Their study reinforced the importance of using multiple stability parameters to obtain a reliable assessment of genotype performance across various environments. Olanrewaju et al. (2021) carried out a study in Nigeria on Bambara groundnut (BGN) genotypes, evaluating the extent of G × E interaction effects across four environments. Their analysis revealed that TVSu-1589, TVSu-1905, and TVSu-2048 were the most stable genotypes, while other genotypes like TVSu-1763 and TVSu-1899 showed high adaptability across environments, demonstrating the role of multivariate stability models in identifying adaptable and stable genotypes for different environmental conditions.

Gupta et al. (2021) focused on lentil genotypes for grain iron (Fe) and zinc (Zn) concentrations, evaluating their stability across three different locations in India. The AMMI model revealed that genotypes IG 49, P 16214, ILL 147, and P 2118 exhibited high stability and better nutrient concentrations. The study emphasized the utility of multivariate stability indices in identifying genotypes with stable performance for specific traits such as nutrient content across diverse environments.

Anuradha et al. (2022) further explored stability analysis in finger millet genotypes, using AMMI, MASV, and MASV1 to identify stable high-yielding genotypes. Their studies reinforced the importance of using multivariate approaches and stability indices for selecting genotypes with consistent performance across various environments. Employing multivariate stability models such as AMMI and GGE biplot analysis, along with stability indices like MASV, MASI, and SSI, has demonstrated significant effectiveness in the evaluation of genotypes. These models assist breeders in pinpointing stable, high-yielding genotypes that are adaptable to diverse environmental conditions, thereby refining the selection process and ultimately fostering enhanced crop production and food security.

9.4.3 Predicting Performance in New Environments

Predicting how genotypes perform in untested or future environments is a critical task in plant breeding, especially under the uncertainties of climate change and evolving agro-ecological conditions. Multivariate statistical tools provide a robust framework to enhance the accuracy and reliability of such forecasts by modeling complex G × E interaction.

(a) *Regression-Based Multivariate Models:* Multivariate regression models extend traditional regression by incorporating multiple traits and/or multiple environments simultaneously. These models help identify environmental covariates (e.g., rainfall, temperature, soil nitrogen) that explain genotype responses. By establishing relationships between genotype performance and environmental variables, breeders can predict how a genotype is likely to perform in a new location that shares similar environmental profiles (Cooper et al. 1996; van Eeuwijk et al. 2005). Techniques like factorial regression and partial least squares regression (PLSR) are widely used in this context (Jarquín et al. 2014).

(b) *Mixed Linear Models (MLM) with Multivariate Extensions:* Mixed models treat genotypic effects as random and environmental effects as fixed (or vice versa), capturing both main effects and G × E interactions efficiently. When extended to multivariate frameworks using REML or Bayesian approaches, these models can handle multiple traits and environments concurrently (Smith et al. 2001; de Campos et al. 2013). This allows breeders to borrow strength from correlated traits and environments, enhancing prediction accuracy. Best Linear Unbiased Prediction (BLUP) and Genomic BLUP (GBLUP) also leverage these structures to forecast performance under novel or low-data environments (Piepho et al. 2008; Crossa et al. 2017).

(c) *Principal Component Analysis (PCA) and GGE Biplots:* Tools like PCA and GGE biplots reduce data dimensionality and visually identify genotype adaptability and stability across environments. The patterns extracted from known environments can be extrapolated to predict behavior in similar, untested environments by positioning them relative to existing environmental vectors (Yan et al. 2000; Yan and Tinker 2006). These tools are particularly valuable for breeders to identify mega-environments and high-performing stable genotypes.
(d) *Factor Analytic Models and AMMI:* Factor analytic (FA) models and additive main effects and multiplicative interaction (AMMI) models decompose G × E into interpretable components. These models are particularly powerful in predicting genotype performance in new environments based on shared interaction structures (Smith et al. 2001; Gauch 2006). They also help identify genotypes with wide or specific adaptability, thus guiding targeted breeding strategies.

Through AMMI, the interaction component here is decomposed into several orthogonal PCs. Among multiple PCs extracted from the data, a few explain pattern and others consist of noise which is unpredictable the choice of extractable PCs that explain signal can be made through approaches like ANOVA, sum of squares, and cross-validation through postdiction (Bernardo 2002). AMMI is not a single model but a family of models designated as AMMI0, AMMI1……AMMIF, depending on the number of components retained in the model diagnosis. AMMI0 is the null model, while AMMIF is the full model with all PCs (Gauch 2013). Once this information is obtained, then prediction of trait performance in new/untested environments can be made. Thus, the parsimonious model that explains most of the predictable signal is more efficient in estimating the trait in new environments.

To determine the best family of model, the number of pcs to be included in the model has to be considered. This can be done by following three ways. First one is Gallob's F test that can be obtained from AMMI model itself. The number of significant PCs in the AMMI ANOVA can be considered for ascertaining the stability of the genotypes. Next, in AMMI ANOVA, the sum of squares due to PCA and signal component can be estimated (Gauch et al. 2008). If the three principal components explain as much are the sum of squares as genotype signal, then three can be considered as optimum number of principal components. Another approach is predictive approach, where the entire data series are divided into training and validation sets, and the optimum number of PCs can be selected through root mean square error (RMSE) or mean absolute errors (MAE) or mean absolute percent errors (MAPE) between observed and predicted values. Studies have shown that the model accuracy *vis-à-vis* success of the prediction is higher across years at a single location than across locations or seasons. Once optimal model is chosen, then AMMI can identify genotypes that consistently perform well across a range of environments depicting stability and viral adaptability.

9.4.4 In Molecular Breeding

Multivariate statistical models have become powerful tools in the analysis of G × E interaction, especially in molecular studies aimed at dissecting the genetic basis of complex traits. These models allow the simultaneous analysis of multiple variables (traits, markers, genes) and environments, enabling a more holistic understanding of plant responses. Key applications include:

(a) *Genomic Selection and Marker-Trait Association:* Multivariate G × E interaction models improve the accuracy of genomic selection (GS) by incorporating multiple environments and traits simultaneously (Jarquín et al. 2014; Montesinos-López et al. 2016). Models like multi-environment GBLUP and Bayesian multi-trait models help identify molecular markers or genomic regions that are consistently associated with performance across environments (Crossa et al. 2017). For example, Identifying stable QTLs/genes for drought or nitrogen use efficiency using multi-trait GWAS (Korte and Farlow 2013).
(b) *Transcriptomic and Proteomic Integration:* PCA, PLS-DA, and canonical correlation analysis (CCA) are applied to large-scale transcriptomic/proteomic datasets to correlate gene expression patterns with phenotypic plasticity across environments (Castro et al. 2008; Jiang et al. 2023). They are useful for identifying stress-responsive genes and regulatory networks contributing to G × E interactions.
(c) *Pathway-Level G × E Studies:* Tools like multivariate analysis of variance (MANOVA) and redundancy analysis (RDA) can be used to explore how entire pathways or gene sets behave across environments (Anderson 2001). They help in identifying molecular pathways that are robust or sensitive to environmental variation.
(d) *Reaction Norm Models:* These models integrate environmental gradients and molecular data to understand how genotypes (at molecular levels) respond to changing environments (Malosetti et al. 2013). They are useful for predicting gene expression or metabolite accumulation in varying conditions.
(e) *Gene Co-expression Network Analysis Under G × E interaction:* Techniques like WGCNA (Weighted Gene Co-expression Network Analysis) are extended to multi-environment trials to understand how co-expression modules vary across environments (Langfelder and Horvath 2008). They help in linking gene networks to adaptive traits influenced by G × E interaction.
(f) *Integration with High-Throughput Phenotyping:* Multivariate G × E interaction models are combined with omics and phenomics data to identify molecular signatures associated with trait performance under environmental stress (Araus et al. 2018). They facilitate machine learning-based prediction of stress responses based on molecular data (Washburn et al. 2019).
(g) *eQTL and pQTL Mapping in Multi-Environments:* Expression QTL (eQTL) and protein QTL (pQTL) mapping under different environments using multivariate models to detect G × E-influenced regulatory loci (Keurentjes et al. 2007; Korte et al. 2012).

9.5 Conclusion

In plant breeding, the G × E interaction analysis is essential for developing crop varieties that are both stable and high-yielding across varying environmental conditions. Multivariate statistical models, such as AMMI, GGE biplot, and stability indices, are highly effective in understanding and quantifying these interactions. These models help dissect the complexity of how genetic and environmental factors interact to influence crop performance, offering breeders valuable understandings into genotype stability and adaptability. Multivariate models enable breeders to identify stable genotypes that perform consistently well across diverse environments. Stability indices such as the AMMI stability value (ASV), modified ASV (MASV), and simultaneous selection index for yield and stability (SSI) allow breeders to rank genotypes based on their yield stability in different environments. This ability is crucial in breeding programs targeting to develop crop resilience to diverse conditions like drought, salinity, or varying soil fertility. By utilizing these models, breeders can select genotypes that exhibit broad adaptability while maintaining high productivity across various environmental stress factors. The growing unpredictability of environmental conditions due to climate change makes identifying stable, high-performing genotypes even more important. Multivariate models assist breeders in selecting genotypes with wide adaptability to multiple environmental scenarios while ensuring they retain high yields. Additionally, these models can help assess the influence of specific environmental factors, enabling breeders to pinpoint favorable growing conditions for each genotype. Multivariate models also facilitate the identification of mega-environments—regions where specific genotypes consistently perform well. By recognizing these regions, breeders can optimize breeding efforts and resources to develop varieties tailored to the unique conditions of each mega-environment. This approach leads to more informed decisions about genotype selection and environmental adaptation, ultimately fostering the development of varieties that are resilient to various environmental stresses. Incorporating multivariate models also aids in uncovering the genetic architecture underlying yield stability. By integrating phenotypic data from multiple environments, these models provide insights into the genetic mechanisms governing trait stability. This understanding paves the way for identifying genetic markers linked to stable performance, which can be utilized in marker-assisted selection (MAS) to accelerate the development of improved crop varieties. Furthermore, multivariate approaches combined with genome-wide association studies (GWAS) can pinpoint specific genetic loci associated with traits of interest, enhancing the breeding programs precision. The integration of molecular techniques, such as proteomics, transcriptomics, and metabolomics, with multivariate models allows breeders to explore the biochemical and molecular pathways that govern to stability of trait across environments. This deeper understanding helps breeders select genotypes that are more resilient to environmental stresses, such as drought or heat, by targeting specific molecular pathways responsible for adaptation. Moreover, multivariate models improve high-throughput breeding by incorporating genomic data from next-generation sequencing (NGS) technologies. By linking genetic variation with phenotypic outcomes across diverse environments, these models provide more accurate predictions of genotype performance, enabling breeders to identify superior genotypes early in the breeding cycle. Molecular markers associated with stability traits can also be tracked using these models, leading to more efficient and targeted selection. One promising development is the use of multivariate models in genomic selection (GS), where genomic data is used to predict the breeding value of genotypes. Incorporating multivariate analysis into GS can enhance the accuracy of predicting yield stability and adaptability, particularly under challenging environmental conditions. As genomic data continues to expand, the combination of multivariate models with genomic selection will become an increasingly powerful tool for improving crop performance and resilience.

9.6 Future Prospects

The future of multivariate statistical models in plant breeding is promising, with several advancements that can enhance their efficiency and applicability. As technological innovations continue to evolve, the integration of high-throughput phenotyping data, molecular markers, and genomic information will provide more accurate and robust insights into G × E interactions. Here are some key areas where future developments are expected:

(a) *Integration with Genomic Data:* The combination of multivariate models with genomic selection methods will allow for the genomic regions identification associated with stable performance over environments. Genomic data can help breeders in understanding the genetic basis of yield stability and provide markers for selecting stable genotypes more efficiently.
(b) *Advanced Phenotyping Techniques:* Employing advanced phenotyping technologies, such as remote sensing, drones, and precision agriculture tools, will allow for the pooling more precise and detailed phenotypic data. These data can be integrated into multivariate models, providing better insights into genotype behavior under changing environmental conditions.
(c) *Climate Change Adaptation*: As climate change continues to pose challenges to agriculture, the need for developing crops that can withstand a range of environmental stresses becomes even more critical. Multivariate models will play a key role in identifying genotypes that are resilient to climate-induced stresses, such as heat, drought, and flooding, aiding to ensure food security in the face of varying climatic conditions.
(d) *Multi-Environment Trials (METs):* The plant breeding future will likely involve even more complex and larger-scale multi-environment trials. With advancements in computational power and machine learning techniques, multivariate models can be applied to vast datasets from global breeding programs. This will enable the identification of stable genotypes across a wider range of environments and accelerate the breeding process.

(e) *Automation and Artificial Intelligence*: Automation and AI will possibly play a significant role in the future of plant breeding. These technologies can help automate the process of data collection, data analysis, and final decision-making, making it faster and more efficient to identify high-yielding and stable genotypes across environments. AI can also assist in the optimization of breeding programs by analyzing large datasets more efficiently than traditional methods.

(f) *Integration of Soil and Environmental Data:* The integration of detailed soil and environmental data into multivariate models will allow for a good understanding of how different soil types, water availability, and climate factors interact with genotype performance. This will enable the development of more precise recommendations for genotype selection based on specific environmental conditions.

(g) *Personalized Crop Breeding:* Advances in precision breeding may lead to more personalized breeding strategies, where crop varieties are developed for specific regions or even individual farmers. This approach will maximize yield stability and adaptability, addressing the unique challenges faced by local farming communities.

References

Acito F (2023) Dimensionality reduction. In: Predictive analytics with KNIME: analytics for citizen data scientists. Springer, Cham, pp 85–103

Agahi K, Ahmadi J, Oghan HA, Fotokian MH, Orang SF (2020) Analysis of genotype × environment interaction for seed yield in spring oilseed rape using the AMMI model. Crop Breed Appl Biotechnol 20(1). https://doi.org/10.1590/1984-70332020v20n1a1

Ajay BC, Aravind J, Abdul Fiyaz R, Bera SK, Kumar N, Gangadhar K et al (2018) Modified AMMI stability index (MASI) for stability analysis. ICAR-DGR Newsl 18:4–5

Ajay BC, Aravind J, Abdul FR, Narendra K, Chuni L, Gangadhar K et al (2020) Evaluation of genotype × environment interaction and yield stability analysis in peanut under phosphorus stress condition using stability parameters of AMMI model. Agric Res 9:477–486. https://doi.org/10.1007/s40003-020-00459-2

Anderson MJ (2001) A new method for non-parametric multivariate analysis of variance. Austral Ecol 26:32–46. https://doi.org/10.1111/j.1442-9993.2001.01070.pp.x

Ansarifard I, Mostafavi K, Khosroshahli M, Reza Bihamta M, Ramshini H (2020) A study on genotype–environment interaction based on GGE biplot graphical method in sunflower genotypes (Helianthus annuus L.). Food Sci Nutr 8:3327–3334. https://doi.org/10.1002/fsn3.1608

Anuradha N, Patro TSK, Singamsetti A, Sandhya RY, Triveni U, Nirmala KA et al (2022) Comparative study of AMMI and BLUP-based simultaneous selection for grain yield and stability of finger millet [Eleusine coracana (L.) Gaertn.] genotypes. Front Plant Sci 12:78–83. https://doi.org/10.3389/fpls.2021.831282

Araus JL, Kefauver SC, Zaman-Allah M, Olsen MS, Cairns JE (2018) Translating high-throughput phenotyping into genetic gain. Trends Plant Sci 23:451–466. https://doi.org/10.1016/j.tplants.2018.02.001

Balfourier F, Oliveira JA, Charmet G, Arbones E (1997) Factorial regression analysis of genotype by environment interaction in ryegrass populations, using both isozyme and climatic data as covariates. Euphytica 98:37–46. https://doi.org/10.1023/A:1003095720009

Baril CP, Denis JB, Wustman R, Van Eeuwijk FA (1995) Analysing genotype by environment interaction in Dutch potato variety trials using factorial regression. Euphytica 82:149–155. https://doi.org/10.1007/BF00027061

Bernardo R (2002) Breeding for quantitative traits in plants. Stemma Press, Woodbury

Bhartiya A, Aditya JP, Kumari V, Kishore N, Purwar JP, Agrawal A et al (2017) GGE biplot and AMMI analysis of yield stability in multi-environment trial of soybean [Glycine max (L.) Merrill] genotypes under rainfed condition of north western Himalayan hills. J Anim Plant Sci 27:227–238

Brancourt-Hulmel M, Denis JB, Lecomte C (2000) Determining environmental covariates which explain genotype environment interaction in winter wheat through probe genotypes and biadditive factorial regression. Theor Appl Genet 100:285–298. https://doi.org/10.1007/s001220050038

Bustos-Korts D, Romagosa I, Borràs-Gelonch G, Casas AM, Slafer GA, van Eeuwijk FA (2018) Genotype by environment interaction and adaptation. In: Meyers RA (ed) Encyclopedia of sustainability science and technology. Springer, New York. https://doi.org/10.1007/978-1-4939-2493-6_1029-1

Carbonell SA, Filho JA, Dias LA, Garcia AA, Morais LK (2004) Common bean cultivars and lines interactions with environments. Sci Agric 61:169–177. https://doi.org/10.1590/S0103-90162004000200009

Castro C, Motto M, Rossi V, Manetti C (2008) Variation of metabolic profiles in developing maize kernels up-and down-regulated for the hda101 gene. J Exp Bot 59:3913–3924. https://doi.org/10.1093/jxb/ern231

Cheloei G, Ranjbar G, Babaeian N, Bagheri N, Nouri M (2020) Using AMMI model and its parameters for yield stability analysis of rice (Oryza sativa L.) advanced mutant genotypes of Tarrom-Mahalli. J Genet 9:70–83

Chobe AC, Ararsa AD (2018) AMMI and GGE biplot analysis of linseed (Linum usitatissimum L.) genotypes in central and south-eastern highlands of Ethiopia. J Plant Breed Genet 6:117–127

Cooper M, Woodruff DR, Phillips IG, Basford KE, Gilmour AR (1996) Genotype-by-management interactions for grain yield and grain protein concentration of wheat. Field Crop Res 47:247–271. https://doi.org/10.1016/0378-4290(96)00018-5

Coulibaly M, Bodjrenou G, Fassinou Hotègni NV, Akohoue F, Agossou CA, Azon CF et al (2024) Genotype by environment interaction and stability analysis of three agronomic traits in Kersting's groundnut (Macrotyloma geocarpum) using factor analytic modeling and environmental covariates. Crop Sci 64:2095–2115. https://doi.org/10.1002/csc2.21210

Crossa J, Vargas M, Joshi AK (2010) Linear, bilinear, and linear-bilinear fixed and mixed models for analysing genotype × environment interaction in plant breeding and agronomy. Can J Plant Sci 90:561–574. https://doi.org/10.4141/CJPS10013

Crossa J, Pérez-Rodríguez P, Cuevas J et al (2017) Genomic selection in plant breeding: methods, models, and perspectives. Trends Plant Sci 22:961–975. https://doi.org/10.1016/j.tplants.2017.08.011

Cruz-Castillo JGS, Ganeshanandam S, Mackay BR, Lawes GS, Woolley DJ (1994) Application of canonical discriminant analysis in horticultural research. HortScience 29:1115–1119. https://doi.org/10.21273/HORTSCI.29.10.1115

Dabessa A, Birru A, Zerihun A, Dagnachew L (2016) Genotype by environment interaction kernel and yield stability of groundnut (Arachis hypogaea L.) varieties in Western Oromia, Ethiopia. J Agric Crops 2:113–120

de los Campos G, Hickey JM, Pong-Wong R, Daetwyler HD, Calus MPL (2013) Whole-genome regression and prediction methods applied to plant and animal breeding. Genetics 193:327–345. https://doi.org/10.1534/genetics.112.143313

Destan S, Yurtseven I (2012) Canonical and typological analysis of the relationship between soil and trees—a step toward ecosystem management. Afr J Agric Res 7:1566–1576. https://doi.org/10.5897/AJAR11.1376

Fängström I, Willén E (1987) Clustering and canonical correspondence analysis of phytoplankton and environmental variables in Swedish lakes. Vegetatio 71:87–95. https://doi.org/10.1007/BF00044815

Farshadfar E (2011) Chromosomal localization of the genes controlling adaptation in Agropyron elongatum using a new AMMI–based simultaneous selection index of yield and yield stability. Int J Plant Breed 5:80–83

Farshadfar E, Mahmodi N, Yaghotipoor A (2011) AMMI stability value and simultaneous estimation of yield and yield stability in bread wheat (Triticum aestivum L.). Aust J Crop Sci 5:514–524

Fisher RA (1938) The statistical utilization of multiple measurements. Ann Eugenics 8:376–386. https://doi.org/10.1111/j.1469-1809.1938.tb02189.x

Freeman GH (1973) Statistical methods for the analysis of genotype-environment interactions. Heredity 31:339–354. https://doi.org/10.1038/hdy.1973.90

Gabur I, Simioniuc DP, Snowdon RJ, Cristea D (2022) Machine learning applied to the search for nonlinear features in breeding populations. Front Artif Intell 5:876578. https://doi.org/10.3389/frai.2022.876578

Gauch HG (2006) Statistical analysis of yield trials by AMMI and GGE. Crop Sci 46:1488–1500. https://doi.org/10.2135/cropsci2005.07-0193

Gauch HG (2013) A simple protocol for AMMI analysis of yield trials. Crop Sci 53:1860–1869. https://doi.org/10.2135/cropsci2013.04.0241

Gauch HG, Zobel RW (1988) Predictive and postdictive success of statistical analyses of yield trials. Theor Appl Genet 76:1–10. https://doi.org/10.1007/BF00288824

Gauch HG, Piepho HP, Annicchiarico P (2008) Statistical analysis of yield trials by AMMI and GGE: further considerations. Crop Sci 48:866–889. https://doi.org/10.2135/cropsci2007.09.0513

Gollob HF (1968) A statistical model which combines features of factor analytic and analysis of variance techniques. Psychometrika 33:73–115. https://doi.org/10.1007/BF02289676

Gupta S, Das S, Dikshit HK, Mishra GP, Aski MS, Bansal R et al (2021) Genotype by environment interaction effect on grain iron and zinc concentration of Indian and Mediterranean lentil genotypes. Agronomy 11:1761. https://doi.org/10.3390/agronomy11091761

Hai-Ping T, Gao-Ming J, Xin-Shi Z (1999) Application of discriminant analysis in distinguishing plant photosynthetic types—a case study in Northeast China transect (NECT) area. J Integr Plant Biol 41:1–10

Hébert Y, Plomion C, Harzic N (1995) Genotype × environment interaction for root traits in maize, as analysed with factorial regression models. Euphytica 81:85–92. https://doi.org/10.1007/BF00022460

Hotelling H (1933) Analysis of a complex of statistical variables into principal components. J Educ Psychol 24:417–441. https://doi.org/10.1037/h0071325

Houmanat K, Douaik A, Charafi J, Hssaini L, El Fechtali M, Nabloussi A (2021) Appropriate statistical methods for analysis of safflower genetic diversity using agglomerative hierarchical cluster analysis through combination of phenotypic traits and molecular markers. Crop Sci 61:4164–4180. https://doi.org/10.1002/csc2.20620

Hunter JD (2007) Matplotlib: A 2D graphics environment. Computing in Science & Engineering 9(3):90–95. https://doi.org/10.1109/MCSE.2007.55

Jambhulkar NN, Bose LK, Singh ON (2014) AMMI stability index for stability analysis. CRRI Newsl 35(1):15

Jarquín D, Crossa J, Lacaze X et al (2014) A reaction norm model for genomic selection using high-dimensional genomic and environmental data. Theor Appl Genet 127:595–607. https://doi.org/10.1007/s00122-013-2243-1

Jiang MZ, Aguet F, Ardlie K, Chen J, Cornell E, Cruz D et al (2023) Canonical correlation analysis for multi-omics: application to cross-cohort analysis. PLoS Genet 19:e1010517. https://doi.org/10.1371/journal.pgen.1010517

Jolliffe IT, Cadima J (2016) Principal component analysis: a review and recent developments. Philos Trans R Soc A 374:20150202. https://doi.org/10.1098/rsta.2015.0202

Kanbar A, Toorchi M, Motohashi T, Kondo K, Shashidhar HE (2010) Evaluation of discriminant analysis in identification of deep and shallow rooted plants in early segregating generation of rice (Oryza sativa L.) using single tiller approach. Aust J Basic Appl Sci 4:3909–3916

Kassambara A, Mundt F (2020) factoextra: Extract and Visualize the Results of Multivariate Data Analyses. R package version 1.0.7. https://CRAN.R-project.org/package=factoextra

Katsenios N, Sparangis P, Leonidakis D, Katsaros G, Kakabouki I, Vlachakis D et al (2021) Effect of genotype × environment interaction on yield of maize hybrids in Greece using AMMI analysis. Agronomy 11:479. https://doi.org/10.3390/agronomy11030479

Kempton RA (1984) The use of biplots in interpreting variety by environment interactions. J Agric Sci 103:123–135. https://doi.org/10.1017/S0021859600043392

Keurentjes JJB, Fu J, Terpstra IR et al (2007) Regulatory network construction in Arabidopsis by using genome-wide gene expression quantitative trait loci. Proc Natl Acad Sci USA 104:1708–1713. https://doi.org/10.1073/pnas.0610429104

Keymeulen R, Beeckman H (1990) Species-environment relationships in three woodlands using canonical correspondence analysis. Silva Gandavensis 55:1–18

Korte A, Farlow A (2013) The advantages and limitations of trait analysis with GWAS: a review. Plant Methods 9:29. https://doi.org/10.1186/1746-4811-9-29

Korte A, Vilhjálmsson BJ, Segura V, Platt A, Long Q, Nordborg M (2012) A mixed-model approach for genome-wide association studies of correlated traits in structured populations. Nat Genet 44:1066–1071. https://doi.org/10.1038/ng.2376

Krishnappa G, Khan H, Amaresh, Vinayaka, Pathy TL, Gupta A, Mahendru-Singh A, Ahlawat AK, Jasavantlal Manilal Patel JM, Biradar SS, Mamrutha HM, Khobra R, Pandey V, Singh GP, Tiwari R (2025). Deciphering low nitrogen tolerance of wheat in mega-environments using integrated multivariate approaches. European Journal of Agronomy 173,127919. https://doi.org/10.1016/j.eja.2025.127919

Kumar N, Ramya V, Sameer Kumar CV, Raju T, Kumar N, Seshu G et al (2021) Identification of pigeonpea genotypes with wider adaptability to rainfed environments through AMMI and GGE biplot analyses. Indian J Genet Plant Breed 81:63–73. https://doi.org/10.31742/ISGPB.81.1.7

Lal C, Ajay BC, Rupapara KV (2021) AMMI and GGE models indicating seasonal variations as major source of variations for nodulation related characters in peanut. Indian J Genet Plant Breed 81:277–288. https://doi.org/10.31742/IJGPB.81.2.11

Langfelder P, Horvath S (2008) WGCNA: an R package for weighted correlation network analysis. BMC Bioinform 9:559. https://doi.org/10.1186/1471-2105-9-559

Lê S, Josse J, Husson F (2008) FactoMineR: An R package for multivariate analysis. Journal of Statistical Software, 25(1):1–18. https://doi.org/10.18637/jss.v025.i01

Lee JS, Son DH, Lee SH, Myeong HH, Cho JS, Lee JC et al (2020) Canonical correspondence analysis ordinations and competitor, stress tolerator, and ruderal strategies of coastal dune plants in South Korea. J Coast Res 36:528–535. https://doi.org/10.2112/JCOASTRES-D-19-00059.1

Makarenkov V, Legendre P (2002) Nonlinear redundancy analysis and canonical correspondence analysis based on polynomial regression. Ecology 83:1146–1161. https://doi.org/10.1890/0012-9658(2002)083[1146:NRAACC]2.0.CO;2

Malosetti M, Ribaut JM, van Eeuwijk FA (2013) The statistical analysis of multi-environment data: modeling genotype-by-environment interaction and its genetic basis. Front Physiol 4:44. https://doi.org/10.3389/fphys.2013.00044

Mandel J (1969) The partitioning of interaction in analysis of variance. J Res Natl Bur Stand 73B:309–328. https://doi.org/10.6028/jres.073B.026

Mandel J (1971) A new analysis of variance model for non-additive data. Technometrics 13:1–18. https://doi.org/10.1080/00401706.1971.10488751

de Mendiburu F (2023) agricolae: Statistical procedures for agricultural research (Version 1.3-7) [R package]. Comprehensive R Archive Network (CRAN). https://CRAN.R-project.org/package=agricolae

Montesinos-López OA, Montesinos-López A, Crossa J et al (2016) A genomic Bayesian multi-trait and multi-environment model. G3 Genes|Genomes|Genetics 6:2725–2744. https://doi.org/10.1534/g3.116.032359

Montesinos-López OA, Chavira-Flores M, Kismiantini, Crespo-Herrera L, Saint Pierre C, Li H et al (2024) A review of multimodal deep learning methods for genomic-enabled prediction in plant breeding. Genetics 228:161. https://doi.org/10.1093/genetics/iyae161

Niazian M, Niedbała G (2020) Machine learning for plant breeding and biotechnology. Agriculture 10:436. https://doi.org/10.3390/agriculture10100436

Olanrewaju OS, Oyatomi O, Babalola OO, Abberton M (2021) GGE biplot analysis of genotype × environment interaction and yield stability in Bambara groundnut. Agronomy 11:1839. https://doi.org/10.3390/agronomy11091839

Oliveira JP, Moreira WN Jr, Duarte JB, Chaves LJ, Pinheiro JB (2003) Genotype-environment interaction in maize hybrids: an application of the AMMI model. Crop Breed Appl Biotechnol 3:185–192. https://doi.org/10.12702/1984-7033.v03n03a02

Olivoto T, Lúcio AD (2020) metan: An R package for multi-environment trial analysis. Methods in Ecology and Evolution 11(6):783–789. https://doi.org/10.1111/2041-210X.13384

Pearson K (1901) On lines and planes of closest fit to systems of points in space. Philos Mag 2:559–572. https://doi.org/10.1080/14786440109462720

Pedregosa F, Varoquaux G, Gramfort A, Michel V, Thirion B, Grisel O, Blondel M, Prettenhofer P, Weiss R, Dubourg V, Vanderplas J, Passos A, Cournapeau D, Brucher M, Perrot M, Duchesnay É (2011) Scikit-learn: Machine learning in Python. Journal of Machine Learning Research 12: 2825–2830. https://jmlr.org/papers/v12/pedregosa11a.html

Piepho HP, Blancon J (2023) Extending Finlay–Wilkinson regression with environmental covariates. Plant Breed 142:621–631. https://doi.org/10.1111/pbr.13122

Piepho HP, Möhring J, Melchinger AE, Büchse A (2008) BLUP for phenotypic selection in plant breeding and variety testing. Euphytica 161:209–228. https://doi.org/10.1007/s10681-007-9449-8

Przystalski M, Osman A, Thiemt E, Rolland B, Ericson L, Østergård H et al (2008) Comparing the performance of cereal varieties in organic and non-organic cropping systems in different European countries. Euphytica 163:417–433. https://doi.org/10.1007/s10681-008-9715-4

Purchase JL, Hatting H, Van Deventer CS (2000) Genotype × environment interaction of winter wheat (Triticum aestivum L.) in South Africa: II. Stability analysis of yield performance. S Afr J Plant Soil 17:101–107. https://doi.org/10.1080/02571862.2000.10634878

Raman R, Kumar V, Pillai BG, Rabadiya D, Patre S, Meenakshi R (2024) The impact of enhancing the k-means algorithm through genetic algorithm optimization on high dimensional data clustering outcomes. In: 2024 International conference on knowledge engineering and communication systems, vol 1. IEEE, pp 1–5. https://doi.org/10.1109/IC-KECS57263.2024.10748020

Rao PJM, Kishore NS, Sandeep S, Neelima G, Rao PM, Das DM et al (2020) Evaluation of performance of yield stability analysis based on AMMI and GGE-Biplot in promising pigeonpea [Cajanus cajan (L.) Millspaugh] genotypes. Legum Res 45:1414–1420. https://doi.org/10.18805/LR-4591

Seabold S, Perktold J (2010) Statsmodels: Econometric and statistical modeling with Python. In 9th Python in Science Conference (Vol. 57, p. 61). https://doi.org/10.25080/Majora-92bf1922-011

Sharifi P, Aminpanah H, Erfani R, Mohaddesi A, Abbasian A (2017) Evaluation of genotype × environment interaction in rice based on AMMI model in Iran. Rice Sci 24:173–180. https://doi.org/10.1016/j.rsci.2016.11.005

Sharma RC, Morgounov AI, Braun HJ, Akin B, Keser M, Kaya Y et al (2012) Yield stability analysis of winter wheat genotypes targeted to semi-arid environments in the International Winter Wheat Improvement Program. Int J Plant Breed 6:1–10

Sharma N, Kumar M, Sharma N, Singh NK (2025) Integration of artificial intelligence (AI) and machine learning (ML) in plant breeding. In: Plant stress tolerance. CRC Press, pp 110–117

Smith AB, Cullis BR, Thompson R (2001) Analysing variety by environment data using multiplicative mixed models and adjustments for spatial field trend. Biometrics 57:1138–1147. https://doi.org/10.1111/j.0006-341X.2001.01138.x

Smith AB, Borg LM, Gogel BJ, Cullis BR (2019) Estimation of factor analytic mixed models for the analysis of multi-treatment multi-environment trial data. J Agric Biol Environ Stat 24:573–588. https://doi.org/10.1007/s13253-019-00367-1

van Eeuwijk FA, Malosetti M, Yin X, Struik PC, Stam P (2005) Statistical models for genotype by environment data: from conventional ANOVA models to eco-physiological QTL models. Aust J Agric Res 56:883–894. https://doi.org/10.1071/AR05155

Van Eeuwijk FA, Bustos-Korts DV, Malosetti M (2016) What should students in plant breeding know about the statistical aspects of genotype × environment interactions? Crop Sci 56:2119–2140. https://doi.org/10.2135/cropsci2015.06.0375

Verma A, Singh GP (2020) Simultaneous use of AMMI model and yield for stability analysis of wheat genotypes evaluated under central zone of the India. Eur J Agric Food Sci 2(6). https://doi.org/10.24018/ejfood.2020.2.6.113

Vinayaka, Lakshmi Pathy T, Gopalareddy K, Kumari S, Murali P, Govindaraj P, Prasad PRC, Vinaykumar LN (2025a) CANE: Comprehensive groups of experiments analysis for numerous environments. R package version 0.1.1. Comprehensive R Archive Network (CRAN). https://doi.org/10.32614/CRAN.package.CANE

Vinayaka, Krishnapriya V, Lakshmi Pathy T, Amaresh, Gopalareddy K, Suresha GS, Govindaraj P (2025b) PhysioIndexR: Physiological and stress indices for crop evaluation. R package Version 0.1.0. Comprehensive R Archive Network (CRAN). https://doi.org/10.32614/CRAN.package.PhysioIndexR

Washburn JD, Mejia-Guerra MK, Ramstein G, Kremling KAG, Valluru R, Buckler ES (2019) Evolutionarily informed deep learning methods for predicting transcript abundance from DNA sequence. Proc Natl Acad Sci USA 116:5542–5549. https://doi.org/10.1073/pnas.1814551116

Yan W, Tinker NA (2006) Biplot analysis of multi-environment trial data: principles and applications. Can J Plant Sci 86:623–645. https://doi.org/10.4141/P05-169

Yan W, Hunt LA, Sheng Q, Szlavnics Z (2000) Cultivar evaluation and mega-environment investigation based on the GGE biplot. Crop Sci 40:597–605. https://doi.org/10.2135/cropsci2000.403597x

Yoosefzadeh Najafabadi M, Hesami M, Eskandari M (2023) Machine learning-assisted approaches in modernized plant breeding programs. Genes 14:777. https://doi.org/10.3390/genes14030777

Zali H, Farshadfar E, Hosein S, Sabaghpour SH, Karimizadeh R (2012) Evaluation of genotype × environment interaction in chickpea using measures of stability from AMMI model. Ann Biol Res 3:3126–3136

Zhang N, Xu Y, Akash M, McCouch S, Oard JH (2005) Identification of candidate markers associated with agronomic traits in rice using discriminant analysis. Theor Appl Genet 110:721–729. https://doi.org/10.1007/s00122-004-1892-5

Zobel RW, Wright MJ, Gauch HG (1988) Statistical analysis of a yield trial. Agron J 80:388–393. https://doi.org/10.2134/agronj1988.00021962008000030002x

A Comparative Study Among Multivariate Statistical Methods for Analyzing G × E Interaction

10

Lonishree Dutta, Twahira Begum, Anindita Gogoi, Tanmita Gupta, and Mohan Lal

Abstract

Genotype × environment (G × E) interaction plays a crucial role in agricultural research, particularly in plant breeding, where selecting stable and high-performing genotypes across multiple environments is a key objective. Multivariate statistical methods provide powerful tools to dissect these interactions and improve genotype selection. This chapter presents a comparative analysis of widely used multivariate approaches, including cluster analysis, principal component analysis (PCA), canonical correlation analysis (CCA), multivariate analysis of variance (MANOVA), additive main effects and multiplicative interaction (AMMI) model, and genotype + genotype-by-environment interaction (GGE) biplot, along with newer methods such as the multi-trait genotype-ideotype distance index (MGIDI) and multi-trait stability index (MTSI). The strengths, limitations, and suitability of these methods in handling G × E interactions are evaluated based on interpretability, computational efficiency, and ability to model complex interactions. Traditional methods like AMMI and GGE biplots remain fundamental for visualizing interactions, whereas MGIDI and MTSI provide advanced multi-trait selection strategies. The findings emphasize the importance of selecting appropriate statistical tools based on breeding objectives and dataset complexity to enhance genetic improvement programs.

L. Dutta · T. Begum · A. Gogoi · T. Gupta · M. Lal (✉)
Agrotechnology and Rural Development Division, CSIR-North East Institute of Science and Technology, Jorhat, Assam, India

R. Mohammadi et al. (eds.), *Genotype x Environment Interactions and its Implications for Plant Breeding*, https://doi.org/10.1007/978-981-95-5664-9_10

10.1 Introduction

The study of genotype × environment (G × E) interaction is crucial in many scientific disciplines, particularly in agricultural and ecological research. It refers to the phenomenon where the performance of different genotypes (G) varies under different environmental conditions (E). In crop breeding, for example, understanding the nature of G × E interactions is essential for identifying genotypes that perform consistently across multiple environments or those that excel under specific conditions. Interactions between genotype and environment restrict a cultivar's genetic potential and its capacity to thrive in a given environment (Yan et al. 2007; Yan and Kang 2002; Yan 2014). Moreover, analyzing G × E interaction can be challenging due to the complexity of the data involved.

Researchers understand the value of considering the full spectrum of genetic diversity instead of just GEI (Malosetti et al. 2013). Modified models are useful for efficiently separating the GEI component from total genetic variation (Hudson et al. 2022). Multivariate statistical methods are powerful tools that help in understanding and interpreting G × E interactions. These methods allow researchers to handle multiple traits or variables simultaneously, providing a holistic view of the relationships among genotypes, environments, and their interactions. This chapter focuses on comparing some of the most widely used multivariate methods for analyzing G × E interaction, namely:

1. Cluster analysis
2. Principal component analysis (PCA)
3. Canonical correlation analysis (CCA)
4. Multivariate analysis of variance (MANOVA)
5. Additive main effects and multiplicative interaction (AMMI) model
6. Genotype-by-environment interaction (GGE) biplot

We are examining the theoretical foundations, applications, and performance of these methods, along with their merits and demerits.

10.2 The Role of G × E Interaction in Agricultural Research

For agricultural research, and especially for crop breeding, it is essential to find stable genotypes under different environmental conditions. Breeding programs seeking to guarantee good output across all environmental situations can benefit from stable genotypes, which exhibit constant performance across environments.

The extent of G × E interaction can vary by traits (e.g., grain yield, disease resistance, drought tolerance), and understanding this interaction helps researchers decide whether they should select genotypes that are adapted to specific environments or genotypes that perform well across a range of environments.

When studying G × E interactions, the key challenge is to deal with complex datasets where multiple factors (genotype, environment, and interaction) influence the observed outcomes. This is where multivariate statistical methods come in they help dissect these relationships in a way that univariate methods (which analyze one trait at a time) often cannot.

10.3 Multivariate Statistical Methods for G × E Interaction Analysis

10.3.1 Cluster Analysis

Cluster analysis is a powerful tool used to group genotypes, environments, or genotype-environment combinations based on similarities in their response patterns. By categorizing genotypes or environments with similar performance traits, researchers can identify key clusters of interest and gain insights into how specific environmental factors influence genetic expression (Everitt et al. 2011; Jain et al. 1999).

In the context of G × E interaction, cluster analysis can be applied in various ways:

1. *Clustering Genotypes by Performance Across Environments*: Cluster analysis can group genotypes based on how their performance varies across multiple environmental conditions. For instance, genotypes that consistently perform well in drought-prone environments might be clustered together (Mohammadı et al., 2012). This helps identify genotypes with stable, desirable traits under specific environmental stresses.
2. *Clustering Environments by Genotype Response*: Environments can be clustered based on how different genotypes perform within them. This approach is useful for identifying environments that are more favorable to specific genotypes (Neisse et al. 2018). For example, a cluster of environments with similar soil types or climatic conditions may reveal that particular genotypes perform better in those settings.
3. *Clustering G × E Interaction Combinations*: A more sophisticated approach involves clustering the combinations of genotypes and environments (Yan et al. 2002). By examining how specific genotype–environment pairs interact, researchers can uncover patterns that would otherwise be difficult to detect. For instance, certain genotype–environment combinations may show strong performance, indicating a genotype that thrives under specific conditions.

10.3.1.1 History

The mathematical foundations of cluster analysis began to take shape in the early 1900s with the rise of statistical theory. One of the earliest approaches to grouping data was the work of *R.A. Fisher* in 1936 (Fisher 1936). Fisher, best known for his contributions to statistics, introduced the concept of "discriminant analysis," a precursor to clustering techniques. Discriminant analysis, like cluster analysis, aims to classify objects into predefined categories, but it requires labeled data, while clustering works with unlabeled data. The next important development came with *Edward Titchener*'s work on "grouping variables" in psychology during the early twentieth century. His approach involved using correlation techniques to find similar characteristics in psychological traits, a fundamental idea behind cluster analysis that would be expanded upon later.

The formalization of cluster analysis as a distinct field began in the 1930s and 1940s. One of the earliest systematic methods used for clustering was the hierarchical cluster analysis method, introduced by *Alexander Prentiss* in 1938 (Everitt et al. 2011). This method sought to create a tree-like structure, known as a dendrogram, to represent the relationships between groups of objects. Hierarchical clustering was an early attempt at understanding how to combine objects or groups based on their similarities.

Another significant advancement came from Sokal and Michener (1958) when they introduced a method for clustering biological species based on a similarity coefficient. Their approach became a standard for hierarchical clustering methods and formed the basis for the Jaccard coefficient, which is still used today to measure similarity between sets.

10.3.1.2 Cluster analysis and Euclidean Distance

Cluster analysis is a technique used to group data points into clusters based on their similarities. One of the most common ways to measure the similarity (or dissimilarity) between data points is through distance metrics, and Euclidean distance is one of the most widely used distance measures in clustering (Jain et al. 1999). Understanding the relationship between cluster analysis and Euclidean distance is crucial for effective application in various fields like data mining, machine learning, and pattern recognition. In this section, we will explore the concept of Euclidean distance, how it is used in cluster analysis, and the impact it has on the clustering process.

10.3.1.3 What Is Euclidean Distance?

Euclidean distance is the straight-line distance between two points in a multidimensional space. It is a measure of the "ordinary" distance between two points, which can be thought of as the shortest path that connects them. Euclidean distance works in any number of dimensions, so whether we are dealing with a two-dimensional (2D) plane, three-dimensional (3D) space, or a higher-dimensional feature space, this formula can be used.

10.3.1.4 How Euclidean Distance Is Used in Cluster Analysis?

In the context of cluster analysis, Euclidean distance is often used to measure the similarity (or dissimilarity) between data points. The closer the Euclidean distance between two data points, the more similar they are considered to be, and they are more likely to be grouped together in the same cluster. Cluster analysis algorithms, such as K-means clustering and Hierarchical clustering, rely on distance measures like Euclidean distance to define clusters.

1. *K-Means Clustering*

   In K-means clustering, the goal is to partition a dataset into k clusters, with each cluster having a centroid (the mean of all the points in that cluster). The algorithm iterates between two main steps:

   - *Assignment step*: Each data point is assigned to the cluster whose centroid is the closest, based on Euclidean distance.
   - *Update step*: The centroids of the clusters are recalculated by taking the mean of all the data points in each cluster.

   The Euclidean distance is used in the first step to measure how far each data point is from each centroid (Mohammadı et al. 2012). The point is assigned to the cluster whose centroid minimizes the Euclidean distance.

2. *Hierarchical Clustering*

   In hierarchical clustering, the goal is to build a tree of clusters (a dendrogram) by either agglomerating (bottom-up) or dividing (top-down) clusters iteratively. Euclidean distance plays a critical role in how clusters are merged or split (Murtagh and Contreras 2017).

   In agglomerative hierarchical clustering:

   - Initially, each data point is treated as a separate cluster.
   - At each iteration, the two closest clusters (based on Euclidean distance) are merged to form a new cluster.
   - This process continues until all points are grouped into a single cluster or until a stopping criterion is met.

   The distance between clusters can be defined in different ways, such as:

   - *Single linkage*: The distance between the closest pair of points from two clusters.
   - *Complete linkage*: The distance between the farthest pair of points from two clusters.
   - *Average linkage*: The average distance between all pairs of points from two clusters.

   However, in all cases, Euclidean distance is used to calculate the distance between individual points, which then influences the merging process.

10.3.1.5 Impact of Euclidean Distance on Clustering

While Euclidean distance is a simple and widely-used distance measure, its use in clustering has certain advantages and limitations.

Advantages:

1. *Intuitive and Easy to Understand*: The concept of straight-line distance in Euclidean space is simple and intuitive, making it easy to understand and implement.
2. *Scalability*: Euclidean distance scales well with data in low- to moderate-dimensional spaces and works well with many common clustering algorithms, such as K-means and hierarchical clustering.
3. *Efficiency*: Computationally, calculating Euclidean distance is relatively fast, and algorithms like K-means are efficient when using it to assign points to clusters.

Limitations:

1. *Sensitivity to Scaling*: Euclidean distance is sensitive to the scale of the features. If the features have different units or vary widely in magnitude, the Euclidean distance may not reflect true similarity. For example, a feature representing number of tillers for a plant might range from 1 to 200 (approx) while another feature representing days to maturity might range from 90 to more than 300 days (approx). In such cases, feature scaling (e.g., normalization or standardization) is often necessary before applying clustering (Jain et al. 1999).
2. *Assumes Spherical Clusters*: Euclidean distance works well when clusters are spherical (i.e., clusters are roughly circular or globular in shape). However, it may not perform well when clusters have irregular shapes, as it measures distance in straight lines and might fail to capture complex geometries (Everitt et al. 2011).
3. *Outlier Sensitivity*: Euclidean distance is highly sensitive to outliers. A single outlier with extreme values can significantly affect the clustering process, pulling the centroid of a cluster away from the majority of the data points.

10.3.1.6 When to Use Euclidean Distance in Cluster Analysis

Euclidean distance is often the default distance measure when performing clustering, but it is particularly suitable for scenarios where:

- The data has a continuous numerical structure.
- The clusters are expected to have a roughly spherical or globular shape.
- The features are on similar scales or have been properly scaled (e.g., using standardization).

10.3.1.7 Alternatives to Euclidean Distance

In some cases, Euclidean distance may not be the most appropriate measure for clustering. Depending on the nature of the data, other distance measures might be more suitable:

1. *Manhattan Distance (L1 Norm)*: Instead of the straight-line distance, Manhattan distance measures the sum of the absolute differences between the coordinates of two points. This is useful in high-dimensional spaces or when dealing with grid-like structures.
2. *Cosine Similarity*: For text data or data involving angles, cosine similarity, which measures the cosine of the angle between two vectors, can be more effective than Euclidean distance. It focuses on the direction of vectors rather than their magnitude.
3. *Minkowski Distance*: A generalized form of both Euclidean and Manhattan distance, where the exponent ppp allows flexibility in distance calculations. If $p = 2p = 2p = 2$, it becomes Euclidean distance; if $p = 1p = 1p = 1$, it becomes Manhattan distance.

4. *Correlation Distance*: When the focus is on finding clusters based on correlation rather than absolute distance, correlation distance can be more appropriate, especially for time-series data.

10.3.2 Principal Component Analysis (PCA)

PCA is a dimension-reduction technique that is used to transform a set of correlated variables into a smaller number of uncorrelated variables called principal components (PCs) that capture the most significant features of the original data (Jolliffe 2011). This is achieved through linear combinations of the original variables. In the context of G × E interaction, PCA is useful for summarizing the variability in genotype performance across multiple environments and traits, thereby reducing the complexity of the data (Yan and Kang 2002).

PCA has become a core technique in data science, machine learning, and various scientific disciplines, allowing researchers to simplify complex data while retaining its essential patterns. The history of PCA is deeply rooted in the development of multivariate statistics, and it has evolved over time to become one of the most powerful and commonly used methods in the analysis of large datasets (Jolliffe and Cadima 2016).

PCA can trace its roots back to Karl Pearson, a British mathematician, who is often credited with being one of the first to develop the concept of principal components in 1901. In his paper titled "On Lines and Planes of Closest Fit to Systems of Points in Space," Pearson introduced the method of principal components as a way to extract the most significant directions of variation from a set of points in a multidimensional space. His original idea was to find a linear combination of variables that would minimize the least squares errors in representing the data. Pearson's method aimed at finding the line of best fit for a set of data points in multidimensional space, effectively identifying the direction in which the data varied the most.

Pearson's approach was based on the concept of eigenvalues and eigenvectors, which would later become central to the development of PCA. However, it was not until the following decades that the technique would be fully formalized and extended.

Although Pearson's work laid the groundwork for PCA, it was Harold Hotelling, an American mathematician and statistician, who formally developed the modern version of PCA in the early 1930s. In his groundbreaking 1933 paper, "Analysis of a Complex of Statistical Variables into Principal Components," Hotelling provided a comprehensive mathematical framework for PCA. He expanded on Pearson's work by formalizing the concept of transforming a set of correlated variables into a new set of uncorrelated variables (the principal components).

Hotelling's key contribution was recognizing that the problem of finding principal components could be interpreted as finding the eigenvectors of the covariance matrix of the data. The eigenvectors represent the directions of maximum variance in the data, and the corresponding eigenvalues represent the amount of variance explained by each principal component. This development made it clear that PCA was not just a method of finding a line of best fit, but a generalizable technique for transforming multidimensional data into a new coordinate system.

Hotelling's formalization of PCA provided the theoretical basis for the method to be widely used in multivariate statistics, and his work is considered the foundation for modern PCA (Jolliffe 2011).

Merits of PCA:

- *Data Reduction*: PCA reduces the dimensionality of the dataset, enabling easier interpretation and visualization (Jolliffe and Cadima 2016).
- *Pattern Identification*: PCA can identify patterns in the data, such as clusters of genotypes that perform similarly across environments (Yan and Kang 2002).
- *Variance Explanation*: The principal components explain the maximum variance in the dataset, ensuring that the most important features are retained (Jolliffe 2011).

Demerits of PCA:

- *Linear Assumptions*: PCA assumes that the relationships between variables are linear, which may not always be the case in biological systems (Jolliffe and Cadima 2016).
- *Interpretation Complexity*: While PCA reduces data complexity, interpreting the principal components can sometimes be difficult, especially if the components do not have clear real-world interpretations (Jolliffe 2011).
- *Lack of Statistical Testing*: PCA does not provide a direct mechanism for hypothesis testing (e.g., testing the significance of G × E interactions) (Yan and Kang 2002).

10.3.3 Canonical Correlation Analysis (CCA)

CCA is a multivariate method used to explore the relationships between two sets of variables. In G × E analysis, one set might represent genotype characteristics, and the other might represent environmental conditions. CCA identifies the linear relationships between these sets of variables by maximizing the correlation between the linear combinations of each set (Hardoon et al. 2004).

Before the formal introduction of CCA, the foundations for multivariate statistical methods were being established in the early twentieth century. Researchers were developing tools to analyze complex data involving multiple variables simultaneously.

- *Correlation Analysis (Late Nineteenth Century)*: The earliest form of multivariate statistical analysis can be traced back to Francis Galton and Karl Pearson, who developed the concept of correlation in the late nineteenth century. Pearson's work on linear correlation was the basis for understanding how two variables might be related (Pearson 1896). However, Pearson's correlation only dealt with the relationship between two variables and was limited in its ability to analyze more complex relationships involving multiple variables.

- *Multiple Regression (Early Twentieth Century)*: As the need to analyze relationships between more than two variables arose, the concept of multiple regression was developed. Multiple regression, popularized by Sir Ronald A. Fisher, allowed for modeling the relationship between a dependent variable and multiple independent variables (Fisher 1922). However, this technique still focused on the relationship between a single dependent variable and a set of independent variables, rather than two sets of multivariate data.
- *Factor Analysis (1900s–1930s)*: The development of factor analysis in the early twentieth century by Charles Spearman and Louis Thurstone allowed statisticians to analyze the underlying structure in a set of correlated variables (Spearman 1961). Factor analysis aimed to reduce data dimensionality and identify latent variables or factors that explain the observed correlations. However, factor analysis was not designed for directly analyzing the relationships between two sets of variables, which would become the core objective of canonical correlation analysis.

These early multivariate techniques laid the groundwork for more advanced statistical methods, including CCA, which sought to go beyond understanding relationships within a single set of variables and instead explore the associations between two distinct sets of variables.

The development of canonical correlation analysis (CCA) is primarily attributed to the work of Harold Hotelling, an American statistician, in the mid-1930s. The formal introduction of CCA can be traced back to 1936, when Harold Hotelling published his seminal paper, "Relations Between Two Sets of Variates." In this paper, Hotelling introduced the concept of canonical correlation, which aimed to measure the strength of the relationship between two sets of variables by finding pairs of linear combinations of these variables that are maximally correlated (Hotelling 1936). Each pair of linear combinations is referred to as a canonical variate, and the correlation between them is the canonical correlation.

Hotelling's work expanded on the concepts of correlation and regression, developing a method that could identify and quantify the relationships between two multivariate sets, which was a significant step beyond traditional techniques that dealt only with univariate or multivariate responses.

- *Mathematical Formulation*: Hotelling's approach involved solving an optimization problem to find linear combinations of the variables in the two sets that maximize the correlation between the two sets. The resulting pairs of canonical variates represent the dimensions in each dataset that are most strongly related. This mathematical formulation, based on eigenvalue decomposition, laid the foundation for how CCA would be applied in statistical research (Mardia et al. 2024).
- *Significance of Hotelling's Contribution*: Hotelling's introduction of CCA was revolutionary because it offered a method for directly assessing the relationships

between two sets of variables, opening the door to a wealth of applications in various fields. It also introduced the idea of canonical correlation coefficients, which quantify the strength of these relationships and form the basis for testing the significance of the associations (Thompson 1984).

Merits of CCA:

- *Simultaneous Analysis*: CCA enables the simultaneous analysis of multiple traits and environmental variables, making it useful for understanding complex genotype–environment relationships (Hair et al. 2019).
- *Correlations*: It provides direct insight into the correlation between genotype performance and environmental factors (Stewart and Love 1968).
- *Visual Interpretation*: CCA generates canonical variables that can be plotted for better understanding and visual representation of relationships (Thompson 1984).

Demerits of CCA:

- *Linear Relationships Assumption*: Like PCA, CCA assumes linearity, which can limit its application to real-world data where relationships might be nonlinear (Mardia et al. 2024).
- *High Dimensionality*: When the number of variables is large, CCA can be computationally expensive and may require dimensionality reduction techniques to be effective (Hardoon et al. 2004).
- *Sensitivity to Outliers*: CCA is sensitive to outliers in the data, which could skew the results (Thompson 1984).

10.3.4 Multivariate Analysis of Variance (MANOVA)

MANOVA is an extension of the traditional analysis of variance (ANOVA), used when there are multiple dependent variables. It tests the hypothesis that the mean vectors of different groups (e.g., genotypes) are equal across several traits or environments. MANOVA is particularly useful in testing the significance of G × E interactions by considering multiple traits simultaneously (Rencher 2002).

The roots of MANOVA can be traced back to the development of analysis of variance (ANOVA) in the early twentieth century. Sir Ronald A. Fisher, a British statistician, is credited with the development of ANOVA in the 1920s as a method for testing hypotheses about differences between group means in experimental data (Fisher 1925). Fisher's work laid the foundation for what would later become multivariate methods.

In univariate ANOVA, the analysis focuses on one dependent variable and assesses whether the means of different groups differ significantly. However, as data complexity increased and researchers began to work with multiple related outcomes, univariate approaches became insufficient. For example, in biological experiments, researchers might be interested in multiple traits such as height, weight, and yield, all of which might be influenced by similar factors. This necessitated the development of statistical methods capable of simultaneously analyzing multiple outcomes.

The idea of multivariate analysis emerged in the 1930s and 1940s as statisticians began to develop techniques that could handle multiple variables at once. A key milestone in this progression was the development of multivariate normal distribution theory, which laid the theoretical foundation for the study of multivariate data (Anderson-Cook 2004).

- In 1937, the statistician Harold Hotelling introduced the Hotelling's T-squared statistic, which is a generalization of the t-test for multiple variables. This statistic became one of the key tools in multivariate analysis and would later be incorporated into various multivariate techniques, including MANOVA (Hotelling 1937).
- During the same period, other techniques like principal component analysis (PCA) and factor analysis were developed, all of which addressed multivariate data in different ways. However, these techniques were primarily concerned with reducing dimensionality and understanding the relationships between variables, rather than directly testing group differences as MANOVA would (Mardia et al. 2024).
- MANOVA as a formal method was first introduced in the early 1950s as an extension of ANOVA to multivariate contexts. The central goal of MANOVA was to test whether the mean vectors of multiple dependent variables differed between two or more groups, while accounting for the potential correlation between the dependent variables (Huberty and Olejnik 2007).
- In 1951, R.A. Fisher's work on multivariate analysis influenced the formal development of methods for analyzing multiple dependent variables simultaneously. Fisher's theoretical contributions to statistical methods in the context of agricultural experiments helped lay the groundwork for what would later become MANOVA (Fisher 1951).
- In 1957, Roy's largest root test was proposed by S. N. Roy, which is an important statistical test used in MANOVA to test hypotheses about the difference in group means. Roy's test provided a way to assess the statistical significance of the multivariate effects of treatments, making it an important part of MANOVA's theoretical foundation (Roy 1957).

Merits of MANOVA:

- *Multivariate Testing*: MANOVA tests multiple dependent variables simultaneously, offering a more comprehensive analysis than univariate tests (Tabachnick and Fidell 2019).
- *Interaction Detection*: It can detect significant interactions between genotypes and environments, which is the core interest of $G \times E$ analysis (Rencher 2002).
- *Multivariate Significance*: By analyzing covariance structures, MANOVA can provide insights into the overall significance of $G \times E$ interactions for all traits at once (Anderson 2004).

Demerits of MANOVA:

- *Assumptions*: MANOVA requires multivariate normality and homogeneity of variance-covariance matrices. If these assumptions are violated, the results may not be reliable (Huberty and Olejnik 2007).
- *Complexity in Interpretation*: MANOVA tests the overall significance of differences, but understanding the specific nature of the differences can be challenging (Tabachnick and Fidell 2019).
- *Less Visual Appeal*: While it provides statistical significance, MANOVA does not directly provide intuitive visualizations of interactions like PCA or AMMI (Mardia et al. 2024).

10.3.5 Additive Main Effects and Multiplicative Interaction (AMMI) Model

The additive main effects and multiplicative interaction (AMMI) model is a statistical approach used to analyze G × E interactions in agriculture, particularly in crop breeding. It combines the additive effects of genotypes and environments (main effects) with the multiplicative interaction between genotypes and environments, providing a powerful tool to study how genotypes perform across different environmental conditions (Gauch 1992).

The AMMI model is particularly useful because it allows researchers to visualize G × E interactions and to identify stable genotypes (those that perform consistently across different environments) and adapted genotypes (those that perform well in specific environments). The results are often visualized in biplots, which make it easier to interpret complex interactions. The model is often employed to aid decision-making in plant breeding, enabling the selection of genotypes that maximize yields or resilience across varying environmental condition (Gauch 1988).

The AMMI model was first introduced in the 1970s by G.N. Gauch, a prominent figure in the field of agricultural statistics, to address the challenges in the analysis of G × E interactions.

Prior to AMMI, researchers used analysis of variance (ANOVA) and multivariate methods to examine G × E interactions. However, these methods were limited because they did not fully account for the complexity of G × E interactions. The introduction of the AMMI model revolutionized the way G × E interactions were studied.

In traditional ANOVA, G × E interactions are treated as a residual error term, but this does not provide a clear view of how these interactions are structured. The AMMI model, on the other hand, breaks down the G × E interaction into two parts: an additive main effect (which considers the independent effects of genotypes and environments) and a multiplicative interaction effect (which captures the interactions between genotypes and environments) (Gauch 2013).

The AMMI model was first developed as a way to analyze multi-environment trial (MET) data, which is commonly encountered in crop breeding programs. Gauch's method, which initially appeared in his seminal papers, has since been refined and adapted, becoming one of the most widely used models in modern agricultural statistics.

Structure of AMMI Model The AMMI model can be represented as a combination of two components:

1. *Additive Main Effects*: This part of the model captures the independent effects of genotypes and environments.
   - The genotypic main effect reflects how each genotype performs across all environments.
   - The environmental main effect reflects how each environment affects the performance of all genotypes.
2. *Multiplicative Interaction Effects*: This part captures the interaction between genotypes and environments, identifying how the performance of genotypes changes under different environmental conditions. The interaction is modeled as the product of genotype and environment effects, typically using principal components analysis (PCA) to identify the underlying factors that drive the interaction (Gauch 2013).

The AMMI model is mathematically expressed as:

$$Y_{ij} = \mu + G_i + E_j + \sum_{k=1}^{x} \left(\lambda_k \sigma_{ik} \gamma_{jk}\right) + e_{ij}\Big)$$

Where,

$Y_{ijr} =$ Observation of the rt⍰ replicate of the ith genotype in the jth environment
$\mu =$ the overall mean,
$G_i =$ main effect of the genotype,
$E_j =$ main effect of the $j^{t⍰}$ environment,
$x =$ matrix rank$\{ge\}_{ij}$,
$\lambda_k =$ the singular value for principal component k,
$\sigma_{ik} =$ the eigen vector score for genotype I and component k,
$\gamma_{jk} =$ the eigen vector score for environment j and component k,
$e_{ij} =$ the average of the corresponding random errors.

Merits of AMMI:

- *Interaction Visualization*: AMMI is particularly useful for visualizing G × E interactions through biplots, which display both genotype and environmental effects clearly. It helps researchers and breeders quickly assess patterns in the data.
- *Effective in Plant Breeding*: It has been extensively used in plant breeding to identify stable genotypes across different environments.

- *Efficiency in Modelling Interactions*: The AMMI model is efficient in handling G × E data, especially when the interaction is strong.
- *Clear Decomposition of Effects*: AMMI provides a clear separation between main effects and interaction effects, making it easier to interpret genotype-environment relationships.
- *Handling of Complex Interactions*: AMMI is well-suited for handling complex, large datasets common in multi-environment trials and crop breeding programs.

Demerits of AMMI:

- *Complex Interpretation of Interaction*: While biplots help visualize the interactions, understanding the exact nature of these interactions can be complex.
- *Linear Assumption*: The AMMI model assumes that the interaction effects are linear, which may not always reflect real-world complexities in G × E interactions.
- *Choice of Principal Components*: The selection of the number of principal components to retain can be subjective and influence the interpretation of results.
- *Model Assumptions*: The AMMI model assumes that G × E interactions can be captured by a linear additive term and a multiplicative term. This might not always be true for more complex interactions.
- *Sensitivity to Data Structure*: The effectiveness of AMMI depends on the quality of data and the number of genotypes and environments considered.

10.3.6 GGE Biplot

The GGE (genotype + genotype × environment) biplot method was introduced by Yan et al. in the early 2000s as an improvement over conventional statistical methods for analyzing G × E interaction (Yan and Kang 2002). It builds upon the additive main effects and multiplicative interaction (AMMI) model but provides a more visually intuitive and informative graphical representation. The approach has gained widespread acceptance in plant breeding and agronomic studies for evaluating genotypic stability and adaptability across multiple environments.

Why GGE and Not GE?

Traditional statistical models separate the variation into three main components:

1. *G (Genotypic effect)*—The mean performance of each genotype.
2. *E (Environmental effect)*—The impact of different locations or seasons.
3. *G × E (Genotype-by-environment interaction)*—The differential response of genotypes across environments.

However, the GGE biplot approach focuses on G + GE, not just GE, because:

- E is usually large and not relevant for genotype selection.
- GGE captures the real pattern of genotype performance across environments.

Thus, GGE biplots remove the environmental main effect (E) and focus on G + GE, making it more informative for genotype selection.

10.3.6.1 History:

The GGE biplot method was developed by Dr. Weikai Yan in the early 2000s as an improvement over traditional models like AMMI (additive main effects and multiplicative interaction) for analyzing multi-environment trials (METs) (Yan et al. 2000). The concept evolved as a response to the limitations of conventional methods, particularly their inability to visually and effectively separate genotypic performance from environmental influence (Gauch 2006).

Since its development, GGE biplot has been widely adopted in plant breeding, agronomy, and ecology, mainly for evaluating genotypic stability and adaptability across different environments (Yan and Kang 2002).

10.3.6.2 Mathematical Representation of GGE Biplot

The GGE model is derived from singular value decomposition (SVD) and principal component analysis (PCA). It decomposes the genotype + genotype-by-environment interaction (G + GE) matrix into principal components (Yan 2001):

$$Y_{ij} - \mu - \beta_j = \sum_{k=1}^{k} \lambda_k \xi_{ik} \eta_{jk} + \varepsilon_{ij}$$

Where:

Y_{ij} = Observed yield (or any trait) of genotype i in environment j.
μ = Grand mean.
β_j = Environmental mean.
λ_k = Singular value for the k-th principal component.
ξ_{ik} = PC1 score for genotype i.
η_{jk} = PC2 score for environment j.
ε_{ij} = Residual term.

Key Insights from the Equation

1. *Data Decomposition*: The equation decomposes the observed data into components that explain the genotypic effect and G × E interaction while removing the environmental main effect (βj) (Yan et al. 2007).
2. *Dimension Reduction*: The principal component analysis (PCA) reduces the high-dimensional data into two primary axes (PC1 and PC2), which form the biplot representation (Yan and Tinker 2005).
3. *Graphical Interpretation*: By plotting the scores (ξik for genotypes and ηjk for environments) in a two-dimensional space, researchers can visually analyze which genotype performs best in which environment (Gauch et al. 2008).

Practical Application

- The first principal component (PC1) explains the largest proportion of variation, often linked to genotypic performance (Yan et al. 2000).
- The second principal component (PC2) accounts for interaction patterns, helping to distinguish genotypic stability across environments (Gauch 2013).

Interpretation of GGE Biplot

GGE biplots can be used for multiple purposes:

1. *"Which-Won-where" Pattern (Mega-Environment Analysis)*
   - The polygon view of a GGE biplot helps determine which genotypes perform best in specific environments (Yan 2002).
   - Each vertex of the polygon represents the most responsive genotype in at least one environment.
   - Environments are grouped into mega-environments, helping plant breeders develop region-specific genotypes (Yan and Kang 2002).
2. *Ranking Genotypes Based on Performance and Stability*
   - The average-environment coordination (AEC) view helps rank genotypes based on their mean performance and stability (Yan and Holland, 2010).
   - Genotypes closer to the AEC axis and farther along the positive direction are superior.
   - Stable genotypes have a small projection onto the AEC ordinate (smaller perpendicular distance) (Gauch 2013).
3. *Discriminativeness vs. Representativeness of Environments*
   - GGE biplots also show how well an environment differentiates genotypes (discriminative ability) and how well it represents the overall target environment (representativeness) (Yan and Tinker 2005).
   - This helps in selecting the best test locations for breeding trials.

Merits of GGE Biplot

1. *Comprehensive Visualization*—GGE biplots graphically display genotype performance across multiple environments, making it easy to identify superior genotypes and favorable environments (Yan 2001).
2. *Focus on G + GE*—Unlike PCA, which includes environmental effects, GGE biplot isolates genotype and G × E interaction, which is more relevant for breeding programs (Yan 2002).
3. *Efficiency in Interpretation*—By reducing multidimensional data into two principal components (PC1 and PC2), GGE biplots allow rapid visual assessment of genotype rankings (Gauch 2013).
4. *Identification of Mega-Environments*—The method is useful in dividing the testing locations into distinct mega-environments, which helps in targeted breeding and recommendation (Yan and Tinker 2005).

5. *Stability Analysis*—It allows the identification of genotypes with high-yield stability across different environments, helping breeders make informed decisions (Yan and Kang 2002).

Demerits of GGE Biplot

1. *Assumption of Biplot Approximation*—GGE biplots rely on principal component analysis, which may not always capture the full complexity of G × E interactions if higher-dimensional interactions are strong (Gauch 2006).
2. *Data Sensitivity*—The interpretation of a GGE biplot can be affected by data quality, missing values, and outliers, which may skew the analysis (Yan et al. 2007).
3. *Limited to Multiplicative Models*—While powerful, GGE biplots assume a multiplicative model, which may not be the best fit for all datasets, especially those with nonlinear interactions (Yan and Tinker 2005).
4. *Potential Over-Interpretation*—Users may over-rely on visual interpretation without considering statistical rigor, leading to misclassification of genotypes or environments (Gauch et al. 2008).

10.3.7 Other Models for Analyzing GE Interaction: An Overview

10.3.7.1 Multi-Trait Genotype-Ideotype Distance Index (MGIDI)

The multi-trait genotype–ideotype distance index (MGIDI) is an advanced selection index used in plant breeding to evaluate and rank genotypes based on multiple traits simultaneously. It is designed to identify genotypes that closely resemble an "ideotype," which represents the optimal combination of desired traits (Olivoto and Nardino 2021). Unlike traditional selection indices that rely on subjective weight assignments, MGIDI is based on multivariate statistical techniques, allowing for a more comprehensive assessment of genotype performance across multiple environments.

MGIDI considers genotype-by-environment (GE) interactions, making it particularly useful in multi-environment trials (METs). By accounting for variations across different environments, the index helps breeders identify genotypes with high performance and stability (Olivoto et al. 2022, 2019a, b).

History

The MGIDI was introduced as an improvement over traditional selection indices that typically struggle to balance multiple traits effectively. Historically, selection indices such as the Smith-Hazel index have been used in plant breeding, but these indices require pre-assigned economic weights, which may not always reflect real breeding goals (Smith 1936; Hazel 1943). MGIDI, on the other hand, eliminates the need for subjective weight assignments by utilizing factor analysis and Euclidean distance to rank genotypes objectively (Olivoto and Nardino 2021; Olivoto and Lúcio 2020).

Since its introduction, MGIDI has been widely adopted in breeding programs for crops such as wheat, maize, and soybean, proving useful in both stress and non-stress environments (Olivoto et al. 2022; Olivoto and Nardino 2021).

Application in G × E Interaction

MGIDI is particularly effective in handling G × E interactions, which are crucial in determining genotype adaptability across different environmental conditions. Its applications include:

1. *Multi-Trait Selection*: MGIDI facilitates the simultaneous selection of genotypes for multiple traits, such as yield, disease resistance, drought tolerance, and grain quality, without the limitations of traditional selection indices (Olivoto and Nardino 2021).
2. *Genotype Stability Analysis*: By incorporating GE interactions, MGIDI helps identify genotypes with consistent performance across diverse environments, an essential factor in breeding for adaptability (Yan and Kang 2002).
3. *Breeding Program Optimization*: The index aids in prioritizing genotypes that closely resemble the ideotype, thereby improving breeding efficiency and reducing the number of selections needed (Olivoto et al. 2022).
4. *Data-Driven Decision-Making*: MGIDI utilizes multivariate analysis, ensuring that genotype selection is based on robust statistical principles rather than arbitrary weight assignments (Cruz 2006).
5. *Handling Complex GE Interactions*: By assessing and ranking genotypes under varying environmental conditions, MGIDI helps breeders select stable and superior genotypes for specific or broad adaptation (Olivoto and Lúcio 2020).

Interpretation

The MGIDI score represents the Euclidean distance between a given genotype and the ideal genotype. A lower MGIDI score indicates a genotype that is closer to the ideotype and is thus more desirable (Olivoto and Nardino, 2021). Additionally, the contribution of different traits to MGIDI can be analyzed to determine which traits influence genotype ranking the most, allowing breeders to adjust selection strategies accordingly.

In GE interaction studies, MGIDI facilitates comparisons of genotype rankings across multiple environments, enabling the selection of either broadly adapted or specifically adapted genotypes based on breeding objectives (Yan and Kang, 2002).

10.3.7.2 Multi-Trait Stability Index (MTSI) Model

The multi-trait stability index (MTSI) is a selection index used in plant breeding to identify genotypes that combine high performance with stability across multiple traits and environments. It is based on the multi-trait genotypeideo type distance index (MGIDI) but focuses specifically on stability, making it a useful tool for selecting genotypes that perform consistently under varying environmental conditions (Olivoto et al. 2022).

MTSI utilizes factor analysis and G × E interaction data to assess both the mean performance and stability of genotypes. By considering multiple traits simultaneously, it provides a comprehensive assessment that avoids the limitations of single-trait selection methods (Cruz et al. 2006). The index ranks genotypes based on their distance from an ideotype, where a lower MTSI score indicates a more desirable genotype with superior performance and stability across environments (Olivoto and Nardino 2021; Olivoto and Lúcio 2020).

History

MTSI was developed to overcome the limitations of traditional stability indices that often focus on a single trait at a time, such as the AMMI stability value (ASV) and the genotype and genotype-by-environment (GGE) stability index (Yan and Kang 2002). Traditional methods failed to capture the complexity of multi-trait selection, leading to the development of MTSI as a more holistic approach (Olivoto et al. 2019a, b).

The index is an extension of the MGIDI index, which evaluates genotypes based on multiple traits, but MTSI specifically incorporates stability analysis to ensure that selected genotypes maintain their desirable characteristics across different environments (Olivoto et al., 2019a; Olivoto and Lúcio 2020). Since its introduction, MTSI has been widely used in plant breeding programs for crops such as maize, wheat, and soybean to improve selection accuracy and breeding efficiency (Olivoto et al. 2019b; Olivoto and Nardino 2021).

Application in GE Interaction

MTSI has several applications in genotype selection and breeding, particularly in evaluating GE interactions:

1. Multi-Trait Stability Selection: MTSI allows breeders to select genotypes with high stability across multiple traits, ensuring consistent performance across different environmental conditions (Olivoto et al. 2019a, b).
2. Genotype Ranking: The index provides a ranking system that integrates mean performance and stability, allowing breeders to identify the most promising genotypes efficiently (Cruz 2006).
3. Adaptability Assessment: MTSI helps determine whether genotypes are broadly adapted (stable across all environments) or specifically adapted to certain conditions (Yan and Kang 2002).
4. Breeding Program Optimization: By focusing on stability, MTSI reduces the risk of selecting genotypes that perform well only in specific environments but fail under variable conditions (Olivoto and Nardino 2021).
5. Improved Selection Strategies: The index supports data-driven decision-making, integrating multiple traits into a single assessment for better selection accuracy (Olivoto et al. 2019a; Olivoto and Nardino 2021).

Interpretation
The MTSI score represents the Euclidean distance between a genotype and the ideotype in a multi-trait stability context. A lower MTSI score indicates a genotype that is both high-performing and stable across environments, making it a more desirable candidate for selection (Olivoto et al. 2019a; Olivoto and Lúcio 2020).

In G × E interaction studies, MTSI is particularly useful for:

- Identifying genotypes that maintain high performance across diverse environments.
- Differentiating between broadly and specifically adapted genotypes.
- Balancing the trade-off between high-yield potential and stability.

By ranking genotypes based on MTSI, breeders can prioritize those with the best combination of stability and trait performance, ensuring more resilient and productive cultivars in breeding programs (Olivoto et al. 2019a, b).

10.3.8 Comparison of Multivariate Methods

To better understand the strengths and weaknesses of each method, we compare the eight techniques across several important criteria (Table 10.1):

Multivariate methods are essential tools for analyzing genotype × environment (G × E) interactions, as they allow for a more detailed and holistic understanding of how genotypes respond across diverse environments. The methods compared in the table vary in their ability to handle multiple traits, interaction effects, and computational efficiency. This comprehensive analysis explores their strengths, limitations, and best-use scenarios.

1. *Cluster Analysis*

Overview:
Cluster analysis is an unsupervised classification technique that groups genotypes based on similarity across multiple traits or environmental responses. Various algorithms (e.g., k-means, hierarchical clustering) and distance metrics (e.g., Euclidean, Mahalanobis) are used for genotype classification.

Strengths:
- *Flexible*: Can be applied to diverse datasets with different clustering algorithms.
- *Handles multiple traits*: Useful in grouping genotypes with similar performance.
- *Easy visualization*: Dendrograms and heatmaps facilitate interpretation.

Limitations:
- *Limited interpretability*: Clusters do not provide direct inference on G × E interactions.

Table 10.1 Comprehensive analysis of different multivariate methods

Method	Interpretability	Flexibility	Computational efficiency	Handling of interaction
Cluster analysis	Moderate (depends on method used)	High (various distance metrics and clustering approaches)	Moderate to high (depends on algorithm)	Limited (primarily groups data, does not explicitly model interactions)
Principal component analysis (PCA)	High (for dimensionality reduction)	Low (linear relationships assumed)	High (efficient for large datasets)	Limited (does not test interactions directly)
Canonical correlation analysis (CCA)	Moderate (relationships between sets)	High (works well with correlated sets)	Moderate (computationally intensive)	Moderate (linear relationships)
Multivariate analysis of variance (MANOVA)	Moderate (complex interaction)	Moderate (linear assumptions)	Moderate (handles large datasets)	High (tests overall significance)
Additive main effects and multiplicative interaction model (AMMI)	High (visualization through biplots)	High (models both additive and interactive effects)	High (efficient for large datasets)	Very high (visualizes G × E interactions)
Genotype + genotype-by-environment interaction (GGE) biplot	High (clear visualization of G × E interactions)	Moderate (assumes specific models)	High (efficient for large datasets)	Very high (focuses on genotype and environment interaction)
Multi-trait genotype–ideotype distance index (MGIDI)	High (integrates multiple traits and ranks genotypes)	High (no need for pre-assigned weights)	High (efficient multivariate approach)	Very high (considers multiple traits under G × E interaction)
Multi-trait stability index (MTSI)	High (focuses on stability and performance)	High (suitable for multiple traits and environments)	High (computationally efficient for large datasets)	Very high (accounts for both mean performance and stability across environments)

- *No explicit modelling of interactions*: Merely groups data without testing effects.
- *Choice of distance metric impacts results:* Requires careful parameter selection.

Best Use Case:
- Grouping genotypes with similar responses across environments.
- Preliminary exploration of genotype relationships before applying other analytical methods.

2. *Principal Component Analysis (PCA)*

Overview:
PCA is a dimensionality reduction technique that identifies the principal components (PCs) capturing the most variance in a dataset. It is widely used for visualizing high-dimensional data in fewer dimensions.

Strengths:
- *Efficient for large datasets*: Reduces dimensionality without losing significant information.
- *High interpretability*: PCs provide insight into major patterns in the data.
- *Data-driven selection*: Avoids subjective weight assignments.

Limitations:
- *Linear assumptions: Assumes relationships between variables are linear.*
- *Does not explicitly model G × E interactions: Simply reduces dimensionality rather than analyzing interaction.*
- *Component interpretation can be difficult*: PCs are abstract and require careful interpretation.

Best Use Case:
- *Exploratory data analysis* before applying more complex models.
- *Trait correlation analysis* to reduce redundancy in highly correlated traits.

3. *Canonical Correlation Analysis (CCA)*

Overview:
CCA is used to examine the relationships between two sets of variables, making it useful for studying the association between genotype traits and environmental variables.

Strengths:
- Handles multiple correlated traits effectively.
- Useful for understanding trait-environment relationships.
- Captures cross-variable relationships.

Limitations:

- *Computationally intensive*: More complex than PCA.
- *Linear relationships assumed*: Performance declines with nonlinear interactions.
- *Difficult interpretation*: Results require deeper statistical expertise.

Best Use Case:

- Understanding how multiple traits interact with environmental variables.
- Identifying correlated traits across genotypes and environments.

4. *Multivariate Analysis of Variance (MANOVA)*

Overview:
MANOVA is an extension of ANOVA that tests differences among genotype means for multiple traits while considering their correlations.

Strengths:

- Tests significance of G × E effects across multiple traits.
- Considers trait correlations, improving power over separate ANOVAs.
- Handles large datasets efficiently.

Limitations:

- Assumes linear relationships among traits.
- Less visual interpretability compared to biplots.
- Sensitive to outliers and multicollinearity.

Best Use Case:

- *Statistical testing of G × E interactions* for multiple correlated traits.
- *Confirmatory analysis after exploratory multivariate methods.*

5. *Additive Main Effects and Multiplicative Interaction (AMMI) Model*

Overview:
AMMI combines ANOVA (for additive effects) and PCA (for interaction effects), making it one of the most widely used methods in G × E studies.

Strengths:

- Effectively captures G × E interactions.
- Uses biplots for clear visualization of genotype and environment performance.
- Efficient for large datasets.

Limitations:
- Requires balanced data for optimal performance.
- Assumes additive and multiplicative components.
- Interpretation can be complex for high-order interactions.

Best Use Case:
- G × E interaction studies in plant breeding.
- Selection of genotypes with both high performance and stability.

6. *Genotype + Genotype-by-Environment (GGE) Biplot Analysis*

Overview:
GGE biplots focus on genotype effects (G) and G × E interactions, making them highly effective for visualizing multi-environment trials (METs).

Strengths:
- Highly interpretable and widely used in plant breeding.
- Focuses on genotypic stability and performance.
- Efficient even for large datasets.

Limitations:
- Requires careful interpretation of biplots.
- Assumes certain statistical models for representation.
- Not suited for testing statistical significance.

Best Use Case:
- Multi-environment trials (METs).
- Genotype ranking and stability analysis.

7. *Multi-Trait Genotype-Ideotype Distance Index (MGIDI)*

Overview:
MGIDI ranks genotypes based on their distance from an ideal genotype using multiple traits and environments. It is an extension of multivariate selection indices that account for both performance and adaptability.

Strengths:
- Considers multiple traits simultaneously.
- No subjective weight assignment needed.
- Computationally efficient for large datasets.
- Highly effective for G × E interaction studies.

Limitations:

- Requires advanced statistical knowledge for interpretation.
- Less widely used compared to AMMI and GGE biplots.

Best Use Case:

- Multi-trait selection in breeding programs.
- Simultaneous selection for performance and stability.

8. *Multi-Trait Stability Index (MTSI)*

Overview:

MTSI extends MGIDI by focusing on genotype stability while maintaining good performance across multiple traits.

Strengths:

- Simultaneously considers stability and performance.
- Highly efficient in multi-environment trials.
- No need for arbitrary weight assignment.
- Strong statistical foundation.

Limitations:

- Requires computational resources for large datasets.
- Interpretation can be complex for non-experts.

Best Use Case:

- Selecting genotypes with stable performance across environments.
- Breeding programs aiming for robust cultivars.

MGIDI and MTSI provide the most comprehensive solutions for selecting superior genotypes in multi-trait, multi-environment trials. However, AMMI and GGE biplots remain indispensable visualization tools for breeding program (Table 10.2).

Table 10.2 Final comparison and recommendations

Scenario	Recommended method
Exploratory analysis	Cluster analysis, PCA
Trait correlation analysis	CCA, PCA
Testing significance of G × E	MANOVA
Visualizing G × E interactions	AMMI, GGE biplot
Selecting high-performing genotypes	MGIDI
Selecting stable and high-performing genotypes	MTSI

10.3.9 Conclusions

This study provides a comparative analysis of various multivariate statistical methods for evaluating G × E interactions. Each method offers unique strengths and limitations, making their suitability dependent on the research objectives, data structure, and the complexity of interactions present in the dataset (Gauch 2006; Yan and Kang 2002).

Principal component analysis (PCA) effectively reduces dimensionality but assumes linear relationships and does not explicitly model interactions (Jolliffe, 2011). Canonical correspondence analysis (CCA) is useful for analyzing relationships between multiple sets of variables but can be computationally intensive (Ter Braak 1986). Multivariate analysis of variance (MANOVA) provides a robust statistical framework for testing overall significance but assumes linearity and requires large datasets (Rencher 2002). The AMMI (additive main effects and multiplicative interaction) model effectively captures both additive and interaction effects, providing powerful visualization through biplots (Gauch 2013). Cluster analysis groups genotypes based on overall similarity patterns but does not explicitly account for genotype-specific responses across environments or model G × E interactions (Kaufman and Rousseeuw 2009), whereas GGE biplot analysis excels in visualizing and interpreting genotype performance and stability across environments (Yan 2001).

Among these methods, AMMI and GGE biplot stand out as particularly effective for understanding G × E interactions due to their ability to visualize genotype performance and stability in multi-environment trials (Yan and Tinker 2005). While AMMI provides a balance between additive and interaction effects, GGE biplot is more focused on genotypic performance and helps in identifying mega-environments (Yan 2002). However, GGE biplot's reliance on principal component analysis (PCA) can lead to oversimplification in cases of complex interactions (Gauch et al. 2008).

Recently, the multi-trait genotype ideotype distance index (MGIDI) and multi-trait stability index (MTSI) have emerged as powerful tools for multi-trait selection and stability assessment in G × E interaction studies. MGIDI ranks genotypes based on their proximity to an ideal genotype, allowing for the simultaneous selection of multiple traits without subjective weight assignment, making it highly effective for multi-trait breeding programs (Olivoto and Nardino 2021). Meanwhile, MTSI extends this approach by incorporating stability analysis, ensuring that selected genotypes exhibit both high performance and consistency across environments (Olivoto et al. 2019b). These indices have demonstrated higher efficiency in identifying superior and stable genotypes compared to traditional methods like AMMI and GGE biplot, particularly in multi-environment trials where multiple traits must be optimized simultaneously (Olivoto et al., 2019a; Olivoto and Lúcio 2020). As the demand for improved genotype selection increases, MGIDI and MTSI provide breeders with robust, flexible, and computationally efficient alternatives for multi-trait selection in complex environments.

References

Anderson MJ (2014) Permutational multivariate analysis of variance (PERMANOVA). Wiley statsref: statistics reference online 1–15

Anderson-Cook CM (2004) An Introduction to Multivariate Statistical Analysis. Journal of the American Statistical Association 99(467):907–909

Cruz CD (2006) Modelos biométricos aplicados ao melhoramento genético. Editora UFV, Viçosa

Everitt BS, Landau S, Leese M, Stahl D (2011) Cluster analysis, 5th edn. Wiley, Chichester

Fisher RA (1922) The goodness of fit of regression formulae, and the distribution of regression coefficients. J R Stat Soc 85(4):597–612

Fisher RA (1925) Statistical methods for research workers. Oliver and Boyd, Edinburgh

Fisher RA (1936) The use of multiple measurements in taxonomic problems. Ann Eugenics 7(2): 179–188. https://doi.org/10.1111/j.1469-1809.1936.tb02137.x

Fisher RA (1951) The design of experiments, 6th edn. Hafner, New York

Gauch HJ (1992) Statistical analysis of regional yield trials: AMMI analysis of factorial designs (pp. xi+-278)

Gauch HG (1988) Model selection and validation for yield trials with interaction. Biometrics 44(3): 705–715. https://doi.org/10.2307/2531585

Gauch HG (2006) Statistical analysis of yield trials by AMMI and GGE. Crop Sci 46(4):1488–1500

Gauch HG (2013) A simple protocol for AMMI analysis of yield trials. Crop Sci 53(5):1860–1869

Gauch HG, Piepho HP, Annicchiarico P (2008) Statistical analysis of yield trials by AMMI and GGE: further considerations. Crop Sci 48(3):866–889

Hair JF, Black WC, Babin BJ, Anderson RE (2019) Multivariate data analysis, 7th edn. Pearson, Upper Saddle River

Hardoon DR, Szedmak S, Shawe-Taylor J (2004) Canonical correlation analysis: an overview with application to learning methods. Neural Comput 16(12):2639–2664

Hazel LN (1943) The genetic basis for constructing selection indexes. Genetics 28(6):476–490

Hotelling H (1933) Analysis of a complex of statistical variables into principal components. J Educ Psychol 24(6):417–441

Hotelling H (1936) Relations between two sets of variates. Biometrika 28(3/4):321–377

Hotelling H (1937) Multivariate quality control. Biometrika 28(3/4):321–377

Huberty CJ, Olejnik S (2007) Applied MANOVA and discriminant analysis, 2nd edn. Wiley, Hoboken

Hudson AI (2022) Analysis of genotype-by-environment interactions in a maize mapping population. G3 Genes Genomes Genet 12(3)

Jain AK, Murty MN, Flynn PJ (1999) Data clustering: a review. ACM Comput Surv 31(3): 264–323. https://doi.org/10.1145/331499.331504

Jolliffe IT (2011) Principal component analysis, 2nd edn. Springer, New York

Jolliffe IT, Cadima J (2016) Principal component analysis: a review and recent developments. Philos Trans R Soc A 374(2065):20150202

Kaufman L, Rousseeuw PJ (2009) Finding groups in data: an introduction to cluster analysis. Wiley, Hoboken

Malosetti M, Ribaut J, van Eeuwijk F (2013) The statistical analysis of multi environment data: modeling genotype-by-environment interaction and its genetic basis. Front Physiol 4:44. https://doi.org/10.3389/fphys.2013.00044

Mardia KV, Kent JT, Bibby JM (2024) Multivariate analysis. John Wiley & Sons.

Mohammadı M, Karımızadeh R, Sabaghnıa N, Shefazadeh MK (2012) Genotype Ã environment interaction and yield stability analysis of new improved bread wheat genotypes. Turkish Journal of Field Crops 17(1):67–73

Murtagh F, Contreras P (2017) Algorithms for hierarchical clustering: an overview, II. Wiley Interdisciplinary Reviews: Data Mining and Knowledge Discovery 7(6),e1219

Neisse AC, Kirch JL, Hongyu K (2018) AMMI and GGE Biplot for genotype× environment interaction: a medoid–based hierarchical cluster analysis approach for high–dimensional data. Biometrical Letters 55(2):97–121

Olivoto T, Lúcio AD, da Silva JA, Marchioro VS, de Souza VQ, Jost E (2019a) Mean performance and stability in multi-environment trials I: combining features of AMMI and BLUP techniques. Agronomy journal 111(6):2949-2960

Olivoto T, Nardino M, Carvalho IR, Follmann DN, Szareski VJ, Souza VQ (2019b) Mean performance and stability in multi-environment trials II: selection based on multiple traits. Agron J 111(6):1–14. https://doi.org/10.2134/agronj2019.03.0220

Olivoto T, Lúcio AD (2020) Metan: an R package for multi-environment trial analysis. Methods Ecol Evol 11(6):783–789

Olivoto T, Nardino M (2021) MGIDI: Toward an effective multivariate selection in biological experiments. Bioinformatics 37(9):1383–1389

Olivoto T, Diel MI, Schmidt D, Lúcio AD (2022) MGIDI: A powerful tool to analyze plant multivariate data. Plant Methods 18:121

Pearson K (1896) Mathematical contributions to the theory of evolution—III. Regression, heredity, and panmixia. Philos Trans R Soc A 187:253–318

Pearson K (1901) On lines and planes of closest fit to systems of points in space. Philos Mag 2(11): 559–572

Rencher AC (2002) Methods of multivariate analysis, 2nd edn. Wiley, New York

Roy SN (1957) Some aspects of multivariate analysis

Smith HF (1936) A discriminant function for plant selection. Ann Eugenics 7(3):240–250

Sokal RR, Michener CD (1958) A statistical method for evaluating systematic relationships. Univ Kansas Sci Bull 38:1409–1438

Spearman C (1961) "General Intelligence" Objectively Determined and Measured. In J. J. Jenkins & D. G. Paterson (Eds.), Studies in individual differences: The search for intelligence (pp. 59–73). Appleton-Century-Crofts. https://doi.org/10.1037/11491-006

Stewart DW, Love WA (1968) A general canonical correlation index. Psychol Bull 70(3):160–163

Tabachnick BG, Fidell LS (2019) Using multivariate statistics, 7th edn. Pearson, Boston

Ter Braak CJF (1986) Canonical correspondence analysis: a new eigenvector technique for multivariate direct gradient analysis. Ecology 67(5):1167–1179

Thompson B (1984) Canonical correlation analysis: uses and interpretation. SAGE, Beverly Hills

Yan WK, Hunt, LA, Sheng OL, Szlavnics Z (2000) Cultivar evaluation and mega-environment investigation based on the GGE biplot. Crop Science 40(3):597–605

Yan W (2001) GGE biplot a windows application for graphical analysis of multi-environment trial data and other types of two-way data. Agron J 93(5):1111–1118

Yan W (2002) Singular-value partitioning in biplot analysis of multi-environment trial data. Agron J 94(5):990–996

Yan W (2014) Crop variety trials: data management and analysis. Wiley, Chichester

Yan W, Holland JB (2010) A heritability-adjusted GGE biplot for test environment evaluation. Euphytica 171(3):355–369

Yan W, Kang MS (2002) GGE biplot analysis: a graphical tool for breeders, geneticists, and agronomists. CRC Press, Boca Raton

Yan W, Tinker NA (2005) An integrated biplot analysis system for displaying, interpreting, and exploring genotype-by-environment interactions. Crop Sci 45(3):1004–1016

Yan W, Kang MS, Ma B, Woods S, Cornelius PL (2007) GGE biplot vs. AMMI analysis of genotype-by-environment data. Crop Sci 47(2):643–653

Does Drought Tolerance Enhance Crop Stability Performance?

11

Uğur Sesiz, Melike Biçici, Harun Bektas, Kanat Yermekbayev, Liu Qier, Jiansheng Chen, and Hakan Özkan

Abstract

Global food security is threatened by climate change, especially worsening droughts, which disrupt crop production. This review closely examines the association between drought tolerance and yield stability, investigating whether breeding for drought tolerance naturally improves performance consistency across different environments. It discusses key drought-response mechanisms—escape, avoidance, and tolerance—along with their physiological trade-offs and relevance in various agroecological settings. A primary focus is on genotype-by-environment (G × E) interactions, particularly crossover effects, complicating selection choices and underscoring the necessity for both specific and broad

U. Sesiz
Department of Field Crops, Faculty of Agriculture, Şırnak University, Şırnak, Turkey

M. Biçici
Department of Field Crops, Faculty of Agriculture, Çukurova University, Adana, Turkey

H. Bektas
Department of Agricultural Biotechnology, Faculty of Agriculture, Siirt University, Siirt, Turkey

K. Yermekbayev
Zhetysu University named after I.Zhansugurov, Institute of Biotechnology and Ecology, Taldykorgan, Kazakhstan

L. Qier · J. Chen
State Key Laboratory of Wheat Improvement, College of Agronomy, Shandong Agricultural University, Tai'an, Shandong, China

H. Özkan (✉)
Department of Field Crops, Faculty of Agriculture, Çukurova University, Adana, Turkey

State Key Laboratory of Wheat Improvement, College of Agronomy, Shandong Agricultural University, Tai'an, Shandong, China
e-mail: hozkan@cu.edu.tr

R. Mohammadi et al. (eds.), *Genotype x Environment Interactions and its Implications for Plant Breeding*, https://doi.org/10.1007/978-981-95-5664-9_11

adaptation strategies. The review compiles evidence from multi-environment trials, demonstrating how advanced tools like genomic selection, envirotyping, and high-throughput phenotyping can identify genotypes that blend drought resilience with stable yield potential. Additionally, case studies of major crops including maize, wheat, sorghum, rice, and legumes showcase successful breeding programs that have improved both traits. Methodological frameworks such as regression-based stability indices, genotype + genotype-by-environment interaction (GGE) biplots, and additive main effects and multiplicative interaction (AMMI) models are presented as crucial instruments for measuring performance under varying conditions. This chapter concludes by suggesting future breeding directions, stressing the integration of climate models, genomic tools, and de novo domestication to address the complex challenge of ensuring food production in water-scarce and increasingly unpredictable climates.

Keywords

Drought tolerance · Yield stability · Genotype × environment interactions · Climate resilience · Stress adaptation · Crop improvement

11.1 Introduction

Climate change has increased the frequency and severity of droughts, posing a significant threat to global agriculture and food security. Droughts reduce crop yields and create greater unpredictability in production (FAO 2018; Lesk et al. 2016; Xiong et al. 2024). For example, between 2005 and 2015, drought accounted for 83% of all agriculture-related economic losses from natural disasters in developing countries, amounting to approximately $29 billion (FAO 2018). On a national scale, droughts and extreme heat can reduce cereal production by an average of 9–10% (Lesk et al. 2016). These effects highlight the urgent need for crop varieties to sustain performance under water scarcity. Plant breeders transitioned from "only yield-quality" breeding programs to "sustainable and climate resilient" plant breeding (Ahmadi et al. 2018; Cairns et al. 2013; Mwadzingeni et al. 2016).

In breeding, a critical question is whether enhancing a plant's drought tolerance (its ability to withstand water deficits) will ultimately lead to greater yield stability across diverse environments, including drought-stressed and favorable conditions (El-Shawy et al. 2017). Recent advances also highlight the importance of integrating genomic prediction models and environmental characterization ("envirotyping") to better match drought tolerance traits with the specific climatic risks of target regions (Cooper and Messina 2023). This enables breeders to simulate and select resilient genotypes under current conditions and future drought scenarios that are expected due to climate change (Xiong et al. 2024).

One of the complicating factors in plant breeding is the occurrence of genotype × environment (G × E) interactions, where different genotypes respond inconsistently to varying environmental conditions (Kang 2002). G × E is a fundamental aspect of agriculture; a genotype's performance is shaped by genetics and environmental factors rather than by either alone (Begna 2020; Egea-Gilabert et al. 2021). For instance, if a drought-tolerant variety performs well in drought conditions but poorly in environments with plentiful water or vice versa, this signifies a crossover G × E interaction (the rank of genotypes shifts between environments). Such interactions pose challenges for breeders, as the "best" genotype in one environment may not be the top performer in another (Kang 2002). Nonetheless, understanding and leveraging G × E interactions are essential for developing resilient crops.

This chapter examines the relationship between drought tolerance and yield stability. We discuss whether breeding for drought tolerance enhances yield stability, the potential trade-offs that may arise, and how breeders can utilize G × E interactions analysis to select drought-tolerant and stable varieties across various environments. The main goal is to clarify how drought tolerance traits and yield stability can be achieved simultaneously to ensure more reliable crop production amid increasing climate variability.

11.2 Defining Stability in Plant Breeding

Stability across environments, seasons, and changing cultural practices is one of the main aims of plant breeding (Harkness et al. 2023) in addition to overall yield (Hassani et al. 2024; Wondaferew et al. 2024). Stability can be defined with two different terms: static and dynamic (Becker 1981). Static stability is the ability of a crop to produce at similar levels under various conditions, while dynamic stability is the ability of a crop variety to improve conditions positively while still keeping its moderate potential under adverse conditions (Becker and Leon 1988). Dynamic or static stability is not superior to each other. One may be advantageous under a certain condition, while the other may be advantageous under different cultural and environmental conditions (Becker and Leon 1988). For climates with radically changing environments, static stability may become advantageous, while dynamic stability becomes favorable under more stable environmental conditions with relatively rare weather changes. Globally distributed varieties, most of which have broad adaptability, generally perform well under diverse environmental conditions; on the other hand, varieties designed for specific conditions have a limited adaptation and they tend to behave as static stability.

Neither concept is inherently superior; the preferred type of stability depends on breeding objectives and target environments. A statically stable variety with moderate yield across all conditions may be favored in highly unpredictable or stress-prone regions. In areas where the climate is generally favorable but occasionally stressful, a dynamically stable genotype that maximizes yield in good years while tolerating poor years can be beneficial. Crucially, both types of stability contribute to food security by reducing year-to-year yield variability.

Stability is closely tied to the concept of adaptation. A broadly adapted genotype performs well across various environments, often indicating dynamic stability, while a specifically adapted genotype excels in a particular environmental niche, such as consistently drought-prone areas (Ceccarelli 1996a). Breeding for broad adaptation or wide stability can sometimes compromise maximum yield in any single

Table 11.1 Comparison of static versus dynamic stability (conceptual overview)

Feature	Static stability	Dynamic stability
Definition	Low yield variance across environments	Yield changes predictably with environmental changes
Yield in favorable environments	Moderate	High
Yield in unfavorable environments	Moderate	Reasonable
Variance across environments	Low	Higher
Breeding goals	Consistent, reliable yield across all conditions	High-yield potential in favorable conditions, acceptable yield in others
Suitable environments	Highly variable, unpredictable conditions	More predictable but sometimes highly favorable conditions

environment, whereas breeding for specific adaptation may diminish performance outside the target niche (Dwivedi et al. 2021). As originally articulated by Allard and Bradshaw (Allard and Bradshaw 1964), breeders must determine whether to pursue genotypes that are versatile or specialists for certain conditions. In practice, modern crop improvement often seeks to balance these objectives, aiming for reasonable yields in unfavorable environments without sacrificing the ability to capitalize on favorable conditions (Table 11.1) (Ceccarelli 1996b).

11.3 Drought Tolerance Mechanisms in Plants

Drought tolerance in plants refers to maintaining function and yield under limited water availability. This complex trait is controlled by numerous morphological, physiological, and genetic factors. Plant drought-response strategies are classified into three categories: drought escape, drought avoidance, and drought tolerance (strictly) that are widely used by current plant breeding schemes (Haghpanah et al. 2024; Levitt 1981).

11.3.1 Drought Escape

Surviving drought by completing the life cycle before severe water stress occurs is often achieved through developmental plasticity, such as early flowering and rapid maturation, enabling seed set to happen before peak drought (Franks 2011; Shavrukov et al. 2017; Burnette and Eckhart 2021). A drought-escaping genotype essentially "outruns" the stress. This strategy is effective in regions with predictable drought timing, like end-of-season drought. However, escape may sacrifice yield potential due to shorter growth duration in years without drought. For instance, fast-maturing wheat or rice varieties can avoid late-season drought but may not yield as much as full-season varieties in normal years, a trade-off acknowledged by breeders (Fahad et al. 2017; Hlaváčová et al. 2022).

11.3.2 Drought Avoidance

Dehydration avoidance involves maintaining high water status by reducing loss or increasing uptake. Key mechanisms include deep root systems to extract deep soil moisture, leaf rolling to limit transpiration, and stomatal closure. These adaptations help the plant postpone dehydration. For example, certain sorghum and pearl millet genotypes possess deeper root systems, enabling access to lower water tables and enhanced drought avoidance compared to shallow-rooted varieties. Another avoidance trait is improved water use efficiency (WUE), which results in more biomass or yield per water unit used. However, some avoidance traits come with growth penalties; for instance, closing stomata conserves water but sacrifices photosynthesis (Haghpanah et al. 2024). Any breeding scheme must balance these effects. Blum (2011a, b) argued that selecting for very high water use efficiency (WUE) can produce smaller, slow-growing plants that survive drought but yield poorly under normal conditions (Lopes and Reynolds 2011). Thus, excessive drought avoidance can limit yield potential, illustrating the need for balance.

11.3.3 Drought Tolerance (*sensu stricto*)

The ability to withstand internal water deficits, that is, to survive and maintain some metabolic activity at low tissue water potential. Tolerance mechanisms come into play once a plant experiences dehydration. Examples include osmotic adjustment (the accumulation of solutes like proline, sugars, and ions in cells to maintain turgor pressure), the production of protective proteins such as dehydrins and heat-shock proteins to stabilize cellular structures, and the activation of antioxidant enzymes (e.g., superoxide dismutase, catalase) to detoxify reactive oxygen species generated by drought stress. These mechanisms enable the plant to physiologically function under water stress instead of merely avoiding it. For instance, osmotic adjustment in sorghum and barley has been linked to improved yield under drought by sustaining cell turgor (and thus growth) during dry periods (Blum 2017; Haghpanah et al. 2024). Drought tolerance traits are typically polygenic and quantitatively inherited, which complicates their direct selection in conventional breeding programs (Shelake 2022). However, with the advent of modern molecular breeding tools numerous quantitative trait loci (QTLs) and candidate genes that are associated with drought tolerance have been identified. This list includes genes involved in osmoprotectant biosynthesis, abscisic acid (ABA) hormone signaling, and root development (Zhao et al. 2019; Varshney et al. 2014). Genomic studies further reveal that drought tolerance arises from complex networks of genes that interact with both developmental timing and environmental stimuli (Liu et al. 2025). It is important to note that these categories are not mutually exclusive; a single genotype can employ a combination of strategies. For example, “stay-green” sorghum avoids drought by retaining green leaves and deep roots (avoidance) while also tolerating dehydration by sustaining photosynthesis longer and utilizing stored stem carbohydrates (tolerance) (Borrell et al. 2014a; Tovignan et al. 2020). Moreover, the effectiveness of a given

Table 11.2 Key physiological mechanisms of drought tolerance

Mechanism	Description	Examples of plant traits involved	Stage of plant development most relevant
Drought escape	Completing life cycle before severe drought	"Early flowering, short growth duration"	Entire life cycle
Drought avoidance	Maintaining high tissue water potential despite water scarcity	"Deep root systems, reduced leaf area, stomatal closure"	Vegetative and reproductive stages
Drought tolerance (sensu stricto)	Functioning and surviving at low tissue water potentials	"Osmotic adjustment, antioxidant production, heat shock proteins"	Cellular level during stress

mechanism depends on the drought scenario: escaping a late-season drought through early flowering is effective in regions with terminal drought patterns, whereas, in a prolonged mid-season drought, deep roots and stomatal closure (avoidance) may be more beneficial. Drought's timing, intensity, and duration determine which strategy is most advantageous. Therefore, breeders must align drought-response traits with their crop's target population of environments (TPE). Emerging research also investigates the role of the plant microbiome in enhancing drought resilience. Certain beneficial root-associated microbes can improve water uptake and stimulate stress-responsive pathways, potentially providing an additional layer of physiological buffering in drought conditions (Cooper and Messina 2023). Breeding programs may increasingly incorporate microbial interactions into trait evaluation for a more holistic approach to drought tolerance (Table 11.2).

Modern breeding programs assess these traits through both field selection (e.g., selecting early flowering lines in drought nurseries) and specialized phenotyping (e.g., measuring canopy temperature or chlorophyll content as indicators of water status) (Araus and Cairns 2014; Tattaris et al. 2016). It is well established that improving drought tolerance is complex (Budak et al. 2015; Langridge and Reynolds 2015). However, numerous success stories exist; for instance, breeding for the stay-green trait has enhanced post-anthesis drought tolerance in sorghum, resulting in higher and more stable yields under water stress (Borrell et al. 2014a). The key challenge is to improve drought tolerance without incurring penalties in yield under non-stress conditions (Haghpanah et al. 2024).

Some traits for drought tolerance inherited from wild relatives, like small leaves and slow growth, may lower yield potential. Therefore, breeders should be careful not to excessively enhance drought tolerance if it compromises productivity. A study focusing on Mediterranean crops found that certain adaptations from wild plants can lead to reduced yields when integrated into high-yield crop varieties. Consequently, a comprehensive breeding strategy is crucial to balance drought tolerance and yield (Haghpanah et al. 2024). In conclusion, drought tolerance is a complex goal: successful breeding must combine escape, avoidance, and tolerance traits well-suited to the crop's growing environments, all while maintaining yield potential (Fleury et al. 2010; Gepstein and Glick 2013; Guadarrama-Escobar et al. 2024).

11.4 Genotype × Environment Interaction (G × E) and Its Implications

Due to varying environmental conditions, plant breeders routinely observe G × E interactions in multi-location trials (Koutis et al. 2012). G × E interactions refers to the inconsistent performance differences between genotypes across different environments (Kang 2002). Formally, G × E interactions are evident when the performance of genotypes "crosses over" between environments or when the magnitude of differences fluctuates across them.

There are two main categories of G × E interactions:

11.4.1 Crossover Interaction (Qualitative G × E)

Crossover interaction refers to the phenomenon where genotype rankings differ from one environment to another. In one location, genotype A may outperform genotype B, while in a different location, B may outyield A. Graphically, the reaction norms (performance curves) of the two genotypes intersect across environments. Crossover G × E interaction is significant because it suggests that no single genotype is optimal under all circumstances. This scenario often necessitates breeding for specific adaptation, which involves developing distinct varieties suited to different environmental types (Kang 2002; Singh et al. 1999). For example, one maize hybrid may be superior under well-watered conditions, while another hybrid may excel during drought; a crossover G × E interaction suggests one hybrid for irrigated farms and a different one for rainfed farms.

11.4.2 Non-crossover Interaction (Quantitative G × E)

The rank order of genotypes remains consistent across environments, but the magnitude of differences varies. Here, if genotype A yields higher than genotype B in all tested environments, we have a consistent winner; however, the yield gap between A and B may be larger in one environment and smaller in another. Graphically, performance lines do not intersect; they may diverge or converge across environments without changing their ranks. Non-crossover G × E is easier for breeders to manage—one genotype is generally superior, even if its advantage varies across environments. In this scenario, breeding can focus on the best genotype (aiming for broad adaptation), allowing one variety to serve multiple regions (Fig. 11.1).

Significant G × E interactions have practical implications; they reduce the genetic correlation between performances in different environments, making it more difficult

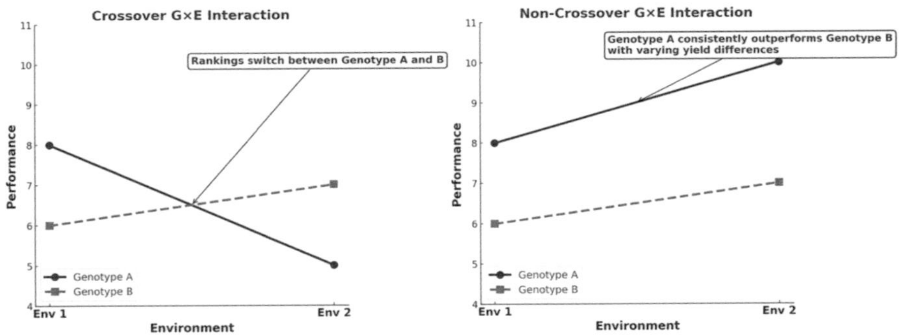

Fig. 11.1 Genotype × environment (G × E) interactions. (The figure presents two conceptual models of G × E interactions: *Left panel (crossover interaction)*: This scenario illustrates a qualitative G × E interaction, where the performance ranking of genotypes changes across environments. In this example, genotype A outperforms genotype B in one environment but underperforms in another. The intersection of performance lines indicates the presence of crossover G × E, suggesting that no single genotype is superior across all environments. This type of interaction necessitates environment-specific breeding strategies. *Right panel (non-crossover interaction)*: This represents a quantitative G × E interaction, where genotype rankings remain consistent across environments, although the magnitude of performance differences varies. Genotype A consistently yields higher than genotype B, but the performance gap differs between environments. Such interactions support the selection of broadly adapted genotypes suitable for diverse growing conditions

to predict a variety's performance in untested conditions (Kang 2002; Robertson 1959). Breeding selections based on one environment may not be applicable in another due to G × E interactions. To address this, plant breeders conduct multi-environment trials (METs) to evaluate genotypes across a range of target conditions, aiming to identify those that perform well on average and/or are specifically adapted to particular niches (Argaw et al. 2025; Lee et al. 2023). METs may span several years and locations, often involving optimal and stress-applied conditions (e.g., managed drought and non-stress trials). Notably, G × E can be quite substantial relative to the main effects. In some studies, the G × E variance for yield has exceeded the variance due to genotype or environment alone. For instance, a multi-site trial of drought-tolerant mung beans in Ethiopia revealed that the G × E interactions component was approximately 4.4 times the genotype component for seed yield, indicating significant differences in genotype rankings across sites (Ganta et al. 2022). Similarly, classic research in wheat estimated that the environment accounted for about 50% of yield variation, genotype approximately 20–30%, and G × E interactions the remaining 20–30%. When G × E interaction represents such a significant portion of variability, breeders must carefully consider it in their selection decisions. Understanding G × E interactions in a breeding program is crucial for guiding strategy. If crossover interactions are the primary focus, it is necessary to develop distinct cultivars tailored for specific environments, such as one for drought-prone regions and another for areas with high rainfall. Conversely, if non-crossover interactions predominantly occur, where genotypes show consistent performance but with varying degrees, breeders can concentrate on a single, broadly adapted genotype and implement management practices to address these differences (Kang 2002).

In practice, plant breeders often encounter both crossover and non-crossover G × E interactions. Breeders exploit G × E interactions by identifying traits or alleles that confer specific adaptations. For instance, a drought tolerance allele that enhances yield under water stress but is neutral under irrigation contributes to crossover G × E interactions, making carriers superior in drought conditions while performing averagely elsewhere (Singh et al. 1999). Conversely, a genotype that consistently performs well across environments, without crossover, can be promoted as a broadly adapted variety (Kang 1997). Understanding G × E interactions is particularly crucial in drought breeding, as water availability can vary dramatically across seasons and locations. Some G × E interactions in drought trials arise from differences in the timing or severity of drought stress. For example, a genotype with early flowering (a form of drought escape) may excel in late-onset drought environments but underperform if drought occurs early, exemplifying G × E interaction (Blum 2010; Craufurd and Wheeler 2009). To effectively interpret G × E interactions, breeders characterize target environments using tools like enviromics or envirotyping, which profile environmental variables (e.g., rainfall patterns, soil moisture, temperature) across trial sites. By grouping environments with similar stress patterns, breeders can better understand G × E; for instance, genotype A may excel in environments with terminal drought, while genotype B thrives under mid-season drought conditions (Cooper and Messina 2021; Xu 2016). Modern approaches integrate environmental data into genomic prediction models, enabling genomic selection (GS) to forecast a genotype's performance in specific environments. This integration of genetics and environmental modeling represents an advanced method through which breeders leverage their understanding of G × E (Jubair and Domaratzki 2023). Breeders are now utilizing climate clustering based on long-term environmental data to refine genotype targeting, and defining breeding zones and mega-environments. These clusters enhance the precision of G × E interpretation and assist breeders in optimizing the deployment of both specific and broadly adapted varieties (Costa-Neto et al. 2021). Additionally, breeders have begun integrating environmental simulation models (e.g., APSIM, DSSAT) into G × E analysis. These models facilitate virtual testing of genotypes using years of weather data, improving predictions of genotype performance under various drought patterns and informing more resilient breeding decisions (Cooper and Messina 2023; Tan et al. 2024).

In conclusion, G × E should not be regarded as mere "noise" complicating selection; it provides essential insights into genetic adaptation. Plant breeders address G × E by conducting tests in representative environments, sometimes creating distinct lines for various niches or developing robust lines that perform well based on observed patterns. Under drought conditions, a comprehensive examination of G × E enables breeders to create varieties that either excel specifically during drought (specific adaptation) or function effectively in both drought and non-drought conditions (broad adaptation), as necessary.

11.5 Does Drought Tolerance Improve Yield Stability?

A central question is whether selecting drought tolerance traits leads to more stable yields across variable environments. Intuitively, one might expect that a drought-tolerant genotype, by withstanding water stress, would have an advantage in drought years and at least maintain equal performance in non-stress years, thus being more stable overall. Indeed, many studies report positive associations between drought tolerance and yield stability. For example, drought-tolerant wheat lines that maintain a higher grain set under water stress also tend to experience less yield fluctuation across wet and dry seasons since they avoid complete failure during dry seasons (Senapati et al. 2019). In maize, breeding programs have shown that hybrids selected for drought tolerance using managed stress trials demonstrate improved year-to-year yield consistency under rainfed conditions compared to older hybrids (Nemali et al. 2015). These correlations suggest that drought tolerance can contribute to yield stability in drought-prone environments.

However, the relationship is complex and not automatic. There are potential trade-offs to consider. A genotype heavily optimized for drought conditions might possess traits that are neutral or even disadvantageous in high-yield, non-stress environments. For instance, a plant that invests in an extensive root system or robust osmotic adjustment mechanisms might divert resources from maximizing shoot growth under optimal water conditions. In a well-watered environment, such a genotype could lag behind a less "stress-prepared" counterpart that allocates more resources to grain production rather than stress defenses. This is sometimes observed: certain drought-bred lines yield slightly less than elite irrigated lines under ideal conditions, even though they excel in drought situations. As Blum (2011a, b) pointed out, traits such as reduced stomatal conductance or smaller leaves conserve water but can limit photosynthesis, thereby reducing top-end yield potential in water-abundant conditions (Lopes and Reynolds 2011). Thus, a highly drought-tolerant genotype may not be the highest-yielding in a favorable environment, which can affect stability metrics depending on how stability is defined.

Another important factor is the specific drought tolerance mechanism used by a genotype, as different mechanisms lead to varying stability outcomes. For example, a drought-escaping variety (which matures early) may sacrifice yield in regions with a longer growing season and adequate moisture. If planted in a year without drought, its yield could be less than optimal because of its early maturation, resulting in a wider yield fluctuation between drought and non-drought years (Craufurd and Wheeler 2009). In contrast, a variety that effectively avoids drought (through traits like deep roots and efficient water use) may perform well under moderate stress and remain competitive in wetter conditions, thus achieving more stable yields across varying rainfall situations (Tardieu 2012). Despite these complexities, a thorough examination of breeding results shows that improving drought resilience generally helps maintain yields in conditions with limited moisture, thus reducing the likelihood of crop failure and enhancing overall stability. In wheat, modeling studies predict that incorporating traits for drought tolerance during the reproductive phase can increase yield potential and stability under future climate conditions. For instance, Senapati et al. (2019) found that a theoretical wheat genotype exhibiting greater drought tolerance during flowering could yield about 13% more on average, with less variability under projected 2050 climate conditions in Europe. Similarly, in chickpeas, Arif et al. (2021) demonstrated that certain genotypes maintained stable yields across varying environmental conditions, underscoring the importance of drought tolerance traits in achieving yield stability. These findings illustrate that traits related to drought tolerance can help shield yields from climate variability.

On the other hand, if a region rarely experiences drought, aggressively pursuing drought tolerance might unnecessarily compromise yield potential or other traits. Conversely, in chronically drought-prone regions, some loss of maximum yield may be acceptable for significant gains in reliability. Therefore, plant breeders aim for genotypes that combine drought tolerance with high yield in non-stress conditions—essentially trying to "have it both ways." The best modern cultivars often minimize the trade-off. An example is maize in Eastern and Southern Africa: new hybrids from the Water Efficient Maize for Africa (WEMA) project were bred for drought tolerance but also tested under optimal conditions. Some of these hybrids (e.g., DroughtTEGO® varieties) yield as well as commercial checks in favorable seasons and significantly better in drought-stricken seasons (Obunyali et al. 2019), thereby achieving superior stability.

In conclusion, drought tolerance plays a vital role in performance in water-limited environments and helps maintain yield stability despite fluctuations in rainfall. However, it does not guarantee stability in every situation. A drought-tolerant genotype may be outperformed in high-productivity environments by a more vigorous but less tolerant genotype. Therefore, breeders must focus on both drought tolerance and yield potential simultaneously, ensuring that advancements in stress conditions do not adversely impact performance under non-stress scenarios (Haghpanah et al. 2024). Comprehensive multi-environment assessments are necessary to ensure that a seemingly drought-tolerant line is truly "stable"—meaning it performs well under drought conditions without significant compromises elsewhere. Often, the most advantageous genotypes exhibit both high average yield and resilience, frequently referred to as "win-win" genotypes. The ultimate goal in breeding for drought-prone regions is to identify these varieties (Lee et al. 2023; Supriadi et al. 2024). Recent studies build on this foundation by emphasizing the strategic application of the extended breeder's equation and G × E × M interactions (genotype × environment × management) to create climate-resilient crop ideotypes. This approach integrates envirotyping and enviromics to map environmental gradients while optimizing genotype selection via reaction-norm modeling in water-limited environments (Cooper and Messina 2023). Additionally, advanced phenotyping technologies like drone-based thermal imaging and hyperspectral sensors are utilized to effectively capture physiological traits associated with drought tolerance and yield stability. These methods help identify genotypes that sustain favorable canopy temperatures, photosynthetic activity, and water usage patterns under fluctuating conditions, thereby enhancing the precision and scalability of stability evaluations (Araus and Cairns 2014).

11.6 Leveraging G × E in Breeding for Drought Tolerance and Stability

Considering the challenges outlined, plant breeders leverage insights from G × E to develop strategies aimed at enhancing drought resistance and achieving yield consistency. They typically adopt two complementary approaches:

11.6.1 Specific Adaptation Breeding

Employing G × E to create genotypes suited for specific environments, particularly drought-influenced ones. For example, breeders may focus on developing a group of lines tailored for semi-arid areas frequently experiencing drought, intensively selecting under drought conditions to enhance beneficial traits (such as deep roots and osmotic adjustment). While these lines might not perform well in wetter climates, they thrive in their adapted environment. By examining G × E, breeders can determine which genotypes are the most suitable for environments with distinct stress characteristics. One example is the breeding of chickpeas in India: one breeding program focuses on post-rainy (terminal drought) environments in central India, selecting lines that flower early and withstand heat, thereby producing cultivars specifically adapted to that stress pattern, which would not be necessary in cooler, irrigated northern India. G × E analysis helps by clustering trial sites into mega-environments—for instance, drought-prone versus high-rainfall—and then selecting winners for each cluster (Berger et al. 2006).

11.6.2 Broad Adaptation (Wide Stability) Breeding

Selecting genotypes that perform consistently well across a diverse range of environments, including both drought-stressed and non-stressed conditions. The goal here is to accumulate trait combinations that confer robustness. Breeders look for lines that have acceptable yields under drought and still have high yields when water is abundant. This often involves index selection for yield in stress and non-stress or selecting on multi-environment average performance. By leveraging G × E data, one can identify genotypes with high mean yield and low variance across environments. Many modern breeding programs explicitly aim for such broadly adapted varieties to buffer farmers against climate uncertainty. For example, tropical maize program of the International Maize and Wheat Improvement Center (CIMMYT) evaluates lines under optimal and drought conditions, advancing those that rank near the top. The result is hybrids that, while perhaps not the absolute highest in perfect conditions, never fall apart during a drought year, providing a desirable stability profile for smallholders in Sub-Saharan Africa (Johnson et al. 1986; McMillen et al. 2022).

In implementing these approaches, breeders are increasingly utilizing environmental characterization and data-driven tools. Envirotyping, the systematic characterization of test environments (such as weather and soil), enables breeders to quantify environmental similarities and more precisely target their selection for specific adaptation (Xu 2016). For instance, by analyzing long-term climate data, breeders can categorize target environments into those with "late-season drought," "occasional drought," or "adequate moisture," and then interpret G × E accordingly (Cooper and Messina 2021). If a crossover G × E is observed mainly between the drought and non-drought groups, it signals an opportunity to breed specialized lines for each. Conversely, if top genotypes rank highly in both groups (non-crossover), breeders may concentrate on those broadly adapted lines (Corlouer et al. 2024).

Advances in genomics and prediction models have enhanced our ability to leverage G × E. For instance, genomic selection (GS) models can incorporate genotype data and environmental covariables to predict performance in new environments. Cooper and Messina (2023) emphasize that integrating genomic and environmental data is essential for predicting how genotypes will perform in future, potentially warmer and drier climates. Marker-assisted breeding is also utilized: if certain alleles are known to confer drought tolerance that is only advantageous under stress, those alleles can be employed in varieties intended for stress-prone areas (thereby enhancing specific adaptation). Conversely, breeders may avoid alleles that lead to yield drag under no stress. An example is the "QTL-hotspot" in chickpeas, a cluster of QTLs for root traits and drought tolerance. The introgression of this hotspot increased yield under drought by approximately 10–15% and had neutral to slight effects under irrigated conditions, making it a valuable tool for improving the broad adaptation of chickpeas (Varshney et al. 2014). By understanding these effects through G × E tests (drought vs. irrigated trials), breeders used the QTL to enhance chickpea lines for areas with variable rainfall.

Environmental simulation models and crop growth models are valuable tools. By simulating how a genotype with specific trait values performs across numerous years of weather data, breeders can evaluate its yield stability in silico. For instance, modeling suggests that a wheat genotype with deeper roots may yield higher in 7 out of 10 years (during mild droughts) but show slightly lower yields in 3 out of 10 wet years, providing insights into the long-term stability benefits or trade-offs associated with that trait (Manschadi et al. 2006). This information guides the decision to incorporate that trait based on the breeder's risk preference and the climate of the target region.

The choice between pursuing specific versus broad adaptation often hinges on the predictability of the target environment. In highly variable climates (e.g., the West African Sahel, where rainfall is erratic), broad adaptation and resilience are prioritized; farmers prefer a variety that provides at least moderate yields every year rather than one that yields very high in some years but fails in others (Beyene et al. 2016; Fisher et al. 2015). In more predictable environments or irrigated systems, specific adaptation can be beneficial (Reynolds et al. 2007; Trnka et al. 2015). For example, developing a high-yield variety for well-watered systems and a separate drought-tolerant variety for purely rainfed areas is important. Often, breeding programs pursue both avenues through different product pipelines or breeding streams. For instance, an international rice program might have one breeding stream for upland rice (drought-prone, specific adaptation) and another for lowland irrigated rice (where drought tolerance is less crucial, so the focus is on maximizing yield and disease resistance). Recent studies have also shown the effectiveness of marker-assisted breeding in developing rice genotypes that are tolerant to multiple stresses like drought, submergence, and disease (Mohanavel et al. 2024).

In summary, breeders utilize G × E knowledge as a tool: by analyzing multi-environment data, they determine whether to focus on broad versus specific adaptation, identify which drought tolerance traits provide advantages in specific environments, and ensure that selection decisions align with the environmental profiles of the target regions. The ultimate goal is to produce crop varieties that excel in challenging environments or perform robustly across diverse conditions. Both outcomes represent forms of "stability"—one reflects stability within a narrow range, while the other signifies stability across a wide range. Modern breeding, through envirotyping and genomic prediction, is increasingly effective in achieving the desired type of stability alongside drought tolerance.

11.7 Methodologies for Evaluating Drought Tolerance and Stability

Robust evaluation methods are crucial for developing drought tolerance and stability. Plant breeders depend on field trial designs and statistical analyses to measure genotype performance, genotype × environment interactions, and stability indices.

Multi-environment trials (METs) serve as the cornerstone of evaluation. In a MET, a set of genotypes (breeding lines or varieties) is tested across various environments that reflect the target conditions. For drought tolerance breeding, METs often include both managed drought stress trials (e.g., rainout shelters or deficit irrigation to impose drought at a specific growth stage) and non-stress trials, along with multiple locations and years. This setup provides data on both performance under drought and overall adaptability. METs are typically organized in randomized complete block or alpha-lattice designs to control experimental error within each environment. Proper blocking is essential because, within any given trial, soil heterogeneity can impact yields; blocking (and replication) assist in ensuring that observed differences are genetic rather than positional (Alvarado et al. 2020; Fischer 2003).

Once data is collected from METs, several statistical tools are used:

11.7.1 Analysis of Variance (ANOVA)

A combined ANOVA across environments is typically conducted first. This separates variance into components attributed to genotype, environment, and G × E interaction (as well as error). A significant G × E term suggests that genotypes are reacting differently across environments, highlighting the importance of assessing stability. ANOVA also facilitates the estimation of genotype means in each environment and overall. Although ANOVA itself doesn't specify which genotypes are the most stable, it acts as a foundational analysis for understanding the variation involved and determining whether G × E is statistically significant (Anshori et al. 2024).

11.7.2 Regression and Univariate Stability Parameters

A classical approach to quantifying stability is the Finlay–Wilkinson/Eberhart–Russell regression analysis, where each genotype's yield is regressed on an environmental index (typically the mean yield of all genotypes in that environment) (Eberhart and Russell 1966; Finlay and Wilkinson 1963). The slope of this regression (denoted bi) indicates responsiveness: bi $\approx$ 1 means average responsiveness, bi <1 indicates below-average response (stable, less sensitive to the environment), and bi >1 indicates above-average responsiveness (more sensitive, performs well in good environments but poorly in bad ones) (Eberhart and Russell 1966). The regression method, introduced by Finlay and Wilkinson (1963), also uses the deviation from regression (S^2d) as an index of unexplained variability; a smaller deviation means the genotype's performance is predictable (good stability). In practice, a stable genotype is often defined as one with *b* close to 1 and low S^2d (Shukla 1972). These parameters are useful for comparing varieties; many breeders still examine regression slopes as an intuitive measure of stability. For example, if a drought-breeding line has bi $= 0.8$ and low deviation, it suggests that it yields relatively well even as overall conditions worsen (a sign of stability, though it might not achieve the highest yields in the best environment). Other univariate indices include the coefficient of variation (CV) of yield, Shukla's stability variance (which measures a genotype's contribution to G × E variance), and Wricke's ecovalence (another measure of contribution to G × E). Each of these provides a scalar value per genotype that reflects stability and often complements one another. It is noted that different statistics may sometimes rank genotypes differently, so breeders frequently examine multiple indices.

11.7.3 AMMI and GGE Biplot Analyses

To gain a deeper understanding of G × E patterns, multivariate analysis is utilized. The additive main effects and multiplicative interaction (AMMI) model integrates ANOVA for main effects with principal component analysis (PCA) applied to the G × E residuals (Gauch 2006). This effectively breaks down the G × E interaction into principal component axes. An AMMI biplot can then display genotypes and environments together, often uncovering clusters of environments and specifically adapted genotypes. AMMI is effective in summarizing complex data; it allows the identification of which genotypes are broadly stable (near the origin in the biplot) as opposed to those that display significant interactions (far from the origin, indicating they are specifically adapted).

The GGE biplot (genotype + G × E) is an approach that focuses on genotype and G × E together, often after subtracting the means of environments (Yan and Kang 2002). GGE biplots provide a visual analysis of "which-won-where," displaying which genotype performs best in each sector of the multivariate environment space. This is extremely useful for drought breeders. For instance, a GGE biplot might demonstrate that a particular drought-tolerant line excels in all environments on the drier side of the plot, while another line succeeds in wetter environments, revealing a clear crossover interaction and suggesting two recommendation domains. These graphical tools have become standard in MET analysis for breeding and assist breeders in determining whether one variety can serve all targets or if multiple varieties are needed (Kang 2002).

11.7.4 New Stability Indices (BLUP-Based)

In recent years, mixed-model approaches utilizing best linear unbiased predictors (BLUPs) have become prevalent in multi-environment trials (MET) analysis. From these mixed models, one can derive genotype BLUPs for each environment along with an overall assessment. A novel metric known as the weighted average of absolute scores (WAASB) was introduced to summarize stability derived from the AMMI model's Interaction Principal Component (IPC) scores in conjunction with BLUPs. Simply put, WAASB provides a single index that penalizes genotypes for significant interaction effects while considering their mean performance. Olivoto et al. (2019), who proposed WAASB, demonstrated that it can be used to rank genotypes based on both yield and stability simultaneously. For example, a genotype with a high yield but considerable instability might receive a lower WAASB score than a genotype with a slightly lower yield but greater stability, depending on the weighting. Breeders have the flexibility to adjust the weights to prioritize yield versus stability according to their objectives (Olivoto and Lúcio 2020). WAASB and comparable indices (such as the Yield Stability Index (YSi), which integrates yield and Shukla's variance) indicate a shift towards compound selection indices, facilitating selection for multifaceted breeding goals in one step (Shukla 1972).

Direct drought tolerance indices: In parallel, when evaluating drought trials specifically, breeders utilize indices such as the stress tolerance index (STI), yield stability index (YSI), and stress susceptibility index (SSI), which compare genotype performance under both stress and non-stress conditions (Mehraban et al. 2018). For example, STI combines yield data under both stress and non-stress conditions to identify genotypes that excel in both scenarios. These indices aid in selecting drought-tolerant lines (performing well under stress) and incur minimal yield penalties in non-stress conditions. A wheat study may calculate STI for each line; a high STI indicates above-average yield in both irrigated and drought trials—traits that are desirable for stability. Correlations among these indices and yield can unveil how tolerance relates to stability metrics.

In addition to performance metrics, evaluating drought tolerance per se often involves measuring specific physiological traits (PTs). Traits such as canopy temperature (an indicator of transpiration cooling), chlorophyll content or "stay-green" score, carbon isotope discrimination (a proxy for transpiration efficiency), and osmotic adjustment capacity are widely used in field trials (Reynolds et al. 2005; Blum 2011a, b). Modern high-throughput phenotyping (HTPP) tools—including spectral reflectance sensors, thermal imaging drones, and ground-based phenomics platforms—enable breeders to quickly screen large populations for these traits (Fiorani and Schurr 2013; Yang et al. 2015). For instance, an infrared thermal camera mounted on a drone can map the canopy temperature of thousands of plots in just a few minutes; cooler canopies in drought conditions often correlate with deeper roots or better hydration (Araus and Cairns 2014). When combined with yield data, this information helps identify genotypes with functional drought tolerance mechanisms. Araus and Cairns (2014) emphasized that these phenotyping advancements are revolutionizing selection for complex traits like drought tolerance, as breeders can now quantify trait expression in the field and select accordingly. By measuring PTs, breeders gain additional confidence that a "stable high yielder" is achieving stability due to genuine drought-response traits rather than, for instance, escaping via early maturity (Blum 2011a, b). Recent developments highlight the potential of an image-based transpiration efficiency (iTE) index for more accurately screening drought tolerance in wild relatives. This iTE index integrates hyperspectral estimates of carbon fixation capacity (Vcmax) and thermal imaging-derived crop water stress indices (CWSI). Combined with de novo domestication strategies, this approach provides a robust path for incorporating underutilized drought tolerance mechanisms into elite breeding lines (Guadarrama-Escobar et al. 2024). Recent advances also involve integrating machine learning algorithms with high-throughput phenotyping data to enhance predictions regarding drought tolerance and yield stability. These data-driven approaches facilitate the identification of complex trait patterns across various environments and improve the efficiency of genotype selection (Araus and Cairns 2014).

Ultimately, no single method offers a comprehensive perspective. Breeders often use a combination of analyses to inform their decisions. For instance, they may employ GGE biplots to identify broadly versus specifically adapted lines, regression analysis to verify the stability of top lines, and physiological measurements to ensure that a line possesses drought tolerance traits. Integrating results leads to more informed selections—a genotype chosen based on multiple criteria (yield, stability indices, and trait data) has a higher likelihood of real-world success. This multifaceted approach is evident in many breeding programs. For example, in a tropical maize trial, researchers identified lines with high BLUP yield, low WAASB (stable), and high chlorophyll retention under stress, advancing those as candidate drought-tolerant, stable varieties (Supriadi et al. 2024).

In summary, assessing drought tolerance and stability involves (1) designing appropriate MET trials to capture essential data, (2) conducting statistical analyses of performance (ANOVA, G × E, stability parameters, biplots, etc.), and (3) performing specialized phenotyping for traits related to drought response. Together, these methodologies allow breeders to dissect complex patterns of genotype performance and select lines that effectively combine high performance with remarkable stability. The synergy between advanced statistical models and modern phenotyping enhances breeders' capability to identify superior genotypes for challenging conditions.

11.8 Case Studies: Breeding Successes in Drought Tolerance and Stability

Despite the challenges, many breeding programs have successfully developed crop varieties that offer improved drought tolerance without compromising, and often enhancing, yield stability across diverse environments. Below, we showcase examples from several crops:

11.8.1 Maize (*Zea mays*)

Maize (*Zea mays*) is highly susceptible to drought, particularly during flowering. Breeders have concentrated on traits such as a short anthesis-silking interval (ASI) to facilitate pollination under stress, deeper roots, and improved water-use efficiency. The CIMMYT-led DTMA/WEMA projects in Africa have introduced over 200 drought-tolerant hybrids. These hybrids (e.g., those marketed as "DroughtTEGO" or "TELA") were thoroughly tested in both stress and optimal conditions. They generally show yield advantages of 20–30% over older varieties in drought conditions while yielding comparably to commercial checks during normal rainfall years (Fisher et al. 2015; Edge et al. 2017). Multi-location trials of some hybrids developed through doubled-haploid (DH) technology demonstrated superior grain yield and stability in both drought-stress and favorable environments. For example, recent studies have identified specific DH-derived hybrids that ranked in the top yield group in environments ranging from well-watered (yield >10 t/ha) to severe drought (yield ~3 t/ha), indicating broad adaptability. The use of genomic selection and rapid cycling (selection in managed drought, followed by testing in multiple environments) has accelerated these gains (Beyene et al. 2016). As a result, farmers in drought-prone areas of Kenya, Zimbabwe, and elsewhere have adopted these varieties, experiencing greater yield stability—the bad years are less bad, while the good years remain strong. Maize breeding demonstrates that careful selection can produce hybrids that tolerate drought and respond well to inputs, bridging the gap between stress-prone and high-potential environments.

11.8.2 Wheat (*Triticum aestivum* and *T. durum*)

The extensive cultivation of wheat means it encounters diverse climates, from irrigated plains to rainfed drylands. Breeding for drought tolerance in wheat has focused on traits such as canopy temperature depression (cool canopies under stress), osmotic adjustment, and deeper roots, as well as flowering time for escape.

A key finding highlights the significance of reproductive-stage drought tolerance for yield stability in future climates. Modeling and field studies (e.g., in Europe and Australia) indicate that genotypes maintaining grain filling under heat/drought (through traits like stay-green or higher stem reserves) achieve both higher yield potential and more stable yields under variable rainfall (Senapati et al. 2019). Breeders have integrated drought-tolerant germplasm from wild relatives (e.g., *Aegilops species*) and landraces. For instance, the introgression of segments from wild wheat (*Triticum dicoccoides*) has led to deeper roots and early vigor, producing new durum wheat lines that outperform checks in dry years and yield comparably in wet years (Araus and Cairns 2014). In India, varieties such as "HI 1500" (drought-tolerant bread wheat) have displayed remarkably low yield variability in multiyear trials, attributed to their efficient root systems and cooler canopies. International Center for Agricultural Research in the Dry Areas (ICARDA)'s breeding approaches also yielded good results. Participatory breeding projects, which aim to direct the contribution of farmers, resulted in the effective use of local genetic diversity. As a result of these, increased drought tolerance due to ABA pathways has been achieved in some regions (Ceccarelli and Grando 2020). These successes in wheat highlight that combining drought tolerance genes, such as *ERA1*, which improves ABA sensitivity, or selecting for the stay-green trait, can indeed enhance yield stability under both current and projected climate scenarios (Senapati et al. 2019).

11.8.3 Sorghum (*Sorghum bicolor*)

Sorghum, a crop well known for its ability to withstand drought, resulted in high-yielding varieties. It has a long stay-green capacity under adverse environmental conditions, which allows farmers in sub-Saharan Africa to achieve high and sustainable yields. Farmers that grew these enhanced sorghums had higher average yields and less volatility from year to year. An additional characteristic, osmotic adjustment, has been chosen in sorghum: lines that store solutes (thereby maintaining cell turgor) show better grain filling during dry times. Such lines tend to have a yield advantage in dry situations and attain equivalent yield in wet environments, resulting in increased stability (Borrell et al. 2014a). Many countries, including India, Australia, and Sub-Saharan Africa, have introduced sorghum varieties that are termed "drought-proof" (Bantte et al. 2022; Borrell et al. 2014b). While no crop is entirely drought-resistant, sorghum comes near; it can produce some grain even in severe droughts, and modern cultivars are more dependable than traditional ones in giving yields in both moderate and high rainfall zones (Menamo et al. 2023).

11.8.4 Rice (*Oryza sativa*)

Rice is commonly cultivated in flooded conditions, but extensive areas of rice in South Asia and Africa depend on rainfall and are vulnerable to drought (Singh et al. 2021; Zu et al. 2017; Comas et al. 2013). Breeding drought-tolerant rice (e.g., for upland and rainfed lowland ecosystems) has historically lagged; however, substantial progress was achieved in the 2000s and 2010s. A significant breakthrough was the identification of the QTL *qDTY12.1* for yield under drought conditions in rice, and its introgression into popular varieties (Mishra et al. 2013). Lines with DTY QTLs (such as the variety "Sahbhagi Dhan" in India) have shown a yield advantage of 0.8–1.2 t/ha under drought and no yield penalty under normal conditions, thereby clearly improving stability. Multi-location trials of Sahbhagi Dhan across India demonstrated yield stability: it achieved superior yields in drought years and yielded comparably to the mega-variety Swarna in non-drought years (Dar et al. 2020).

Other approaches for rice include breeding for deeper roots (e.g., incorporating traits from *aus* rice varieties) and utilizing aerobic rice for water-limited uplands (Singh et al. 2021; Uga et al. 2011, 2013). The adoption of drought-tolerant rice by farmers (e.g., in Eastern India, Nepal, and Bangladesh) has been significant—farmers report that these varieties "rescue" yields during dry spells and continue to provide good harvests in normal rainfall, effectively stabilizing production fluctuations (Dar et al. 2020). Ongoing efforts by International Rice Research Institute (IRRI) and national programs are expanding drought-tolerant rice portfolios by integrating novel donors (wild *Oryza* species) for improved root and stress signaling traits. The case of rice illustrates that even in a water-loving crop, targeted breeding can enhance drought resilience and yield stability in rainfed systems.

11.8.5 Grain Legumes

Many pulse crops, such as chickpea (*Cicer arietinum*), cowpea (*Vigna unguiculata*), and pigeon pea (*Cajanus cajan*), are often cultivated on residual soil moisture and are susceptible to terminal drought. Breeding efforts in these crops have focused on reducing crop duration (drought escape) and enhancing root traits.

In chickpea, the development of extra-early varieties (e.g., maturing in less than 90 days) has enabled the crop to evade end-of-season drought in South Asia (Varshney et al. 2014; Muriuki et al. 2020). Varieties such as "Jodha" and "Sadabahar" combine early maturity with decent yield potential, ensuring more stable yields by preventing catastrophic failure in drought years and allowing flexibility in sowing time during favorable years. Additionally, a "QTL-hotspot" for drought tolerance, containing genes for root depth and shoot biomass, was introduced from the donor ICC 4958 into elite chickpea cultivars using marker-assisted backcrossing. This resulted in lines that yielded 10–15% more under drought conditions and had slightly higher yields in non-stress conditions as well. One such line, BGM 10216, was released as "Pusa Chickpea 10216" for commercial cultivation in India, demonstrating enhanced drought tolerance and yield stability (Bharadwaj et al. 2021).

In cowpea breeding, efforts in West Africa have produced varieties like IT97K-499-35 that cope well with drought and consistently yield. A study of 112 cowpea accessions revealed significant variation in drought and heat tolerance; the best lines, such as IT97K-499-35, exhibited high pod yields in both drought-stressed and non-stressed trials, indicating that selection for drought-related traits, including delayed leaf senescence and vigorous early growth, yielded positive results in terms of stability (Sansa et al. 2024).

Pigeon pea, a drought-hardy legume, has benefitted from hybrid breeding. New hybrids, such as ICPH 2671, demonstrate improved root systems and maintain yields under erratic rainfall, providing smallholders with more stable returns. These hybrids have shown significant yield advantages over traditional inbred cultivars, highlighting the potential of hybrid breeding in enhancing drought resilience and yield stability in pigeon pea (Vales et al. 2012). These examples illustrate that even in relatively neglected crops, focused breeding can produce new cultivars that improve minimum yields in bad seasons while still achieving top yields in good seasons, thereby stabilizing production. These results are increasingly backed by integrative breeding pipelines that combine conventional hybridization with speed breeding, marker-assisted selection, and genomic prediction. This convergence of tools is particularly evident in programs developing dual-purpose lines—those that ensure performance in both stress and non-stress conditions—accelerating real-world impact.

11.8.6 Minor Cereals and Root Crops

Beyond the major staples, breeding for drought tolerance helps stabilize the yields of several other food security crops. For instance, tef (*Eragrostis tef*), an indigenous cereal of Ethiopia, is adapted to drought, but breeding has enhanced its resilience further. Recent work identified landraces tolerant to both drought and waterlogging (as tef is often grown in both conditions)—crosses produced progeny such as Genotype G1 and G3, which exhibited not only dual tolerance but also high mean grain yields with low variance across sites in Ethiopia (Girma and Natol 2024). These have been released as climate-resilient tef varieties, safeguarding farmers against unpredictable rainfall. Cassava (*Manihot esculenta*), a root crop, is inherently drought-tolerant, but breeding has combined drought tolerance with higher dry matter and disease resistance. Some new cassava clones (e.g., those bred by International Institute of Tropical Agriculture (IITA)) show stable root yields across wet and dry zones; farmers in Nigeria have observed that these clones produce reliable harvests even when the rains are delayed, unlike older varieties that might drastically lose yield (Codjia et al. 2025). Similarly, sweet potato breeding in East Africa has introduced drought-tolerant traits (deep storage roots, efficient stomatal control) into high-yielding, orange-fleshed lines, resulting in varieties that provide consistent yields and nutrition under variable rainfall (Low et al. 2020).

These case studies reinforce several key points: achieving both drought tolerance and yield stability typically requires a multifaceted breeding approach. Breeders generally combine conventional methods (hybridization, selection in target environments) with modern tools (marker-assisted selection, genomic selection) and extensive phenotyping. They also utilize diverse germplasm—including wild relatives and landraces—to introduce novel stress-tolerance alleles. The result of

Table 11.3 Examples of drought-tolerant, stable varieties

	Variety (or line)	Key drought-tolerance traits	Yield stability outcomes	Source/ reference
Maize	DH-derived hybrid "A" (example)	Short anthesis-silking interval (ASI); deep roots; DH breeding	High grain yield under both optimal and drought conditions; top performer in 20 environments	Beyene et al. (2016)
Wheat	"Sokoll" (durum line)	Reproductive-stage tolerance; stay-green phenotype	15% higher yield than standard under drought; comparable in rainfed; low G × E interaction	Senapati et al. (2019)
Sorghum	Line E (G × E study in Ethiopia)	Stay green (post-flowering); high chlorophyll retention	Among the most stable for grain and biomass across 5 locations in Ethiopia	Gebre et al. (2024)
Mung bean	NLLP-MGC-24 (G6)	Selected under drought; robust root system	High and stable yield across 6 environments; recommended for release	Ganta et al. (2022)
Tef	Local selection "G3"	Tolerant to both drought and waterlogging	Highest mean yield; low yield variance in multiyear trials in Western Oromia	Girma and Natol (2024)
Chickpea	"Sahbhagi Chana" (experimental)	Early flowering; introgressed QTL-hotspot	10% higher yield under drought; equal yield under irrigation; high STI and stability index	Varshney et al. (2014), Mannur et al. (2019)
Rice	Sahbhagi Dhan (IR74371-70-1-1)	Deep root system; DTY QTLs; medium duration	~4.5 t/ha under normal conditions; ~2.5 t/ha under drought vs. 1–1.5 t/ha for checks	Basu et al. (2017), Rameshkumar et al. (2016)

successful programs is varieties that consistently outperform existing ones across a range of environments, not just in drought plots. Notably, many of these programs involve collaborations (national programs, Consultative Group on International Agricultural Research (CGIAR) centers, universities) and sometimes include participatory variety testing with farmers to ensure the selected varieties meet practical stability needs (e.g., acceptable in both high- and low-input fields) (Table 11.3).

11.9 Future Perspectives and Challenges

As climate change accelerates, breeding for both drought tolerance and yield stability will become increasingly important (Kilian et al. 2021; Massel et al. 2021; Snowdon et al. 2021). Future climates are expected to bring not only more frequent droughts but also greater variability in rainfall distribution and temperature extremes (Xiong et al. 2024). This presents new challenges and opportunities for innovation in breeding:

11.9.1 Deeper Genetic and Physiological Understanding

Ongoing research aims to unravel the complex genetic architectures underlying drought tolerance and stability. Advances in quantitative genetics and functional genomics are identifying key genes and regulatory networks that enable plants to cope with water stress (Savitri 2023; Khalid et al. 2018; Millet et al. 2016). For example, genome-wide association studies (GWAS) in crops such as maize and rice pinpoint loci associated with yield under stress compared to non-stress conditions (some loci contribute specifically to stress yield, while others affect yield in all circumstances) (Mohanavel et al. 2024; Ribeiro et al. 2023; Wu et al. 2022; He et al. 2014; Verma et al. 2023). Understanding how these genes interact—whether through pleiotropy or trade-offs—is crucial. Similarly, systems biology approaches are helping to connect molecular changes, such as transcriptomic and metabolomic shifts under drought, to whole-plant performance and, ultimately, to yield stability. A deeper understanding will guide breeders in stacking beneficial alleles while avoiding detrimental linkages. It may also reveal novel traits. For example, recent findings about the role of the microbiome in plant drought resilience open the possibility of breeding plants that attract or harbor drought-supportive microbes, adding another layer to stability (Lareen et al. 2016; Rolli et al. 2015).

11.9.2 Advanced Phenotyping and Envirotyping

Phenotyping remains a bottleneck for complex traits (Mwadzingeni et al. 2016; Comas et al. 2013; Urfan et al. 2022; Duncan and Topp 2022). Developing more precise, high-throughput phenotyping (HTP) methods is a priority. We anticipate greater use of drones, sensors, and machine learning to measure plant responses (e.g., real-time monitoring of turgor via thermal imagery or root growth using tomography). Remote sensing and image analysis can quickly estimate stress indicators (such as wilting or senescence) across fields, enhancing selection accuracy for drought tolerance (Mir et al. 2019; Wasson et al. 2016; Walter et al. 2015; Schneider et al. 2012). Even phenomics in controlled environments (e.g., drought simulators and lysimeter facilities) can complement field data to dissect trait differences (Urfan et al. 2022; Hacisalihoglu and Armstrong 2023; Furbank and Tester 2011). On the envirotyping side, breeders will increasingly utilize climate data analytics to classify and even predict target environmental scenarios. This may involve using seasonal forecasts or climate change models to determine which breeding lines to advance (e.g., emphasizing lines that perform well under projected future drought scenarios). Integrating these tools will enhance selection efficiency and targeting (Xiong et al. 2024).

11.9.3 Genomic Selection and Gene Editing

The genomic revolution of the past decade will continue to accelerate drought breeding. Genomic selection (GS), already utilized in crops such as maize, wheat, and rice, allows breeders to predict performance (including under drought conditions) based on DNA markers, thereby shortening breeding cycles (Jubair and Domaratzki 2023; Pankaj and Wani 2023; Maulana et al. 2021; Duhnen et al. 2017). As more phenotypic data under stress accumulates, GS models will improve, potentially enabling direct predictions of stability metrics. For polygenic traits like yield stability, GS is particularly useful because it captures the small effects of numerous genes. Marker-assisted selection (MAS) remains relevant for major QTLs; for instance, marker haplotypes for known drought QTLs can routinely be incorporated into new breeding lines to ensure a baseline of tolerance (Singh 2023; Raza et al. 2023; Eltaher et al. 2023; Alahmad et al. 2019; Chukwu et al. 2019). Beyond conventional breeding, gene editing (CRISPR/Cas) offers exciting possibilities for directly tweaking genes for drought adaptation (Massel et al. 2021; Razzaq et al. 2022; Kim et al. 2017). Researchers have identified candidate genes (e.g., transcription factors like DREB, NF-YB, and enzymes such as LOS5/ABA3 that affect ABA biosynthesis) that could be edited to enhance drought response. In rice, CRISPR knockout of a negative regulator of root growth has resulted in plants with deeper roots and improved drought yield (one study showed that OSAUX1 edited lines had 20% more yield under drought) (Sun et al. 2022). Gene editing can also eliminate unwanted sensitivities—modifying a flowering-time gene, for example, to avoid premature flowering under mild stress (thus preventing over-escape). By combining editing with rapid breeding techniques (such as doubled haploids or speed breeding), we may create isogenic lines that differ only in specific drought-responsive genes to confirm their effects on stability (Watson et al. 2018; Repkova and Nedelnik 2014). If better understood, epigenetic modifications and priming might also be leveraged (Oğuz et al. 2023; Balusamy et al. 2022; Zulfiqar et al. 2021; Mal et al. 2019; Sher et al. 2019); some experiments indicate that plants can "remember" drought and respond more quickly later, and harnessing this potential could enhance resilience (Gully 2019).

11.9.4 Crop Modeling and Climate Resilience Planning

Crop simulation models (e.g., DSSAT, APSIM) will be more closely integrated into breeding. These models can simulate how a given genotype (with traits X, Y, and Z) performs over many years and locations. Breeders can utilize these virtual experiments to prioritize trait combinations that yield the most stable outcomes. For example, modeling may indicate that combining early vigor with deep roots provides more stability than pairing deep roots with high transpiration efficiency in a specific region, thus guiding breeding focus (Schäfer et al. 2022; Kalogiros et al. 2016; Daly et al. 2016). In the context of climate change, models can also investigate future scenario performance, essentially "future-proofing" current breeding lines by selecting those predicted to cope with future drought patterns. This is particularly urgent given that breeding cycles span years to decades—varieties being developed now will face the climates of the 2030s and 2040s. There is a growing demand for climate-resilient ideotypes (idealized plant types) (Rao et al. 2021; van der Bom et al. 2020; Gouache et al. 2015; Lynch 2013), which models suggest would be both high-yielding and stable under future conditions (such as the wheat ideotype with reproductive drought tolerance discussed by Senapati et al., 2019). Breeding programs may explicitly target these model-derived ideotypes. Future breeding pipelines may increasingly depend on de novo domestication, wherein significant domestication alleles are introduced into drought-tolerant wild accessions, preserving their resilience traits while enhancing agronomic suitability. This strategy unlocks the full potential of wild germplasm by integrating novel stress tolerance mechanisms with contemporary yield and quality traits (Kumar et al. 2022; Guadarrama-Escobar et al. 2024). Looking ahead, breeding pipelines will likely utilize predictive ideotyping—employing crop models and genomic data to develop ideotypes that are climate-resilient and suited to regional stress patterns. Such ideotypes are currently being investigated for wheat, maize, and rice to enhance both productivity and stability under 2050 climate projections (Senapati et al. 2019).

11.9.5 Despite Optimism, Significant Challenges Remain

Drought tolerance and yield stability are complex, multi-genic traits, and improving them can still be slow and resource-intensive. G × E will continue to complicate selection as environments become more erratic—a variety might perform well in most situations, but some unforeseen combination of stress (e.g., drought + heat + new pathogen) could expose weaknesses. Breeding must increasingly consider multi-stress tolerance (drought is often accompanied by heat or soil nutrient stresses) for true stability. Another challenge is the necessity for long-term breeding programs and investments: developing climate-resilient crops is not a one-time effort but an ongoing process that demands sustained funding and interdisciplinary collaboration. Integrating expertise from physiology, molecular biology, climatology, and socioeconomics will be essential. Finally, capacity building is vital: many regions requiring drought-resistant crops (e.g., parts of Africa and South Asia) need trained breeders and modern facilities. Initiatives to train young scientists and establish breeding infrastructure (phenotyping platforms and seed systems for distributing new varieties) are as crucial as the technical innovations in breeding. Encouragingly, international networks (such as the Heat and Drought Wheat Consortium or the Excellence in Breeding Platform) are concentrating on these issues.

In conclusion, the future of breeding for drought tolerance and stability will likely involve a convergence of advanced technologies and classical breeding wisdom. By integrating high-throughput data, predictive models, and genomic tools with extensive field evaluation, breeders will move closer to the ideal of crops that "thrive, when possible, survive when necessary." These efforts are critical for sustaining agriculture in the face of climate change and feeding a growing global population.

11.10 Conclusion

Climate is changing and extreme weather events, including drought and heat, are becoming more and more common. The stability of a crop variety under changing environmental conditions is linked to its drought tolerance/avoidance ability. Therefore, crop stability is affected by drought tolerance, but it is not solely limited to the drought tolerance mechanism. A variety may yield well under low input and/or adverse conditions, but may not benefit enough from the optimal conditions, or vice versa. Breeding objectives must address specific goals expected from a specific variety under specific conditions. A well-designed" crop variety is expected to withstand adverse conditions while taking maximum advantage of optimum inputs. These kinds of objectives require multiyear–multi environment trials. That demonstrates genotypic potential as well as G–E interactions. Local targeted breeding programs as well as the ones with wide adaptability objectives, both help food security, sustainability, and stability. Any additional knowledge on the needs and abilities of a region or specific target environment helps breeders to work towards specific and objective-oriented breeding projects.

The use of long-term knowledge and modern tools such as marker-assisted breeding made yield and stability possible under diverse conditions for crops such as wheat, maize, and sorghum. These varieties are more consistent and gained the trust of the farmers for their stability. There will be a need for multi-disciplinary approaches to cope with extreme weather events to keep and increase yield and stability within the current climate change and population increase estimates.

The use of biotechnological tools and collaborations with various disciplines will enhance our gain and improve stability and so sustainability. Drought and heat tolerance, as well as other biotic and abiotic stressors, need to be challenged not one by one, but as complementary parts of a single target, yield, and overall stability across various conditions. Aiming to improve multiple traits should be an expected outcome. The collaboration with local players and benefiting from the unexplored gene pools would speed up the process.

References

Ahmadi J et al (2018) Screening wheat germplasm for seedling root architectural traits under contrasting water regimes: potential sources of variability for drought adaptation. Arch Agron Soil Sci 64:1–15. https://doi.org/10.1080/03650340.2018.1432855

Alahmad S et al (2019) A major root architecture QTL responding to water limitation in durum wheat. Front Plant Sci 10:436. https://doi.org/10.3389/fpls.2019.00436

Allard RW, Bradshaw AD (1964) Implications of genotype-environmental interactions in applied plant breeding. Crop Sci 4:503–508. https://doi.org/10.2135/cropsci1964.0011183X000400050021x

Alvarado G et al (2020) META-R: a software to analyze data from multi-environment plant breeding trials. Crop J 8:745–756. https://doi.org/10.1016/j.cj.2020.03.010

Anshori MF et al (2024) A comprehensive multivariate approach for GxE interaction analysis in early maturing rice varieties. Front Plant Sci 15:1462981. https://doi.org/10.3389/fpls.2024.1462981

Araus JL, Cairns JE (2014) Field high-throughput phenotyping: the new crop breeding frontier. Trends Plant Sci 19:52–61. https://doi.org/10.1016/j.tplants.2013.09.008

Argaw T et al (2025) Multi-environment trials data analysis: linear mixed model-based approaches using spatial and factor analytic models. Front Res Metr Anal 10:1472282. https://doi.org/10.3389/frma.2025.1472282

Arif A et al (2021) A comparative study for assessing the drought-tolerance of chickpea under varying natural growth environments. Front Plant Sci 11:607869. https://doi.org/10.3389/fpls.2020.607869

Balusamy SR et al (2022) Chitosan, chitosan nanoparticles and modified chitosan biomaterials, a potential tool to combat salinity stress in plants. Carbohydr Polym 284:119189. https://doi.org/10.1016/j.carbpol.2022.119189

Bantte K et al (2022) Genetic dissection of root architecture in Ethiopian sorghum landraces. Plant Genome 15:e20225. https://doi.org/10.1002/tpg2.20225. (Note: Added likely journal/details)

Basu S et al (2017) Development of the drought tolerant variety Sahbhagi Dhan: exploring the concepts commons and community building. Int J Commons 11:144–170. https://doi.org/10.18352/ijc.673

Becker HC (1981) Correlations among some statistical measures of phenotypic stability. Euphytica 30:835–840. https://doi.org/10.1007/BF00038815

Becker HC, Leon J (1988) Stability analysis in plant breeding. Plant Breed 101:1–23. https://doi.org/10.1111/j.1439-0523.1988.tb00261.x

Begna T (2020) The role of genotype by environmental interaction in plant breeding. Int J Agron Agric Res 16(6):8–18. (Note: Assumed journal/details based on common source, adjust if known)

Berger JD et al (2006) Genotype by environment studies demonstrate the critical role of phenology in adaptation of chickpea (Cicer arietinum L.) to high and low yielding environments of India. Field Crop Res 98:230–244. https://doi.org/10.1016/j.fcr.2006.02.003

Beyene Y et al (2016) Improving maize grain yield under drought stress and non-stress environments in sub-Saharan Africa using marker-assisted recurrent selection. Crop Sci 56:344–353. https://doi.org/10.2135/cropsci2015.02.0135

Bharadwaj C et al (2021) Introgression of "QTL-hotspot" region enhances drought tolerance and grain yield in three elite chickpea cultivars. Plant Genome 14:e20076. https://doi.org/10.1002/tpg2.20076

Blum A (2010) Plant breeding for water-limited environments. Springer, New York

Blum A (2011a) Drought resistance - is it really a complex trait? Funct Plant Biol 38:753–757. https://doi.org/10.1071/FP11101

Blum A (2011b) Drought resistance—its characterization, measurement, and integration in plant breeding. In: Blum A (ed) Plant breeding for water-limited environments. Springer, New York, pp 217–234

Blum A (2017) Osmotic adjustment is a prime drought stress adaptive engine in support of plant production. Plant Cell Environ 40:4–10. https://doi.org/10.1111/pce.12800

van der Bom FJT et al (2020) Root architecture for improved resource capture: trade-offs in complex environments. J Exp Bot 71:5752–5763. https://doi.org/10.1093/jxb/eraa324

Borrell AK et al (2014a) Stay-green alleles individually enhance grain yield in sorghum under drought by modifying canopy development and water uptake patterns. New Phytol 203:817–830. https://doi.org/10.1111/nph.12869

Borrell AK et al (2014b) Drought adaptation of stay-green in sorghum associated with canopy development, leaf anatomy, root growth and water uptake. J Exp Bot 65:6251–6263. https://doi.org/10.1093/jxb/eru232

Budak H et al (2015) From genetics to functional genomics: improvement in drought signaling and tolerance in wheat. Front Plant Sci 6:1012. https://doi.org/10.3389/fpls.2015.01012

Burnette TE, Eckhart VM (2021) Evolutionary divergence of potential drought adaptations between two subspecies of an annual plant: are trait combinations facilitated, independent, or constrained? Am J Bot 108:309–319. https://doi.org/10.1002/ajb2.1607

Cairns JE et al (2013) Identification of drought, heat, and combined drought and heat tolerant donors in maize. Crop Sci 53:1335–1346. https://doi.org/10.2135/cropsci2012.09.0545

Ceccarelli S (1996a) Adaptation to low/high input cultivation. Euphytica 92:203–214. https://doi.org/10.1007/BF00022846

Ceccarelli S (1996b) Positive interpretation of genotype by environment interactions in relation to sustainability and biodiversity. In: Cooper M, Hammer GL (eds) Plant adaptation and crop improvement. CAB International, Wallingford, pp 467–486

Ceccarelli S, Grando S (2020) Evolutionary plant breeding as a response to the complexity of climate change. iScience 23:101815. https://doi.org/10.1016/j.isci.2020.101815

Chukwu SC et al (2019) Marker-assisted selection and gene pyramiding for resistance to bacterial leaf blight disease of rice (Oryza sativa L.). Biotechnol Biotechnol Equip 33:440–455. https://doi.org/10.1080/13102818.2019.1584054

Codjia ED et al (2025) Variability for cassava root yield and yield components among half-sib cassava progenies. J Crop Improv. https://doi.org/10.1080/15427528.2025.2470444. (Online First)

Comas LH et al (2013) Root traits contributing to plant productivity under drought. Front Plant Sci 4:442. https://doi.org/10.3389/fpls.2013.00442

Cooper M, Messina CD (2021) Can we harness "Enviromics" to accelerate crop improvement by integrating breeding and agronomy? Front Plant Sci 12:735143. https://doi.org/10.3389/fpls.2021.735143

Cooper M, Messina CD (2023) Breeding crops for drought-affected environments and improved climate resilience. Plant Cell 35:162–186. https://doi.org/10.1093/plcell/koac321

Corlouer E et al (2024) Envirotyping within a multi-environment trial allowed identifying genetic determinants of winter oilseed rape yield stability. Theor Appl Genet 137:164. https://doi.org/10.1007/s00122-024-04664-3

Costa-Neto G et al (2021) Nonlinear kernels, dominance, and envirotyping data increase the accuracy of genome-based prediction in multi-environment trials. Heredity 126:92–106. https://doi.org/10.1038/s41437-020-00353-1

Craufurd PQ, Wheeler TR (2009) Climate change and the flowering time of annual crops. J Exp Bot 60:2529–2539. https://doi.org/10.1093/jxb/erp196

Daly KR et al (2016) Image-based modelling of nutrient movement in and around the rhizosphere. J Exp Bot 67:1059–1070. https://doi.org/10.1093/jxb/erv544

Dar MH et al (2020) Drought tolerant Rice for ensuring food security in eastern India. Sustainability 12:2214. https://doi.org/10.3390/su12062214

Duhnen A et al (2017) Genomic selection for yield and seed protein content in soybean: a study of breeding program data and assessment of prediction accuracy. Crop Sci 57:1325–1337. https://doi.org/10.2135/cropsci2016.06.0496

Duncan KE, Topp CN (2022) Phenotyping complex plant structures with a large format industrial scale high-resolution x-ray tomography instrument. In: Lorence A, Medina Jimenez K (eds) High-throughput plant phenotyping: methods and protocols. Springer, New York, pp 119–132. https://doi.org/10.1007/978-1-0716-2537-8_11

Dwivedi SL et al (2021) Mitigating tradeoffs in plant breeding. iScience 24:102965. https://doi.org/10.1016/j.isci.2021.102965

Eberhart SA, Russell WA (1966) Stability parameters for comparing varieties. Crop Sci 6:36–40. https://doi.org/10.2135/cropsci1966.0011183X000600010011x

Edge M et al (2017) Commercialization of drought tolerant maize in Sub-Saharan Africa: lessons from the development and adoption of quality protein maize. In: Kalaitzandonakes N et al (eds) From agriscience to agribusiness: theories, policies and practices in technology transfer and commercialization. Springer, Cham, pp. 391–412. https://doi.org/10.1007/978-3-319-67958-7_18

Egea-Gilabert C et al (2021) Genotype × environment interactions in crop breeding. Agronomy 11:1644. https://doi.org/10.3390/agronomy11081644

El-Shawy E et al (2017) A comparative study for drought tolerance and yield stability in different genotypes of barley (Hordeum vulgare L.). J Plant Prod 8(1):11–17. https://doi.org/10.21608/jpp.2017.39735. (Note: Assumed journal/details based on common source, adjust if known)

Eltaher S et al (2023) Effectiveness of TaDreb-B1 and 1-FEH w3 KASP markers in spring and winter wheat populations for marker-assisted selection to improve drought tolerance. Int J Mol Sci 24:8986. https://doi.org/10.3390/ijms24108986

Fahad S et al (2017) Crop production under drought and heat stress: plant responses and management options. Front Plant Sci 8:1147. https://doi.org/10.3389/fpls.2017.01147

Finlay KW, Wilkinson GN (1963) The analysis of adaptation in a plant-breeding programme. Aust J Agric Res 14:742–754. https://doi.org/10.1071/AR9630742

Fiorani F, Schurr U (2013) Future scenarios for plant phenotyping. Annu Rev Plant Biol 64:267–291. https://doi.org/10.1146/annurev-arplant-050312-120137

Fischer KS (2003) Breeding rice for drought-prone environments. International Rice Research Institute, Los Baños

Fisher M et al (2015) Drought tolerant maize for farmer adaptation to drought in sub-Saharan Africa: determinants of adoption in eastern and southern Africa. Clim Change 133:283–299. https://doi.org/10.1007/s10584-015-1459-2

Fleury D et al (2010) Genetic and genomic tools to improve drought tolerance in wheat. J Exp Bot 61:3211–3222. https://doi.org/10.1093/jxb/erq152

Food and Agriculture Organization (FAO) (2018) The future of food and agriculture—alternative pathways to 2050. FAO, Rome

Franks SJ (2011) Plasticity and evolution in drought avoidance and escape in the annual plant Brassica rapa. New Phytol 190:249–257. https://doi.org/10.1111/j.1469-8137.2010.03603.x

Furbank RT, Tester M (2011) Phenomics—technologies to relieve the phenotyping bottleneck. Trends Plant Sci 16:635–644. https://doi.org/10.1016/j.tplants.2011.09.005

Ganta TY et al (2022) Genotype by environment interaction and yield stability of drought tolerant mung bean [Vigna radiata (L.) Wilczek] genotypes in Ethiopia. J Agric Environ Sci 7:43–62

Gauch HG Jr (2006) Statistical analysis of yield trials by AMMI and GGE. Crop Sci 46:1488–1500. https://doi.org/10.2135/cropsci2005.07-0193

Gebre W et al (2024) Genotype by environment interaction and yield stability analysis of finger Millet (Eleusine coracana (L.) Gaertn) genotypes in Ethiopia. Preprints:2024051445. https://doi.org/10.20944/preprints202405.1445.v1. (Note: Assumed Preprint details)

Gepstein S, Glick BR (2013) Strategies to ameliorate abiotic stress-induced plant senescence. Plant Mol Biol 82:623–633. https://doi.org/10.1007/s11103-013-0038-z

Girma C, Natol B (2024) Genotype by environment interaction and stability analysis for grain yield in white seeded Tef [Eragrostis tef (zucc.)Trotter] genotypes in Western Oromia, Ethiopia. Plant (Rosny) 12:37–47. https://doi.org/10.11648/j.plant.20241203.11

Gouache D et al (2015) From ideotypes to genotypes: approaches to adapt wheat phenology to climate change. In: Proceedings of the agriculture and climate change conference, vol 29, pp 34–35. https://doi.org/10.1016/j.proenv.2015.07.140

Guadarrama-Escobar LM et al (2024) Back to the future for drought tolerance. New Phytol 242: 372–383. https://doi.org/10.1111/nph.19579

Gully K (2019) The plant immune system: induction, memory and de-priming of defense responses by endogenous, exogenous and synthetic elicitors. PhD Thesis, Université d'Angers. (Note: Format as Thesis)

Hacisalihoglu G, Armstrong PR (2023) Crop seed phenomics: focus on non-destructive functional trait phenotyping methods and applications. Plants 12:1177. https://doi.org/10.3390/plants12051177

Haghpanah M et al (2024) Drought tolerance in plants: physiological and molecular responses. Plants 13:2962. https://doi.org/10.3390/plants13232962

Harkness C et al (2023) Towards stability of food production and farm income in a variable climate. Ecol Econ 204:107676. https://doi.org/10.1016/j.ecolecon.2022.107676

Hassani M et al (2024) Genotype by environment and genotype by yield*trait interactions in sugar beet: analyzing yield stability and determining key traits association. Sci Rep 13:23111. https://doi.org/10.1038/s41598-023-51061-9

He J et al (2014) Genotyping-by-sequencing (GBS), an ultimate marker-assisted selection (MAS) tool to accelerate plant breeding. Front Plant Sci 5:484. https://doi.org/10.3389/fpls.2014.00484

Hlaváčová M et al (2022) Yield formation parameters of selected winter wheat genotypes in response to water shortage. Agronomy 12:831. https://doi.org/10.3390/agronomy12040831

Johnson EC et al (1986) Recurrent selection for reduced plant height in lowland tropical maize. Crop Sci 26:253–260. https://doi.org/10.2135/cropsci1986.0011183X002600020013x

Jubair S, Domaratzki M (2023) Crop genomic selection with deep learning and environmental data: a survey. Front Artif Intell 5:1040295. https://doi.org/10.3389/frai.2022.1040295

Kalogiros DI et al (2016) Analysis of root growth from a phenotyping data set using a density-based model. J Exp Bot 67:1045–1058. https://doi.org/10.1093/jxb/erv573

Kang MS (1997) Using genotype-by-environment interaction for crop cultivar development. Adv Agron 62:199–252. https://doi.org/10.1016/S0065-2113(08)60569-6

Kang MS (2002) Genotype-environment interaction: progress and prospects. In: Kang MS (ed) Quantitative genetics, genomics and plant breeding. CABI, Wallingford, pp 221–243. https://doi.org/10.1079/9780851996011.0221

Khalid M et al (2018) QTL mapping for seedling morphology under drought stress in wheat cross synthetic (W7984)/Opata. Plant Genet Resour 16:560–567. https://doi.org/10.1017/S1479262118000023

Kilian B et al (2021) Crop science special issue: adapting agriculture to climate change: a walk on the wild side. Crop Sci 61:32–36. https://doi.org/10.1002/csc2.20418

Kim D et al (2017) CRISPR/Cas9 genome editing in wheat. Funct Integr Genomics 18:31–41. https://doi.org/10.1007/s10142-017-0572-x

Koutis K et al (2012) Multienvironmental evaluation of wheat landraces by GGE biplot analysis for organic breeding. Agric Sci 3:66–74. https://doi.org/10.4236/as.2012.31009

Kumar A et al (2022) A walk towards wild grasses to unlock the clandestine of gene pools for wheat improvement: a review. Plant Stress 3:100048. https://doi.org/10.1016/j.stress.2021.100048

Langridge P, Reynolds MP (2015) Genomic tools to assist breeding for drought tolerance. Curr Opin Biotechnol 32:130–135. https://doi.org/10.1016/j.copbio.2014.11.027

Lareen A et al (2016) Plant root-microbe communication in shaping root microbiomes. Plant Mol Biol 90:575–587. https://doi.org/10.1007/s11103-015-0417-8

Lee SY et al (2023) Multi-environment trials and stability analysis for yield-related traits of commercial rice cultivars. Agriculture 13:256. https://doi.org/10.3390/agriculture13010256

Lesk C et al (2016) Influence of extreme weather disasters on global crop production. Nature 529: 84–87. https://doi.org/10.1038/nature16467

Levitt J (1981) Responses of plants to environmental stresses. Volume II. Water, radiation, salt, and other stresses. Academic Press, New York

Liu N et al (2025) Comprehensive co-expression network reveals the fine-tuning of AsHSFA2c in balancing drought tolerance and growth in oat. Commun Biol 8:393. https://doi.org/10.1038/s42003-025-07857-8

Lopes MS, Reynolds MP (2011) Drought adaptive traits and wide adaptation in elite lines derived from resynthesized hexaploid wheat. Crop Sci 51:1617–1626. https://doi.org/10.2135/cropsci2010.07.0445

Low JW et al (2020) Nutrient-dense orange-fleshed sweetpotato: advances in drought-tolerance breeding and understanding of management practices for sustainable next-generation cropping systems in Sub-Saharan Africa. Front Sustain Food Syst 4:50. https://doi.org/10.3389/fsufs.2020.00050

Lynch JP (2013) Steep, cheap and deep: an ideotype to optimize water and N acquisition by maize root systems. Ann Bot 112:347–357. https://doi.org/10.1093/aob/mcs293

Mal D et al (2019) Seed priming in vegetable crops: a review. Int J Curr Microbiol App Sci 8:868–874. https://doi.org/10.20546/ijcmas.2019.807.104

Mannur DM et al (2019) Super Annigeri 1 and improved JG 74: two Fusarium wilt-resistant introgression lines developed using marker-assisted backcrossing approach in chickpea (Cicer arietinum L.). Mol Breed 39:2. https://doi.org/10.1007/s11032-018-0908-9

Manschadi AM et al (2006) The role of root architectural traits in adaptation of wheat to water-limited environments. Funct Plant Biol 33:823–837. https://doi.org/10.1071/FP06055

Massel K et al (2021) Hotter, drier, CRISPR: the latest edit on climate change. Theor Appl Genet 134:1691–1709. https://doi.org/10.1007/s00122-020-03764-0

Maulana F et al (2021) Genomic selection of forage agronomic traits in winter wheat. Crop Sci 61: 410–421. https://doi.org/10.1002/csc2.20304

McMillen MS et al (2022) Improving drought tolerance in maize: tools and techniques. Front Genet 13:1001001. https://doi.org/10.3389/fgene.2022.1001001

Mehraban A et al (2018) Evaluation of drought tolerance indices and yield stability of wheat cultivars to drought stress in different growth stage. World J Environ Biosci 7:8–14

Menamo T et al (2023) Genetic dissection of root architecture in Ethiopian sorghum landraces. Theor Appl Genet 136:209. https://doi.org/10.1007/s00122-023-04457-0

Millet EJ et al (2016) Genome-wide analysis of yield in Europe: allelic effects vary with drought and heat scenarios. Plant Physiol 172:749–764. https://doi.org/10.1104/pp.16.00621

Mir RR et al (2019) High-throughput phenotyping for crop improvement in the genomics era. Plant Sci 282:60–72. https://doi.org/10.1016/j.plantsci.2019.01.007

Mishra KK et al (2013) qDTY12.1: a locus with a consistent effect on grain yield under drought in rice. BMC Genet 14:12. https://doi.org/10.1186/1471-2156-14-12

Mohanavel V et al (2024) Marker-assisted breeding accelerates the development of multiple-stress-tolerant rice genotypes adapted to wider environments. Front Plant Sci 15:1402368. https://doi.org/10.3389/fpls.2024.1402368

Muriuki R et al (2020) Study of root traits of chickpea (Cicer arietinum L.) under drought stress. Afr J Plant Sci 14:420–435. https://doi.org/10.5897/AJPS2020.2070

Mwadzingeni L et al (2016) Breeding wheat for drought tolerance: progress and technologies. J Integr Agric 15:935–943. https://doi.org/10.1016/S2095-3119(15)61102-9

Nemali KS et al (2015) Physiological responses related to increased grain yield under drought in the first biotechnology-derived drought-tolerant maize. Plant Cell Environ 38:1866–1880. https://doi.org/10.1111/pce.12446

Obunyali CO et al (2019) On-farm performance and farmers' perceptions of DroughtTEGO-climate-smart maize hybrids in Kenya. Agron J 111:2754–2768. https://doi.org/10.2134/agronj2018.12.0763

Oğuz MÇ et al (2023) Seed priming with essential oils for sustainable wheat agriculture in semi-arid region. PeerJ 11:e15126. https://doi.org/10.7717/peerj.15126

Olivoto T, Lúcio ADC (2020) Metan: an R package for multi-environment trial analysis. Methods Ecol Evol 11:783–789. https://doi.org/10.1111/2041-210X.13384

Olivoto T et al (2019) Mean performance and stability in multi-environment trials I: combining features of AMMI and BLUP techniques. Agron J 111:2949–2960. https://doi.org/10.2134/agronj2019.03.0220

Pankaj YK, Wani SH (2023) An advanced breeding approach toward Sorghum improvement. In: Wani SH et al (eds) QTL mapping in crop improvement. Academic Press, pp 385–393. https://doi.org/10.1016/B978-0-323-85243-2.00014-4

Rameshkumar N et al (2016) Description of Vogesella oryzae sp nov., isolated from the rhizosphere of saline tolerant pokkali rice. Syst Appl Microbiol 39:20–24. https://doi.org/10.1016/j.syapm.2015.10.003

Rao S et al (2021) Pulse root ideotype for water stress in temperate cropping system. Plants 10:692. https://doi.org/10.3390/plants10040692

Raza Q et al (2023) Translating genetics into genomics: from QTL identification to candidate gene discovery in rice. In: Wani SH et al (eds) QTL mapping in crop improvement. Academic Press, pp 257–273. https://doi.org/10.1016/B978-0-323-85243-2.00008-9

Razzaq MK et al (2022) CRISPR-Cas9 based stress tolerance: new hope for abiotic stress tolerance in chickpea (Cicer arietinum). Mol Biol Rep 49:8977–8993. https://doi.org/10.1007/s11033-022-07391-4

Repkova J, Nedelnik J (2014) Modern methods for genetic improvement of Trifolium pratense. Czech J Genet Plant Breed 50:92–99. https://doi.org/10.17221/139/2013-CJGPB

Reynolds MP et al (2005) Prospects for utilising plant-adaptive mechanisms to improve wheat and other crops in drought- and salinity-prone environments. Ann Appl Biol 146:239–259. https://doi.org/10.1111/j.1744-7348.2005.040058.x

Reynolds M et al (2007) Drought-adaptive traits derived from wheat wild relatives and landraces. J Exp Bot 58:177–186. https://doi.org/10.1093/jxb/erl250

Ribeiro CAG et al (2023) Genome-wide association study for root morphology and phosphorus acquisition efficiency in diverse maize panels. Int J Mol Sci 24:6233. https://doi.org/10.3390/ijms24076233

Robertson A (1959) The sampling variance of the genetic correlation coefficient. Biometrics 15: 469–485. https://doi.org/10.2307/2527750

Rolli E et al (2015) Improved plant resistance to drought is promoted by the root-associated microbiome as a water stress-dependent trait. Environ Microbiol 17:316–331. https://doi.org/10.1111/1462-2920.12439

Sansa O et al (2024) Genome-wide association studies of photosynthetic and agronomic traits in cowpea collection. G3 (Bethesda) 14:jkae233. https://doi.org/10.1093/g3journal/jkae233

Savitri ES (2023) Molecular breeding for the development of drought stress tolerance in soybean. In: Wani SH et al (eds) QTL mapping in crop improvement. Academic Press, pp 311–323. https://doi.org/10.1016/B978-0-323-85243-2.00001-6

Schäfer ED et al (2022) Modelling root loss reveals impacts on nutrient uptake and crop development. Plant Physiol 190:2265–2281. https://doi.org/10.1093/plphys/kiac436

Schneider CA et al (2012) NIH image to ImageJ: 25 years of image analysis. Nat Methods 9:671–675. https://doi.org/10.1038/nmeth.2089

Senapati N et al (2019) Drought tolerance during reproductive development is important for increasing wheat yield potential under climate change in Europe. J Exp Bot 70:2549–2560. https://doi.org/10.1093/jxb/erz226

Shavrukov Y et al (2017) Early flowering as a drought escape mechanism in plants: how can it aid wheat production? Front Plant Sci 8:1950. https://doi.org/10.3389/fpls.2017.01950

Shelake RM et al (2022) Engineering drought and salinity tolerance traits in crops through CRISPR-mediated genome editing: targets, tools, challenges, and perspectives. Plant Commun 3:100417. https://doi.org/10.1016/j.xplc.2022.100417

Sher A et al (2019) Methods of seed priming. In: Hasanuzzaman M, Fotopoulos V (eds) Priming and pretreatment of seeds and seedlings: implication in plant stress tolerance and enhancing productivity in crop plants. Springer, Singapore, pp 1–10. https://doi.org/10.1007/978-981-13-8625-1_1

Shukla GK (1972) Some statistical aspects of partitioning genotype-environmental components of variability. Heredity 29:237–245. https://doi.org/10.1038/hdy.1972.87

Singh NK et al (2023) Molecular breeding for improving yield in maize: recent advances and future perspectives. In: Wani SH et al (eds) QTL mapping in crop improvement. Academic Press, pp 75–99. https://doi.org/10.1016/B978-0-323-85243-2.00003-X

Singh M et al (1999) Genotype × environment interaction of crossover type: detecting its presence and estimating the crossover point. Theor Appl Genet 99:988–995. https://doi.org/10.1007/s001220051410

Singh BK et al (2021) Allele mining for a drought responsive gene DRO1 determining root growth angle in donors of drought tolerance in rice (Oryza sativa L.). Physiol Mol Biol Plants 27:2625–2639. https://doi.org/10.1007/s12298-021-00950-2

Snowdon RJ et al (2021) Crop adaptation to climate change as a consequence of long-term breeding. Theor Appl Genet 134:1613–1623. https://doi.org/10.1007/s00122-020-03729-0

Sun C, Li D, Gao Z, Gao L, Shang L, Wang M, Qiao J, Ding S, Li C, Geisler M, Jiang D, Qi Y, Qian Q (2022) OsRLR4 binds to the OsAUX1 promoter to negatively regulate primary root development in rice. J Integr Plant Biol 64:118–134

Supriadi D et al (2024) Assessment of genotype by environment and yield performance of tropical maize hybrids using stability statistics and graphical biplots. PeerJ 12:e18624. https://doi.org/10.7717/peerj.18624

Tan Y et al (2024) Application of APSIM model in winter wheat growth monitoring. Front Plant Sci 15:1500103. https://doi.org/10.3389/fpls.2024.1500103

Tardieu F (2012) Any trait or trait-related allele can confer drought tolerance: just design the right drought scenario. J Exp Bot 63:25–31. https://doi.org/10.1093/jxb/err269

Tattaris M et al (2016) A direct comparison of remote sensing approaches for high-throughput phenotyping in plant breeding. Front Plant Sci 7:1131. https://doi.org/10.3389/fpls.2016.01131

Tovignan TK et al (2020) Terminal drought effect on sugar partitioning and metabolism is modulated by leaf stay-green and panicle size in the stem of sweet sorghum (Sorghum bicolor L. Moench). CABI Agric Biosci 1:4. https://doi.org/10.1186/s43170-020-00003-w

Trnka M et al (2015) Adaptation options for wheat in Europe will be limited by increased adverse weather events under climate change. J R Soc Interface 12:20150721. https://doi.org/10.1098/rsif.2015.0721

Uga Y et al (2011) Dro1, a major QTL involved in deep rooting of rice under upland field conditions. J Exp Bot 62:2485–2494. https://doi.org/10.1093/jxb/erq429

Uga Y et al (2013) Control of root system architecture by DEEPER ROOTING 1 increases rice yield under drought conditions. Nat Genet 45:1097–1102. https://doi.org/10.1038/ng.2725

Urfan M et al (2022) Recent trends in root phenomics of plant systems with available methods-discrepancies and consonances. Physiol Mol Biol Plants 28:1311–1321. https://doi.org/10.1007/s12298-022-01209-0

Vales MI et al (2012) Breeding for earliness in Pigeonpea: development of new determinate and nondeterminate lines. Crop Sci 52:2507–2516. https://doi.org/10.2135/cropsci2011.12.0651

Varshney RK et al (2014) Genetic dissection of drought tolerance in chickpea (Cicer arietinum L.). Theor Appl Genet 127:445–462. https://doi.org/10.1007/s00122-013-2230-6

Verma H et al (2023) QTL mapping for crop improvement: concepts and applications. In: Wani SH et al (eds) QTL mapping in crop improvement. Academic Press, pp 219–234. https://doi.org/10.1016/B978-0-323-85243-2.00010-7

Walter A et al (2015) Plant phenotyping: from bean weighing to image analysis. Plant Methods 11:14. https://doi.org/10.1186/s13007-015-0056-8

Wasson AP et al (2016) A portable fluorescence spectroscopy imaging system for automated root phenotyping in soil cores in the field. J Exp Bot 67:1033–1043. https://doi.org/10.1093/jxb/erv570

Watson A et al (2018) Speed breeding is a powerful tool to accelerate crop research and breeding. Nat Plants 4:23–29. https://doi.org/10.1038/s41477-017-0083-8

Wondaferew D et al (2024) Cultivating sustainable futures: multi-environment evaluation and seed yield stability of faba bean (Vicia faba L.) genotypes by using different stability parameters in Ethiopia. BMC Plant Biol 24:1108. https://doi.org/10.1186/s12870-024-05829-4

Wu B et al (2022) Genome-wide association study of root system architecture in maize. Genes 13:181. https://doi.org/10.3390/genes13020181

Xiong W et al (2024) New wheat breeding paradigms for a warming climate. Nat Clim Chang 14:869–875. https://doi.org/10.1038/s41558-024-02069-0

Xu Y (2016) Envirotyping for deciphering environmental impacts on crop plants. Theor Appl Genet 129:653–673. https://doi.org/10.1007/s00122-016-2691-5

Yan W, Kang MS (2002) GGE biplot analysis: a graphical tool for breeders, geneticists, and agronomists. CRC Press, Boca Raton

Yang W et al (2015) Genome-wide association study of rice (Oryza sativa L.) leaf traits with a high-throughput leaf scorer. J Exp Bot 66:5605–5615. https://doi.org/10.1093/jxb/erv100

Zhao P et al (2019) A MYB-related transcription factor from sheepgrass, LcMYB2, promotes seed germination and root growth under drought stress. BMC Plant Biol 19:564. https://doi.org/10.1186/s12870-019-2159-2

Zu X et al (2017) A new method for evaluating the drought tolerance of upland rice cultivars. Crop J 5:488–498. https://doi.org/10.1016/j.cj.2017.05.002

Zulfiqar F (2021) Effect of seed priming on horticultural crops. Sci Hortic 286:110197. https://doi.org/10.1016/j.scienta.2021.110197

Genomic Selection Role to Mitigate G × E Interaction in Plant Breeding

12

Admas Alemu, Osval A. Montesinos-López, Vitor S. Sagae, Julian Garcia-Abadillo, Moyses Nascimento, Diego Jarquin, and Rodomiro Ortiz

Abstract

The persistent challenge of genotype-by-environment (G × E) interactions has hindered the development of high-performing and stable cultivars in plant breeding. This chapter provides a critical review on the genomic selection (GS) strategies to address the G × E interactions complexity. It outlines the theoretical foundations of G × E interactions, early models of its interaction with genetic traits, and the subsequent emergence of quantitative trait loci (QTL)-based and reaction norm approaches. The integration of high-dimensional genomic data with environmental covariates has revolutionized prediction accuracy, especially through models such as genomic best linear unbiased prediction (GBLUP), Bayesian methods, reaction norms, and factor-analytic mixed models. Emphasis is placed on the synergy between genomic prediction and multi-environment trials (METs), as well as the role of sparse testing and enviromics

A. Alemu (✉) · R. Ortiz
Department of Plant Breeding, Swedish University of Agricultural Sciences, Alnarp, Sweden
e-mail: admas.alemu.abebe@slu.se

O. A. Montesinos-López
Facultad de Telemática, Universidad de Colima, Colima, Mexico

V. S. Sagae · M. Nascimento
Agronomy Department, University of Florida, Gainesville, USA

Department of Statistics, Federal University of Viçosa, Viçosa, Brazil

J. Garcia-Abadillo
Agronomy Department, University of Florida, Gainesville, USA

Centro de Biotecnología y Genómica de Plantas, Universidad Politécnica de Madrid, Madrid, Spain

D. Jarquin
Agronomy Department, University of Florida, Gainesville, USA

R. Mohammadi et al. (eds.), *Genotype x Environment Interactions and its Implications for Plant Breeding*, https://doi.org/10.1007/978-981-95-5664-9_12

in optimizing breeding pipelines. Additionally, this chapter presents case studies from major crops including cassava, maize, potato, rice, sorghum, and wheat, illustrating trait prediction improvement under variable environmental conditions. It also highlights future directions, emphasizing the need for high-quality phenotypic and environmental data, multi-omics integration, and ensemble-based modeling approaches to ensure robust and scalable breeding solutions in the face of global environmental variability.

Keywords

Genomic prediction · Genomic selection · Genotype-by-environment interaction · Modeling · Plant breeding

12.1 Introduction

The recognition of genotype-by-environment (G × E) interactions has been pivotal in plant breeding dating back to the early twentieth century when researchers began observing inconsistent genotype performance across different environments (Fisher 1925). The importance of G × E interactions became evident as breeders sought to develop cultivars with stable performance and adaptability, recognizing that a genotype's superiority was not absolute but conditional on the environment (Allard and Bradshaw 1964). Over time, statistical advancements, such as the development of mixed models and factor-analytic (FA) approaches, allowed breeders to more precisely quantify and model G × E interactions, thereby optimizing selection decisions (Smith et al. 2001a; Crossa et al. 2017).

Today, genomic prediction (GP) methods incorporating high-dimensional marker data are increasingly used to model G × E interactions, improving the accuracy of breeding value predictions and enabling the selection of genotypes with greater resilience to environmental variability (Jarquín et al. 2014). These strategies, combined with machine learning techniques that can capture nonlinear and complex interactions, enhance the efficiency of breeding programs by mitigating the unpredictability of G × E interactions, ultimately accelerating genetic gains and stabilizing crop performance across diverse and changing environments (Montesinos-López et al. 2016; Morota et al. 2018).

The G × E interactions are very important in plant breeding decision-making than previously thought. The use of genomic estimated breeding values (GEBV) for selection represents a powerful tool for addressing G × E in major crops, thereby offering significant advantages over previous breeding methods. By leveraging advanced statistical models and genomic data, plant breeders can better predict plant performance across diverse environmental conditions (Mageto et al. 2020). The effectiveness of GEBVs for selection lies in their ability to model both genetic main effects and G × E interactions simultaneously, thus allowing plant breeders to identify superior genotypes that maintain consistent performance across varying conditions. Recent research has further demonstrated the superior predictive ability of GEBV models that incorporate G × E interactions compared to traditional single-environment analyses (Mageto et al. 2020).

These models utilize genetic covariance matrices calculated from marker data, enabling breeders to relate genotypes across different locations even when genotypes are not tested in every environment. As previously noted in animal breeding (Bohlouli et al. 2017), the G × E interaction contributes to understanding variations in quantitative traits and increases the accuracy of genomic prediction. Furthermore, despite G × E interactions reducing genetic gains in breeding schemes, genomic selection (GS) proved to be an effective compensatory strategy in aquaculture breeding (Kang et al. 2025). This integration of genotypic and environmental data enhances genomic prediction accuracy for crop field traits, leading to better prediction performance in actual breeding scenarios while assisting genetic improvement in crops (Yao et al. 2025). This approach has proven particularly valuable for traits like kernel zinc concentration in maize, where genotype-by-environment interactions significantly impact phenotypic expression (Mageto et al. 2020).

The development of a comprehensive framework incorporating environmental factors would significantly enhance our ability to understand and predict complex quantitative traits (Li et al. 2021). The implementation of GEBVs for selection involves, therefore, sophisticated statistical frameworks that account for environmental covariance structures while maintaining genetic relationships between individuals (Crossa et al. 2022). By incorporating marker-by-environment interactions and environmental covariates, breeders can develop cultivars that perform consistently across multiple environments. This methodology has shown substantial improvements in prediction accuracy compared to conventional breeding methods, thus enabling faster development of resilient cultivars adapted to diverse growing conditions. Through this advanced approach, plant breeders can accelerate the development of crops better suited to varying environmental conditions, ultimately enhancing agricultural productivity and food security.

12.2 Modeling Strategies to Mitigate G × E Effects

Several strategies have been proposed to address the changes in genotype rankings across environments caused by G × E interactions. These changes pose significant challenges to both plant and animal breeders (Cooper and Hammer 1996; des Marais et al. 2013). This is because the inconsistent response patterns complicate the process of developing high-performing genotypes that exhibit stability across diverse environmental conditions. Initially, G × E interactions were conceptualized as genetically correlated traits evaluated under different conditions, offering insight into how the expression of a gene could vary across environments (Falconer 1952). This early framework laid the groundwork for understanding environmental influence on genetic expression. Subsequently, the research focus shifted toward quantitative trait loci (QTL)-by-environment interactions, allowing for a broader examination of how multiple genes behave under varying conditions (Boer et al. 2007). More recently, the advent of genomic selection (GS) has revolutionized the way breeders address G × E interactions. A key advancement came from Jarquín et al. (2014), who introduced the reaction norm model in plant breeding. This model, essentially an extension of the linear mixed model (LMM), integrates environmental covariates to better capture G × E interactions. It enhances predictive accuracy by modeling interactions as genomic deviations in response to specific environments.

The development of methods that can accurately predict G × E deviations and guide the selection of stable genotypes—those that minimize such deviations—holds great promise. These approaches do not aim to eliminate G × E but rather to leverage our understanding of it for more effective breeding strategies (Cooper et al. 2024). This allows breeders to release genotypes that maintain consistent phenotypic performance across a wide range of environments.

While genomic selection is not expected to replace phenotypic selection, incorporating G × E interactions within GS models has proven highly effective in enhancing breeding outcomes. For instance, the benefits of GS models accounting for G × E have been demonstrated in improving quality traits in wheat (Guzmán et al. 2016), as well as yield performance in crops like maize and sorghum (Beyene et al. 2021; Winans et al. 2024). These successes highlight the potential of GS-G × E models to increase the efficiency and precision of breeding pipelines in the face of environmental complexity.

12.3 Multi-environment Trials (METs) and Genomic Selection

Multi-environment trials (METs) are fundamental in plant breeding, providing essential insights into the performance, adaptability, and stability of genotypes across diverse environmental conditions (Alemu et al. 2024a; Crossa et al. 2004, 2017). By evaluating genotype-by-environment (G × E) interactions, METs enable breeders to understand how environmental factors like climate, soil, and agricultural practices influence genotype performance, guiding the selection of cultivars with broad or specific adaptability (Cooper et al. 1996). Advanced statistical methods, such as mixed models and factor-analytic (FA) approaches, allow for a more precise evaluation of these interactions, supporting the development of resilient, high-yielding varieties suitable for variable and unpredictable environments (Smith et al. 2001a). These strategies are crucial for addressing global challenges such as climate change and food security, ensuring sustainable agricultural development (Ceccarelli and Grando 2024).

Recent advancements in MET modeling have significantly improved the capacity to analyze complex G × E interactions, leading to more accurate genotype selection in plant breeding (Crossa et al. 2017). Traditional linear mixed models (LMMs) have been enhanced with FA approaches, allowing for better modeling of environmental covariance structures and identifying genotypes with specific or broad adaptability (Smith et al. 2001b). Additionally, genomic-enabled prediction models incorporating molecular marker data are now used to predict G × E interactions more accurately, maximizing genetic gains and accelerating breeding cycles (Heslot et al. 2015). Machine learning techniques, including random forests and deep learning, are increasingly employed to capture nonlinear and complex G × E interactions, further improving prediction accuracy (Montesinos-López et al. 2019). The integration of high-throughput phenotyping, multi-omics data, and advanced modeling approaches is reshaping MET analysis, enabling breeders to develop crop varieties that can adapt to diverse and changing environments.

The combination of METs and GS has enhanced the prediction accuracy of genotype performance across variable environments, offering a more comprehensive approach to understanding G × E interactions (Crossa et al. 2017). Traditional MET approaches relied primarily on phenotypic data, limiting genetic insights. However, integrating genomic data enables breeders to estimate genetic values for untested genotypes across diverse environments using models like genomic best linear unbiased prediction (GBLUP) and Bayesian methods, accounting for both additive and non-additive genetic effects (Jarquín et al. 2014; de Los Campos et al. 2013). This approach optimizes the allocation of testing resources, reduces the time and cost of field evaluations, and accelerates genetic gains (Alemu et al. 2024b; Burgueño et al. 2012).

Machine learning methods complement this integration by effectively modeling complex, nonlinear G × E interactions that conventional methods may overlook (Montesinos-López et al. 2019). Techniques like deep learning and random forests analyze large-scale genomic and environmental datasets, improving the identification of environment-specific genomic regions associated with trait expression (Morota et al. 2018). Additionally, multi-trait and multi-environment models that assess multiple traits across environments can exploit correlations to enhance prediction accuracy (Rincent et al. 2017). The synergy between MET and GS is vital for developing climate-resilient, high-yielding crops that can withstand unpredictable environmental conditions, contributing to sustainable global food production.

12.3.1 Sparse Testing

Sparse testing is an efficient strategy in plant breeding designed to reduce the number of field trials required while maintaining accurate and reliable predictions of genetic performance. By selectively phenotyping only a subset of genotype-environment combinations, this approach minimizes costs and logistical demands without compromising the quality of selection decisions. Sparse testing is particularly valuable in modern breeding programs, where the integration of genomic data and the scale of multi-environment trials often render comprehensive field testing infeasible. By optimizing the use of limited resources, it enables breeders to evaluate a larger number of genotypes and environments, thereby accelerating genetic gain and enhancing program efficiency (Burgueño et al. 2012). By strategically selecting a subset of environments and genotypes for evaluation, breeders can generate sufficient data to make informed decisions, maximizing the genetic gains while minimizing the resources required (Jarquín et al. 2014; Montesinos-López et al. 2023a, b).

Sparse testing has emerged as a strategic solution in plant breeding to address the increasing complexity and cost of large-scale multi-environment trials. Its core objective is to reduce the number of genotype–environment combinations that require phenotyping, thereby minimizing operational costs, labor, and time investments—without compromising the accuracy of genetic evaluations. This is particularly critical in the context of modern breeding programs, where the expansion of genomic data and the growing number of environments make full-scale field testing both impractical and economically unsustainable (Crespo-Herrera et al. 2021; Montesinos-López et al. 2023a, b). Beyond cost reduction, sparse testing plays a vital role in enhancing the prediction of genotype performance in untested or partially tested environments. This predictive capability is essential for selecting genotypes with high stability and adaptability across diverse and dynamic environmental conditions, a common challenge due to genotype-by-environment (G × E) interactions. By strategically selecting a subset of genotype–environment combinations and applying advanced genomic prediction models, breeders can make informed decisions even in the absence of complete field data (Bernal-Vasquez et al. 2016).

Recent developments in statistical modeling have further strengthened the utility of sparse testing. Techniques that integrate genomic and environmental covariates—alongside models that explicitly account for G × E interactions—have significantly improved prediction accuracy and robustness. These innovations support several critical goals: (a) increasing the efficiency of genomic prediction (GP), (b) allowing the evaluation of a larger number of genotypes within fixed resources, (c) enabling reliable inference in incomplete trials, and (d) promoting precision breeding through the targeted use of environmental information (Montesinos-López et al. 2023a, b, 2024). In summary, sparse testing offers a practical, scalable, and scientifically rigorous framework for accelerating genetic gains in plant breeding. It addresses real-world constraints while enabling more informed selection decisions, positioning it as a cornerstone strategy in the era of genomic-assisted breeding.

Several methods have been proposed to implement sparse testing in genomic prediction and improve the efficiency and accuracy of breeding programs. For instance, Montesinos-López et al. (2023b) introduced a framework based on balanced incomplete block designs (BIBD), which enables structured sparsity in field trials. This approach improves prediction accuracy by preserving connectivity among genotypes, traits, and environments, thereby facilitating effective information sharing across the dataset. The authors validated their method through both simulations and real datasets, demonstrating that BIBD-based designs provide a robust structure for sparse testing in multi-environment trials. In addition, commonly used genomic prediction models such as ridge regression best linear unbiased prediction (RR-BLUP) and genomic best linear unbiased prediction (GBLUP) have been successfully adapted to accommodate incomplete trial data (Crossa et al. 2017). These models integrate genomic and environmental information to strengthen prediction performance under sparsity. When applied to sparse testing, they support targeted genotype recommendations by leveraging environmental covariates to guide the selection of genotypes for phenotyping. This strategy enhances predictive accuracy, particularly for environments not directly represented in the training data, thus offering practical utility for real-world breeding scenarios.

12.3.2 Genotype-Specific Environmental Covariables

Modeling G × E interactions using genomics and environmental covariates is a powerful approach to understand how genetic and environmental factors influence complex traits. By integrating high-throughput genomic data with detailed environmental measurements, researchers can identify specific genetic variants that show differential effects across environments. This approach enhances predictive models in plant and animal breeding, as well as in human health studies, by improving the accuracy of genomic selection under variable conditions (Crossa et al. 2021). Additionally, machine learning and Bayesian methods have been increasingly used to model nonlinear G × E interactions, leading to more robust predictions of phenotypic outcomes (Heslot et al. 2015).

Despite its advantages, modeling G × E interactions presents significant challenges, including the need for large, well-structured datasets that capture both genetic and environmental variability (Cooper et al. 2016). Environmental factors are often difficult to quantify precisely, and their interactions with genetic effects may be highly complex and context-dependent. Moreover, statistical models must balance complexity and interpretability, as overly complex models may overfit data and reduce generalizability (Jarquin et al. 2020). Computational demands are another hurdle, as integrating genomic and environmental data requires significant resources for storage and processing, especially in large-scale studies. Current research in this field is focusing on improving the integration of multi-omics data (e.g., transcriptomics and epigenomics) with environmental variables to refine G × E predictions (Resende et al. 2021). Additionally, advancements in remote sensing and climate modeling are providing more precise environmental covariates, which can improve the accuracy of G × E models (Crossa et al. 2021). Future directions also include the use of deep learning techniques to capture complex nonlinear interactions while maintaining biological interpretability. As researchers continue to refine these models, their application in precision agriculture, personalized medicine, and conservation genetics is expected to expand, leading to more tailored and efficient strategies for managing genetic resources.

12.3.3 Modeling G × E Interaction with Environmental Covariates

Recent advances in genomic prediction have introduced innovative modeling approaches that integrate G × E interaction through the incorporation of environmental covariates, aiming to enhance predictive accuracy in METs (Heslot et al. 2015; Jarquín et al. 2014). One of the widely employed methods in plant breeding to account for G × E interaction is the Finlay–Wilkinson (FW) regression. This approach models the performance of each genotype across different environments as a linear function of the environmental mean, thereby allowing for the identification of both stable and responsive genotypes (Finlay and Wilkinson 1963). The FW regression is particularly valuable for distinguishing genotypes with broad adaptability from those that exhibit specific adaptation to particular environmental conditions. However, despite its simplicity and interpretability, FW regression may not fully capture complex G × E interaction, which is often nonlinear and influenced by multiple environmental and genetic factors. Recent extensions of the FW regression model have incorporated environmental covariates, significantly improving the ability to capture G × E interactions. This enhancement is crucial for genomic prediction in plant breeding programs, where environmental variability plays a major role in determining phenotype expression. By integrating environmental information into genomic models, researchers can better account for differential genetic responses across diverse environmental conditions, leading to more robust and accurate predictions of genetic merit (Crossa et al. 2017).

The inclusion of environmental covariates in FW regression aligns with the growing emphasis on METs, where the interaction between genotypes and environments often dictates the success of selection strategies. The genomic prediction approaches that rely solely on genetic markers may fail to capture these interactions adequately, leading to suboptimal breeding decisions. However, models that incorporate environmental data can enhance predictive performance by explicitly modeling how specific genetic effects vary across environments (Heslot et al. 2015). This is particularly relevant in breeding for climate resilience, where understanding genotype performance under varying climatic conditions is essential for developing stress-tolerant cultivars (Jarquín et al. 2014).

Moreover, integrating environmental covariates improves the adaptability of genomic prediction models by enabling researchers to identify stable and adaptive genotypes across diverse agro-ecological zones. This approach not only refines selection criteria but also facilitates the development of varieties with enhanced yield stability, a critical factor in modern breeding programs facing unpredictable climate patterns (Crossa et al. 2017). By leveraging environmental information, FW regression extensions bridge the gap between statistical modeling and biological realism, ensuring that genomic selection remains effective across heterogeneous environments (de Los Campos et al. 2020).

Factorial regression (FR) is another approach used to explicitly model G × E interactions by incorporating environmental covariates as explanatory variables (Van Eeuwijk et al. 1996). Unlike traditional regression methods that assume a uniform response across environments, FR decomposes the interaction structure into meaningful components, allowing for a more detailed characterization of genotype responses to specific environmental factors (Malosetti et al. 2013). The interaction terms in FR are typically arranged into a matrix, with genotypes as rows and environments as columns, and subsequently analyzed using singular value decomposition (SVD) to extract dominant patterns of variation (Crossa et al. 2017). The first principal component from the SVD often serves as an environmental index, summarizing the most influential environmental gradients affecting genotype performance (Van Eeuwijk et al. 1996). However, despite its analytical power, FR becomes computationally complex when numerous environmental covariates are considered, leading to challenges in model estimation, interpretation, and overfitting risks (Jarquín et al. 2014). To overcome these limitations, recent advancements have focused on integrating environmental covariates in more parsimonious ways to enhance the predictive power of genomic models.

One of the recent approaches for integrating environmental covariates in genomic prediction models is the use of linear kernel regression models (Jarquín et al. 2014) that have been widely used in plant breeding to model G × E interactions by leveraging environmental covariance structures. These models extend the traditional GBLUP framework by incorporating an environmental covariance matrix derived from environmental covariates, analogous to the construction of kinship matrices from genomic data (Crossa et al. 2017). This approach enables a more refined characterization of G × E interactions by capturing the relationships among environments and their influence on genotype performance. One of the key advantages of linear kernel regression is its ability to efficiently integrate high-dimensional environmental data without requiring explicit interaction terms, thereby mitigating the risk of overparameterization and improving predictive accuracy (Heslot et al. 2015). However, the linearity assumption may limit its ability to model complex, nonlinear G × E patterns, which has motivated the exploration of more flexible kernel methods, such as Gaussian kernel regression and deep learning approaches, to enhance predictive performance in genomic selection (Montesinos-López et al. 2019, 2021a, b).

Factor-analytic linear mixed models (Smith et al. 2001a, b) have been widely employed in METs to model and interpret G × E interactions. These models offer a flexible and computationally efficient framework for capturing complex interaction structures by decomposing the G × E variance–covariance matrix into latent factors that represent unobserved environmental influences (Smith et al. 2001a, b). In this approach, the environment is treated as a random factor, with its effects modeled through a factor-analytic variance–covariance structure. This formulation assumes that the environmental effects can be expressed as linear regressions on a small number of underlying latent variables, which effectively reduce the dimensionality of the interaction space while preserving biologically meaningful patterns of genotype response across environments (Oakey et al. 2016). A key advantage of the factor-analytic approach is its ability to accommodate correlations between environments, enabling improved borrowing of information across trials and facilitating genotype evaluation under diverse conditions (Piepho 1997, 1998). Furthermore, it provides a more interpretable representation of G × E interactions compared to purely empirical methods, allowing breeders to identify genotypic stability and adaptability with greater precision (Smith et al. 2001a, b). However, despite its advantages, the choice of the number of latent factors remains a critical consideration, as overfitting may lead to inflated model complexity while underfitting may fail to capture essential interaction patterns (Jarquín et al. 2014). Recent advancements in modeling methodologies have further expanded the applicability of FA models by integrating genomic and environmental information, significantly enhancing predictive accuracy in genomic selection (Burgueño et al.

2012; Smith et al. 2001a, b). FA models are particularly well-suited for capturing genotype-by-environment (G × E) interactions, as they decompose the genetic signal across multiple environments into latent factors that explain shared and specific variation (Smith et al. 2001a, b). This approach not only improves model interpretability but also enables more precise predictions of genotype performance across diverse environmental conditions. The incorporation of environmental covariates into FA models strengthens their ability to characterize the underlying structure of G × E interactions by explicitly linking genetic effects to environmental gradients. The conventional genomic prediction models often assume a uniform genetic effect across environments, which can lead to biased predictions when environmental heterogeneity is substantial (Heslot et al. 2015). FA models mitigate this issue by allowing genetic expression to vary systematically with environmental drivers, thereby improving both predictive accuracy and biological realism (Jarquín et al. 2014). Moreover, the integration of environmental data into FA models facilitates the identification of adaptive traits, enabling breeders to select genotypes that perform optimally across a range of conditions. This is particularly relevant in breeding for climate resilience, where understanding genotype stability and plasticity is critical for developing varieties that maintain high yields under variable stress conditions (Crossa et al. 2017). By leveraging both genetic and environmental information, FA models provide a robust framework for optimizing selection strategies in METs, ultimately leading to more reliable genomic predictions and improved breeding efficiency (Meyer 2009).

12.3.4 G × E-Aided Genomic Selection in Major Crops

12.3.4.1 Maize

Genomic estimated breeding values for selection represent a powerful approach in maize breeding that specifically addresses G × E interactions through sophisticated prediction modeling. Research has demonstrated that by incorporating environmental factors directly into genomic prediction models, maize breeders can significantly improve the accuracy of selecting superior genotypes across diverse growing conditions (Mageto et al. 2020). The effectiveness of genomic selection in mitigating G × E interactions was evident through distinct prediction models (Krause et al. 2020). The most basic model incorporates only environment and line effects, while progressively more advanced versions include genomic information and crucially, the interaction between genomic markers and environmental conditions (Acosta-Pech et al. 2017; Tolley et al. 2023). Research has shown that the model accounting for these interactions consistently achieves higher prediction accuracies, with average prediction abilities reaching 0.39 and 0.44 for newly developed lines, and impressive values of 0.71 and 0.51 for lines tested across multiple locations (Mageto et al. 2020). A compelling demonstration on the power of genomic selection comes from research focused on kernel zinc concentration in tropical maize germplasm (Mageto et al. 2020). By implementing these predictive models across three distinct environments, scientists observed significant improvements in identifying superior genotypes. The inclusion of G × E interaction terms in the prediction models proved particularly valuable, enabling breeders to identify cultivars that maintained desirable traits across varying environmental conditions. Furthermore, multi-environment trial data along with machine learning methods in genomic prediction seem to be an innovative tool that enables maize breeders to predict phenotypic characteristics of single cross hybrids without requiring actual field testing (Barreto et al. 2024). This approach leverages advanced genomics and statistical modeling to forecast how specific hybrid combinations would perform across different environments, effectively streamlining the breeding process by eliminating the need for physical evaluation of every potential cross. Through sophisticated algorithms and genomic markers, breeders can now predict hybrid performance without conducting extensive field trials, thus revolutionizing the efficiency and scope of modern breeding programs. The practical implications of this approach are significant for maize breeding programs. By accounting for genotype-by-environment interactions through selection based on GEBVs, maize breeders can accelerate the development of resilient cultivars that perform consistently across diverse growing conditions. This represents a significant advancement over traditional breeding methods, which often struggle to predict cultivar performance in different environments. The integration of genomic prediction with environmental interaction analysis allows for more informed decision-making in breeding programs, thereby contributing to improved crop adaptation and yield stability. Through this sophisticated approach, genomic selection also effectively bridges the gap between genetic potential and environmental expression, providing breeders with powerful tools to develop maize cultivars that maintain consistent performance regardless of growing conditions. The success of these methods has been validated for various characteristics through extensive testing across multiple populations and environments, thus demonstrating their reliability and effectiveness in modern maize breeding programs.

12.3.4.2 Potato

The implementation of GEBVs for selection in potato breeding involves several key mechanisms. First, researchers employ specialized statistical models that explicitly account for G × E interactions, allowing for the identification of genetic variants that contribute to environmental adaptation (Ortiz et al. 2022; Agha et al. 2024). These models enable breeders to distinguish between genetic factors that confer broad adaptability versus those that lead to local adaptation, which is a critical distinction for potato breeding programs aiming to develop cultivars for specific regions or conditions. Recent advances in genomic prediction methods have significantly enhanced the effectiveness of G × E interaction modeling. Techniques such as kernel regression models and deep learning approaches have improved the accuracy of predictions across multiple environments (Cuevas et al. 2023; Ortiz et al. 2023). These methods particularly benefit potato breeding due to the crop's complex tetraploid nature and the extensive G × E interactions observed for yield and quality traits (Ortiz 2023). The practical implementation of these approaches has demonstrated promising results. Research has shown that GEBVs for selection can effectively identify genetic variants associated with environmental responses, enabling breeders to make informed decisions about variety development for specific target environments (Alemu et al. 2024b). Furthermore, the integration of reaction norm models allows researchers to predict how genotypes will respond across environmental gradients, providing valuable insights for potato breeding focusing on either broad adaptation or environment-specific improvement (Ortiz et al. 2024).

In conclusion, the success of genomic selection in mitigating G × E interactions for potato breeding stems from its ability to capture both major genes and multigenic effects contributing to environmental adaptation. The comprehensive approach provided by GEBV for selection enables potato breeders to balance the need for broad adaptability with the potential benefits of environment-specific improvements, ultimately leading to more efficient and effective cultivar development for diverse growing conditions.

12.3.4.3 Rice

By utilizing genome-wide markers to predict breeding values, rice breeders understand and manage better how rice cultivars perform across different environmental conditions (Nguyen et al. 2023). Recent advances in multi-environment genomic prediction models have significantly enhanced the effectiveness of GEBVs for selection in rice breeding (Bančič et al. 2023). These sophisticated models incorporate environmental covariates such as weather and soil information, allowing for more accurate predictions of variety performance across different growing conditions. Research has consistently shown that models accounting for G × E interactions achieve higher prediction accuracy compared to traditional single-environment analyses, thus enabling rice breeders to identify cultivars that will perform consistently across diverse environmental conditions. Some GEBV multi-environment models in rice integrated genomic markers (DNA information), phenotypic data (observed traits), and environmental covariates (climate and soil measurements, which account for G × E, and improve prediction accuracy. Multi-environment models demonstrated comparable predictive accuracy to single environment models for unobserved phenotypes in untested rice breeding germplasm, with variations in predictive ability ranging from −6% to 4% across different traits and statistical models. However, when evaluating rice germplasm under both alternate wetting and drying and continuous flooding management systems, multi-environment models showed superior performance compared to single environment models, achieving approximately 30% greater predictive ability (Ben Hassen et al. 2018). Hence, the GEBV for selection approach helps accelerate genetic gain in rice breeding while reducing breeding cycle length, ultimately contributing to improved rice productivity and food security.

12.3.4.4 Sorghum

Significant improvement in prediction accuracy was achieved by incorporating G × E interaction through advanced statistical models, thus allowing sorghum breeders to better identify germplasm that perform consistently across diverse environments (Maulana et al. 2023; Kent et al. 2025). Likewise, Velazco et al. (2020) indicated that genomic prediction models must incorporate G × E and locus-specific genetic variances to better predict how different sorghum cultivars will perform across diverse environmental conditions. Winans et al. (2024) have shown that envirotype data provide valuable insights that effectively delineate established sorghum production regions, thereby enabling breeders to allocate resources more strategically. Furthermore, the findings demonstrate the practical utility of implementing sparse hybrid trials within sorghum breeding programs that utilize genomic predictions. As a result, incorporating envirotype data into genomic prediction models enhances the informational resources available to sorghum breeding programs while facilitating more efficient hybrid trial design methodologies. The effectiveness of genomic selection in mitigating genotype-by-environment interaction is particularly evident in sorghum hybrid breeding programs, where prediction models incorporating population structure and G × E interaction effects have proven highly effective. Research have shown that these approaches can effectively predict hybrid performance while accounting for environmental variation, though success depends on factors such as trait repeatability, genetic architecture, parental similarity, and training set composition. Furthermore, advanced genomic-enabled prediction models that incorporate general and specific combining abilities have demonstrated improved prediction accuracy across different environments, thereby enabling breeders to develop cultivars with broader adaptability. The implementation of genomic selection in sorghum breeding programs offers practical advantages, including reduced breeding cycles and accelerated genetic gains. As indicated by Maulana et al. (2023) three cycles of genomic-assisted recurrent selection per year combined with strategic updates to training populations may maximize genetic gains while maintaining prediction accuracy across environments.

12.3.4.5 Wheat

To build effective prediction models for wheat breeding, phenotypic data must come from environments that strongly correlate with target production environments (Ovenden et al. 2018). As noticed by Raffo and Jensen (2022), accounting for both gene × gene (G × G) and G × E in GEBVs for selection can significantly enhance genetic improvement by increasing the rate of genetic gain and adaptation of wheat cultivars. The characterization and modeling G × E in GEBVs for selection demonstrated high predictive ability in wheat breeding (Lado et al. 2016).

12.3.4.6 Other Crops

As shown above for cereal and tuber crops, GEBV appears to be also useful for managing genotype-by-environment interactions in cassava breeding by enabling early-stage selection and improving prediction accuracy across diverse growing conditions in beverage (Adunola et al. 2023), forage (Fois et al. 2021), pulse (Roorkiwal et al. 2018), and root (Ly et al. 2013) crops. Through high-density marker genotyping, GEBV for selection allows breeders of such crops to predict genetic values without extensive phenotypic evaluation, particularly valuable for traits showing strong environmental sensitivity. Prediction accuracy further depends on the complexity of the trait, as demonstrated in cassava by Esuma et al. (2021). For instance, genomic prediction achieved an accuracy of 0.52 for total carotenoid content—a relatively simple trait—compared to just 0.20 for resistance to cassava brown streak disease, a more complex trait. Research also demonstrated that GEBVs based on multi-environmental trials enhance genetic gains while breeding for host plant resistance to both crown and stem rust in ryegrass (Fois et al. 2021). Similarly, Adunola et al. 2023 indicated that genomic prediction can allow selecting coffee germplasm combining high yield and stability across years in the target population of environments. Likewise, GEBV modeling including various environmental variables increases accuracy in chickpea breeding (Roorkiwal et al. 2018), thus improving the ability to predict performance of testing germplasm in environment yet to be used for their assessment.

12.4 Challenges and Future Directions

Challenge 1: Environmental characterization through meaningful covariates (QTL × E interaction; redesign target population of environments using meaningful covariates)

Considering a reaction norm model, environments with similar covariates (temperature, precipitation, radiation, management, or geographical distribution) are expected to share similar G × E patterns, based on the hypothesis that related genotypes will show similar adaptive patterns in related environments. However, while this is relatively easy to prove retrospectively using cross-validation, the challenge remains for collecting meaningful environmental covariates for future environments. Predicting future environments (known locations in future years) is crucial for understanding G × E, because often the interest relies in selecting genotypes that can withstand environmental fluctuations, such as drought, heat, cold or flood stress, and other factors as pests and diseases infestation. Identifying meaningful covariates is crucial to predict future abiotic and abiotic stresses, and this can be done by considering all environment factors which affect the crop development or the occurrence of some disease, holding the potential to redesign the characterization of target population of environments (Cooper and Messina 2023). In addition, the development of improved genotypes should be focused not on the current environmental conditions but on those that will occur in 3–5 years, that is when the genotypes to release will reach the market.

Challenge 2: Multi-omics integration

Advancements in high-throughput phenotyping technologies have enabled access to massive datasets with the potential to enhance breeding decisions in multi-environment trials, allowing the measurement of genotype stability and superiority for a large amounts of individuals. For instance, datasets collected through automated systems, such as those captured by multi- or hyperspectral cameras with unmanned aerial vehicles (UAVs), which are referred as "phenomics" (Rincent et al. 2018). Besides phenomics, there are other types of omics which encompass biological molecules such as transcripts, proteins, and metabolites that can be used to build whole biological systems (Mackay et al. 2009).

Although integrating information from diverse sources holds great promise for improving predictive accuracy in breeding programs (Crossa et al. 2017), it also presents new challenges—particularly in terms of how to manage and interpret the vast volume of heterogeneous data in a transparent and explainable manner (Montesinos-López et al., 2021a). To address this issue, several tools have been developed. One such tool is CHiDO, an open-source, no-code platform designed to facilitate the application of reaction norm methodologies to multi-omics data organized in tabular format. The CHiDO enables users to analyze complex G × E interactions without requiring programming expertise, thus democratizing access to advanced statistical tools in genomics and breeding research (González et al. 2025). The development of these tools allows researchers to evaluate the usefulness of their datasets for specific tasks, whether for addressing theoretical biological hypotheses or applied protocols for releasing superior varieties. Multi-omics integration through the so-called multi-kernel approach is being tested in breeding programs and has shown promising results for crops such as oat (Hu et al. 2021). However, some challenges remain regarding the partition of the explained variability when considering multiple sources of information.

12.5 Future Directions

12.5.1 Importance of High-Quality Data

Genomic selection primarily relies on training statistical models using various approaches, which can be broadly categorized into parametric, semi-parametric, and non-parametric methods (Alemu et al. 2024b). Parametric methods include RR-BLUP and GBLUP, as well as Bayesian alphabet methods such as Bayes A, B, and Cπ. Semi-parametric approaches, such as the Reproducing Kernel Hilbert Space (RKHS), offer greater flexibility by capturing nonlinear relationships without assuming a strict parametric form. Meanwhile, non-parametric methods—including random forests and deep learning algorithms (e.g., neural networks)—can model complex, nonlinear interactions without relying on underlying distributional assumptions. However, all these approaches have in common the need to employ high-quality data to conduct reliable predictions (Montesinos-López 2021). Thus, good data recording on fields is still crucial to achieve GS highest potential. Also, the ongoing development of genotyping platforms and tools to collect other sources of data through high-throughput systems and weather sensors is important not only to make them cheaply available but also to increase our ability to represent environments and genotypes using this information (Crossa et al. 2021). Despite their methodological differences, all GS approaches share a common requirement: the need for high-quality data to enable reliable and accurate predictions (Montesinos-López 2021). Consequently, robust data recording in the field remains essential for realizing the full potential of GS. Moreover, the continued advancement of genotyping platforms, along with the development of high-throughput phenotyping systems and environmental data collection tools (e.g., weather sensors), plays a critical role in reducing costs and enhancing our capacity to accurately characterize both genotypes and environments (Crossa et al. 2021). These innovations contribute significantly to improving the precision and scalability of genomic prediction models.

12.5.2 Dealing with Data Scalability

Integrating high-dimensional omics data—such as genomics, phenomics, and enviromics—and especially modeling their interactions through contrasts or covariance structures, presents significant challenges in terms of computational efficiency (Jarquín et al. 2014; Crossa et al. 2017). Generating an interaction matrix using contrasts requires constructing a matrix with a column for each possible combination between groups of factors. For instance, in a dataset containing 10,000 genetic markers evaluated across 100 environments, the resulting interaction matrix would need to have one column for each marker-by-environment combination. This would lead to a matrix with dimensions of n (number of individuals) × 1,000,000 (10,000 markers × 100 environments), which presents substantial computational and storage challenges when modeling G × E interaction effects.

Employing the reaction norm model can overcome the limitation of contrasts by calculating interaction matrices with dimensions $n \times n$. However, fitting models using datasets with a high number of observations can also be prohibitive. As an alternative, Xavier et al. (2025) presented different approaches to leverage complex interactions through multivariate mixed-model solvers. Another alternative that we visualize is the use of graphic processing units (GPUs) not necessarily through deep learning approaches but parallelizing heavily linear algebra tasks such as matrix inversion or decompositions while implementing the most common solvers via linear mixed models or Bayesian approaches.

12.5.3 Combine State-of-the-Art Methods and New Approaches

Despite the rapid advancement of novel methods, particularly those based on neural networks, studies have reported only marginal or no significant improvements over established state-of-the-art approaches. Moreover, these deep learning models often require careful and computationally intensive hyperparameter tuning, which may limit their practical utility in some breeding contexts (Montesinos-López 2021). This may be an indicator to look at these tools not as a replacement to older methods, but as complements to leverage the strengths of both worlds. For example, neural networks can account for nonlinear patterns through activation functions, being a potential tool to redesign target population of environments to recommend genotypes (Farooq et al. 2024). Still, parametric approaches allow us to obtain important genetic parameters such as genetic variance and heritability, being important measures to evaluate the progress of a breeding program (de Los Campos et al. 2015). Hence, ensemble-based models that combine different approaches to decision support, such as the one proposed by Nascimento et al. (2024) for coffee prediction, are strong candidates to deal with a world under constant change and new challenges to guarantee better data-driven recommendations to drive genetic progress.

12.6 Conclusion

Genomic selection has emerged as a pivotal tool in plant breeding to address the challenges posed by G × E interactions. By leveraging high-throughput genotyping, advanced statistical models, and environmental covariates, breeders can now more accurately predict genotype performance across variable conditions. This enables the identification of stable and high-yielding genotypes, even in the presence of complex environmental heterogeneity. The integration of multi-environment trials, sparse testing strategies, and modern machine learning techniques has further enhanced the scalability and precision of breeding programs. Moreover, the use of reaction norm models, factor analytic approaches, and environmental kernels allows for deeper biological interpretation of G × E effects and facilitates the development of climate-resilient cultivars. Despite significant progress, key challenges has been remained particularly in environmental characterization, data scalability, and the effective integration of multi-omics datasets. Future efforts must prioritize the development of efficient computational tools, the refinement of prediction models, and the collection of robust phenotypic and environmental data. Combining state-of-the-art statistical approaches with flexible, interpretable artificial intelligence (AI) models offers a promising path forward. Eventually, genomic selection enriched by G × E modeling is critical for the future sustainable agriculture intensification, thereby enabling faster genetic gains and improving food security in the face of climate change and evolving agricultural demands.

References

Acosta-Pech R, Crossa J, de Los CG, Teyssèdre S, Claustres B, Pérez-Elizalde S, Pérez-Rodríguez P (2017) Genomic models with genotype × environment interaction for predicting hybrid performance: an application in maize hybrids. Theor Appl Genet 130:1431–1440. https://doi.org/10.1007/s00122-017-2898-0

Adunola P, Ferrão MAG, Ferrão RG, da Fonseca AFA, Volpi PS, Comério M, Verdin Filho AC, Munoz PR, Ferrão LFV (2023) Genomic selection for genotype performance and environmental stability in Coffea canephora. G3 Genes Genom Genet 13:jkad062. https://doi.org/10.1093/g3journal/jkad062

Agha HI, Endelman JB, Chitwood-Brown J, Clough M, Coombs J, De Jong WS, Douches DS, Higgins CR, Holm DG, Novy R, Resende MFR, Sathuvalli V, Thompson AL, Yencho GC, Zotarelli L, Shannon LM (2024) Genotype-by-environment interactions and local adaptation shape selection in the US National Chip Processing Trial. Theor Appl Genet 137:99. https://doi.org/10.1007/s00122-024-04610-3

Alemu A, Åstrand J, Montesinos-López OA, Isidro Y Sánchez J, Fernández-Gónzalez J, Tadesse W, Vetukuri RR, Carlsson AS, Ceplitis A, Crossa J, Ortiz R, Chawade A (2024a) Genomic selection in plant breeding: key factors shaping two decades of progress. Mol Plant 17: 552–578. https://doi.org/10.1016/j.molp.2024.03.007

Alemu A, Singh PK, Chawade A (2024b) Adaptation and grain yield stability analysis of winter wheat cultivars with and without fungicides treatment from national variety trials in Sweden. Agriculture 14:2229. https://doi.org/10.3390/agriculture14122229

Allard RW, Bradshaw AD (1964) Implications of genotype-environmental interactions in applied plant breeding. Crop Sci 4:503–508

Bančič J, Ovenden B, Gorjanc G, Tolhurst DJ (2023) Genomic selection for genotype performance and stability using information on multiple traits and multiple environments. Theor Appl Genet 136:104. https://doi.org/10.1007/s00122-023-04305-1

Barreto CAV, das Graças Dias KO, de Sousa IC, Azevedo CF, Nascimento ACC, Guimarães LJM, Guimarães CT, Pastina MM, Nascimento M (2024) Genomic prediction in multi-environment trials in maize using statistical and machine learning methods. Sci Rep 14:1062. https://doi.org/10.1038/s41598-024-51792-3

Ben Hassen M, Bartholomé J, Valè G, Cao TV, Ahmadi N (2018) Genomic prediction accounting for genotype by environment interaction offers an effective framework for breeding simultaneously for adaptation to an abiotic stress and performance under normal cropping conditions in rice. G3 Genes|Genomes|Genetics 8:2319–2332. https://doi.org/10.1534/g3.118.200098

Bernal-Vasquez AM, Utz HF, Piepho HP (2016) Outlier detection methods for generalized lattices: a case study on the transition from ANOVA to REML. Theor Appl Genet 129:787–804

Beyene Y, Gowda M, Pérez-Rodriguez P, Olsen M, Robbins KR, Burgueño J, Prasanna BM, Crossa J (2021) Application of genomic selection at the early stage of breeding pipeline in tropical maize. Front Plant Sci 12:685488

Boer MP, Wright D, Feng L, Podlich DW, Luo L, Cooper M, van Eeuwijk FA (2007) A mixed-model quantitative trait loci (QTL) analysis for multiple-environment trial data using environmental covariables for QTL-by-environment interactions, with an example in maize. Genetics 177:1801–1813

Bohlouli M, Alijani S, Javaremi AN, König S, Yin T (2017) Genomic prediction by considering genotype × environment interaction using different genomic architecture. Ann Anim Sci 17:683–701. https://doi.org/10.1515/aoas-2016-0086

Burgueño J, de los Campos G, Weigel K, Crossa J (2012) Genomic prediction of breeding values when modeling genotype × environment interaction using pedigree and dense molecular markers. Crop Sci 52:707–719

Ceccarelli S, Grando S (2024) Diversity as a plant breeding objective. Agronomy 14:550

Cooper M, Hammer GL (1996) Plant adaptation and crop improvement. IRRI, Manila

Cooper M, Messina CD (2023) Breeding crops for drought-affected environments and improved climate resilience. Plant Cell 35:162–186

Cooper M, Byth DE, Hammer GL (1996) Understanding plant adaptation to achieve systematic applied crop improvement–a fundamental challenge. In: Cooper M, Hammer GL (eds) Plant adaptation and crop improvement. CAB International, Wallingford, pp 5–23

Cooper M, Technow F, Messina C, Gho C, Totir LR (2016) Use of crop growth models with whole-genome prediction: application to a maize multienvironment trial. Crop Sci 56:2141–2156

Cooper M, van Eeuwijk F, Chapman SC, Podlich DW, Löffler C (2024) Genotype-by-environment interactions under water-limited conditions. In: Ribaut JM (ed) Drought adaptation in cereals. CRC Press, Boca Raton, pp 51–96

Crespo-Herrera L, Howard R, Piepho HP, Pérez-Rodríguez P, Montesinos-Lopez O, Burgueño J, Crossa J (2021) Genome-enabled prediction for sparse testing in multi-environmental wheat trials. Plant Genome 14:e20151

Crossa J, Montesinos-López OA, Pérez-Rodríguez P, Costa-Neto G, Fritsche-Neto R, Ortiz R, Martini JWR, Lillemo M, Montesinos-López A, Jarquin D, Breseghello F, Cuevas J, Rincent R (2022) Genome and environment-based prediction models and methods of complex traits incorporating genotype × environment interaction. In: Ahmadi N, Bartholomé J (eds) Methods in molecular biology, vol 2467. Humana, New York, pp 245–283. https://doi.org/10.1007/978-1-0716-2205-6_9

Crossa J, Fritsche-Neto R, Montesinos-Lopez OA, Costa-Neto G, Dreisigacker S, Montesinos-Lopez A, Bentley AR (2021) The modern plant breeding triangle: optimizing the use of genomics, phenomics, and enviromics data. Front Plant Sci 12:651480

Crossa J, Pérez-Rodríguez P, Cuevas J, Montesinos-López O, Jarquin D, de Los Campos G, Burgueño J, González-Camacho JM, Pérez-Elizalde S, Beyene Y (2017) Genomic selection in plant breeding: methods, models, and perspectives. Trends Plant Sci 22:961–975

Crossa J, Yang RC, Cornelius PL (2004) Studying crossover genotype × environment interaction using linear-bilinear models and mixed models. J Agric Biol Environ Stat 9:362–380

Cuevas J, Reslow F, Crossa J, Ortiz R (2023) Modeling genotype × environment interaction for single and multitrait genomic prediction in potato (Solanum tuberosum L.). G3 Genes Genom Genet 13:jkac322. https://doi.org/10.1093/g3journal/jkac322

de Los Campos G, Hickey JM, Pong-Wong R, Daetwyler HD, Calus MP (2013) Whole-genome regression and prediction methods applied to plant and animal breeding. Genetics 193:327–345

de Los Campos G, Pérez-Rodríguez P, Bogard M, Gouache D, Crossa J (2020) A data-driven simulation platform to predict cultivars performances under uncertain weather conditions. Nat Commun 11:4876

de Los Campos G, Sorensen D, Gianola D (2015) Genomic heritability: what is it? PLoS Genet 11: e1005048

des Marais DL, Hernandez KM, Juenger TE (2013) Genotype-by-environment interaction and plasticity: exploring genomic responses of plants to the abiotic environment. Annu Rev Ecol Evol Syst 44:5–29

Esuma W, Ozimati A, Kulakow P, Gore MA, Wolfe MD, Nuwamanya E, Egesi C, Kawuki RS (2021) Effectiveness of genomic selection for improving provitamin a carotenoid content and associated traits in cassava. G3 Genes Genom Genet 11:jkab160. https://doi.org/10.1093/g3journal/jkab160

Falconer DS (1952) The problem of environment and selection. Am Nat 86:293–298

Farooq MA, Gao S, Hassan MA, Huang Z, Rasheed A, Hearne S, Prasanna B, Li X, Li H (2024) Artificial intelligence in plant breeding. Trends Genet. https://doi.org/10.1016/j.tig.2024.05.006

Finlay KW, Wilkinson GN (1963) The analysis of adaptation in a plant-breeding programme. Aust J Agric Res 14:742–754

Fisher RA (1925) Statistical methods for research workers. Oliver and Boyd, Edinburgh

Fois M, Malinowska M, Schubiger FX, Asp T (2021) Genomic prediction and genotype-by-environment interaction analysis of crown and stem rust in ryegrasses in European multi-site trials. Agronomy 11:1119. https://doi.org/10.3390/agronomy11061119

González F, Garcia-Abadillo J, Jarquin D (2025) Introducing CHiDO—a no code genomic prediction software implementation for the characterization and integration of driven omics. Plant Genom 18:e20519

Guzmán C, Mondal S, Govindan V, Autrique JE, Posadas-Romano G, Cervantes F, Crossa J, Vargas M, Singh RP, Peña RJ (2016) Use of rapid tests to predict quality traits of CIMMYT bread wheat genotypes grown under different environments. LWT Food Sci Technol 69:327–333

Heslot N, Jannink JL, Sorrells ME (2015) Perspectives for genomic selection applications and research in plants. Crop Sci 55:1–12

Hu H, Campbell MT, Yeats TH, Zheng X, Runcie DE, Covarrubias-Pazaran G, Broeckling C, Yao L, Caffe-Treml M, Gutiérrez L (2021) Multi-omics prediction of oat agronomic and seed nutritional traits across environments and in distantly related populations. Theor Appl Genet 134:4043–4054

Jarquín D, Crossa J, Lacaze X, Cheyron P, Daucourt J, Lorgeou J, Piraux F, Guerreiro L, Pérez P, Calus M, de los Campos G (2014) A reaction norm model for genomic selection using high-dimensional genomic and environmental data. Theor Appl Genet 127:595–607

Jarquin D, Howard R, Crossa J, Beyene Y, Gowda M, Martini JWR, Gracia-Romero A, Burgueño J, Gastelo M, Rodrigues F, Vicente FS, Pacheco A, Vallejo G, Pereira HS, Guimarães P, Fritsche-Neto R, Olsen M, Prasanna BM (2020) Genomic prediction enhanced sparse testing for multi-environment trials. G3 Genes Genom Genet 10:2725–2739

Kang Z, Kong J, Li Q, Sui J, Dai P, Luo K, Meng X, Chen B, Cao J, Tan J, Fu Q, Xing Q, Luan S (2025) Genomic selection strategies to overcome genotype by environment interactions in biosecurity based aquaculture breeding programs. Genet Sel Evol 57:2. https://doi.org/10.1186/s12711-025-00949-3

Kent MA, Fonseca JMO, Klein PE, Klein RR, Hayes CM, Rooney WL (2025) Inclusion of genotype by environment interactions into genomic prediction models for sorghum seed parents in hybrid testcrosses. J Crop Improv 39:146–166. https://doi.org/10.1080/15427528.2025.2460571

Krause MD, Dias KOG, dos Santos JPR, de Oliveira AA, Guimarães LJ, Pastina MM, Margarido GRA, Garcia AAF (2020) Boosting predictive ability of tropical maize hybrids via genotype-by-environment interaction under multivariate GBLUP models. Crop Sci 60:3049–3065. https://doi.org/10.1002/csc2.20253

Lado B, Barrios PG, Quincke M, Silva P, Gutiérrez L (2016) Modeling genotype × environment interaction for genomic selection with unbalanced data from a wheat breeding program. Crop Sci 56:2165–2179. https://doi.org/10.2135/cropsci2015.04.0207

Li X, Guo T, Wang J, Bekele WA, Sukumaran S, Vanous AE, McNellie JP, Tibbs-Cortes LE, Lopes MS, Lamkey KR, Westgate ME, McKay JK, Archontoulis SV, Reynolds MP, Tinker NA, Schnable PS, Yu J (2021) An integrated framework reinstating the environmental dimension for GWAS and genomic selection in crops. Mol Plant 14:874–887. https://doi.org/10.1016/j.molp.2021.03.010

Ly D, Hamblin M, Rabbi I, Melaku G, Bakare M, Gauch HG Jr, Okechukwu R, Dixon AGO, Kulakow P, Jannink J-L (2013) Relatedness and genotype × environment interaction affect prediction accuracies in genomic selection: a study in cassava. Crop Sci 53:1312–1325. https://doi.org/10.2135/cropsci2012.11.0653

Mackay TFC, Stone EA, Ayroles JF (2009) The genetics of quantitative traits: challenges and prospects. Nat Rev Genet 10:565–577

Mageto EK, Crossa J, Pérez-Rodríguez P, Dhliwayo T, Palacios-Rojas N, Lee M, Guo R, San Vicente F, Zhang X, Hindu V (2020) Genomic prediction with genotype by environment interaction analysis for kernel zinc concentration in tropical maize germplasm. G3 Genes|Genomes|Genetics 10:2629–2639. https://doi.org/10.1534/g3.120.401172

Malosetti M, Ribaut JM, van Eeuwijk FA (2013) The statistical analysis of multi-environment data: modeling genotype-by-environment interaction and its genetic basis. Front Physiol 4:44

Maulana F, Perumal R, Serba DD, Tesso T (2023) Genomic prediction of hybrid performance in grain sorghum (Sorghum bicolor L.). Front Plant Sci 14:1139896. https://doi.org/10.3389/fpls.2023.1139896

Meyer K (2009) Factor-analytic models for genotype × environment type problems and structured covariance matrices. Genet Sel Evol 41:1–11

Montesinos-López OA, Montesinos-López A, Crossa J, Toledo FH, Pérez-Hernández O, Eskridge KM, Rutkoski J (2016) A genomic Bayesian multi-trait and multi-environment model. G3 Genes|Genomes|Genetics 6:2725–2744. https://doi.org/10.1534/g3.116.032359

Montesinos-López A, Montesinos-López OA, Montesinos-López JC, Flores-Cortes CA, de la Rosa R, Crossa J (2021a) A guide for kernel generalized regression methods for genomic-enabled prediction. Heredity 126:577–596. https://doi.org/10.1038/s41437-021-00412-1

Montesinos-López OA, Montesinos-López A, Pérez-Rodriguez P, Barrón-López JA, Martini JWR, Fajardo-Flores SB, Gaytan-Lugo LS, Santana-Mancilla PC, Crossa J (2021b) A review of deep learning applications for genomic selection. BMC Genomics 22:1–23

Montesinos-López OA, Montesinos-López A, Tuberosa R, Maccaferri M, Sciara G, Ammar K, Crossa J (2019) Multi-trait, multi-environment genomic prediction of durum wheat with genomic best linear unbiased predictor and deep learning methods. Front Plant Sci 10:1311

Montesinos-López OA, Mosqueda-González BA, Salinas-Ruiz J, Montesinos-López A, Crossa J (2023b) Sparse multi-trait genomic prediction under balanced incomplete block design. Plant Genome 16:e20305

Montesinos-López OA, Saint Pierre C, Gezan SA, Bentley AR, Mosqueda-González BA, Montesinos-López A, Mondal S, Crespo-Herrera L, Huerta J, Crossa J (2023a) Optimizing sparse testing for genomic prediction of plant breeding crops. Genes 14:927

Montesinos-López OA, Vitale P, Gerard G, Crespo-Herrera L, Saint Pierre C, Montesinos-López A, Crossa J (2024) Genotype performance estimation in targeted production environments by using sparse genomic prediction. Plants 13:3059

Morota G, Ventura RV, Silva FF, Koyama M, Fernando SC (2018) Machine learning and data mining advance predictive big data analysis in precision animal agriculture. J Anim Sci 96: 1540–1550

Nascimento M, Nascimento ACC, Azevedo CF, de Oliveira ACB, Caixeta ET, Jarquin D (2024) Enhancing genomic prediction with stacking ensemble learning in Arabica coffee. Front Plant Sci 15:1373318

Nguyen VH, Morantte RIZ, Lopena V, Verdeprado H, Murori R, Ndayiragije A, Katiyar SK, Islam MR, Juma RU, Flandez-Galvez H, Glaszmann JC, Cobb JN, Bartholomé J (2023) Multi-environment genomic selection in rice elite breeding lines. Rice 16:7. https://doi.org/10.1186/s12284-023-00623-6

Oakey H, Cullis B, Thompson R, Comadran J, Halpin C, Waugh R (2016) Genomic selection in multi-environment crop trials. G3 Genes Genom Genet 6:1313–1326

Ortiz R, Crossa J, Reslow F, Perez-Rodriguez P, Cuevas J (2022) Genome-based genotype × environment prediction enhances potato (Solanum tuberosum L.) improvement using pseudo-diploid and polysomic tetraploid modeling. Front Plant Sci 13:785196. https://doi.org/10.3389/fpls.2022.785196

Ortiz R, Reslow F, Montesinos-López A, Huicho J, Pérez-Rodríguez P, Montesinos-López OA, Crossa J (2023) Partial least squares enhance multi-trait genomic prediction of potato cultivars in new environments. Sci Rep 13:9947. https://doi.org/10.1038/s41598-023-37169-y

Ortiz R, Reslow F, Vetukuri R, García-Gil MR, Pérez-Rodríguez P, Crossa J (2024) Genomic prediction for inbred and hybrid polysomic tetraploid potato offspring. Agriculture 14:455. https://doi.org/10.3390/agriculture14030455

Ortiz R (2023) Challenges for crop improvement. Emerg Top Life Sci 7:197–205. https://doi.org/10.1042/ETLS20230106

Ovenden B, Milgate A, Wade LJ, Rebetzke GJ, Holland JB (2018) Accounting for genotype-by-environment interactions and residual genetic variation in genomic selection for water-soluble carbohydrate concentration in wheat. G3 Genes|Genomes|Genetics 8:1909–1919. https://doi.org/10.1534/g3.118.200038

Piepho HP (1997) Analyzing genotype-environment data by mixed models with multiplicative terms. Biometrics 53:761–766

Piepho HP (1998) Methods for comparing the yield stability of cropping systems. J Agron Crop Sci 180:193–213

Raffo MA, Jensen J (2022) Gene × gene and genotype × environment interactions in wheat. Crop Sci 63:1779–1793. https://doi.org/10.1002/csc2.20986

Resende RT, Piepho HP, Rosa GJ, Silva-Junior OB, e Silva FF, de Resende MDV, Grattapaglia D (2021) Enviromics in breeding: applications and perspectives on envirotypic-assisted selection. Theor Appl Genet 134:95–112

Rincent R, Charcosset A, Moreau L (2017) Predicting genomic selection efficiency to optimize calibration set and to assess prediction accuracy in highly structured populations. Theor Appl Genet 130:2231–2247

Rincent R, Charpentier J-P, Faivre-Rampant P, Paux E, le Gouis J, Bastien C, Segura V (2018) Phenomic selection is a low-cost and high-throughput method based on indirect predictions: proof of concept on wheat and poplar. G3 Genes Genom Genet 8:3961–3972

Roorkiwal M, Jarquin D, Singh MK, Gaur PM, Bharadwaj C, Rathore A, Howard R, Srinivasan S, Jain A, Garg V, Kale S, Chitikineni A, Tripathi S, Jones E, Robbins KR, Crossa J, Varshney RK (2018) Genomic-enabled prediction models using multi-environment trials to estimate the effect of genotype × environment interaction on prediction accuracy in chickpea. Sci Rep 8:11701. https://doi.org/10.1038/s41598-018-30027-2

Smith A, Cullis B, Thompson R (2001a) Analyzing variety by environment data using multiplicative mixed models and adjustments for spatial field trend. Biometrics 57:1138–1147. https://doi.org/10.1111/j.0006-341x.2001.01138.x

Smith A, Cullis B, Gilmour A (2001b) The analysis of crop variety evaluation data in Australia. Aust N Z J Stat 43:129–145

Tolley SA, Brito LF, Wang DR, Tuinstra MR (2023) Genomic prediction and association mapping of maize grain yield in multi-environment trials based on reaction norm models. Front Genet 14:1221751. https://doi.org/10.3389/fgene.2023.1221751

Van Eeuwijk FA, Denis JB, Kang MS (1996) Incorporating additional information on genotypes and environments in models for two-way genotype by environment tables. In: Kang MS, Gauch HG (eds) Genotype-by-environment interaction. CRC Press, Boca Raton, pp 15–49

Velazco JG, Jordan DR, Hunt CH, Mace ES, van Eeuwijk FA (2020) Genomic prediction for broad and specific adaptation in sorghum accommodating differential variances of single nucleotide polymorphism effects. Crop Sci 60:2328–2342. https://doi.org/10.1002/csc2.20221

Winans ND, Fonseca JMO, Perumal R, Klein PE, Klein RR, Rooney WL (2024) Envirotyping can increase genomic prediction accuracy of new environments in grain sorghum trials depending on mega-environment. Crop Sci 64:2519–2533. https://doi.org/10.1002/csc2.21213

Xavier A, Runcie D, Habier D (2025) Megavariate methods capture complex genotype-by-environment interactions. Genetics 229:iyae179

Yao Z, Yao M, Wang C, Li K, Guo J, Xiao Y, Yan J, Liu J (2025) GEFormer: a genotype-environment interaction-based genomic prediction method that integrates the gating multilayer perceptron and linear attention mechanisms. Mol Plant 18:527–549. https://doi.org/10.1016/j.molp.2025.01.020

Can Selection in Segregating Population Reduce the Effects of G × E Interactions in Breeding Programs?

13

Sanaz Khalifani, Hamid Hatami Maleki, and Reza Darvishzadeh

Abstract

The interaction between genotype and environment is central to crop improvement and the future of plant breeding. In early generations (e.g., F_3), the large number of genotypes often limits testing to a few locations, increasing the risk of losing promising materials due to genotype-by-environment (G × E) interactions. By leveraging advances in statistical modeling, molecular genetics, and multi-environment testing, breeders are increasingly equipped to develop varieties that are both productive and stable under a wide range of environmental conditions. Genomic selection (GS) is a powerful tool for early generations in plant breeding programs. By utilizing genomic markers and limited phenotypic data, GS predicts genomic estimated breeding value (GEBVs) and identifies stable genotypes. Effective management of G × E guarantee enhancing crop adaptability, ensuring consistent performance, and sustaining agricultural productivity amid global challenges.

Keywords

Envirotyping · Genomic selection · Multi-environment trials · Stability analysis · Segregating generations

S. Khalifani · R. Darvishzadeh (✉)
Department of Plant Production and Genetics, Faculty of Agriculture, Urmia University, Urmia, Iran
e-mail: r.darvishzadeh@urmia.ac.ir

H. H. Maleki
Department of Plant Production and Genetics, Faculty of Agriculture, University of Maragheh, Maragheh, Iran

R. Mohammadi et al. (eds.), *Genotype x Environment Interactions and its Implications for Plant Breeding*, https://doi.org/10.1007/978-981-95-5664-9_13

13.1 Introduction

Living organisms are identified neither by their genes nor by their environment; they are the result of the interaction of genes and environment (Crossa 2012). Genotype-by-environment interaction (G × E) refers to the differential performance of genotypes across diverse environments. Indeed, G × E is observed when the relative performance of two or more genotypes varies across environments, resulting in changes in genotype rankings or the magnitude of genetic, environmental, and phenotypic variances (Bowman 1972). G × E is interpreted as differential genotypic sensitivities to environments and produce a great challenge to crop improvement in a target region (De Leon et al. 2016). G × E is a critical point in plant breeding, particularly in selecting genotypes that perform consistently across target environments. For decades, breeders and biometricians have sought to understand and manage G × E and its implications for yield stability (Begna 2020). A significant G × E reduces the predictability of genotype performance across environments, complicating selection decisions. Conversely, a non-significant G × E simplifies selection, as the top-performing genotype in one environment tends to perform similarly in other environments (Begna 2020, 2021).

Phenotypic expression depends on the genotype, the environment, and their interaction (De Leon et al. 2016). G × E contributes to genotype-specific responses, causing unstable performance across environments. A significant interaction may arise from either changes in the magnitude of genotypic differences or changes in genotype rankings (Falconer 1952). In this regard, G × E interactions are classified into crossover (changes in genotype ranking) and non-crossover (changes in performance magnitude) types. These interactions arise because environmental factors influence gene expression, producing a range of phenotypes for the same genotype across different conditions (van Eeuwijk et al. 1996; Begna 2020). G × E interactions are common for quantitative traits such as grain yield, plant height, and biomass, all of which are central to crop breeding objectives (Zhuang et al. 1997). G × E and yield stability analyses remain essential for assessing varietal suitability across seasons and ecological zones (Habib et al. 2024). The degree of G × E is typically more pronounced in environments that differ greatly in stress factors such as drought, temperature extremes, soil fertility, and pathogen pressure. These environmental factors can be internal or external, biotic or abiotic, and may act at micro- and macro-environmental levels (Begna 2020). Plant breeders use statistical models to dissect G × E interactions, separating predictable components from unpredictable ones and identifying stable genotypes for cultivation (Teressa et al. 2021). Several parametric and non-parametric statistical methods have been developed to analyze G × E, to improve selection efficiency and genetic gain across variable environments. The primary goal remains the identification of stable and high-performing genotypes (De Leon et al. 2016). The applied plant breeders now have the ability to manipulate genes identified as mechanistically involved in G × E. Furthermore, understanding the principles of G × E interactions enable breeders to design experiments that efficiently assess the performance of genotypes under varied conditions (Napier et al. 2023). Multi-environment trials (METs) are vital for evaluating genotype performance under varied conditions, enabling plant breeders to select adaptable and resilient varieties (Fernandes Filho et al. 2025).

Throughout the history of crop domestication and modern breeding, managing G × E has been a persistent objective. Today, breeders possess advanced tools, including genomic selection (GS) and marker-assisted selection (MAS), which enable precise identification of alleles associated with favorable responses to environmental variability (Crossa et al. 2017; Paniza 2024). As climate change intensifies environmental variability, understanding G × E interactions becomes increasingly critical for ensuring crop productivity and food security (Chapman et al. 2012). These tools allow for more targeted and efficient breeding strategies, supporting the development of genotypes that are both high-yielding and resilient. As climate change intensifies the variability of agricultural environments, understanding and managing G × E become increasingly critical (Chapman et al. 2012). Breeding programs must anticipate future environmental conditions and integrate adaptive capacity into genotype development. This approach ensures long-term crop resilience and food security.

13.2 Populations Differences in Response to G × E

In plant breeding, qualitative traits can be selected through limited nursery tests, whereas complex traits like yield, heavily influenced by environmental factors, require testing across multiple environments (Richards et al. 2010). In early generations (e.g., F_3), the large number of genotypes often limits testing to a few locations, increasing the risk of losing promising materials due to G × E interactions (Ntare and Aken'Ova 1985). Even in populations with similar reaction norms, differences among populations may reflect historical G × E interactions shaped by distinct evolutionary histories (Saltz et al. 2018). These differences suggest that selection pressures or genetic drift varied, leading to unique reaction norms in ancestral populations.

Moreover, genetic drift, particularly in small or isolated populations, can significantly influence the development of reaction norms. Random fluctuations in allele frequencies may lead to the fixation of traits regardless of their adaptive value, potentially masking historical G × E signals and complicating interpretations of evolutionary trajectories (Star and Spencer 2013). Understanding G × E and local adaptation requires a multifaceted research approach. This includes integrating genetic data with ecological context and considering historical and contemporary environmental interactions. By doing so, researchers can more effectively disentangle the roles of selection, drift, and plasticity in shaping phenotypic diversity. Understanding these dynamics is crucial for evolutionary biology and predicting population responses to future environmental change. Ultimately, historical G × E interactions continue to influence current phenotypic expression, offering valuable insights into the evolutionary potential of plant populations.

13.3 Selection in Plant Breeding

Plant breeding enhances desirable traits to meet producer and consumer needs, relying on genetic diversity as its foundation. Selection is the cornerstone of crop improvement, enabling breeders to enhance performance, quality, and resistance to biotic and abiotic stresses (Vieira et al. 2025). Genetic diversity is critical for improving quantitative traits like yield and disease resistance, which are controlled by multiple genes and influenced by G × E interactions (Buckler et al. 2006). Consequently, the accurate assessment of quantitative variation and the identification of its underlying genetic determinants are essential for achieving substantial and sustained genetic gains.

The initial phase of most breeding programs involves the creation of genetically diverse populations. This diversity is commonly introduced through the crossing of genetically unrelated parents or mutagenesis of existing germplasm (Rauf et al. 2010). This initial genetic variation creates the basis for effective selection by enabling the identification of superior genotypes. Moreover, the planned and strategic enrichment of genetic resources not only helps avoid genetic bottlenecks and the accumulation of deleterious alleles but also accelerates the development of improved cultivars with enhanced resilience and adaptability (Salgotra and Chauhan 2023).

Selection involves isolating elite genotypes from segregating populations, such as F_2, F_3, backcross (BC), recombinant inbred lines (RILs), or doubled haploids (DHs) which are typically derived from well-designed breeding crosses (Acquaah 2009).

The effectiveness of selection largely depends on the breeder's ability to accurately identify superior individuals within a population exhibiting a wide range of genetic variation. Although selection is often based on observable phenotypic traits, it is important to recognize that phenotype is the result of both genetic and environmental factors. This complexity becomes especially pronounced when traits are influenced by recessive alleles, minor-effect genes, or when gene expression is significantly affected by environmental conditions (Vieira et al. 2025). As a result, the efficiency of selection is closely linked to the degree of correspondence between phenotype and genotype. In the past, cultivated plants were primarily shaped by natural selection, which favored genotypes best adapted to prevailing environmental conditions (Ackerly et al. 2000). In contrast, modern plant breeding relies on targeted and intentional selection conducted under controlled conditions to identify and propagate genotypes that possess specific, desirable traits (Lamichhane and Thapa 2022). This shift from natural to artificial selection has significantly enhanced the precision and effectiveness of breeding programs in developing improved crop varieties (Lamichhane and Thapa 2022).

13.4 Integrating Genomic Tools and G × E Models to Improve Selection Accuracy and Crop Resilience

Phenotypic selection (PS) is one of the traditional plant breeding methods that involves selecting parents based on observable traits of the plants. Due to its time-consuming and resource requirements, especially in variable environments, this method may not always be fully sufficient. Therefore, marker-assisted selection (MAS) methods and quantitative trait locus (QTL) analyses are employed to identify genetic markers associated with traits, enhancing both the speed and precision of the selection process. With advances in genetics, marker-assisted selection (MAS) methods have been developed, utilizing QTL studies to identify gene loci that affect quantitative traits, enabling indirect selection in early stages of plant growth (Hospital and Charcosset 1997). MAS is effective for traits with simple genetic control and major-effect QTLs; however, its efficiency is limited for complex traits with polygenic control and significant environmental effects (Bernardo and Charcosset 2006). To overcome the limitations of MAS, GS (genomic selection) was introduced, which uses comprehensive genomic estimated breeding values (GEBVs) of genotypes (Sabadin et al. 2022; Gholamiasl et al. 2025). Unlike MAS, which focuses on specific QTLs, GS considers the entire genome and is applicable to complex traits with low heritability. This method offers a more accurate and rapid assessment of the genetic composition of plants, particularly for traits with lower heritability that are influenced by environmental factors. Among its advantages are reduced breeding time, preservation of genetic diversity, and increased productivity.

From a genetic perspective, parameters such as heritability, genetic variance, and population structure directly influence the success of GS. The use of models (such as GOCS and OPV[1]) that can maintain genetic diversity while optimizing the contribution of parents has proven to be highly effective in preserving long-term stability and increasing productivity. Researches by da Silva et al. (2021) and Covarrubias-Pazaran et al. (2022) have shown that G × E interactions can have a significant impact on genetic progress, especially in traits affected by environmental fluctuations. Therefore, employing more precise models to understand these effects and fine-tune selection strategies is crucial.

Multi-trait selection, especially for independent and complex traits, benefits from multivariate Bayesian models or multi-objective optimization frameworks such as MOOB,[2] which enhance prediction accuracy and help maintain a balance between various traits. These methods are particularly useful when traits are correlated with each other or when it is difficult to assign an economic weight to each trait (Akdemir et al. 2019).

---

[1]GOCS (genomic optimal cross-selection) and OPV (optimal population value) selection are population-based genomic selection methods aimed at improving breeding program efficiency.

[2]MOOB (multi-objective optimized genomic breeding) leverages genomic selection and optimization techniques to enhance plant breeding programs by considering multiple traits simultaneously, accelerating genetic gain, and improving breeding strategies.

Finally, one of the most important factors in the success of breeding programs is considering G × E interactions. Buntaran et al. (2022) demonstrated that GS models that account for environmental effects can predict the performance of genotypes more accurately in specific environments. Using G × E models that incorporate genotype-year and genotype-location interactions, and focusing on traits with medium to high heritability, plays a significant role in developing resilient and stable germplasm.

Overall, advances in genetic tools, statistical models, and data-collection technologies have propelled plant breeding toward more precise, faster, and more sustainable selection. The application of these methods can not only improve target traits but also ensure the long-term preservation of genetic diversity and sustainability.

13.5 Optimizing Selection Strategies in Early Segregating Generations for Crop Improvement

Segregating generations play a pivotal role in crop improvement programs, as they provide plant breeders with the opportunity to develop superior commercial cultivars through continued selection and advancement (Gibely 2021). The F_2 generation is particularly important in this context, as it exhibits the highest level of genetic segregation available for selection. Heterozygosity peaks in the F_2 generation and decreases by approximately 50% in each subsequent self-pollinated generation. Therefore, selection in the F_2 generation is typically performed on individual plants, while in the F_3 generation, it is conducted at the family level (Acquaah 2009).

Model-based studies suggest that selection in early generations can be effective, especially when non-genetic variation (such as G × E interaction) is not excessively high. This is critical for the success of breeding programs. In such models, phenotype is regarded as the sum of genetic and environmental effects, and dominance variance is generally disregarded due to its rapid decline under selfing. Thus, only additive and additive-type epistatic variances are considered in the analysis (Yang 2009).

In early generations such as F_2 and F_3, due to the relatively low genetic variance, increasing the number of replications is necessary to accurately identify superior genotypes. Two major selection strategies are typically evaluated (Table 13.1): simple selection (among lines or sub-lines) and combined selection, in which information from multiple generations is integrated using weighted indices. Additionally, a two-stage selection strategy has been examined, wherein initial selection occurs among F_2 lines followed by selection among sub-lines in subsequent generation (Wang et al. 2003).

Literature review indicates that despite the lower genetic variance in early generations, effective selection is still possible if non-genetic variance—especially that arising from G × E or year effects—remains within acceptable limits. In addition, combined and two-stage selection strategies, when properly designed

Table 13.1 Comparison of different selection strategies in plant breeding in relation to G × E interaction and associated tools

Selection strategy	Description	Advantages	Limitations	Impact on G × E	Complementary tools
Simple selection	Selection of genotypes based on phenotype in F_2	Simple, low-cost	Sensitive to G × E	Limited reduction	Traditional phenotyping
Combined selection	Use of data from multiple generations with weighted indices	Improved accuracy	Requires multi-year data	Moderate reduction	AMMI, GV
Two-stage selection	Initial selection in F_2, then in F_3 with MET	Reduces MET costs	Complex design	Significant reduction	MET, advanced phenotyping
GS	Prediction of GEBV using genomic markers	High accuracy	Requires genomic infrastructure	Strong reduction	Speed-GS, gBLUP
Multi-trait selection (GYT/MTSI)	Simultaneous evaluation of multiple traits	Considers trait correlations	Computational complexity	Effective reduction	GYT, MTSI

(e.g., optimal number of lines and sub-lines, appropriate index weighting), outperform simple selection (Cochran 1951).

GS is a powerful tool for early generations in plant breeding programs (Table 13.1). By utilizing genomic markers and limited phenotypic data, GS predicts GBVs and identifies stable genotypes. A simulation study on wheat breeding programs demonstrated that GS leads to greater genetic gain compared to phenotypic selection (PS). Specifically, the use of genomic information enhanced selection accuracy and increased genetic gain up to threefold (Tessema et al. 2020). The observed increase in genetic gain with GS was approximately 300%. Similarly, a previous study by Gaynor et al. (2017) reported that implementing GS in a wheat breeding program resulted in a 1.21-fold increase in genetic gain. Bonnett et al. (2022) also highlighted that GS in early wheat generations (e.g., the F_2 generation) can significantly enhance genetic gain, although its success depends on the appropriate choice of statistical methods for predicting GEBVs. Furthermore, GS has shown acceptable accuracy in predicting traits related to baking quality, and by enabling early selection, it can double the selection response, thereby facilitating the development of high-yielding and high-quality wheat cultivars (Michel et al. 2018). In addition, the use of multi-trait and multi-environment GS models, which account for G × E interactions, has been shown to increase GEBV prediction accuracy by up to 9.6%, particularly for traits with low heritability (Ward et al. 2019).

In another study, Sousa et al. (2019) evaluated the accuracy of GS in the F_2 generation of Arabica coffee and found that GS could reduce the selection cycle time by 50%, thereby increasing genetic gain per unit time.

In advanced stages of plant breeding, data from METs play a crucial role in identifying superior genotypes (Table 13.1). In this context, the multi-trait stability index (MTSI) has been introduced as a novel approach for selecting genotypes based on both mean performance and stability (Koundinya et al. 2021). Although MET analyses often focus on a single trait, considering multiple traits simultaneously can significantly improve the accuracy of genotype selection (Koundinya et al. 2021). MTSI is based on genotype performance and stability across diverse environments, where lower index values indicate higher genotype stability across multiple traits (Olivoto et al. 2019a). This method is implemented using the weighted average of absolute scores (WAASB), derived from the singular value decomposition of the BLUP matrix. It has been successfully applied in various crops including, soybeans under drought and salinity stress (Zuffo et al. 2020), drought-tolerant cassava (Koundinya et al. 2021), black beans for cultivar recommendation (Ambrósio et al. 2024), Norwegian spruce (Alexandru et al. 2023), and sunflower (Akbari and Darvishzadeh 2025). Due to its ability to account for the correlation structure among traits, MTSI is considered a robust and practical tool for selecting high-performing and stable genotypes under multi-environment conditions. Collectively, these findings highlight that in breeding programs—especially during early yield testing stages—GS models that incorporate G × E and multiple traits are effective tools for predicting the performance of new genotypes, even when phenotypic data from some environments are incomplete.

13.6 Genomic Selection Strategies to Mitigate the Effects of G × E in Early Generations

GS, as an advanced plant breeding method, utilizes genomic marker data and phenotypic information to estimates GBVs, thereby accelerating genetic gain (Crossa et al. 2017). However, G × E, particularly in early breeding generations such as F_1 and F_2, poses significant challenges. This interaction, caused by environmental heterogeneity—such as variations in soil and climate conditions—can reduce the accuracy of GEBV prediction and affect genotype performance stability (Lado et al. 2013). G × E refers to differential genotype performance across environments, which is exacerbated in early generations due to limited phenotypic data and environmental variability. Such heterogeneity can increase experimental error and reduce GEBV prediction accuracy. Furthermore, early parent selection is constrained by delays in phenotypic data availability, limiting prediction accuracy at early stages. To mitigate the effects of G × E and enhance the accuracy of GS in early generations, the implementation of METs, combined with environmental typing, enables more accurate estimation of environmental effects and G × E interactions. Spatially adjusted statistical models, such as those proposed by Lado et al. (2013), can correct for environmental heterogeneity and improve GEBV prediction accuracy. This approach helps reduce errors caused by environmental variability and ensures genotype performance stability.

Advanced phenotyping technologies, including aerial imaging and high-resolution sensors, facilitate large-scale, high-precision phenotypic data collection (Pandey et al. 2016). These technologies improve the quality of input data for GEBV prediction, minimize experimental error, and enable more accurate selection. Incorporating phenotypic data from early generations (e.g., F_2) into the training population enhances GEBV accuracy during initial breeding cycles. This strategy, through continuous phenotypic data updates, mitigates the negative effects of G × E and supports the preservation of genetic diversity. Accurate parent selection using haplotype-based breeding panels and genomic BLUP (gBLUP) can reduce G × E effects while maintaining genetic diversity. The Speed-GS method, which integrates GS with speed breeding (SB), accelerates the production of homozygous lines and facilitates genotype evaluation across diverse environments (Voss-Fels et al. 2019). Avoiding selection of closely related individuals and incorporating exotic germplasm also help preserve genetic diversity.

Generation interval reduction methods—such as rapid generation advance (RGA), speed breeding, and doubled haploid (DH) techniques—expedite the breeding process (Araus et al. 2018). These methods enable early genotype evaluation in target environments by rapidly producing homozygous lines and decreasing the impact of G × E through increased breeding cycles. Effective population size directly influences the mitigation of G × E effects and the maintenance of genetic diversity. Larger populations reduce genetic drift and preserve favorable alleles, resulting in more effective selection responses (Meuwissen 2009). However, managing large populations can be costly and may increase experimental errors. The use of DNA markers for genomic typing and targeted sampling strategies (e.g., terminal sampling) can address these challenges and improve selection efficiency.

Reducing the impact of G × E interactions in early breeding generations requires integrating GS strategies with advanced technologies and precise population management. Multi-environment trials, high-throughput phenotyping, incorporation of early-generation phenotypes into the training set, precise parental selection, and shortened breeding cycles are among the most effective strategies to enhance GEBV accuracy and genotype performance stability. These approaches, through preserving genetic diversity and optimizing selection intensity, ensure long-term genetic gain and contribute to the development of resilient and high-yielding cultivars (Sinha et al. 2021).

13.7 Impact of Climate Change on G × E Interaction in Segregating Populations

Climate change, through increasing the frequency of extreme events such as drought and heat stress, has intensified the heterogeneity of agricultural environments, thereby complicating G × E. This complexity is particularly challenging during the early stages of plant breeding, when genotypes are still segregating and phenotypic data are sparse. The increased occurrence of crossover interactions across environments alters genotype rankings, reducing selection accuracy, and limiting genetic gain. Consequently, breeding programs may be forced to partition target regions and lower selection intensity (Xiong et al. 2022). To manage this complexity, approaches such as GS, performance modeling, and envirotyping have proven useful in enhancing selection efficiency within segregating populations and improving yield stability under unstable climatic conditions (Xiong et al. 2022). As temporal and spatial environmental variability increases, the predictability of genotype performance significantly declines, making selection decisions more difficult. Furthermore, this variability reduces trait measurement repeatability, leading to lower heritability and diminished selection accuracy in early generations (Crossa et al. 2017). In response to escalating climate change, breeding programs must adapt their selection strategies, particularly in segregating populations. Identifying repeatable patterns of G × E remains essential. When such patterns are present, the target region can be divided into mega-environments, enabling the use of genotype-specific cultivars to enhance heritability and accelerate genetic gain. Otherwise, the entire region must be treated as a single unit, and G × E must be managed through robust experimental designs (Yan et al. 2023).

In this context, envirotyping) supported by phenomics technologies and controlled platforms(has emerged as a powerful tool in modern breeding. It facilitates the identification of environmental factors influencing G × E, classification of environments, optimization of trial sites, and a deeper understanding of gene–environment interactions (Xu 2016). Historically, breeding relied mainly on phenotypic data. However, with the advent of "tri-typing" technologies,[3] combining

[3]Tri-typing refers to a combined approach in plant breeding that considers genotype, phenotype, and envirotype (environmental factors) to understand and predict plant performance.

genotypic, phenotypic, and environmental (G-P-E) components across time and space has become crucial for predicting genotype performance across diverse conditions (Xu et al. 2022). Large-scale implementation of breeding programs through METs demands advanced computational tools, robust algorithms, and efficient data management systems (Xu et al. 2022). To utilize these complex datasets effectively, standardized ontologies, uniform protocols, high-precision instrumentation, and high-quality data collected under realistic field conditions are essential. The integration of genotypic, phenotypic, and environmental data via modern platforms such as data warehouses and data lakes significantly enhances prediction accuracy and selection efficiency (Harfouche et al. 2019). Given the ongoing climate-related challenges, selection strategies in segregating populations must increasingly incorporate predictive models that integrate genomic data with precise environmental covariates through envirotyping. Early-stage METs combined with high-throughput phenotyping (HTP) allow breeders to capture relevant G × E patterns and identify genotypes with superior performance and adaptability. When combined with GS, these tools can accelerate genetic gain even under climatic uncertainty.

13.8 The Role of Environmental Indices in Genomic Prediction Modeling for Plant Breeding

In recent years, environmental indices (EIs) have gained considerable importance in genomic prediction modeling for plant breeding (Table 13.2). These indices utilize environmental data to enhance the analysis of G × E, particularly in evaluating genotype sensitivity to environmental conditions (Costa-Neto et al. 2021). EI-based models contribute to optimizing METs and can help mitigate or manage G × E (Piepho 2022). EIs also serve as novel predictors of genotype performance across environments, enabling more sophisticated regression models that better simulate genetic–environmental relationships (Li et al. 2022). Machine learning algorithms such as neural networks, deep learning, and support vector machines (SVM) are increasingly employed in prediction models to capture nonlinear interactions among genomic, phenotypic, and environmental variables (Montesinos López et al. 2022).

Advancements in proximal and remote sensing technologies have enabled more precise measurement of environmental variables, significantly improving prediction accuracy for genotype selection under diverse climate scenarios (Kusmec et al. 2021). This trend is especially impactful in selecting climate-resilient genotypes, thereby enhancing crop performance under variable environmental conditions (Beans 2020; Cooper et al. 2014).

Ultimately, incorporating environmental indices into genomic prediction models holds great potential for accelerating genetic progress and improving genotype selection in variable and challenging environments.

Table 13.2 Practical applications of environmental indexing (EI) in plant breeding and genomic prediction

Axis	Practical application description
G × E management in plant breeding	Using EI allows breeders to more accurately select genotypes adapted to specific environments at early stages. This leads to cost reduction and increased efficiency.
Optimizing trial site selection	By leveraging GIS– G × E or kernel-based models, sites can be selected that maximize environmental diversity or desired genetic/environmental similarity.
Predicting performance in new environments	Combining sensor data and satellite imagery with prediction models enables forecasting genotype performance under future conditions or in previously untested locations.
Modeling G × E × M interactions	Including M (management) as an environmental variable in models allows predicting genotype responses to different management practices, which is key for precision agriculture planning.
Improving GS accuracy	Incorporating EIs into GS models can increase prediction accuracy of breeding values, especially under stress conditions or diverse environments.
Identifying genes adapted to specific climates	Combining climate data with genotypic information enables identification of genes or QTLs associated with adaptation to specific climates, which is critical for climate-resilient breeding.

13.9 Measuring G × E Interaction

The interaction of genotype with the environment is one of the main and most complex issues in plant breeding programs for developing high-yielding, compatible, and stable cultivars. To identify cultivars with high and stable yields for different regions with diverse climatic conditions, multiple trials across multi-years and multi-locations have been proposed (Reckling et al. 2021). In such experiments, it is necessary to address the degree of stability of cultivars in addition to performance criteria. Several statistical methods have been used to increase awareness of the interaction of genotype × environment and its relationship with sustainability (Table 13.3). A suite of univariate as well as multivariate statistical methods has been introduced for the interpretation of G × E (Flores et al. 1998). One prevalence approach exists to analyze genotype stability is additive main effects and multiplicative interaction (AMMI) (Gauch Jr 1988). The AMMI model in combination with analysis of variance (ANOVA) outputs AMMI1 and AMMI2 biplots, with the first displaying the genotypic means and their relationship to the first principal component analysis (PCA) and the second showing genotypes' relationships to the first two PCAs (Bocianowski et al. 2019). An alternative method for assessing stability is by the use of genotype × genotype-by-environment (GGE) biplot analysis, as described by Yan and Kang (2002). The GGE biplot offers a visual depiction of G × E effects, enabling the detection of trends, correlations, and outliers. This facilitates the interpretation and discussion of data. The AMMI and GGE models are well

Table 13.3 Statistical methods for G × E interaction analysis

Statistical method	Mathematical formula/model	Software or implementation tool	Reference
AMMI	$\gamma_{\text{ger}}=\mu+\alpha_g+\beta_e+\sum_n \lambda_n\gamma_{gn}\delta_{en}+\rho_{ge}+K_{r(e)}+\varepsilon_{ger}$	AMMISOFT, R (Packages: agricolae, metan, ammistability)	Zobel and Gauch (1996)
GGE Biplot	$\widehat{Y}_{ij}-u-\beta_j=\lambda_1\xi_{I1}\eta_{1J+}\lambda_2\xi_{i2}\eta_{2j}+\varepsilon_{ij}$	GEA-R Software, R (packages: GGEBiplotGUI, metan)	Yan and Kang (2002)
WAASB_g	$\text{WAASB}_g=\frac{\sum_{n=1}^{p}\left\lvert \text{IPCA}_{gn}\times \text{EP}_n\right\rvert}{\sum_{n=1}^{p}\text{EP}_n}$	metan (R package)	Olivoto et al. (2019b)
WAASBY	$\text{WAASBY}_g=\frac{\left(\text{rG}_g+\theta_Y\right)+\left(\text{rW}_g+\theta_S\right)}{\theta_Y+\theta_S}$		
MTSI	$\text{MTSI}_i=\left[\sum_{j=1}^{f}\left(F_{ij}-F_j\right)^2\right]^{0.5}$		
HMGV	$\text{HMGV}=\frac{E}{\sum_{J=1}^{E}\frac{1}{\text{GV}_{ij}}}$	Selegen-REML/BLUP (standalone Software by Resende); also in R using packages breedR, lme4	De Resende (2007)
RPGV	$\text{RPGV}=\frac{1}{E}\sum_{j=1}^{E}\text{CV}_{ij}/\mu_j$		
HMRPGV	$\text{HMRPGV}=\frac{E}{\sum_{j=1}^{E}\frac{1}{\text{GV}_{ij}/\mu_j}}$		

recognized as two prominent stability approaches, particularly when two principal components (PCs) hold major importance. Recently, stability analysis approaches based on mixed models (REML/BLUP) are currently in wide use in plant breeding programs (de Sousa et al. 2020). The REML (restricted maximum likelihood) procedure estimates the variance components and genetic parameters, and BLUP (best linear unbiased prediction) is the ideal selection procedure (de Resende 2016). In the context of mixed models, some stability parameters such as harmonic mean of the genotypic values (HMGV), relative performance of genotypic values (RPGV), and harmonic mean of the relative performance of the genotypic values (HMRPGV) can be calculated to simultaneously evaluate yield stability and adaptability (de Resende 2016). Another new method introduced for analysis of multi-trials in multi-environments is WAASB (weighted average absolute scores of BLUPs) which was described by (Olivoto et al. 2019b). This method profits from the advantages of both AMMI and BLUP methods. Briefly, in WAASB, singular value decomposition was applied to the BLUP matrix to analyze G × E generated by a linear mixed model. Then, the biplot of WAASB × dependent variable (Y) which is defined as WAASBY was used to jointly interpret stability and dependent variable productivity. But the notable clue is that selection based on only one attribute such as yield (as an economic part of the plant) is not the most appropriate strategy, because yield could be influenced by other plant characteristics. Therefore, plant breeders are going to incorporate several traits into an identified stable genotype. To solve this challenge, two approaches, including genotype by yield × trait (GYT) (Yan and Frégeau-Reid 2018) and MTSI (Olivoto et al. 2019a), were proposed to study yield stability considering multi-traits simultaneously. In the GYT method, there are graphical representations which make it an effective and comprehensive method and also superiority index (SI) value to assess genotypes based on yield × trait combinations. However, the MTSI index is calculated based on the distance from the ideal genotype estimated through factor analysis using all measured traits simultaneously.

13.10 Analysis of G × E and Modern Tools in Plant Breeding

G × E represents one of the major challenges in selecting superior genotypes in plant breeding programs. Various statistical models have been developed to analyze this complex interaction, each offering unique features and advantages (Abdi et al. 2025a, b). One of the most effective models is the additive main effects and multiplicative interaction (AMMI) model, which integrates analysis of variance (ANOVA) for the main effects of genotype and environment with principal component analysis (PCA) for the G × E interaction component. This model facilitates the identification of stable and adaptable genotypes (Zobel et al. 1988). In early generations, where environmental heterogeneity and intense G × E interactions reduce selection accuracy, the AMMI model serves as a valuable tool for breeders through visual aids such as AMMI1 and AMMI2 biplots. For instance, in a study on spring barley involving the thousand grain weight of 32 genotypes across six locations, 24.5% of the total variance was attributed to G × E. The AMMI model effectively identified highly stable genotypes (Bocianowski et al. 2019). Similarly, in maize, this model successfully differentiated F_1 hybrids and inbred parents in terms of performance across environments (Abuali et al. 2014).

The GGE biplot model, which focuses on genotype main effects and G × E interaction (excluding environmental effects), provides a visual and interpretable approach for multi-environment trial (MET) analysis. Particularly valuable in early generations with high genetic variability, this model enables the identification of genotypes with either specific adaptability or general stability (Yan and Kang 2002). For example, in a study on 488 maize hybrids across multiple environments in South Africa, the GGE biplot efficiently identified superior genotypes in terms of stability and adaptability (Musundire et al. 2021). Moreover, in an analysis of canola parents and their F_1 hybrids, the model contributed to identifying effective heterotic combinations (Payghamzadeh and Oghan 2024).

In recent years, the weighted average of absolute scores from BLUP (WAASB) has emerged as an innovative tool for analyzing G × E, combining the AMMI model with mixed models (REML/BLUP). By utilizing BLUP estimates, this method enhances prediction accuracy in unbalanced or incomplete datasets (Olivoto et al. 2019b). For example, in a study involving 16 chickpea genotypes across 10 locations, AMMI, BLUP, and WAASB models were simultaneously applied, and genotype G1 was identified as superior (Arshad et al. 2024). Similarly, in an evaluation of 32 soybean genotypes across eight environments over 2 years, the WAASB and WAASBY indices identified MACS 1620 as a stable and high-yielding genotype, while the MTSI index highlighted DSb 33 for its earliness and high-yield potential (Nataraj et al. 2021). In another study on 28 sunflower genotypes across semi-arid environments, AMMI analysis revealed strong G × E interaction. WAASB and WAASBY indices identified GMU-494, LTRR-341, EC-601901, and GMU-481 as stable genotypes, while the MTSI index identified EC-198078 as superior in terms of yield, oil quality, and resistance to downy mildew (SDM) (Dudhe et al. 2024). Additionally, in a wheat trial, genotypes G12 and G17 were identified as the most drought- and heat-tolerant using MGIDI, WAASB, and WAASBY indices (Al-Ashkar et al. 2023).

With technological advances, the integration of genomics, phenomics, and enviromics has emerged as a modern approach for analyzing complex traits and implementing breeding strategies (Crossa et al. 2024). In the early stages of breeding, genomic data are employed to select parents and design crosses, using GS indices (GS index), multivariate Bayesian decision theory, and genetic diversity data (Villar-Hernández et al. 2018). Cross-predictions are also conducted using simulation algorithms based on marker and phenotypic data from parents (Mohammadi et al. 2015). Since genotyping in early generations (e.g., F_2) is costly, machine learning (ML) models are increasingly applied for phenotypic screening. In the F_4 generation, family selection is based on A-matrix and HTP data, while in the F_5 generation, GS based on GEBVs is used to eliminate weak lines (Crossa et al. 2024). Krause et al. (2020) demonstrated that vegetation indices extracted from HTP at early growth stages could significantly enhance wheat yield prediction. To accelerate breeding programs, early sparse testing is implemented in intermediate generations, aiming to allocate genotypes to selected environments within the target population of environments (TPE). In this phase, genomic prediction models are used for unobserved environments. Montesinos-López et al. (2023) evaluated four sparse testing strategies and reported that multi-trait models outperformed single-trait models due to their higher accuracy and robustness against data reduction. Moreover, breeding cycles can be shortened using rapid-cycle recurrent GS (RCRGS). In a study at The International Maize and Wheat Improvement Center (known – even in English – by its Spanish acronym CIMMYT for Centro Internacional de Mejoramiento de Maíz y Trigo) (CIMMYT) on bread wheat, three cycles of RCRGS resulted in a 12% (0.28 t/ha) average increase in grain yield (Volpato et al. 2021). A similar study in tropical maize also reported enhanced genetic gain and shorter generation intervals (Zhang et al. 2017).

In this context, genomic prediction models, particularly multi-trait multi-environment (MTME) models that account for G × E interactions, play a pivotal role. MTME models improve prediction accuracy by leveraging trait correlations across environments and reducing the need for extensive field phenotyping (Dreisigacker et al. 2021). Montesinos-López et al. (2018) showed that combining item-based collaborative filtering (IBCF) with MTME models improved across-nursery prediction accuracy and was suitable for large-scale datasets. In F_1 hybrid breeding programs, artificial neural networks (ANNs) have been effectively used for predicting parental combinations. Khaki et al. (2020) demonstrated that ANN combined with matrix decomposition outperformed other prediction models.

These models, capable of capturing nonlinear relationships, offer superior performance over traditional methods and are particularly useful in analyzing genotypic stability and G × E interactions, especially when integrated with HTP (Harfouche et al. 2019; Niazian et al. 2018). Recently, machine learning models such as SVM, random forest, and ANN have been widely used in GS. These models effectively capture complex relationships among variables and have been instrumental in predicting traits like disease resistance, yield, and stress tolerance (González-Camacho et al. 2018; Montesinos-López et al. 2019). For example, SVM has shown high accuracy in predicting wheat rust resistance, and ANNs have excelled in predicting yield in maize and wheat (Alves et al. 2018; Han et al. 2022; Yildirim et al. 2022; Cvejić et al. 2023).

13.11 Perspectives and Challenges of Genomic Prediction in Plant Breeding with a Focus on G × E Interaction

The G × E interaction plays a crucial and determining role in the performance of crop plants, especially in the face of challenges posed by climate change and the increasing global demand for agriculture products (Bailey-Serres et al. 2019). While G × E was previously considered a limitation in breeding programs, it is now increasingly viewed as a valuable opportunity to enhance plant adaptability across diverse and changing environments (Bailey-Serres et al. 2019).

Genomic prediction in agriculture, particularly under the complex interplay of G × E, requires models capable of capturing these interactions multiplicatively and beyond simple linear relationships (Onogi et al. 2021). The use of multi-trait multi-environment (MTME) models has improved prediction accuracy; however, challenges such as varying trait correlations across different environments still remain (Gill et al. 2021). The first step in advancing this field involves designing comprehensive frameworks that integrate whole-genome sequencing data with multi-environment field trials. These datasets can be analyzed using tools such as environmental covariate search affecting genetic correlations (ECGC) algorithms and advanced methods like "mash" (multivariate adaptive shrinkage) to identify environmental gradients and genomic regions associated with G × E (Urbut et al. 2019).

Environmental data serve as a critical bottleneck for accurate prediction and effective selection of genotypes in unknown or future environments (Montesinos-López et al. 2022). Employing methods like principal component regression (PCR[4]), partial least squares (PLS) regression, and sophisticated machine learning algorithms helps overcome challenges related to model interpretability and optimizes the use of large-scale environmental datasets (Montesinos-López et al. 2022). The integration of genomic, phenomic, and enviromic data, combined with machine learning

[4]Principal components regression (PCR) reduces a large number of explanatory variables in a regression model down to a small number of principal components.

models, offers an innovative and powerful approach to enhance phenotypic prediction and increase breeding efficiency. This integrative strategy is especially beneficial in early breeding generations and variable environments, significantly accelerating both the speed and accuracy of selection (Crossa et al. 2024).

Single-cell transcriptomics also provides detailed insights into cellular responses to environmental stresses, bridging the gap between phenotypic traits and underlying regulatory mechanisms (Y. Wang et al. 2021). Moreover, genome editing technologies such as CRISPR/Cas9 empower researchers to manipulate genomic regions involved in G × E interactions, facilitating improved stress resilience in crops (Soyk et al. 2017).

Furthermore, studies on the costs and constraints of genetic plasticity reveal that while plasticity aids environmental adaptation, it can incur energetic costs and informational complexity (Napier et al. 2023). Phenomena such as conditional neutrality, where genes exhibit contrasting effects across environments, can be addressed through advanced breeding tools to optimize trait performance (Napier et al. 2023).

As we enter the era of big data, breeders must develop skills to analyze, integrate, and interpret vast datasets from diverse sources, as only through advanced statistical methods and machine learning algorithms can complex data be transformed into actionable breeding decisions.

Finally, although high-throughput phenotyping (HTP) data have enhanced phenotypic prediction accuracy, challenges such as low interpretability and the presence of non-genetic sources of variation remain significant. Effective utilization of these data demands a combination of deep biological knowledge, precise modeling, and advanced data processing technologies to achieve rapid, accurate, and sustainable genetic gains in breeding programs (Crossa et al. 2024).

13.12 Conclusion

In plant breeding, managing G × E interactions are pivotal for developing stable, high-yielding cultivars, particularly through selection in early segregating generations like F_2 and F_3. Targeted selection strategies, such as combined and two-stage selection, effectively reduce G × E effects by integrating multi-generational data and optimizing selection indices, outperforming traditional phenotypic methods. GS stands out as a transformative approach, leveraging whole-genome markers to predict GBVs with high accuracy, minimizing the impact of environmental variability and accelerating genetic gain. The integration of HTP and envirotyping enhances this process by providing detailed phenotypic and environmental data, enabling precise identification of genotypes resilient to diverse conditions. Advanced statistical models, including AMMI, GGE Biplot, WAASB, and the MTSI, offer robust tools to dissect G × E interactions and select for both performance and stability across multiple traits and environments. In the face of climate change, which exacerbates environmental heterogeneity, METs and environmental indices further refine selection by capturing repeatable G × E patterns.

Looking ahead, emerging technologies like CRISPR-based genome editing promise to fine-tune genetic responses to environmental challenges. Collectively, these integrated strategies empower breeders to mitigate G × E effects, enhance selection efficiency in segregating populations, and develop adaptable cultivars that ensure sustainable agricultural productivity amidst global environmental uncertainties.

References

Abdi N, Bordbar M, Darvishzadeh R, Rabiei B, Alipour H, Soufimaleky S, Hatami Maleki H, Jabbari M (2025a) A review of stability analysis methods in plant breeding with an emphasis on cereals I: a study of nonparametric and univariate parametric approaches. Cereal Res. (in press). https://doi.org/10.22124/cr.2025.29968.1857

Abdi N, Bordbar M, Darvishzadeh R, Rabiei B, Alipour H, Soufimaleky S, Hatami Maleki H, Jabbari M (2025b) A review of stability analysis methods in plant breeding with an emphasis on cereals II: an exploring of multivariate techniques and future prospects. Cereal Res. (in press). https://doi.org/10.22124/cr.2025.29969.1858

Abuali AI, Abdelmula AA, Khalafalla MM, Hamza NB, Abdalla AH, Idris AE (2014) Assessment of yield stability and adaptability of parental inbred lines and F1-hybrids of grain maize (Zea mays L.) using AMMI analysis. Br Biotechnol J 4:339–349

Ackerly DD, Dudley SA, Sultan SE, Schmitt J, Coleman JS, Linder CR, Sandquist DR, Geber MA, Evans AS, Dawson TE (2000) The evolution of plant ecophysiological traits: recent advances and future directions: new research addresses natural selection, genetic constraints, and the adaptive evolution of plant ecophysiological traits. Bioscience 50:979–995

Acquaah G (2009) Principles of plant genetics and breeding. Wiley, Chichester

Akbari N, Darvishzadeh R (2025) Selection of oilseed sunflower genotypes with high and stable seed yield using the multi-trait stability index (MTSI). Iran J Crop Sci. (in press)

Akdemir D, Beavis W, Fritsche-Neto R, Singh AK, Isidro-Sánchez J (2019) Multi-objective optimized genomic breeding strategies for sustainable food improvement. Heredity 122:672–683

Al-Ashkar I, Sallam M, Almutairi KF, Shady M, Ibrahim A, Alghamdi SS (2023) Detection of high-performance wheat genotypes and genetic stability to determine complex interplay between genotypes and environments. Agronomy 13:585. https://doi.org/10.3390/agronomy13020585

Alexandru A-M, Mihai G, Stoica E, Curtu AL (2023) Multi-trait selection and stability in Norway spruce (Picea abies) provenance trials in Romania. Forests 14:456

Alves GR, Teixeira IR, Melo FR, Souza RTG, Silva AG (2018) Estimating soybean yields with artificial neural networks. Acta Sci Agron 40:e35250

Ambrósio M, Daher RF, Santos RM, Santana JGS, Vidal AKF, Nascimento MR, Leite CL, de Souza AG, Freitas RS, Stida WF (2024) Multi-trait index: selection and recommendation of superior black bean genotypes as new improved varieties. BMC Plant Biol 24:525

Araus JL, Kefauver SC, Zaman-Allah M, Olsen MS, Cairns JE (2018) Translating high-throughput phenotyping into genetic gain. Trends Plant Sci 23:451–466

Arshad W, Aleem S, Khan MI, Saqib M, Zeeshan M, Mohsin MU, Malik A, Sharif I, Younas A, Ihsan F (2024) AMMI, BLUP, and WAASB stability models for the selection of stable chickpea (Cicer arietinum L.) genotypes in different agroecological regions of Pakistan. J Food Agric Technol Res 3:66–81

Bailey-Serres J, Parker JE, Ainsworth EA, Oldroyd GED, Schroeder JI (2019) Genetic strategies for improving crop yields. Nature 575:109–118

Beans C (2020) Crop researchers harness artificial intelligence to breed crops for the changing climate. Proc Natl Acad Sci USA 117:27066–27069

Begna T (2020) The role of genotype by environmental interaction in plant breeding. Am J Plant Sci 11:1483–1496
Begna T (2021) Genotype-environment interaction and stability analysis. Int J Agri Biosci 10(2): 120–127
Bernardo R, Charcosset A (2006) Usefulness of gene information in marker-assisted recurrent selection: a simulation appraisal. Crop Sci 46:614–621
Bocianowski J, Warzecha T, Nowosad K, Bathelt R (2019) Genotype by environment interaction using AMMI model and estimation of additive and epistasis gene effects for 1000-kernel weight in spring barley (Hordeum vulgare L.). J Appl Genet 60:127–135
Bonnett D, Li Y, Crossa J, Dreisigacker S, Basnet B, Pérez-Rodríguez P, Alvarado G, Jannink J-L, Poland J, Sorrells M (2022) Response to early generation genomic selection for yield in wheat. Front Plant Sci 12:718611
Bowman JC (1972) Genotype x environment interactions. Ann Génét Sél Anim 4:117–123
Buckler ES, Gaut BS, McMullen MD (2006) Molecular and functional diversity of maize. Curr Opin Plant Biol 9:172–176
Buntaran H, Bernal-Vasquez AM, Gordillo A, Sahr M, Wimmer V, Piepho H-P (2022) Assessing the response to genomic selection by simulation. Theor Appl Genet 135:2891–2905
Chapman SC, Chakraborty S, Dreccer MF, Howden SM (2012) Plant adaptation to climate change—opportunities and priorities in breeding. Crop Pasture Sci 63:251–268
Cochran WG (1951) Improvement by means of selection. In: Neyman J (ed) Proceedings of the second Berkeley symposium on mathematical statistics and probability. University of California Press, Berkeley, pp 449–470
Cooper M, Messina CD, Podlich D, Totir LR, Baumgarten A, Hausmann NJ, Wright D, Graham G (2014) Predicting the future of plant breeding: complementing empirical evaluation with genetic prediction. Crop Pasture Sci 65:311–336
Costa-Neto G, Galli G, Carvalho HF, Crossa J, Fritsche-Neto R (2021) EnvRtype: a software to interplay enviromics and quantitative genomics in agriculture. G3 Genes Genom Genet 11: jkab040
Covarrubias-Pazaran G, Gebeyehu Z, Gemenet D, Werner C, Labroo M, Sirak S, Coaldrake P, Rabbi I, Kayondo SI, Parkes E (2022) Breeding schemes: what are they, how to formalize them, and how to improve them? Front Plant Sci 12:791859
Crossa J (2012) From genotype x environment interaction to gene x environment interaction. Curr Genomics 13:225–244
Crossa J, Montesinos-López OA, Costa-Neto G, Vitale P, Martini JWR, Runcie D, Fritsche-Neto R, Montesinos-López A, Pérez-Rodríguez P, Gerard G (2024) Machine learning algorithms translate big data into predictive breeding accuracy. Trends Plant Sci. https://doi.org/10.1016/j.tplants.2024.04.003
Crossa J, Pérez-Rodríguez P, Cuevas J, Montesinos-López O, Jarquín D, De Los Campos G, Burgueño J, González-Camacho JM, Pérez-Elizalde S, Beyene Y (2017) Genomic selection in plant breeding: methods, models, and perspectives. Trends Plant Sci 22:961–975
Cvejić S, Hrnjaković O, Jocković M, Kupusinac A, Doroslovački K, Gvozdenac S, Jocić S, Miladinović D (2023) Oil yield prediction for sunflower hybrid selection using different machine learning algorithms. Sci Rep 13:17611
De Leon N, Jannink J, Edwards JW, Kaeppler SM (2016) Introduction to a special issue on genotype by environment interaction. Crop Sci 56:2081–2089
De Resende MDV (2007) SELEGEN-REML/BLUP: sistema estatístico e seleção genética computadorizada via modelos lineares mistos. Embrapa Florestas, Colombo
De Resende MDV (2016) Software Selegen-REML/BLUP: a useful tool for plant breeding. Crop Breed Appl Biotechnol 16:330–339
Dreisigacker S, Crossa J, Pérez-Rodríguez P, Montesinos-López OA, Rosyara U, Juliana P, Mondal S, Crespo-Herrera L, Govindan V, Singh RP (2021) Implementation of genomic selection in the CIMMYT global wheat program, findings from the past 10 years. Crop Breed Genet Genom 3:e210007

Dudhe MY, Jadhav MV, Sujatha M, Meena HP, Rajguru AB, Gahukar SJ, Ghodke MK (2024) WAASB-based stability analysis and validation of sources resistant to Plasmopara halstedii race-100 from the sunflower working germplasm for the semiarid regions of India. Genet Resour Crop Evol 71:1435–1452

Van Eeuwijk FA, Denis JB, Kang MS (1996) Incorporating Additional Information on Genotypes and Environments in Models for Two-Way Genotype by Environment Tables. Genotype-by-Environment Interaction, eds. M. S. Kang and H. G. Gauch, Jr., Boca Raton: CRC Press, pp. 15–50

Falconer DS (1952) The problem of environment and selection. Am Nat 86:293–298

Fernandes Filho CC, Lima Barrios SC, Santos MF, Nunes JAR, do Valle CB, Jank L, Rios EF (2025) Assessing genotype adaptability and stability in perennial forage breeding trials using random regression models for longitudinal dry matter yield data. G3 Genes Genom Genet 15: jkae306

Flores F, Moreno MT, Cubero JI (1998) A comparison of univariate and multivariate methods to analyze G×E interaction. Field Crop Res 56:271–286

Gauch HG Jr (1988) Model selection and validation for yield trials with interaction. Biometrics 44: 705–715

Gaynor RC, Gorjanc G, Bentley AR, Ober ES, Howell P, Jackson R, Mackay IJ, Hickey JM (2017) A two-part strategy for using genomic selection to develop inbred lines. Crop Sci 57:2372–2386

Gibely RH (2021) Variability, genetic components and selection response in segregating generations among some cotton crosses. J Sustain Agric Sci 47:1–12

Gill HS, Halder J, Zhang J, Brar NK, Rai TS, Hall C, Bernardo A, Amand PS, Bai G, Olson E (2021) Multi-trait multi-environment genomic prediction of agronomic traits in advanced breeding lines of winter wheat. Front Plant Sci 12:709545

Gholamiasl M, Darvishzadeh R, Fayaz Moghaddam A, Alipour H (2025) Genomic selection for grain yield in maize (Zea mays L.) under non stress and phosphorus deficit conditions. Iran J Crop Sci. (in press)

González-Camacho JM, Ornella L, Pérez-Rodríguez P, Gianola D, Dreisigacker S, Crossa J (2018) Applications of machine learning methods to genomic selection in breeding wheat for rust resistance. Plant Genom 11:170104

Habib MA, Azam MG, Haque MA, Hassan L, Khatun MS, Nayak S, Abdullah HM, Ullah R, Ali EA, Hossain N (2024) Climate-smart rice (Oryza sativa L.) genotypes identification using stability analysis, multi-trait selection index, and genotype-environment interaction at different irrigation regimes with adaptation to universal warming. Sci Rep 14:13836

Han X, Liu F, He X, Ling F (2022) Research on rice yield prediction model based on deep learning. Comput Intell Neurosci 2022:1922561

Harfouche AL, Jacobson DA, Kainer D, Romero JC, Harfouche AH, Mugnozza GS, Moshelion M, Tuskan GA, Keurentjes JJB, Altman A (2019) Accelerating climate resilient plant breeding by applying next-generation artificial intelligence. Trends Biotechnol 37:1217–1235

Hospital F, Charcosset A (1997) Marker-assisted introgression of quantitative trait loci. Genetics 147:1469–1485

Khaki S, Khalilzadeh Z, Wang L (2020) Predicting yield performance of parents in plant breeding: a neural collaborative filtering approach. PLoS One 15:e0233382

Koundinya AVV, Ajeesh BR, Sheela MN, Mohan C, Asha KI (2021) Genetic parameters, stability and selection of cassava genotypes between rainy and water stress conditions using AMMI, WAAS, BLUP and MTSI. Sci Hortic 281:109949

Krause MR, Mondal S, Crossa J, Singh RP, Pinto F, Haghighattalab A, Shrestha S, Rutkoski J, Gore MA, Sorrells ME (2020) Aerial high-throughput phenotyping enables indirect selection for grain yield at the early generation, seed-limited stages in breeding programs. Crop Sci 60: 3096–3114

Kusmec A, Zheng Z, Archontoulis S, Ganapathysubramanian B, Hu G, Wang L, Yu J, Schnable PS (2021) Interdisciplinary strategies to enable data-driven plant breeding in a changing climate. One Earth 4:372–383

Lado B, Matus I, Rodriguez A, Inostroza L, Poland J, Belzile F, del Pozo A, Quincke M, Castro M, von Zitzewitz J (2013) Increased genomic prediction accuracy in wheat breeding through spatial adjustment of field trial data. G3 Genes Genom Genet 3:2105–2114

Lamichhane S, Thapa S (2022) Advances from conventional to modern plant breeding methodologies. Plant Breed Biotechnol 10:1–4

Li M, Zhang Y-W, Zhang Z-C, Xiang Y, Liu M-H, Zhou Y-H, Zuo J-F, Zhang H-Q, Chen Y, Zhang Y-M (2022) A compressed variance component mixed model for detecting QTNs and QTN-by-environment and QTN-by-QTN interactions in genome-wide association studies. Mol Plant 15: 630–650

Meuwissen THE (2009) Accuracy of breeding values of 'unrelated' individuals predicted by dense SNP genotyping. Genet Sel Evol 41:35

Michel S, Kummer C, Gallee M, Hellinger J, Ametz C, Akgöl B, Epure D, Löschenberger F, Buerstmayr H (2018) Improving the baking quality of bread wheat by genomic selection in early generations. Theor Appl Genet 131:477–493

Mohammadi M, Tiede T, Smith KP (2015) PopVar: a genome-wide procedure for predicting genetic variance and correlated response in biparental breeding populations. Crop Sci 55: 2068–2077

Montesinos-López OA, Martín-Vallejo J, Crossa J, Gianola D, Hernández-Suárez CM, Montesinos-López A, Juliana P, Singh R (2019) A benchmarking between deep learning, support vector machine and Bayesian threshold best linear unbiased prediction for predicting ordinal traits in plant breeding. G3 Genes Genom Genet 9:601–618

Montesinos-López OA, Montesinos-López A, Crossa J, Montesinos-López JC, Mota-Sanchez D, Estrada-González F, Gillberg J, Singh R, Mondal S, Juliana P (2018) Prediction of multiple-trait and multiple-environment genomic data using recommender systems. G3 Genes Genom Genet 8:131–147

Montesinos-López OA, Montesinos-López A, Kismiantini, Roman-Gallardo A, Gardner K, Lillemo M, Fritsche-Neto R, Crossa J (2022) Partial least squares enhances genomic prediction of new environments. Front Genet 13:920689

Montesinos-López OA, Saint Pierre C, Gezan SA, Bentley AR, Mosqueda-González BA, Montesinos-López A, van Eeuwijk F, Beyene Y, Gowda M, Gardner K (2023) Optimizing sparse testing for genomic prediction of plant breeding crops. Genes 14:927

Montesinos López OA, Montesinos López A, Crossa J (2022) Multivariate statistical machine learning methods for genomic prediction. Springer, Cham

Musundire L, Derera J, Dari S, Lagat A, Tongoona P (2021) Stability assessment of single-cross maize hybrids using GGE-biplot analysis. J Agric Sci 13:78–92

Napier JD, Heckman RW, Juenger TE (2023) Gene-by-environment interactions in plants: molecular mechanisms, environmental drivers, and adaptive plasticity. Plant Cell 35:109–124

Nataraj V, Bhartiya A, Singh CP, Devi HN, Deshmukh MP, Verghese P, Singh K, Mehtre SP, Kumari V, Maranna S (2021) WAASB-based stability analysis and simultaneous selection for grain yield and early maturity in soybean. Agron J 113:3089–3099

Niazian M, Sadat-Noori SA, Abdipour M, Tohidfar M, Mortazavian SMM (2018) Image processing and artificial neural network-based models to measure and predict physical properties of embryogenic callus and number of somatic embryos in ajowan (Trachyspermum ammi (L.) Sprague). In Vitro Cell Dev Biol Plant 54:54–68

Ntare BR, Aken'Ova ME (1985) Yield stability in segregating populations of cowpea. Crop Sci 25: 208–211

Olivoto T, Lúcio ADC, da Silva JAG, Sari BG, Diel MI (2019a) Mean performance and stability in multi-environment trials II: selection based on multiple traits. Agron J 111:2961–2969

Olivoto T, Lúcio ADC, da Silva JAG, Marchioro VS, de Souza VQ, Jost E (2019b) Mean performance and stability in multi-environment trials I: combining features of AMMI and BLUP techniques. Agron J 111:2949–2960

Onogi A, Sekine D, Kaga A, Nakano S, Yamada T, Yu J, Ninomiya S (2021) A method for identifying environmental stimuli and genes responsible for genotype-by-environment interactions from a large-scale multi-environment data set. Front Genet 12:803636

Pandey MK, Roorkiwal M, Singh VK, Ramalingam A, Kudapa H, Thudi M, Chitikineni A, Rathore A, Varshney RK (2016) Emerging genomic tools for legume breeding: current status and future prospects. Front Plant Sci 7:455

Paniza HM (2024) Challenges in plant breeding under climate change: a review. In: Kumar A, Kumar P, Singh S (eds) Plant quarantine challenges under climate change anxiety. Springer, Singapore, pp 533–556

Payghamzadeh K, Oghan HA (2024) Selection of superior Brassica napus L. hybrids by integrated graphical and numerical diallel analysis by yield× traits: using hayman and biplot diallel approaches. Heliyon 10:e31579

Piepho H-P (2022) Prediction of and for new environments: what's your model? Mol Plant 15:581–582

Rauf S, da Silva JT, Khan AA, Naveed A (2010) Consequences of plant breeding on genetic diversity. Int J Plant Breed 4:1–21

Reckling M, Ahrends H, Chen T-W, Eugster W, Hadasch S, Knapp S, Laidig F, Linstädter A, Macholdt J, Piepho H-P (2021) Methods of yield stability analysis in long-term field experiments. A review. Agron Sustain Dev 41:27

Richards RA, Rebetzke GJ, Watt M, Condon AGT, Spielmeyer W, Dolferus R (2010) Breeding for improved water productivity in temperate cereals: phenotyping, quantitative trait loci, markers and the selection environment. Funct Plant Biol 37:85–97

Sabadin F, DoVale JC, Platten JD, Fritsche-Neto R (2022) Optimizing self-pollinated crop breeding employing genomic selection: from schemes to updating training sets. Front Plant Sci 13: 935885

Salgotra RK, Chauhan BS (2023) Genetic diversity, conservation, and utilization of plant genetic resources. Genes 14:174

Saltz JB, Bell AM, Flint J, Gomulkiewicz R, Hughes KA, Keagy J (2018) Why does the magnitude of genotype-by-environment interaction vary? Ecol Evol 8:6342–6353

da Silva ÉDB, Xavier A, Faria MV (2021) Impact of genomic prediction model, selection intensity, and breeding strategy on the long-term genetic gain and genetic erosion in soybean breeding. Front Genet 12:637133

Sinha P, Singh VK, Bohra A, Kumar A, Reif JC, Varshney RK (2021) Genomics and breeding innovations for enhancing genetic gain for climate resilience and nutrition traits. Theor Appl Genet 134:1829–1843

de Sousa AM d CB, da Silva VB, Lopes ÂC d A, Gomes RLF, Carvalho LCB (2020) Prediction of grain yield, adaptability, and stability in landrace varieties of lima bean (Phaseolus lunatus L.). Crop Breed Appl Biotechnol 20:e295120115

Sousa TV, Caixeta ET, Alkimim ER, Oliveira ACB, Pereira AA, Sakiyama NS, Zambolim L, Resende MDV (2019) Early selection enabled by the implementation of genomic selection in Coffea arabica breeding. Front Plant Sci 9:1934

Soyk S, Lemmon ZH, Oved M, Fisher J, Liberatore KL, Park SJ, Goren A, Jiang K, Ramos A, van der Knaap E (2017) Bypassing negative epistasis on yield in tomato imposed by a domestication gene. Cell 169:1142–1155

Star B, Spencer HG (2013) Effects of genetic drift and gene flow on the selective maintenance of genetic variation. Genetics 194:235–244

Teressa T, Semahegn Z, Bejiga T (2021) Multi environments and genetic-environmental interaction (GxE) in plant breeding and its challenges: a review article. Int J Res Stud Agric Sci 7:11–18

Tessema BB, Liu H, Sørensen AC, Andersen JR, Jensen J (2020) Strategies using genomic selection to increase genetic gain in breeding programs for wheat. Front Genet 11:578123

Urbut SM, Wang G, Carbonetto P, Stephens M (2019) Flexible statistical methods for estimating and testing effects in genomic studies with multiple conditions. Nat Genet 51:187–195

Vieira RA, Nogueira APO, Fritsche-Neto R (2025) Optimizing the selection of quantitative traits in plant breeding using simulation. Front Plant Sci 16:1495662

Villar-Hernández B d J, Pérez-Elizalde S, Crossa J, Pérez-Rodríguez P, Toledo FH, Burgueño J (2018) A Bayesian decision theory approach for genomic selection. G3 Genes Genomes Genet 8:3019–3037

Volpato L, Bernardeli A, Gomez F (2021) Genomic selection with rapid cycling: current insights and future prospects. Crop Breed Appl Biotechnol 21(spe):e394721S14

Voss-Fels KP, Herzog E, Dreisigacker S, Sukumaran S, Watson A, Frisch M, Hayes B, Hickey LT (2019) "SpeedGS" to accelerate genetic gain in spring wheat. In: Miedaner T, Korzun V (eds) Applications of genetic and genomic research in cereals. Woodhead Publishing, Duxford, pp 303–327

Wang J, van Ginkel M, Podlich D, Ye G, Trethowan R, Pfeiffer W, DeLacy IH, Cooper M, Rajaram S (2003) Comparison of two breeding strategies by computer simulation. Crop Sci 43:1764–1773

Wang Y, Huan Q, Li K, Qian W (2021) Single-cell transcriptome atlas of the leaf and root of rice seedlings. J Genet Genomics 48:881–898

Ward BP, Brown-Guedira G, Tyagi P, Kolb FL, Van Sanford DA, Sneller CH, Griffey CA (2019) Multienvironment and multitrait genomic selection models in unbalanced early-generation wheat yield trials. Crop Sci 59:491–507

Xiong W, Reynolds M, Xu Y (2022) Climate change challenges plant breeding. Curr Opin Plant Biol 70:102308

Xu Y (2016) Envirotyping for deciphering environmental impacts on crop plants. Theor Appl Genet 129:653–673

Xu Y, Zhang X, Li H, Zheng H, Zhang J, Olsen MS, Varshney RK, Prasanna BM, Qian Q (2022) Smart breeding driven by big data, artificial intelligence, and integrated genomic-enviromic prediction. Mol Plant 15:1664–1695

Yan W, Frégeau-Reid J (2018) Genotype by yield* trait (GYT) biplot: a novel approach for genotype selection based on multiple traits. Sci Rep 8:8242

Yan W, Kang MS (2002) GGE biplot analysis: a graphical tool for breeders, geneticists, and agronomists. CRC Press, Boca Raton

Yan W, Nilsen KT, Beattie A (2023) Mega-environment analysis and breeding for specific adaptation. Crop Sci 63:480–494

Yang R-C (2009) When is early generation selection effective in self-pollinated crops? Crop Sci 49: 2065–2079

Yildirim T, Moriasi DN, Starks PJ, Chakraborty D (2022) Using artificial neural network (ANN) for short-range prediction of cotton yield in data-scarce regions. Agronomy 12:828

Zhang X, Pérez-Rodríguez P, Burgueño J, Olsen M, Buckler E, Atlin G, Prasanna BM, Vargas M, San Vicente F, Crossa J (2017) Rapid cycling genomic selection in a multiparental tropical maize population. G3 Genes Genom Genet 7:2315–2326

Zhuang J-Y, Lin H-X, Lu J, Qian H-R, Hittalmani S, Huang N, Zheng K-L (1997) Analysis of QTL× environment interaction for yield components and plant height in rice. Theor Appl Genet 95:799–808

Zobel RW, Gauch HG (1996) AMMI analysis of yield trials. In: Kang MS, Gauch HG (eds) Genotype-by-environment interaction. CRC Press, Boca Raton, pp 85–122

Zobel RW, Wright MJ, Gauch HG Jr (1988) Statistical analysis of a yield trial. Agron J 80:388–393

Zuffo AM, Steiner F, Aguilera JG, Teodoro PE, Teodoro LPR, Busch A (2020) Multi-trait stability index: a tool for simultaneous selection of soya bean genotypes in drought and saline stress. J Agron Crop Sci 206:815–822

A Comparative Study Among Parametric and Non-parametric Univariate Stability Statistics for Stability Analysis

14

Alireza Pour-Aboughadareh, Jan Bocianowski, and Bita Jamshidi

Abstract

Genotype-by-environment ($G \times E$) interaction is a critical concept in plant breeding and agronomic research, as it reflects the differential responses of genotypes to varying environmental conditions. The presence of significant $G \times E$ interaction poses challenges for the identification and selection of superior genotypes, given that their performance may fluctuate across environments. To assess and interpret $G \times E$ interactions effectively, stability statistics play a pivotal role in quantifying genotype adaptability and performance consistency. Over the past few decades, numerous stability statistics have been developed for the analysis $G \times E$ interactions. These statistics are generally classified into two broad categories: univariate and multivariate approaches. For greater conceptual clarity, this chapter presents all statistics within the frameworks of independent and model-based approaches. Furthermore, a practical example is provided to guide researchers in implementing these methods using R software.

A. Pour-Aboughadareh (✉)
Seed and Plant Improvement Institute, Agricultural Research, Education and Extension Organization (AREEO), Karaj, Iran

J. Bocianowski
Department of Mathematical and Statistical Methods, Poznań University of Life Sciences, Poznań, Poland

B. Jamshidi
Department of Food Security and Public Health, Khabat Technical Institute, Erbil Polytechnic University, Erbil, Iraq

R. Mohammadi et al. (eds.), *Genotype x Environment Interactions and its Implications for Plant Breeding*, https://doi.org/10.1007/978-981-95-5664-9_14

14.1 Introduction

In plant breeding, one of the principal challenges lies in the identification of genotypes that not only exhibit high yield potential but also maintain performance stability across a broad range of environmental conditions. A central concept understanding this challenge is the genotype-by-environment ($G \times E$) interaction, which describes the differential phenotypic responses of genotypes to varying environmental stimuli. This interaction is critical for assessing the adaptability and overall performance of genotypes, thereby playing a pivotal role in selection decisions within in breeding programs. The inherent complexity of $G \times E$ interactions necessitates a rigorous evaluation of genotype stability and adaptability. Breeders are particularly focused on identifying superior genotypes capable of delivering consistent yields despite environmental variability (Becker and Leon 1988). To this end, a range of statistical methods and models have been developed to analyze $G \times E$ interactions, broadly categorized into two parametric and non-parametric approaches. Within the context of breeding programs, the application of stability statistics is indispensable during selection cycles, as it facilitates the identification of genotypes that optimize both yield but and resilience. By integrating stability analyses with $G \times E$ evaluation, breeders are better equipped to make informed decisions that contribute to the development of cultivars with enhanced productivity and environmental adaptability (Kang 2002). Such integration is fundamental for advancing sustainable agricultural and ensuring the long-term success of crop improvement strategies.

This chapter aims to present a comprehensive comparative analysis of both parametric and non-parametric stability statistics within the $G \times E$ interactions framework. By examining the underlying methodologies, practical applications, and implications of these approaches for genotype selection, this chapter elucidates their respective strengths and limitations. The overarching objective is to provide meaningful insights into the most effective statistical strategies for breeding high-yielding, stable genotypes capable of thriving under diverse environmental conditions.

14.2 Non-parametric Stability Statistics

Non-parametric stability statistics are widely employed to estimate genotype stability in multi-environmental trials (METs), particularly in cases where the data do not fulfill the assumptions required for parametric methods. These statistics offer robust alternatives by evaluating genotype performance based on ranks rather than absolute values, thereby eliminating the need for specific distributional assumptions. One of the key advantages of non-parametric approaches is their distribution-free nature, which enhances their applicability across diverse datasets without necessitating normality assumptions. Additionally, they are less influenced by outliers, making them especially useful in scenarios where extreme values could distort parametric analyses. Their simplicity and ease of interpretation further contribute to their utility in breeding programs. Non-parametric statistics are particularly well-suited to situations involving small sample sizes and often provided consistent evaluations across variable environmental conditions. However, despite these advantages, non-parametric methods typically exhibit lower statistical power compared to parametric approaches when the underlying assumptions of normality are met. They may also produce less precise estimates and show reduced sensitivity to subtle differences in genotype performance. Furthermore, non-parametric methods offer fewer analytical options, and some rank-based metrics can be difficult to interpret in complex datasets. Nevertheless, non-parametric stability statistics remain valuable tools for identifying high-yielding, stable genotypes, particularly in contexts where data irregularities compromise the validity of parametric assumptions. The following section outlines several key non-parametric stability statistics commonly used in genotype evaluation.

14.2.1 Huehn's and Nassar and Huehn's Statistics

Huehn (1979) introduced a set of non-parametric stability statistics based on rank-based assessments to evaluate genotype performance in METs. These statistics are instrumental in quantifying the stability of genotypes across varying environmental conditions. The key metrics include:

- $S^{(1)}$: Represents the average of the absolute rank differences for a specific genotype across all tested environments. Lower values indicate greater consistency in performance, regardless of environmental variability.
- $S^{(2)}$: Reflects the variance of genotype ranks across all environments. A smaller value denotes a more stable genotype with minimal fluctuations in rank across different settings.
- $S^{(3)}$: Measures the total absolute deviation of each genotype's rank from the mean rank. This statistic emphasizes the relative consistently of a genotype's performance, with lower values signifying higher stability.
- $S^{(6)}$: This statistic signifies the total sum of squares of ranks for each genotype in relation to the mean rank, providing another layer of insight into the stability of genotypic performance across environments.

To compute these statistics, yield data from each genotype-environment combination are transformed into ranks. Genotypes exhibiting stable performance will maintain consistent ranks across the environments evaluated. Notably, for each of these statistics, lower values indicate greater stability, underscoring a genotype's ability to perform reliably under diverse environmental conditions. Detailed mathematical formulations and computational procedures for these statistics are presented in the subsequent formulas.

$$S^{(1)} = \frac{2\sum_{j=1}^{N-1}\sum_{j'=j+1}^{N}\left|r_{ij} - r_{ij'}\right|}{N(N-1)} \qquad S^{(2)} = \frac{\sum_{j=1}^{N}\left(r_{ij} - \overline{r_{i.}}\right)^2}{N-1}$$

$$S^{(6)} = \sum_{j=1}^{N}\frac{\left|r_{ij} - \overline{r_{i.}}\right|}{\overline{r_{i.}}} \qquad S^{(3)} = \sum_{j=1}^{N}\frac{\left(r_{ij} - \overline{r_{i.}}\right)^2}{\overline{r_{i.}}}$$

where r_{ij} is the rank of the ith genotype in j^{th} environment, $\overline{r}_{i.}$ is the mean rank across all environments for each genotype, and N is the number of test environments.

14.2.2 Kang's Rank

Kang's rank-sum (RK) method, proposed by Kang (1988), offers a robust statistical approach for selecting high-yielding and stable genotypes across a range of environmental conditions. This statistic combines mean yield and Shukla's stability variance (σ_i^2) into a single selection index, thereby enabling the simultaneous assessment of both productivity and adaptability. In application, each genotype is ranked separately for yield and stability variance, with the highest-yielding genotypes and those exhibiting the lowest σ_i^2 assigned a rank of one for each respective criterion. These ranks are then summed for each genotype, and the genotypes with the lowest cumulative rank-sum are considered the most desirable, reflecting superior performance and stability. This method provides plant breeders with a straightforward yet effective tool for genotype selection, supporting more informed and balanced decision-making in the context of multi-environment trials.

14.2.3 TOP-Fox Stability Statistic

Fox et al. (1990) introduced the TOP-Fox method as a simple yet effective non-parametric approach for evaluating genotype adaptability across multiple environments. This stratified ranking technique assessed general adaptability by classifying genotype performance in each test environments into three distinct groups: LOW (bottom third), MID (middle third), and TOP (top third). The fundamental premise of the method is that genotypes frequently ranked in the TOP category—indicating consistent high performance across environments—are considered broadly adapted and suitable for selection. For each genotype, the proportion of environments in which it falls within the LOW, MID, or TOP groups is calculated, yielding a set of non-parametric indicators of stability and adaptability.

A high value indicates that the genotype has demonstrated superior performance across various environments, making it an ideal candidate for widespread cultivation. Due to its simplicity and effectiveness, the TOP-Fox method provides breeders with a quick and reliable tool for identifying genotypes with broad adaptability, bypassing complex statistical assumptions required by parametric models.

14.2.4 Thennarasu's Statistics

When evaluating genotype performance across diverse environments, it is important to recognize that ranking based solely on raw phenotypic values may be insufficient. This limitation underscores the need for an independent assessment of genotype stability that is disentangled from genotypic main effects. To address this issue, a methodology has been developed to refine the ranking patterns of genotypes and environments through the use of adjusted phenotypic values. The $NP^{(1-4)}$ statistics, introduced as a suite of alternative non-parametric stability metrics by Thennarasu (1995), serve this purpose effectively. These statistics are derived from the ranks of adjusted genotype means across multiple environments, offering a more accurate and robust evaluation of genotype stability.

The formulas to compute these statistics are shown below:

$$NP^{(1)} = \frac{1}{N}\sum_{j=1}^{n}\left|r_{ij}^{*} - M_{di}^{*}\right| \qquad NP^{(2)} = \frac{1}{N}\left[\sum_{j=1}^{n}\left|r_{ij}^{*} - M_{di}^{*}\right|/M_{di}\right]$$

$$NP^{(3)} = \frac{\sqrt{\frac{\sum\left(r_{ij}^{*} - \overline{r_{i.}^{*}}\right)^{2}}{N}}}{\overline{r_{i.}}} \qquad NP^{(4)} = \frac{2}{N(N-1)}\left[\sum_{j=1}^{n-1}\sum_{[j'=j+1]}^{n}\left|r_{ij}^{*} - r_{ij'}^{*}\right|/\overline{r_{ij}}\right]$$

In above relations, r_{ij}^{*} is the rank of ith genotype in the jth environment based on adjusted data, $\overline{r_{ij}^{*}}$ is the mean ranks for adjusted data, M_{di}^{*} is the median ranks for adjusted data, while $\overline{r_{i.}}$ and M_{di} are the same statistics obtained from the original data (unadjusted). The low values of these statistics indicate high stability.

14.3 Parametric Stability Statistics

Parametric stability statistics represent an important class of analytical tools used to evaluate genotype performance stability in METs. These methods are grounded in quantitative models designed to interpret $G \times E$ interactions. They generate numerical estimates that how genotypes respond to environmental variables such as temperature, soil quality, and precipitation. A key strength of parametric stability statistics lies in their capacity to provide precise and objective evaluations based on rigorously defined mathematical formulations. However, the effectiveness of these methods is contingent upon the availability of extensive datasets from multi-environment trials, making their application relatively resource-intensive. Moreover, parametric models may introduce bias if relevant environmental factors are not adequately incorporated, highlighting the importance of supplementing them with non-parametric approaches for a more comprehensive stability assessment. The following section outlines several prominent parametric stability statistics, which quantify genotype stability by evaluating yield deviations, environmental responsiveness, and adaptability.

14.3.1 Independent Stability Statistics

14.3.1.1 Environmental Variance (S^2)

Roemer (1917) proposed the use of yield variance across environments as an indicator of genotypes stability. This measure evaluates the consistency of a genotype's performance in response to varying environmental conditions. The formula for this statistic is given by:

$$S^2 = \sum (R_{ij} - m_i)^2/(e-1)$$

where R_{ij} represents the yield of the ith genotype in the jth environment, m_i denotes the grand mean yield across all environments, and e is the total number of environments. Lower values of S^2 indicate greater stability. Derived measures such as the square root of this variance (standard deviation) and the coefficient of variation are also commonly used to assess stability.

14.3.1.2 Mean Variance Component (θ)

Plaisted and Peterson (1959) introduced a statistical measure for evaluating the variance components of $G \times E$ interactions, focusing on interactions between all possible pairs of genotypes. This method assesses the average estimate from various combinations that include a specific genotype, using this average as a measure of stability across environments. The stability statistic is defined by the following equation:

$$\theta = \left[\frac{p}{2(p-1)(q-1)} \sum_{j=1}^{q} \left(X_{ij} - \overline{X_{i.}} - \overline{X_{.j}} + \overline{X_{..}} \right)^2 \right] + \frac{\text{SSGE}}{2(p-2)(q-1)}$$

$$\text{SSGE} = \sum \sum \left(X_{IJ} - \overline{X_{I.}} - \overline{X_{.J}} + \overline{X_{..}} \right)^2$$

In the equation provided, X_{ij} represents the grain yield (GY) of the genotype i^{th} in the environment j^{th}. $\overline{X_{i.}}$ is the average grain yield of the genotype i^{th}, while $\overline{X_{.j}}$ denotes the average grain yield of the environment j^{th}. The grand mean is indicated by $\overline{X_{..}}$. SSGE refers to the sum of squares for genotype-environment interaction. The variables p and q represent the total number of genotypes and environments, respectively. According to this statistic, genotypes exhibiting a lower value for θ are regarded as more stable.

14.3.1.3 GE Variance Component (θ')

This statistic represents a modified approach to assessing genotype stability. The following equation demonstrates that, when the ith genotype is excluded from the complete dataset, the $G \times E$ interaction variance calculated from the resulting subset serves as stability index for the ith genotype (Plaisted 1960).

$$\theta = \left[\frac{-p}{(p-1)(p-2)(q-1)} \sum_{j=1}^{q} \left(X_{ij} - \overline{X_{i.}} - \overline{X_{.j}} + \overline{X_{..}} \right)^2 \right] + \frac{\text{SSGE}}{(p-2)(q-1)}$$

where X_{ij} is the grain yield of genotype i^{th} in environment j^{th}; $\overline{X_{i.}}$ is the mean grain yield of genotype i^{th}; $\overline{X_{.j}}$ is the mean grain yield of the environment j^{th}; $\overline{X_{..}}$ is the grand mean; SSGE is $G \times E$ interaction sum square; and p and q are the numbers of genotypes and environments, respectively. Genotypes with higher values of this stability index are considered more stable.

14.3.1.4 Wricke's Ecovalence (W^2)

Wricke (1962) introduced the concept of ecovalence, a measure that quantified the contribution of each genotype to the overall $G \times E$ interaction. Ecovalence, denotes as W for the ith genotype, is calculated by squaring the interaction deviations of that genotype across environments and summing these squared values across all environments. Genotypes with lower ecovalence values show smaller deviations from the average performance across varying environmental conditions, thereby greater stability. The following equation presents the mathematical formulation of this important stability metric:

$$W^2 = \sum \left(X_{ij} - \overline{X_{i.}} - \overline{X_{.j}} + \overline{X_{..}} \right)^2$$

where X_{ij} is the grain yield of genotype i^{th} in environment j^{th}; $\overline{X_{i.}}$ is the mean grain yield of genotype i^{th}; $\overline{X_{.j}}$ is the mean grain yield of the environment j^{th}; $\overline{X_{..}}$ is the grand mean.

14.3.1.5 Shukla's Stability Variance (σ^2)

Shukla's stability variance quantifies the contribution of each genotype to the $G \times E$ interaction by calculating the variance of its yield performance across environments. A lower stability variance indicates a more stable genotype, whereas a higher variance reflects greater sensitivity to environmental variability (Shukla 1972). This statistic is calculated using the following equation:

$$\sigma^2 = \left[\frac{p}{(p-2)(q-1)} \right] W^2 - \frac{\sum W_i^2}{(p-1)(p-2)(q-1)}$$

In this equation, W^2 denotes Wricke's ecovalence, while p and q denote the number of genotypes and environments, respectively.

14.3.1.6 Pinthus's Coefficient of Determination (R^2)

This stability statistic proposed by Pinthus (1973) serves as a measure of the predictability of genotype responses across environments and is recognized as a key indicator of stability. It quantifies the extent to which variations in mean yield can be attributed to genotype-specific responses under varying environmental conditions. This statistic is mathematically represented by the following equation:

$$R^2 = \frac{b_i^2 \sum \left(\overline{X_{.j}} - \overline{X_{..}}\right)^2}{\sum \left(X_{ij} - \overline{X_{i.}}\right)^2}$$

where b_i, X_{ij}, $\overline{X_{i.}}$, $\overline{X_{.j}}$, and $\overline{X_{..}}$ denote the slope regression, the grain yield of genotype i^{th} in environment j^{th}, the mean grain yield of genotype i^{th}, the mean grain yield of the environment j^{th}, and the grand mean, respectively. A genotype with a higher value of this statistic is considered more stable, indicating greater resilience and consistency across diverse environmental conditions.

14.3.1.7 Joint Regression Analysis (JRA)

The joint regression (JR) model was originally proposed by Yates and Cochran (1938) as a statistical framework for evaluating genotype performance across multiple environments. This model was subsequently revisited and refined by several researchers, including Finlay and Wilkinson (1963), Eberhart and Russell (1966), and Perkins and Jinks (1968), each introducing modifications that enhanced its utility for stability analysis. The primary objective of these adaptations was to effectively identify and characterize genotypes that demonstrated consistent performance under diverse environmental conditions. The foundational structure of the JR model is represented by the following equation:

$$Y_{ij} = \mu + G_i + E_j + b_i E_j + d_{ij} + e_{ij}$$

In this equation, Y_{ij} denotes the average yield for the ith genotype in the jth environment, and μ denotes the overall mean yield of all genotypes combined. G_i corresponds to the effects of the ith genotype, and E_j indicates the effect of the jth environment. The parameter b_i is the regression coefficient associated with the ith genotype, derived from the environmental index, while d_{ij} represents the deviation from the linear regression. The term e_{ij} accounts for the random errors associated with the specific genotype–environment combination. Within the JR model, $G \times E$ interactions are interpreted in terms of genotype sensitivity to environmental changes—where some genotypes exhibit greater responsiveness to favorable environments that others. A notable advantage of the JR model is its capacity to handle unbalanced data within a univariate analytical framework, in contrast to other methods that require balanced dataset. Finlay and Wilkinson (1963) proposed the regression coefficient (b_i) as an effective indicator of genotype stability and relative adaptability. Eberhart and Russell (1966) expanded this concept by incorporating the deviation from linear regression (S_{di}^2) as an additional stability parameter. Consequently, the two primary components of the RJ model are the regression coefficient b_i and the deviation from regression S_{di}^2. Genotypes characterized by $b_i = 1$ and $S_{di}^2 = 0$ are considered highly stable. When such stability is accompanied by high mean performance, the genotype is regarded as generally adaptable. In contrast, when paired with low mean performance, it suggests poor adaptability across environments. Moreover, the interpretation of b_i values provides insights into genotype-specific responses: values greater than 1 ($b_i > 1$) indicate genotypes that are more responsive to favorable (high-yielding) environments, whereas values less than 1 ($b_i < 1$) reflect genotypes better suited to less favorable (low-yielding) conditions.

14.3.1.8 Coefficient of Variance (CV)

Francis and Kannenberg (1987) proposed the coefficient of variation (CV) as a valuable parametric statistic for evaluating yield stability. This statistic combines the standard deviation, mean yield, and environmental variance to provide a comprehensive assessment of genotype performance across environments. By jointly considering these components, researchers can better understand the consistency and reliability of crop yield under varying environmental conditions. This parametric statistic can be described with the following equation:

$$\mathrm{CV} = \frac{S_x}{\overline{X}} \cdot 100$$

In this formulation, S_x denotes the standard deviation of the mean yield of a genotype across different environments, while $\overline{X}$ denotes the overall mean yield. Genotypes characterized by a low CV, low environmental variance, and high average yield are considered the most desirable in terms of stability and performance.

14.3.1.9 Superiority Index (*P*)

The superiority index (*P*) is a widely used metric in stability analysis for evaluating the performance of genotypes across multiple environments (Lin and Binns 1988). It quantifies the mean square of the distance between a genotype's response and the maximum response observed across environments. A lower value of the superiority index indicates greater relative stability, suggesting that the genotype consistently performs close to the best-performing genotype under diverse environmental conditions. The superiority index is mathematically defined as follows:

$$P = \sum_{j=1}^{N} \frac{\left(X_{ij} - M_j\right)^2}{2N}$$

where n is the number of environments, X_{ij} is the yield of the ith genotype in the jth environment, and M_j is the maximum response (yield) among all genotypes in the jth environment.

14.3.2 Model-Based Stability Statistics

Among the key parametric stability statistics derived from various models, those produced by the additive main effects and multiplicative interaction (AMMI) and best linear unbiased prediction (BLUP) models are particularly prominent. These two statistical approaches are fundamental for the effective analysis of $G \times E$

interactions, providing critical insights into genotypes performance across diverse environmental conditions. By incorporating both main effects and interaction components, they facilitate more precise estimation of genotype responses in multi-environment trials (METs). The following sections examine the specific stability statistics associated with each model, emphasizing their significance in the context of MET-based stability analysis.

14.3.2.1 AMMI-Based Stability Statistics

AMMI-Based Stability Parameter (ASTAB)

The *ASTAB* stability statistic is a parameter derived from the AMMI model, designed to evaluate genotype stability in METs. As described by Rao and Prabhakaran (2005), it is calculated as the sum of squared interaction principal component (IPC) scores derived from the AMMI model. This statistic enables researchers to assess the consistency of a genotype's performance across diverse environmental conditions.

$$\mathrm{ASTAB} = \sum_{n=1}^{N} \lambda_n \gamma_{in}^2$$

where λ_n and γ_{in} are the eigenvalue of the n-th IPCA and the eigenvector value for i^{th} genotype. Lower values of this statistic indicate greater stability of the genotype across environments.

AMMI Stability Index (ASI)

The AMMI stability index (ASI) is a statistical measure employed in genotype stability analysis to evaluate the consistency of genotype performance across diverse environments. Originally proposed by Jambhulkar et al. (2014), the ASI is derived from the interaction principal component (IPC) scores generated by the AMMI model. This approach allows researchers to identify genotypes that exhibit both high yield and stability. Lower ASI values correspond to greater stability, indicating that the genotype's performance is less influenced by environmental variability.

$$\mathrm{ASI} = \sqrt{\left[\left(\mathrm{IPCA1} \cdot \theta_1^2\right)^2 + \left(\mathrm{IPCA2} \cdot \theta_2^2\right)^2\right]}$$

In this relation, IPCA1 and IPCA2 denote the scores for the first two principal component interactions. θ_1^2 and θ_2^2 also represent the percentage of the sum of squares accounted for by the effects of the two first IPCA components.

AMMI Stability Value (ASV)

The AMMI stability value (ASV) is a stability metric developed by Purchase et al. (2000) to quantify the $G \times E$ interaction effects by integrating principal component scores derived from the AMMI model. This statistic enables the identification of genotypes with consistent performance across multiple environments. Lower ASV values denote greater stability, indicating that the genotype is less susceptible to environmental fluctuations. ASV is particularly valuable in the selection of genotypes that maintain stable performance under diverse environmental conditions.

$$\mathrm{ASV} = \sqrt{\left[\frac{\mathrm{SS_{IPCA1}}}{\mathrm{SS_{IPCA2}}}(\mathrm{IPCA1})\right]^2 + (\mathrm{IPCA2})^2}$$

In this relation, $\mathrm{SS_{IPCA1}}/\mathrm{SS_{IPCA2}}$ is the ratio between the sum of squares from the first and second IPCAs, and IPCA1 and IPCA2 are the genotypic scores of these components in the AMMI model.

Sum Across Environments of Absolute Value of GEI Modeled by AMMI (AV_{AMGE})

The AV_{AMGE}, proposed by Zali et al. (2012), is a stability statistic employed in the analysis of $G \times E$ interaction within the framework of the AMMI model. It measure the average absolute value of $G \times E$ interaction effects across multiple environments by incorporating all statistically significant IPCs retained in the AMMI model.

$$\mathrm{AV_{(AMGE)}} = \sum_{j=1}^{E} \sum_{n=1}^{N} \left|\lambda_n \gamma_{in} \delta_{jn}\right|$$

where δ_{jn} is the eigenvector for the jth environment from the nth IPCA. N is the number of significant IPCs which retain in the AMMI model via F-test. A genotype exhibiting the lowest AV_{AMGE} value is considered the most stable, as it demonstrates minimal interaction with environmental variability.

Annicchiarico's and Zhang's D Parameter Values (D_a & D_z)

Annicchiarico (1997) and Zhang et al. (1998) proposed two stability statistics that quantify the distance (D_a and D_z, respectively) of a genotype's IPCA score from the origin in a multi-dimensional interaction space. These statistics serve as comprehensive indices that reflect the overall stability of a genotype relative to a set of environments. The fundamental principle underlying these statistics is genotype with smallest D value considered the most stable, as it exhibits the least deviation across varying environmental conditions. The following equations present the mathematical formulation for computing the distance D:

$$D_a = \left[\sum_{n=1}^{N} (\lambda_n \gamma_{in})^2\right]^{0.5}$$

$$D_a = \left[\sum_{n=1}^{N} (\gamma_{in})^2\right]^{0.5}$$

Sums of the Averages of the Squared Eigenvector Values (EV)

The EV, developed by Zobel (1994), is a calculated as the variance of a genotype's yield or response across multiple environments. A lower EV value indicates greater stability, meaning that the genotype exhibits more consistent performance under varying environmental conditions.

$$\mathrm{EV} = \sum_{n=1}^{N} \frac{\gamma_{in}^2}{N}$$

for EV_1, N equals one; for EV_V, N represents the number of IPC retained in the AMMI model based on validation procedures; and for EV_F, N corresponds to the number of IPC retained according to the F-test.

Stability Measure Based on Fitted AMMI Model (FA)

The FA is another stability statistics that quantifies genotype stability by incorporating all significant IPCs retained in the AMMI model (Raju 2002).

$$\mathrm{FA} = \sum_{n=1}^{N} \lambda_n^2 \gamma_{in}^2$$

Modified AMMI Stability Index (MASI)

Unlike the ASI, the Modified AMMI stability index (MASI) incorporates all statistically significant IPCs retained in the AMMI model when assessing genotype stability. This makes *MASI* a more comprehensive and robust metric for evaluating the stability of genotypes across multiple environments. Genotypes with the lowest MASI values are considered the most stable (Ajay et al. 2018a).

$$\mathrm{MASV} = \sqrt{\sum_{n=1}^{N-2} \left[\frac{\mathrm{SS}_{\mathrm{IPCA}_n}}{\mathrm{SS}_{\mathrm{IPCA}_{n+1}}}\right]^2 + \left(\mathrm{IPCA}_N\right)^2}$$

Sums of the Absolute Value of the IPC Scores (SIPC)

The sums of the absolute value of interaction principal component scores (SIPC) incorporates all statistically significant IPCs to evaluate genotype stability. A lower SIPC value indicates a higher degree of stability, implying that the genotype is less affected by environmental fluctuations (Sneller et al. 1997). Such stability is essential for predicting consistent genotype performance across a range of environments.

$$\mathrm{SIPC} = \sum_{n=1}^{N} \left|\lambda_n^{0.5} \gamma_{in}\right|$$

Absolute Value of the Relative Contribution of IPCs to the Interaction (Z_a)

The Z_a statistic, proposed by Zali et al. (2012), represents the absolute value of the relative contribution of each IPCA and serves as an alternative metric for assessing genotype stability. Lower Z_a values indicate greater stability, reflecting minimal interaction effects across environments.

$$Z_a = \sum_{i=1}^{N} |\theta_n \gamma_{in}|$$

where θ_n is the percentage of the total sum of squares explained by the nth IPCA.

Weighted Average of Absolute Scores (WAAS)

The weighted average of absolute scores (WAAS) statistic, proposed by Olivoto et al. (2019), enables researchers to evaluate the consistency of genotype performance under varying environmental conditions while accounting for $G \times E$ interactions. This statistic is computed as a weighted average of the absolute values of interaction scores across all significant IPCs, with weights based on the proportion of variance each IPCA explains. Lower WAAS values indicate greater genotype stability, signifying reduced sensitivity to environmental variability.

$$\text{WAAS} = \frac{\sum_{k=1}^{p} |\text{IPCA}_{ik} \cdot \text{EP}_k|}{\sum_{k=1}^{p} \text{EP}_k}$$

where EP_k denotes the proportion of variance explained by the kth IPCA. A genotype with the lowest WAAS value is considered the most stable according to this metric.

Genotype Stability Index (GSI)

Farshadfar (2008) proposed a statistic approach for evaluating genotype performance by comparing two key metrics: the rank of the mean yield of each genotype across diverse environmental conditions (denoted as RY) and the rank of the genotype's ASV value (denoted as RASV). This composite statistic, known as the genetic stability index (GSI), allows for the identification of genotypes that combine high yield with stability. A genotype exhibiting the lowest GSI is considered to possess superior yield stability across environments, reflecting its consistency and reliability in performance.

$$\text{GSI} = \text{RASV} + \text{RY}$$

14.3.2.2 BLUP-Based Stability Statistics

Three stability statistics derived from the BLUP model were proposed by Resende (2007) to facilitate the identification of genotypes that combine high yield with stability across METs.

Relative Performance of Genotypes in Various Environments (RPGV)

The relative performance of genotypic values (RPGV) quantifies the performance of genotypes by accounting for their yield or trait expression across multiple environments. Higher RPGV values indicate superior adaptability and overall performance under diverse environments. In essence, this metric evaluates the relative performance of genotypes across environments and serves as an indicator of phenotypic adaptability.

$$\mathrm{RPGV} = \frac{1}{E}\sum_{j=1}^{E} \mathrm{GV}_{ij}/\mu_j$$

In all the formulas mentioned above, GV_{ij} represents the genotypic values (BLUP) of the ith genotype in the jth environment, μ_j denotes the overall mean for each environment j, and E indicates the total number of environments.

Homogeneity Mean Genotype Value (HMGV)

This statistic assesses both the stability and overall performance of genotypes across multiple environments. The *HMGV* is calculated as the harmonic mean of a genotype's performance across environments. Higher HMGV values indicate greater stability, meaning that the genotype maintains consistent performance across different environmental conditions. Thus, this metric primarily serves as an indicator of stability.

$$\mathrm{HMGV} = \frac{E}{\sum_{j=1}^{E} \frac{1}{\mathrm{GV}_{ij}}}$$

Harmonic Mean of the Relative Performance of Genotypic Values (HMRPGV)

This statistic integrates the concepts of HMGV and RPGV to provide a comprehensive assessment of both genotypic stability and relative performance. It offers a balanced measure of genotype stability and adaptability, with higher HMRPGV values indicating genotypes that combine high yield with stability across diverse environments.

$$\mathrm{HMRPGV} = \frac{E}{\sum_{j=1}^{E} \frac{1}{\mathrm{GV}_{ij}/\mu_j}}$$

According to Resende (2007), the optimal selections correspond to the highest values of these parameters.

14.4 Concepts of Stability in the $G \times E$ Interaction Analysis

In $G \times E$ interaction analysis, stability is typically conceptualized as either static or dynamic, each offering a distinct perspective on genotype performance. Static stability refers to genotypes that maintain consistent rankings across environments, regardless of yield level. Such genotypes exhibit minimal interaction with environmental conditions, reflected by low correlation with environmental conditions. In contrast, dynamic stability emphasizes genotypes that not only achieve high yields but also respond positively to environmental variation, demonstrating a strong correlation between performance and environmental factors. In plant breeding programs, dynamic stability is generally preferred, as it identifies genotypes capable of consistently high performance across varying environmental conditions. Nonetheless, both concepts offer valuable insights and inform genotype selection strategies tailored to specific agricultural contexts. According to previous studies (Ahmadi et al. 2015; Khalili and Pour-Aboughadareh 2016; Vaezi et al. 2019; Pour-Aboughadareh et al. 2022), stability statistics that shows a positive and significant correlation with mean yield is indicative of dynamic stability. Therefore, correlation analysis plays crucial role in categorizing stability metrics and guiding their interpretation.

14.5 How Can We Calculate Non-parametric and Parametric Stability Statistics?

Advancements in computer science and programming languages have facilitated the development of various scripts and software tools that enable breeders to efficiently analyze large-scale experimental datasets. Among the available analytical platforms, the R programming environment and the STABILITYSOFT application are widely recognized for their accessibility and comprehensive functionality in computing both parametric and non-parametric stability statistics. STABILITYSOFT, a user-friendly web-based application, estimates several parametric and non-parametric stability statistics (Pour-Aboughadareh et al. 2019). In addition, it provides genotype ranking patterns and generates heatmaps based on Pearson's and Spearman's correlation coefficients. Further information, including a user guide, is available at https://manzik.com/stabilitysoft/. To avoid redundancy in presenting results, the subsequent example focuses on the use of R software. Although multiple R packages are available for estimating stability statistics—such as '*ammistability*' (Ajay et al. 2018b), '*toolStability*' (Wang and Chen 2023), '*agrostab*'(Cheshko 2019), and '*metan*' (Olivoto and Lucio 2020)—this example employs the '*metan*' package due to its versatility and broad application in multi-environment trial analysis.

14.6 Practical Example

This practical example is designed to demonstrate the calculation of stability statistics and to explore their interrelationships using an empirical dataset. The data were obtained from a comprehensive MET involving 19 barley genotypes. The trial was conducted across five diverse experimental research stations in the warm climate of Iran, including Ahvaz, Zabol, Gonbad, Moghan, and Darab. Genotypic evaluation was carried out over 3 years at the Ahvaz, Darab, Gonbad, and Moghan stations, while at the Moghan station, assessments were additionally performed over two consecutive years. All trials were laid out in a randomized complete block design (RCBD) with three replications per environment. As a result of this extensive experimental design, the dataset comprises 14 distinct environments.

To conduct the analysis, the '*metan*' package in R is used. It can be installed using one of the following commands:

```
install.packages("metan")
```

or

```
devtools::install_github("TiagoOlivoto/metan")
# To build the HTML vignette use
devtools::install_github("TiagoOlivoto/metan", build_vignettes =
TRUE)
> library(metan)
```

To ensure complete display of all columns and rows in the resulting output tables, the following options should be set:

```
> options(tibble.width = Inf) # displays all columns.
> options(tibble.print_max = Inf) # to show all the rows.
```

Before proceeding with the estimation of stability statistics, a combined analysis of variance will be conducted:

```
> j_an <- anova_joint(data, env = ENV, gen = GEN, rep = REP, resp = c(GY))
```

Note In this script, '*data*' refers to the name of the dataset imported into R.

Based on this result (Table 14.1), there were significant effect of E, G, and their interaction ($G \times E$) on grain yield (GY). These findings underscore the necessity of analyzing the $G \times E$ interaction and estimating relevant stability parameters to identify stable and high-performing genotypes.

Table 14.1 Joint analysis of variance for grain yield in 19 barley genotypes across 14 test environments

S.O.V.	*df*	SS	MS	*F*-value	Pr (>*F*)
ENV	13	933.5	71.807	51.92	4.43E-16
REP(ENV)	28	38.7	1.383	2.84	2.95E-06
GEN	18	46.3	2.572	2.07	7.71E-03
GEN:ENV	234	291.1	1.244	2.55	1.32E-18
ENV/GEN	247	1224.6	4.958	10.17	9.87E-105
ENV/1	13	112.8	8.678	17.79	4.99E-34
ENV/2	13	63.3	4.869	9.98	8.91E-19
ENV/3	13	86.7	6.67	13.68	3.36E-26
ENV/4	13	98.1	7.543	15.47	1.16E-29
ENV/5	13	66.3	5.098	10.45	9.71E-20
ENV/6	13	66.5	5.118	10.5	7.97E-20
ENV/7	13	50.5	3.883	7.96	1.44E-14
ENV/8	13	51	3.924	8.05	9.55E-15
ENV/9	13	64	4.923	10.1	5.26E-19
ENV/10	13	46.2	3.557	7.29	3.65E-13
ENV/11	13	59.7	4.596	9.42	1.28E-17
ENV/12	13	39.5	3.042	6.24	6.22E-11
ENV/13	13	54.7	4.207	8.63	5.85E-16
ENV/14	13	59.9	4.607	9.45	1.15E-17
ENV/15	13	75.8	5.829	11.95	8.83E-23
ENV/16	13	72.2	5.551	11.38	1.25E-21
ENV/17	13	60.5	4.654	9.54	7.23E-18
ENV/18	13	38.6	2.972	6.09	1.24E-10
ENV/19	13	58.2	4.477	9.18	4.09E-17
Residuals	504	245.8	0.488		
CV (%)	17.12				

To estimate a comprehensive set of stability statistics—including rank-based, univariate, and model-based approaches—the following command is used:

```
> model <- ge_stats(data, ENV, GEN, REP, GY)
> model
```

The obtained results are presented in Table 14.2. Additionally, specific stability statistics can be estimated individually using the corresponding functions, as demonstrated below:

 # Huehn's stability statistics

```
> out <- Huehn (data, ENV, GEN, GY)
> out
```

Table 14.2 Estimated stability statistics for 19 investigated genotypes of barley across 14 environments

Code	S1	S1_R	S2	S2_R	S3	S3_R	S6	S6_R	N1	N1_R	N2	N2_R	N3	N3_R	N4	N4_R	FOX
G1	0.0769	14.5	41.4	18	54.3	17	7.38	17	5.21	16	0.579	13	0.616	13	0.008	14	1
G2	0.0659	13	27.6	7	30.3	3	4.75	2	4.14	6	0.592	14	0.622	14	0.008	16.5	3
G3	0.022	4.5	31.6	11	11.4	1	2.85	1	4.54	11	0.907	18	0.936	19	0.004	9.5	3
G4	0.022	4.5	45.2	19	37.8	12	5.97	9	6	19	1.33	19	0.916	18	0.003	7	4
G5	0.011	1.5	32.9	12	34.2	9	5.62	8	4.71	13	0.725	17	0.645	15	0.001	2	1
G6	0.044	10.5	38.1	15	43	15	6.51	13	5.54	17	0.692	16	0.677	16	0.005	11	4
G7	0.0549	12	24.1	4	34.2	8	5.46	5	3.86	2	0.386	5	0.463	4	0.005	12	2
G8	0.0989	18.5	22.9	2	35	10	6.22	12	3.71	1	0.31	3	0.372	2	0.008	15	1
G9	0.0879	16.5	29.8	9	33	6	5.57	6	4.46	8	0.406	6	0.483	7	0.008	16.5	1
G10	0.022	4.5	21.7	1	22.7	2	4.99	3	3.93	3	0.462	10	0.464	5	0.002	4	0
G11	0.033	8	37.8	14	68.2	19	10.6	19	4.96	14	0.31	4	0.466	6	0.003	5.5	2
G12	0.033	8	23.3	3	35.4	11	6.8	15	4	4	0.308	2	0.372	3	0.003	5.5	2
G13	0.0769	14.5	39.6	16	51.6	16	7.35	16	5	15	0.455	9	0.59	12	0.008	13	2
G14	0.0879	16.5	28.2	8	38.2	13	6.05	10	4.36	7	0.545	12	0.573	11	0.010	18	2
G15	0.011	1.5	27.6	6	31.4	5	6.12	11	4.64	12	0.464	11	0.488	8	0.001	1	2
G16	0.0989	18.5	35.1	13	31.3	4	5.34	4	4.5	9.5	0.643	15	0.689	17	0.012	19	5
G17	0.033	8	29.9	10	33.1	7	5.58	7	4.5	9.5	0.409	7	0.516	9	0.003	8	3
G18	0.022	4.5	25	5	38.4	14	6.77	14	4.07	5	0.281	1	0.357	1	0.002	3	0
G19	0.044	10.5	41.3	17	56.2	18	8.49	18	5.64	18	0.418	8	0.534	10	0.004	9.5	4

Code	CV	CV_R	Shukla	Shukla_R	Ecoval	Ecoval_R	bij	Sij	Sij_R	R2	R2_R	Pi_a	Pi_a_R	Gai	Gai_R
G1	41.50	19	0.70	16	25.40	16	1.39	0.335	13	0.841	6	1.2	14	3.79	13
G2	29.90	7	0.30	8	11.20	8	1.03	0.147	9	0.824	8	0.757	3	4.08	4
G3	32.30	15	0.36	10	13.30	10	1.25	0.124	7	0.881	5	0.388	1	4.39	1
G4	35.80	18	0.73	18	26.20	18	1.23	0.493	17	0.759	13	0.853	4	4.18	3
G5	31.30	13	0.38	11	14.20	11	1.03	0.229	12	0.787	12	1.04	9	3.96	7
G6	31.00	11	0.54	13	19.70	13	0.98	0.383	14	0.705	14	1.04	10	4	6
G7	28.20	4	0.26	5	9.80	5	0.91	0.0995	6	0.813	10	1.09	11	3.89	9
G8	29.20	5	0.13	2	5.27	2	0.97	−0.0177	2	0.898	3	1.2	13	3.76	14
G9	31.80	14	0.16	3	6.24	3	1.09	0.00037	1	0.908	2	0.975	8	3.84	12
G10	26.80	2	0.10	1	4.28	1	0.93	−0.051	4	0.913	1	0.93	7	3.94	8
G11	34.50	17	0.71	17	25.60	17	0.85	0.518	18	0.59	19	1.95	19	3.38	19
G12	26.40	1	0.40	12	14.90	12	0.75	0.166	10	0.701	15	1.44	17	3.69	17
G13	30.10	8	0.65	15	23.60	15	0.82	0.446	16	0.599	18	1.31	15	3.76	15
G14	29.40	6	0.34	9	12.60	9	0.98	0.187	11	0.79	11	0.881	5	4.03	5
G15	34.30	16	0.26	6	9.96	6	1.17	0.0748	5	0.887	4	0.916	6	3.85	11
G16	31.00	10	0.76	19	27.20	19	0.96	0.591	19	0.624	17	0.516	2	4.2	2
G17	31.00	9	0.28	7	10.70	7	1.01	0.136	8	0.822	9	1.11	12	3.85	10
G18	26.80	3	0.22	4	8.39	4	0.81	0.0202	3	0.83	7	1.53	18	3.58	18
G19	31.20	12	0.62	14	22.50	14	0.86	0.438	15	0.629	16	1.32	16	3.72	16

Code	ASTAB	ASTAB_R	ASI	ASI_R	ASV	ASV_R	AVAMGE	AVAMGE_R	DA	DA_R	DZ	DZ_R	EV	EV_R	FA	FA_R
G1	1.90	17	0.255	14	1.42	14	7.76	15	2.76	15	0.713	17	0.084	17	7.61	15
G2	0.16	1	0.024	2	0.138	2	2.21	1	0.723	1	0.225	1	0.008	1	0.522	1
G3	2.23	18	0.219	13	1.23	13	8.39	19	2.91	18	0.78	18	0.101	18	8.47	18
G4	0.93	6	0.284	15	1.59	15	5.47	11	2.16	12	0.457	6	0.034	6	4.67	12
G5	1.33	14	0.395	19	2.21	19	7.85	16	2.77	16	0.485	7	0.039	7	7.65	16
G6	1.21	11	0.102	6	0.571	6	5.18	9	1.97	10	0.622	14	0.064	14	3.89	10
G7	0.93	7	0.120	9	0.673	9	5.06	6	1.74	6	0.558	9	0.051	9	3.02	6
G8	2.39	19	0.082	5	0.463	5	7.94	17	2.97	19	0.808	19	0.109	19	8.79	19
G9	0.94	8	0.170	11	0.95	11	5.08	7	1.78	8	0.566	12	0.053	12	3.17	8
G10	0.57	5	0.114	7	0.639	7	3.49	4	1.36	4	0.444	5	0.032	5	1.85	4
G11	1.75	16	0.319	17	1.78	17	8.29	18	2.72	14	0.695	16	0.080	16	7.41	14
G12	0.97	9	0.047	3	0.266	3	5.11	8	1.77	7	0.562	11	0.052	11	3.13	7
G13	1.01	10	0.158	10	0.885	10	5.46	10	1.96	9	0.531	8	0.047	8	3.84	9
G14	1.65	15	0.369	18	2.06	18	7.7	14	2.83	17	0.62	13	0.064	13	8.01	17
G15	1.30	12	0.120	8	0.672	8	5.83	12	2.06	11	0.652	15	0.070	15	4.24	11
G16	1.31	13	0.297	16	1.66	16	6.65	13	2.46	13	0.560	10	0.052	10	6.06	13
G17	0.50	4	0.173	12	0.969	12	4.1	5	1.5	5	0.350	4	0.020	4	2.25	5
G18	0.27	2	0.068	4	0.383	4	2.55	2	0.979	3	0.293	2	0.014	2	0.958	3
G19	0.27	3	0.019	1	0.111	1	2.69	3	0.856	2	0.330	3	0.018	3	0.733	2

Code	MASI	MASI_R	MASV	MASV_R	SIPC	SIPC_R	ZA	ZA_R	WAAS	WAAS_R	HMGV	HMGV_R	RPGV	RPGV_R	HMRPGV	HMRPGV_R
G1	0.270	14	1.95	15	3.12	18	0.249	19	0.578	19	3.65	14	0.991	10	0.977	13
G2	0.049	2	0.54	1	0.85	1	0.0553	2	0.118	2	3.8	8	1	8	1	8
G3	0.24	13	1.94	14	3.13	19	0.226	18	0.500	15	3.36	19	0.905	19	0.886	19
G4	0.288	15	1.71	12	1.90	9	0.178	12	0.441	13	3.59	17	0.953	17	0.947	17
G5	0.395	19	2.22	17	1.48	5	0.185	13	0.487	14	3.65	13	0.975	14	0.964	15
G6	0.140	7	1.49	11	2.02	10	0.141	9	0.312	7	3.85	5	1.030	5	1.020	5
G7	0.136	6	1.26	6	1.89	8	0.135	7	0.302	6	3.70	11	0.991	11	0.985	10
G8	0.222	12	2.32	19	2.65	16	0.19	14	0.416	12	4.01	2	1.070	2	1.050	3
G9	0.177	11	1.28	7	1.85	7	0.139	8	0.325	9	3.72	10	0.988	12	0.983	11
G10	0.124	5	1.03	4	1.60	6	0.109	4	0.246	4	3.51	18	0.930	18	0.926	18
G11	0.329	17	2.12	16	2.73	17	0.222	17	0.532	17	3.63	16	0.969	15	0.957	16
G12	0.117	4	1.35	8	2.13	11	0.132	6	0.280	5	3.91	4	1.040	4	1.030	4
G13	0.172	9	1.36	9	2.14	12	0.164	10	0.373	10	4.14	1	1.100	1	1.100	1
G14	0.373	18	2.24	18	2.40	14	0.219	16	0.542	18	3.99	3	1.070	3	1.060	2
G15	0.153	8	1.45	10	2.54	15	0.172	11	0.376	11	3.81	7	1.020	7	1.010	7
G16	0.302	16	1.88	13	2.37	13	0.217	15	0.528	16	3.83	6	1.030	6	1.020	6
G17	0.176	10	1.10	5	1.43	4	0.130	5	0.316	8	3.77	9	0.995	9	0.990	9
G18	0.079	3	0.71	3	1.18	3	0.0852	3	0.191	3	3.64	15	0.967	16	0.965	14
G19	0.042	1	0.60	2	0.925	2	0.0478	1	0.096	1	3.70	12	0.986	13	0.983	12

R is the rank of genotypes based on each stability statistic

 # TOP-Fox stability statistic

```
> out1 <- Fox(ch14, ENV, GEN, GY)
> out
```

 # Thennarasu's stability statistics

```
> out <- Thennarasu(data_ge, ENV, GEN, GY)
> out
```

 # Wricke's model

```
> out <- ecovalence(data,
env = ENV,
gen = GEN,
rep = REP,
resp = GY)
> out
```

 # Shukla's stability variance statistic

```
> out <- shukla(data,
env = ENV,
gen = GEN,
rep = REP,
resp = GY)
> out
```

 # Lin and Binns' superiority index

```
> out <- superiority(data, ENV, GEN, GY)
> out
```

 # AMMI-based stability statistics

```
> model <- performs_ammi(data, env = ENV, gen = GEN, rep= REP, GY)
> model_indexes <- ammi_indexes(model)
> model_indexes
```

 # BLUP-based stability statistics

```
> res_ind <- waasb(ch14, env = ENV, gen = GEN, rep = REP, resp = c(GY),
verbose = FALSE)
> model_indexes <- blup_indexes(res_ind)
> model_indexes
```

For instance, genotypes G3, G4, G16, G2, and G14 showed the highest grain yield among all tested genotypes. According to Shukla's and ecovalence stability statistics, genotypes G7, G8, G9, G10, G17, and G18 were identified as the most stable. All AMMI-based stability statistics identified genotypes G2, G18, and G19 as the most stable genotypes than others. Moreover, genotypes G2, G6, G8, G12, G13, and G14 were selected as high-yielding and stable genotypes using the BLUP-based stability statistics. To investigate the relationships among the various stability statistics, the following script offers various graphical perspectives on their correlation.

One of the most widely used R packages for visualizing correlation coefficients matrices is '*corrplot*' (Wei and Simko 2024). The analysis continues using this package.

```
> install.packages("metan")
> library(corrplot)
```

Note Before analyzing the data, please import the data file without the genotype column.

```
  > cor <- read.csv(file.choose(), header = T)
> M = cor(cor)
```

The following commands can be used to generate various types of correlation heatmaps:

```
corrplot(M, method = 'number'
> corrplot(M, method = 'color', order = 'alphabet')
> corrplot(M, method = 'circle')
> corrplot(M, method = 'shade', order = 'AOE', diag = FALSE)
> corrplot(M, method = 'square', order = 'FPC', type =
'lower', diag = FALSE)
> corrplot(M, method = 'ellipse', order = 'AOE', type = 'upper')
> corrplot.mixed(M, order = 'AOE')
> corrplot.mixed(M, lower = 'shade', upper = 'pie', order = 'hclust')
> corrplot(M, method = 'square', diag = FALSE, order = 'hclust',
addrect = 3, rect.col = 'blue', rect.lwd = 3, tl.pos = 'd')
```

In this example, the following command was used to examine the relationships among the stability metrics and to visually cluster them:

```
> corrplot(M, method = 'circle')
```

Based on the results (Fig. 14.1), all AMMI-based stability statistics were positively correlated with one another. Grain yield (GY) showed a significant positive association with the b_i, GAI, N^2, N^3, and Top-Fox statistics. Similarly, all BLUP-based statistics metrics were positively correlated and also displayed a positive correlation with the N^4 statistic.

Another key step involves clustering the stability statistics. To perform this task, several R packages can be utilized, including '*tidyheatmap*' (Mangiola and Papenfuss 2020), '*gplots*' (Gregory et al. 2024), and '*pheatmap*' (Kolde 2019). In this example, we employed the '*gplots*' package:

```
> library(gplots)
```

We start by standardizing the data to make statistics (as variable):

```
df <- scale(data)
```

Note data is the name of the imported data.

```
> heatmap.2(df, scale = "column", trace = "none", density.info =
"none")
```

Based on the results (Fig. 14.2), all stability statistics were classified into two major groups. The first group comprises all BLUP-based statistics, as well as S^1, S^2, S^3, S^6, N^1, N^4, S_i, Shukla, and Ecovalence. From the perspective of stability theory, GY was clustered alongside GAI, N^2, N^3, and R^2 within a distinct subgroup of the second main cluster. These indicates that these statistics show a *dynamic* interpretation of stability.

In addition, the corresponding heatmap categorized all evaluated genotypes into two main clusters. The first cluster consisted of genotypes G1, G3, G4, G5, G6, G8, G13, G14, and G16, while the second cluster included G2, G7, G9, G10, G11, G12, G15, G17, G18, and G19. These results suggest that the first cluster predominantly contains the most stable genotypes.

14.7 Summary and Perspective for the Future

The stability of genotype yields regardless of environmental conditions is very important in the breeding process. On the other hand, the adaptation of genotypes to specific environmental conditions also provides valuable information to

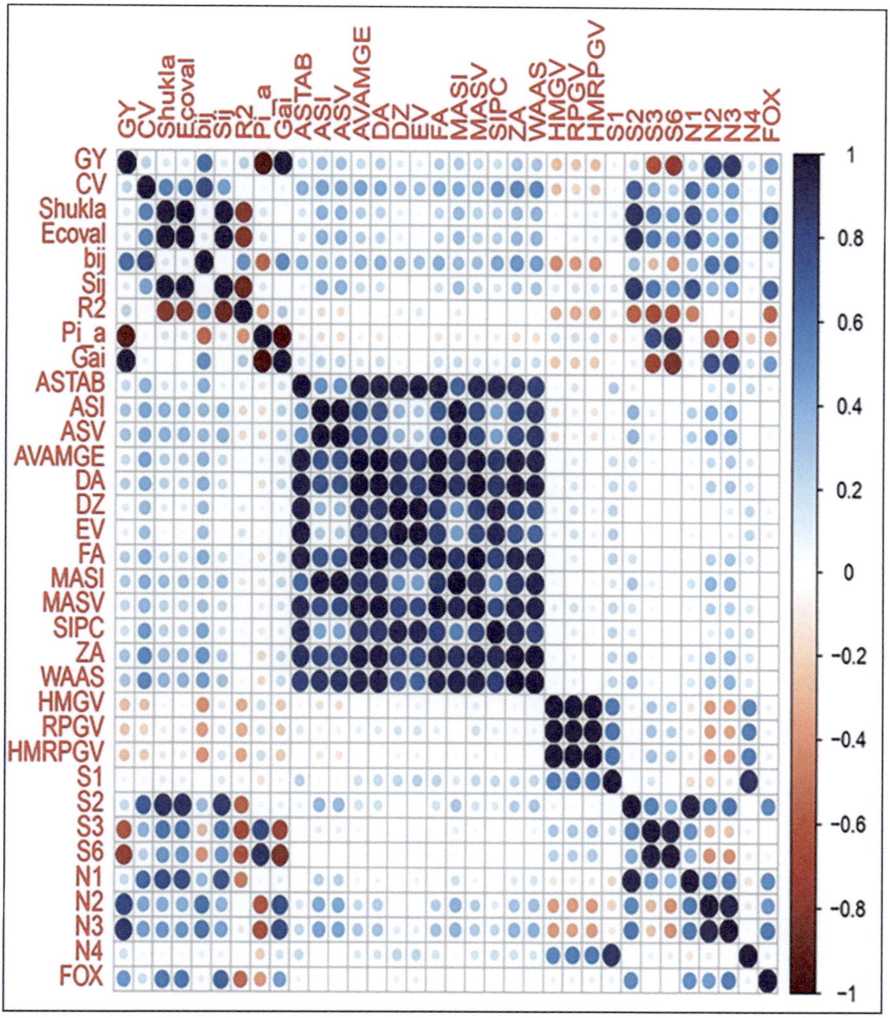

Fig. 14.1 A heatmap based on Pearson's correlation coefficients to study the relationships among stability statistics

experimenters. In order to assess the stability and adaptation of genotypes in genotype–environment interaction studies, a number of statistical methods and models have been developed. These methods can broadly be classified into parametric and non-parametric approaches, each of which may be further subdivided into independent and dependent models. The use of the presented parameters supports breeders in selecting genotypes that are stable due to different environmental conditions or adaptive to specific environmental conditions. This can result in increased efficiency of breeding programs and a better understanding of the mechanisms of genotype–environment interaction. Stability statistics are typically

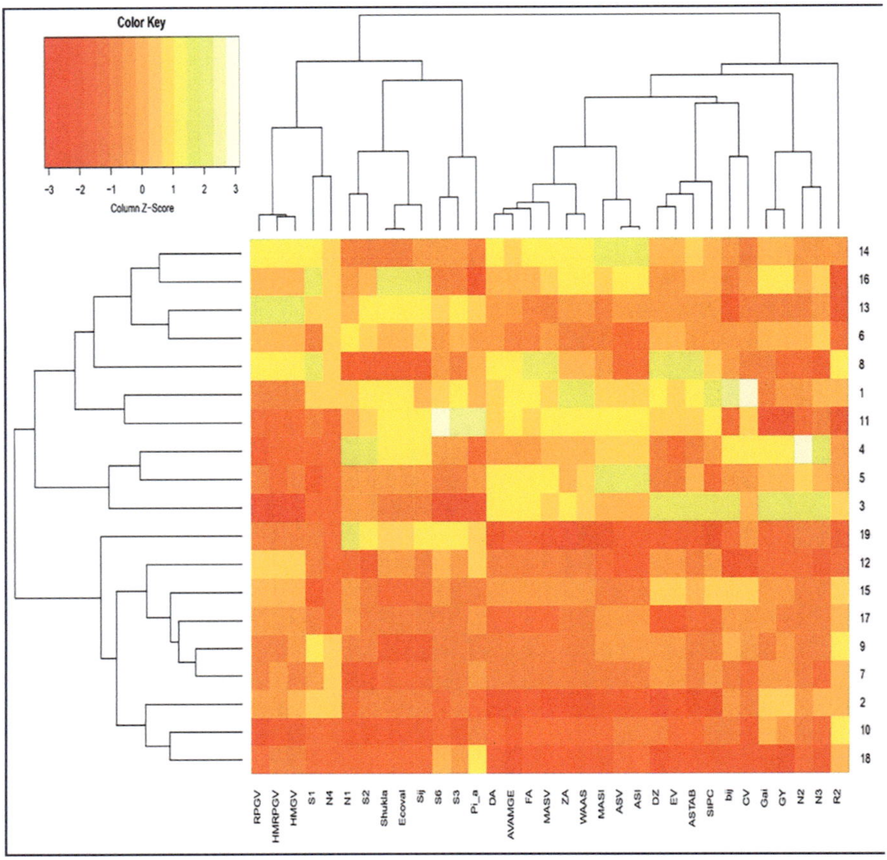

Fig. 14.2 Two-way dendrogram illustrating the clustering of stability statistics and the genotypes under using Ward's method

categorized into overarching concepts: "static" and "dynamic" stability. Static stability refers to genotypes whose performance remains relatively invariant across different environmental conditions. In this context, a statically stable genotype exhibits minimal fluctuation in yield or other agronomic traits regardless of environmental variability. The central premise is largely unaffected by environmental conditions, leading to consistent results under both favorable and stressful conditions. This form of stability is particularly desirable in stress-prone or unpredictable environments, where reliable performance is prioritized over maximizing yield in specific environments. Statistically, measures such as environmental variance or the coefficient of variation are commonly employed, with lower values indicating greater static stability. In contrast, dynamic stability recognizes that genotype performance may vary with environmental conditions but considers these changes "stable" if they are systematic and proportional to environment quality. Under this framework, genotypes are assessed based on their responsiveness to an environmental index. A dynamically stable genotype demonstrates predictable increases in yield under favorable conditions and decreases under unfavorable conditions, typically in a linear or near-linear manner. This concept is especially relevant in breeding programs that aim to exploit the yield potential of genotypes while maintaining performance predictability. Methods such as Finlay and Wilkinson's joint regression analysis and the Eberhart and Russell model operationalize this concept by evaluating the regression coefficient (b_i) of genotype performance against an environmental index. Genotypes with regression coefficients close to 1 and low deviation from the regression line are considered dynamically stable, as their performance closely aligns with environmental conditions. Generally, stability statistics positively correlated with mean yield are indicative of "dynamic" stability.

In summary, when the breeding objective is to ensure consistent performance—especially in marginal or high-risk environments—genotypes with high static stability are preferable. Alternatively, in favorable or well-characterized environments where maximizing genetic potential is the primary goal, dynamically stable genotypes that exhibit predictable responses to environmental variation are more advantageous.

References

Ahmadi J, Vaezi B, Shaabani A, Khademi K, Fabriki Ourang S, Pour-Aboughadareh A (2015) Non-parametric measures for yield stability in grass pea (*Lathyrus sativus* L.) advanced lines in semi warm regions. J Agric Sci Tech 17:1825–1838

Ajay BC, Aravind J, Abdul Fiyaz R, Bera SK, Narendra K, Gangadhar K, Kona P (2018a) Modified AMMI stability index (MASI) for stability analysis. ICAR-DGR Newsl 18:4–5

Ajay BC, Aravind J, Fiyaz AR (2018b) Ammistability: additive main effects and multiplicative interaction model stability parameters. https://CRAN.R-project.org/package=ammistability

Annicchiarico P (1997) Joint regression vs AMMI analysis of genotype-environment interactions for cereals in Italy. Euphytica 94:53–62

Becker HC, Leon J (1988) Stability analysis in plant breeding. Plant Breed 101:1–23

Cheshko A (2019) agrostab: stability analysis for agricultural research. Available at: https://cran.r-project.org/web/packages/agrostab/index.html

Eberhart SAT, Russell WA (1966) Stability parameters for comparing varieties. Crop Sci 6:36–40

Farshadfar E (2008) Incorporation of AMMI stability value and grain yield in a single non-parametric index (GSI) in bread wheat. Pak J Biol Sci 11:1791–1796

Finlay KW, Wilkinson GN (1963) Adaptation in a plant breeding programme. Aust J Agric Res 14: 742–754

Fox PN, Skovmand B, Thompson BK, Braun J, Cormier R (1990) Yield and adaptation of hexaploid spring triticale. Euphytica 47:57–64

Francis TR, Kannenberg LW (1987) Yield stability studies in short-season maize: I. A descriptive method for group-ing genotypes. Can J Plant Sci 58:1029–1034

Gregory RW, Bolker B, Bonebakker L, Gentleman R, Huber W, Liaw A, Lumley T, Maechler M, Magnusson A, Moeller S, Schwartz M, Venables B, Galili T (2024) gplots: various R programming tools for plotting data. Available at: https://cran.r-project.org/web/packages/gplots/index.html

Huehn VM (1979) Beitrage zur erfassung der phanotypischen stabilitat. EDV Med Biol 10:112–117

Jambhulkar NN, Bose LK, Singh ON (2014) AMMI stability index for stability analysis. In: Mohapatra T (ed) Cen-tral Rice Research Institute, Cuttack, Orissa, p 15

Kang MS (1988) A rank-sum method for selecting high yielding stable corn genotypes. Cereal Res Comm 16:113–115

Kang MS (2002) Genotype-environment interaction: progress and prospects. Agron J 94:1130–1135

Khalili M, Pour-Aboughadareh A (2016) Parametric and nonparametric measures for evaluating yield stability and adaptability in barley doubled haploid lines. J Agric Sci Tech 18:789–803

Kolde R (2019) pheatmap: pretty heatmaps. Available at: https://cran.r-project.org/web/packages/pheatmap/index.html

Lin CS, Binns MR (1988) A superiority measure of cultivar performance for cultivar x location data. Can J Plant Sci 68:193–198

Mangiola S, Papenfuss AT (2020) tidyHeatmap: an R package for modular heatmap production based on tidy principles. J Open Source Softw 5:2472

Olivoto T, Lucio ADC (2020) Metan: an R package for multi-environment trial analysis. Methods Ecol Evol 11:783–789

Olivoto T, Lucio ADC, da Silva JAG, Sari BG, Diel MI (2019) Mean performance and stability in mul-ti-environment trials II: selection based on multiple traits. Agron J 111:2961–2969

Perkins JM, Jinks JL (1968) Environment and genotype-environmental components of variability. Heredity 23:339–356

Pinthus JM (1973) Estimate of genotype value: a proposed method. Euphytica 22:121–123

Plaisted RI, Peterson LC (1959) A technique for evaluating the ability of selection to yield consistently in different locations or seasons. Am Potato J Res 36:381–385

Plaisted RL (1960) A shorter method for evaluating the ability of selections to yield consistently over locations. Am Potato J Res 37:166–172

Pour-Aboughadareh A, Barati A, Koohkan SA, Jabari M, Marzoghian A, Gholipoor A, Shahbazi-Homonloo K, Zali H, Poodineh O, Kheirgo M (2022) Dissection of genotype-by-environment interaction and yield stability analysis in barley using AMMI model and stability statistics. Bull Natl Res Cent 46:19

Pour-Aboughadareh A, Yousefian M, Moradkhani H, Poczai P, Siddique KHM (2019) STABILITYSOFT: a new online program to calculate parametric and non- parametric stability statistics for crop traits. Appl Plant Sci 7:e1211

Purchase JL, Hatting H, Van Deventer CS (2000) Genotype × environment interaction of winter wheat in South Africa: II. Stability analysis of yield performance. S Afr J Plant Soil 17:101–107

Raju BMK (2002) A study on AMMI model and its biplots. J Indian Soc Agric Stat 55:297–322

Rao AR, Prabhakaran VT (2005) Use of AMMI in simultaneous selection of genotypes for yield and stability. J Indian Soc Agric Stat 59:76–82

Resende MDV (2007) SELEGEN-REML/BLUP: Sistema estatístico e seleção genética computadorizada via modelos lineares mistos. Embrapa Florestas, Colombo, p 359

Roemer J (1917) Sinde die ertagdreichen sorten ertagissicherer? Mitt DLG 32:87–89

Shukla GK (1972) Some statistical aspects of partitioning genotype–environmental components of variability. Heredity 28:237–245

Sneller CH, Kilgore-Norquest L, Dombek D (1997) Repeatability of yield stability statistics in soybean. Crop Sci 37:383–390

Thennarasu K (1995) On certain non-parametric procedures for studying genotype-environment interactions and yield stability. PhD thesis,. PJ School, Indian Agricultural Research Institute, New Delhi, India

Vaezi B, Pour-Aboughadareh A, Mohammadi R, Mehraban A, Hossein-Pour T, Koohkan E, Ghasemi S, Moradkhani H, Siddique KH (2019) Integrating different stability models to investigate genotype × environment interactions and identify stable and high-yielding barley genotypes. Euphytica 215:63

Wang TC, Chen TW (2023) toolStability: tool for stability indices calculation. Available at: https://cran.r-project.org/web/packages/toolStability/index.html

Wei T, Simko V (2024) R package "Corrplot": visualization of a correlation matrix. Available at: https://cran.r-project.org/web/packages/corrplot/vignettes/corrplot-intro.html#introduction

Wricke G (1962) Übereine Methode zur Erfassung der ökologischen Streubreite in Feldversuchen. Zeitschrift für Pflan-zenzüchtung 47:92–96

Yates F, Cochran WG (1938) The analysis of groups of experiments. J Agric Sci 28:556–580

Zali H, Farshadfar E, Sabaghpour SH, Karimizadeh R (2012) Evaluation of genotype × environment interaction in chickpea using measures of stability from AMMI model. Annal Biol Res 3: 3126–3136

Zhang Z, Lu C, Xiang Z (1998) Analysis of variety stability based on AMMI model. Acta Agron Sin 24:304–309

Zobel RW (1994) Stress resistance and root systems. In: Proceedings of the workshop on adaptation of plants to soil stress. University of Nebraska-Lincoln, Institute of Agriculture and Natural Resources, pp 80–99

An Introductory Guide to the Usage of Statistical Softwares for Investigating G × E Interaction: From Theory to Application

15

Alireza Pour-Aboughadareh, Jan Bocianowski, and Bita Jamshidi

Abstract

Genotype-by-environment (G × E) interactions represent a critical component of plant breeding, as they significantly affect crop yield and stability across diverse cultivation conditions. Understanding these interactions requires advanced statistical analysis methods and appropriate software tools. This chapter presents a theoretical overview of G × E interactions and demonstrates the application of selected statistical tools, including Genstat and R softwares, for their analysis. Practical examples from breeding programs are also presented, emphasizing the advantages of employing modern analytical approaches.

The original version of the chapter has been revised. A correction to this chapter can be found at https://doi.org/10.1007/978-981-95-5664-9_16

A. Pour-Aboughadareh (✉)
Seed and Plant Improvement Institute, Agricultural Research, Education and Extension Organization (AREEO), Karaj, Iran

J. Bocianowski
Department of Mathematical and Statistical Methods, Poznań University of Life Sciences, Poznań, Poland

B. Jamshidi
Department of Food Security and Public Health, Khabat Technical Institute, Erbil Polytechnic University, Erbil, Iraq

R. Mohammadi et al. (eds.), *Genotype x Environment Interactions and its Implications for Plant Breeding*, https://doi.org/10.1007/978-981-95-5664-9_15

15.1 Introduction

Genotype-by-environment (G × E) interactions play a pivotal role in plant breeding, as they facilitate the identification of genotypes that are well-adapted to specific environmental conditions or exhibit stability across a wide range of environmental variability (Pour-Aboughadareh et al. 2022). However, pronounced G × E interactions can complicate genotype evaluation and hinder the effective recommendation of new varieties. To address these challenges, a range of statistical methodologies and computational tools have been developed to enhance the analysis of G × E interactions. This study aims to introduce the application of statistical software in G × E research, illustrate its practical applications, and underscore the advantages of incorporating advanced analytical approaches in modern plant breeding.

15.2 Theoretical Foundations of G × E Interactions

G × E interaction arises when genotypes exhibit differential responses across varying environmental conditions. This can manifest in the following ways:

- *Rank change*: A situation where genotype A yields higher than genotype B in one environment but lower in another.
- *Scale change*: When the range of genotype responses varies depending on the environment.

The analysis of these interactions requires the application of statistical models such as standard analysis of variance (ANOVA), mixed models, and multivariate methods. These models not only facilitate the identification of significant interactions but also support the evaluation of yield stability and genotype-specific adaptation.

15.3 Statistical Tools for Analyzing G × E Interactions

Modern statistical software offers a broad range of tools for analyzing G × E interactions. Several platforms and packages—including SAS, GGEbiplot, Genstat, ASReml, and various R packages—have been specifically developed for analyzing G × E interaction. Among these, Genstat and R are widely recognized for their user-friendly interfaces and robust analytical capabilities, making them particularly valuable for researches investigating G × E interactions.

15.3.1 Genstat

A commercial software package dedicated to agricultural research, offering a wide range of tools for advanced analysis of G × E interactions and genotype stability. In Genstat, G × E interaction analysis can be performed using various procedures, including the `ANOVA`, `fitting model (FIT)` or `AMMI` directives, each providing different perspectives on the interaction effects and their interpretation:

15.3.1.1 ANOVA

In a completely random design (no blocks):`TREATMENTSTRUCTURE Genotype*Environment`

```
ANOVA [PRINT=aovtable;fprobability=yes;pse=lsd] Trait
```

In a block design:`BLOCKSTRUCTURE Blocks`

```
TREATMENTSTRUCTURE Genotype*Environment
ANOVA [PRINT=aovtable;fprobability=yes;pse=lsd] Trait
```

The ANOVA directive in Genstat is employed for the analysis of balanced experimental designs, encompassing a wide array of commonly utilized layouts such as randomized blocks designs, Latin squares, split-plot, and other orthogonal designs. Additionally, it accommodates designs with balanced confounding, including balanced lattices and balanced incomplete block designs. Moreover, many partially balanced designs can be accommodated, provided they satisfy the criterion of first-order balance, as defined algorithmically by Wilkinson (1970) and Payne and Wilkinson (1977), and mathematically by James and Wilkinson (1971) and Payne and Tobias (1992). Prior to conducting an ANOVA, it is essential to specify the model to be fitted. This involves defining the treatment structure using the TREATMENTSTRUCTURE directive, which identifies the experimental factors of interest, and the BLOCKSTRUCTURE directive, which outlines the hierarchical or nested arrangement of blocks and other sources of variation in the experimental design.

The PRINT option specifies the components of the output to be displayed.

aovtable: Provides an analysis-of-variance table, summarizing information on aliasing, non-orthogonality, and any significant residuals.
covariates: Reports estimates of covariate regression coefficients.
residuals: Presents tables of estimated residuals.
means: Displays predicted means for treatment terms.
cbeffects: Shows estimated effects of treatment terms by aggregating information across all strata where each term is evaluated.
cbmeans: Provides predicted means for treatment terms by combining information from all relevant strata.
stratumvariances: Lists the estimated variances of units within each stratum along with the stratum variance components.
%cv: Reports coefficients of variation and standard errors for individual units.
missingvalues: Displays estimates for missing values.

The PSE option controls the type of Standard errors to be printed with tables of means. By default, it is set to differences, which display the standard errors of pairwise differences between means. Alternatively, setting it to means provides the standard errors of the individual means, while LSD produces least significant differences (LSDs). Specifying PSE = * suppresses the display of standard errors entirely. The LSDLEVEL option allows users to modify the significance level used in the calculation of LSDs, with the default value set at 5%.

The ANOVA directive in Genstat was used to assess the interactions among others by Weber et al. (2006), Golinski et al. (2010), Seidler-Łożykowska et al. (2013a, b), Bocianowski et al. (2015), Niemann et al. (2018), and Liersch et al. (2024).

15.3.1.2 FIT

In a completely random design (no blocks):`MODEL Trait`

```
FIT [PRINT=accumulated;fprobability=yes] Genotype*Environment
```

A FIT statement must always be preceded by a corresponding MODEL statement, although they are not required to appear consecutively within the code. Multiple FIT statements can follow a single MODEL statement, allowing for the evaluation of different explanatory variables. The FIT directive specifies the explanatory variables included in the model and may also utilize a model formula to represent interactions and functional relationships among variables. Interactions between factors are interpreted analogously to those specified in the TREATMENTSTRUCTURE directive. The PRINT option controls the output generated by the analysis, and multiple settings can be specified simultaneously to produce reports on different aspects of the analysis. The model provides a detailed description of the model, including both response and explanatory variables. When the summary setting is used, the output provides a summary analysis of variance or deviance (for generalized linear models), including F-probabilities if the option FPROBABILITY = yes is specified. It is important to note, however, that the interpretation of F-probabilities depends on the fulfillment of standard assumptions underlying regression analysis. The FIT directive in Genstat was used to assess the interactions among others by Chełkowski et al. (2000), Seidler-Łożykowska et al. (2013a, b), and Szulc et al. (2018).

15.3.1.3 AMMI

```
AMMI        [PRINT=aovtable,genotype,environment,estimates,envtable,
cluster,stability; NROOTS=3; PLOT=ipca,mean] Trait; GENOTYPES=Gen;
ENVIRONMENTS=Env; REPLICATES=Repl
```

The AMMI (additive main effects and multiplicative interaction) model provides a powerful framework for exploratory analysis of G × E interactions (Chełkowski et al. 2000). The AMMI procedure is a robust statistical approach for analyzing data classified by two factors. It integrates the additive main effects from ANOVA with the multiplicative interaction effects derived through principal component analysis (PCA) (Gauch 1992). Specifically, a principal components model is applied to the residuals from the ANOVA, generating interaction PCA (IPCA) scores for both genotypes and environments. Data for analysis are provided through the DATA parameter, with genotype and environment factors defined using the GENOTYPES and ENVIRONMENTS parameters, respectively. Additionally, the REPLICATES parameter allows specification of replicate factors within environments. The AMMI model assumes that replicates are derived from a randomized block design within each environment. It is important to note that missing values are not permitted, and equal replication is required for valid analysis. The NROOTS option can be employed to define the number of principal component axes (roots, sets of scores) to be retained and interpreted. This analytical flexibility makes the AMMI model a valuable tool for capturing complex G × E interaction patterns and enhancing the interpretation of multi-environment trial data.

The PRINT option enables the selection of specific results to be requested through designated settings.

aovtable: Analysis-of-variance table summarizing the contribution of each component to the interaction term.
genotype: Genotype means, interaction scores, and stability indices.
environment: Environment-specific means and interaction scores.
envtable: Table displaying environment means and variances.
estimates: Genotypic estimates for each environment.
cluster: Hierarchical clustering of AMMI genotype estimates across environments, employing the average linkage method and Euclidean distance for similarity assessment.
stability: AMMI stability values following the methodology of Purchase et al. (2000).

The PLOT option controls the visualization of biplots generated during the analysis. When set to mean, it produces a biplot that displays genotype and environment means plotted against their corresponding IPCA (interaction principal component analysis) scores. Alternatively, the IPCA setting generates a biplot exclusively based on the IPCA scores. The scaling of the IPCA biplot is regulated using the SCALING option. The available settings are as followed:

Environment—multiplies the environment scores by their corresponding eigenvalues; genotype—multiplies the genotype scores by their corresponding eigenvalues; symmetric—multiplies both environment and genotype scores by the square roots of their respective eigenvalues.

The STABILITY parameter stores the AMMI stability values as defined by Purchase et al. (2000). These stability values represent the distances of genotypes from zero in the two-dimensional plot of genotype scores, with an additional weighting for the IPCA1 scores to account for their more substantial contribution to the G × E interaction.

The AMMI model was used to assess the interactions among others by Abbasi and Bocianowski (2021), Barati et al. (2025), Bocianowski et al. (2015, 2018,

2019a, b, c, d, e, 2020, 2021a, b, 2022, 2024a, b), Bocianowski and Prażak (2022), Ghazvini et al. (2024), Jedzura et al. (2023), Liersch et al. (2020), Pour-Aboughadareh et al. (2022, 2023, 2024), Rahmati et al. (2024), and Nowosad et al. (2016, 2017, 2018, 2023).

If Environments is declared as Years and Locations interaction, then GYL interaction can be analyzed using:

ANOVA

In a completely random design (no blocks):
```
TREATMENTSTRUCTURE Genotype*Year*Location
ANOVA [print=aovtable;fprobability=yes;pse=lsd] Trait
```

In a block design:
```
BLOCKSTRUCTURE Blocks
TREATMENTSTRUCTURE Genotype*Year*Location
ANOVA [print=aovtable;fprobability=yes;pse=lsd] Trait
```

or

FIT
```
MODEL Trait
FIT [PRINT=accumulated; fprobability=yes] Genotype*Year*Location
```

15.3.1.4 Practical Example

The following example is a practical example of G × E interaction analysis using Genstat software. The experimental data relates to grain yield (GY) of 19 barley genotypes evaluated across 14 environments (a combination of various years and locations) in the warm climate of Iran.

To obtain information on G × E interaction in a completely randomized design using the ANOVA procedure, we write the lines with commands as given below:
```
TREATMENTSTRUCTURE Gen*Env
ANOVA [PRINT=aovtable;fprobability=yes;pse=lsd] GY
```

In the output window, we get a table with the results of the two-way analysis of variance (Table 15.1) and the means (overall, for individual levels of both factors and for the combination of levels of both factors) and the values of the least significant differences (LSD) at the declared significance level.

To obtain information about G × E interaction in a block design using the ANOVA procedure, we write the lines with commands as given below:

Table 15.1 Results of two-way analysis of variance in a completely random design (no blocks) for GY obtained after applying the ANOVA procedure

Source of variation	d.f.	s.s.	m.s.	*F-value*	F pr.
Gen	18	46.2944	2.5719	4.81	<0.001
Env	13	933.4221	71.8017	134.29	<0.001
Gen.Env	234	291.1198	1.2441	2.33	<0.001
Residual	532	284.4461	0.5347		
Total	797	1555.282			

d.f. the number of degrees of freedom; *s.s.* sum of squares; *m.s.* mean squares; *v.r.* value of the test *F*-statistic; *F pr.* probability of rejecting the null hypothesis of no differences

Table 15.2 Results of two-way analysis of variance in a block design for GY obtained after applying the ANOVA procedure

Source of variation	d.f.	s.s.	m.s.	v.r.	F pr.
Rep stratum	2	4.5473	2.2737	4.31	
Rep.*Units* stratum					
Gen	18	46.2944	2.5719	4.87	<0.001
Env	13	933.4221	71.8017	135.96	<0.001
Gen.Env	234	291.1198	1.2441	2.36	<0.001
Residual	530	279.8987	0.5281		
Total	797	1555.282			

```
BLOCKSTRUCTURE Rep
TREATMENTSTRUCTURE Gen*Env
ANOVA [PRINT=aovtable;fprobability=yes;pse=lsd] GY
```

The output includes a table presenting the results of the two-way analysis of variance (Table 15.2), along with the means (overall, for individual levels of both factors and for the combination of levels of both factors) and the corresponding least significant differences (LSDs) values calculated at the specified significance level.

Analysis of variance can also be performed in Genstat using regression analysis. To obtain information on G × E interaction using the `FIT` procedure, we write the command lines as given below:

```
MODEL GY
FIT [PRINT=accumulated;fprobability=yes] Gen*Env
```

In the output window, we get a table with the results of the two-way analysis of variance (Table 15.3).

Analysis of variance can also be performed in Genstat by using the AMMI model. To investigate G × E interaction using the `AMMI` procedure, the analysis is performed by executing the following command lines:

Table 15.3 Results of two-way analysis of variance in a block design for GY obtained after applying the FIT procedure

Change	d.f.	s.s.	m.s.	v.r.	F pr.
+ Gen	18	46.2944	2.5719	4.81	<.001
+ Env	13	933.4221	71.8017	134.29	<.001
+ Gen.Env	234	291.1198	1.2441	2.33	<.001
Residual	532	284.4461	0.5347		
Total	797	1555.2824	1.9514		

Table 15.4 Analysis of variance of main effects and interactions for GY of 19 genotypes obtained after applying the AMMI procedure

Source	d.f.	s.s.	m.s.	v.r.	F pr
Total	797	1555.3	1.951		
Treatments	265	1270.8	4.796	9.84	<0.001
Genotypes	18	46.3	2.572	5.28	<0.001
Environments	13	933.4	71.802	51.92	<0.001
Block	28	38.7	1.383	2.84	<0.001
Interactions	234	291.1	1.244	2.55	<0.001
IPCA 1	30	100.9	3.364	6.9	<0.001
IPCA 2	28	52.1	1.86	3.82	<0.001
IPCA 3	26	41.5	1.596	3.27	<0.001
Residuals	150	96.6	0.644	1.32	0.0143
Error	504	245.7	0.488		

```
AMMI [PRINT=aovtable,genotype,environment,estimates,envtable,
cluster,stability; NROOTS=3; PLOT=ipca,mean] GY; GENOTYPES=Gen;
ENVIRONMENTS=Env; REPLICATES=Rep
```

In the output window, we get a table with the results of the two-way analysis of variance from AMMI model (Table 15.4).

Additionally, the use of the AMMI procedure makes it possible to obtain a set of tables containing various characteristics: genotype means and scores from AMMI procedure (Table 15.5), environment means and AMMI scores (Table 15.6), AMMI stability values from AMMI procedure (Table 15.7), environment means and variances from AMMI procedure (Table 15.8), AMMI-estimates per each environment, and table of first four AMMI selections per environment (Table 15.9).

The AMMI procedure additionally provides AMMI estimates per each environment. This information is very extensive. The program offers a summary table with first four AMMI selections per environment (Table 15.9).

The Genstat program, through the application of the AMMI procedure, also provides graphical representation of the results. These visualizations include three key biplots: a genotype by environment interaction biplot illustrating both primary and secondary interaction components (Fig. 15.1), a biplot displaying the primary

Table 15.5 Genotype means and scores from AMMI procedure

Genotype	Number	Mean	IPCAg1	IPCAg2	IPCAg3
G1	1	4.101	0.55638	0.93481	0.51075
G2	2	4.258	0.12348	0.11085	−0.60452
G3	3	4.609	0.26816	0.71366	−0.38803
G4	4	4.432	1.06566	−0.11873	0.00115
G5	5	4.16	0.21064	0.53608	0.51672
G6	6	4.207	0.81491	−0.52376	0.2908
G7	7	4.032	0.45161	−0.41766	0.1837
G8	8	3.918	0.14672	0.25689	0.16667
G9	9	4.03	0.05538	−0.02706	0.00603
G10	10	4.066	0.04589	0.10655	−0.25804
G11	11	3.593	−0.18101	−1.17292	0.55108
G12	12	3.817	−0.81786	−0.15113	0.16757
G13	13	3.932	−1.14084	0.09278	−0.02083
G14	14	4.221	−0.2579	0.27713	−0.16976
G15	15	4.067	0.26584	−0.43832	0.13718
G16	16	4.388	0.10967	−0.41739	−1.41633
G17	17	4.021	−0.46728	0.29333	−0.02229
G18	18	3.711	−0.3268	−0.07937	0.08479
G19	19	3.909	−0.92266	0.02426	0.26337

Table 15.6 Environment means and scores from AMMI procedure

Environment	Number	Mean	IPCAe1	IPCAe2	IPCAe3
1	1	2.032	−0.41085	−0.28079	0.14603
2	2	3.48	−0.62928	0.04958	0.36806
3	3	3.596	−0.59981	−1.09111	−1.21084
4	4	3.85	−0.01332	−0.26367	0.14794
5	5	4.482	−0.3877	0.09787	0.82718
6	6	4.173	0.22225	−0.61284	0.39679
7	7	6.267	1.68389	0.50477	−0.32866
8	8	6.025	−0.00321	−0.08964	0.28771
9	9	4.63	−0.16098	0.15004	0.19083
10	10	4.722	−1.13295	1.30364	−0.40704
11	11	3.296	0.40302	−0.25616	0.61681
12	12	3.873	0.27451	−0.24355	−0.35046
13	13	2.945	0.44234	0.18845	−0.11562
14	14	3.712	0.31209	0.54342	−0.56874

interaction component in relation to the mean value of the observed trait (Fig. 15.2), and a biplot depicting the secondary interaction component alongside the trait mean (Fig. 15.3).

Table 15.7 AMMI stability values from AMMI procedure

Rank	Genotype	Number	Stability (ASV)	Mean
1	G9	9	0.1107	4.03
2	G10	10	0.1388	4.066
3	G2	2	0.2637	4.258
4	G8	8	0.3831	3.918
5	G16	16	0.4684	4.388
6	G14	14	0.5714	4.221
7	G18	18	0.6381	3.711
8	G5	5	0.6737	4.16
9	G15	15	0.6763	4.067
10	G3	3	0.8828	4.609
11	G17	17	0.9517	4.021
12	G7	7	0.9696	4.032
13	G11	11	1.2242	3.593
14	G1	1	1.4269	4.101
15	G12	12	1.5918	3.817
16	G6	6	1.6635	4.207
17	G19	19	1.7878	3.909
18	G4	4	2.0681	4.432
19	G13	13	2.2123	3.932

Table 15.8 Environment means and variances from AMMI procedure

Env	N observed	Mean	Variance
1	57	2.032	0.2862
2	57	3.48	0.7837
3	57	3.596	2.0517
4	57	3.85	0.4191
5	57	4.482	0.42
6	57	4.173	0.3182
7	57	6.267	1.7587
8	57	6.025	0.513
9	57	4.63	0.4959
10	57	4.722	1.7122
11	57	3.296	0.3653
12	57	3.873	0.9332
13	57	2.945	0.3408
14	57	3.712	0.7067
Margin	798	4.077	1.9514

15.3.2 R software

R is a flexible and widely used open-source programming language and software environment specifically designed for statistical analysis and data visualization. It is extensively employed by statisticians, data scientists, and researchers for tasks such as data manipulation, statistical modeling, visualization, and machine learning.

Table 15.9 First four AMMI selections per environment

Number	Environment	Mean	Score	1	2	3	4
7	7	6.267	1.6839	G4	G3	G1	G6
13	13	2.945	0.4423	G3	G4	G16	G1
11	11	3.296	0.403	G4	G6	G7	G5
14	14	3.712	0.3121	G3	G16	G2	G4
12	12	3.873	0.2745	G16	G4	G3	G2
6	6	4.173	0.2223	G6	G4	G11	G7
8	8	6.025	−0.0032	G4	G3	G6	G5
4	4	3.85	−0.0133	G4	G6	G3	G16
9	9	4.63	−0.161	G3	G5	G14	G1
5	5	4.482	−0.3877	G5	G19	G1	G13
1	1	2.032	−0.4108	G13	G19	G16	G3
3	3	3.596	−0.5998	G16	G2	G13	G11
2	2	3.48	−0.6293	G13	G19	G12	G14
10	10	4.722	−1.133	G3	G13	G14	G17

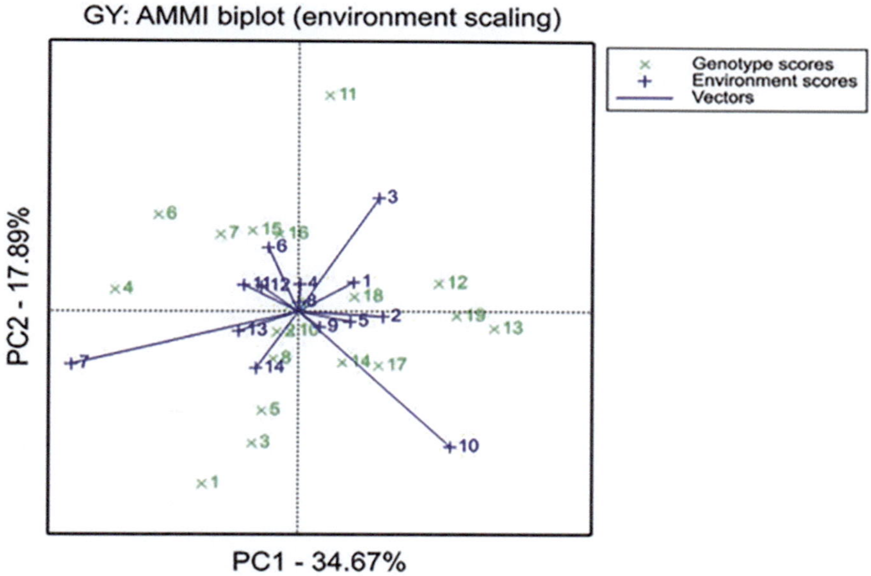

Fig. 15.1 Biplot for genotype by environment interaction of 19 genotypes in 14 environments, showing the effects of primary and secondary components (IPC1 and IPC2, respectively)

Several R packages, such as GEInfo (Wang et al. 2022), aomisc (Onfri 2024), and metan (Olivoto and Lucio 2020), have developed for analyzing multi-environments trials (MET) data and interpret the G × E interaction effects. Based on a review of the literature, the 'metan' package appears to be among the most widely adopted tools

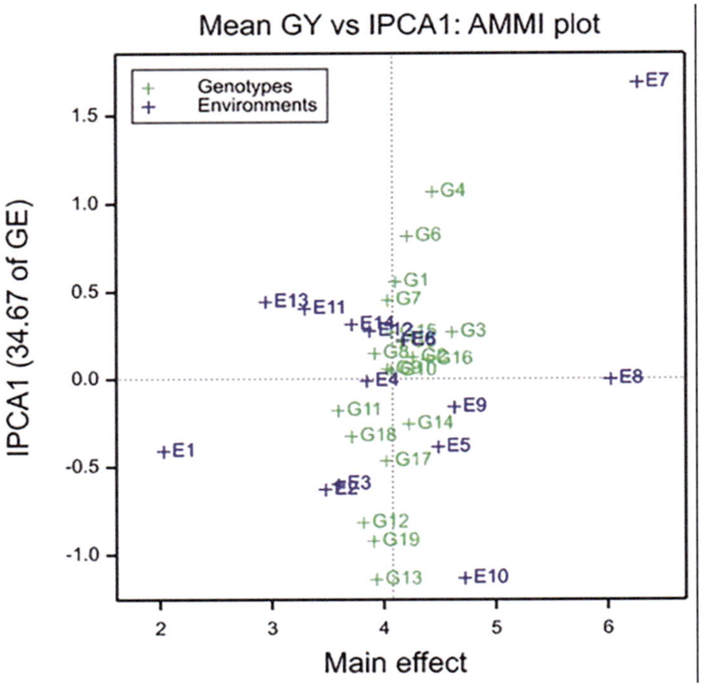

Fig. 15.2 Biplot for the primary component of interaction (IPC1) and the average GY. Vertical line at the center of biplot is the general grand mean

for G × E analysis. Therefore, this section introduces the 'metan' package and presents a practical example using essential R scripts.

15.3.2.1 *'metan'*

This package is a comprehensive and versatile tool designed for the analysis of MET data, with particular emphasis on applications in agronomy and plant breeding. It provides a wide array of functions to support both parametric and non-parametric approaches to MET analysis. Specifically, the package facilitates the computation of descriptive statistics, the execution of within-environment and combined analyses of variance, and the implementation of AMMI (additive main effects and multiplicative interaction) and GGE (genotype +genotype by environment) biplot analyses. Additionally, it enables the estimation of various stability indices and the application of the BLUP (best linear unbiased prediction) methodology, along with its associated stability indices. More information about this package is available in Olivoto and Lucio (2020).

15.3.2.2 Installation and Open Library

As with other R packages, the target package must first be installed prior to use. The installation can be performed using one of the following two command lines:

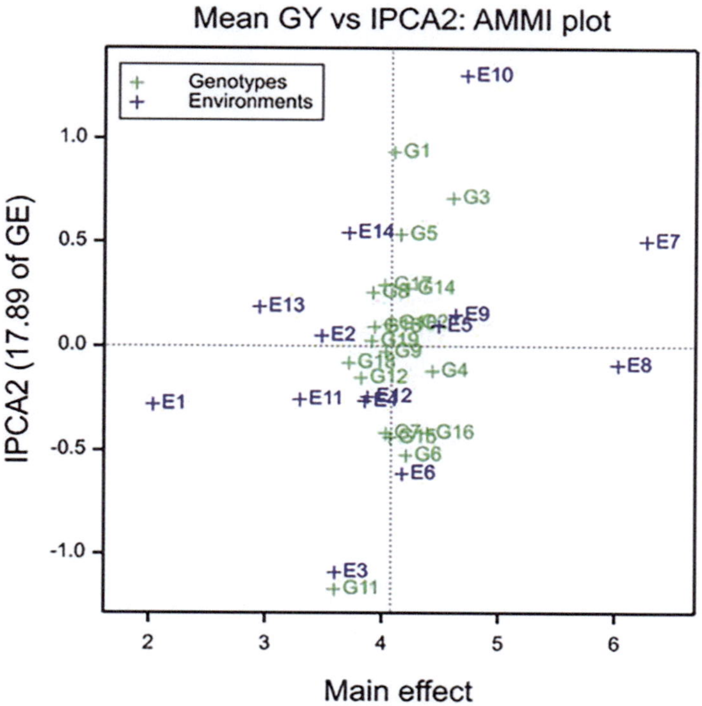

Fig. 15.3 Biplot for the secondary component of interaction (IPC2) and the average GY. Vertical line at the center of biplot is the general grand mean

```
install.packages("metan")
```

or

```
devtools::install_github("TiagoOlivoto/metan")
# To build the HTML vignette use
devtools::install_github("TiagoOlivoto/metan", build_vignettes =
TRUE)
```

Subsequent to the package installation and preceding the commencement of the statistical analysis, the *metan* package is activated via the following command:

```
> library(metan)
```

In order to ensure the complete display of all columns and rows within the extracted tables, the following script is employed:

```
> options(tibble.width = Inf) # displays all columns.
> options(tibble.print_max = Inf) # to show all the rows.
```

15.3.2.3 Scripts for the Analysis of Variance and Stability Models

Joint-Analysis of Variance

Before conducting stability analyses, it is essential to initiate a combined analysis of variance. This preliminary step is crucial for evaluating the significance of G × E interaction effects. This analysis can be carried out using the following script:

```
> j_an <- anova_joint(data, env = ENV, gen = GEN, rep = REP, resp = c
(GY))
```

This script performs a joint analysis of variance to detect the presence of G × E interaction, accommodating both randomized complete block and alpha-lattice designs. In all scripts, GY, ENV, GEN, REP, and data indicate grain yield trait, environments, genotypes, replications, and data source, respectively.

Practical Example

To illustrate the joint analysis of variance, the experimental dataset described in Sect. 15.3.1.4 is utilized (Table 15.10).

```
> j_an <- anova_joint(data, env = ENV, gen = GEN, rep = REP, resp = c
(GY))
```

The results indicate a significant effect of E, G, and their interaction (G × E) on grain yield (GY). More information about descriptive statistics (Table 15.11) can be shown using the following script:

```
> get_model_data(j_an, "details")
```

For example, the lowest and highest grain yield are related to G5 in E3 (1.11 ton ha^{-1}) and G16 in E3 (9.29 ton ha^{-1}), respectively.

AMMI Model

Using the AMMI model, the G × E interaction effect can be divided into several interaction principal components (IPCAs). Hence, this model can be computed through the following script:

```
> model <- performs_ammi (data, ENV, GEN, REP, resp = c(GY))
```

Table 15.10 Joint analysis of variance for grain yield in 19 barley genotypes across 14 test environments

S.O.V.	Df	SS	MS	*F*-value	Pr (> *F*)
ENV	13	933.5	71.807	51.92	4.43E-16
REP(ENV)	28	38.7	1.383	2.84	2.95E-06
GEN	18	46.3	2.572	2.07	7.71E-03
GEN:ENV	234	291.1	1.244	2.55	1.32E-18
ENV/GEN	247	1224.6	4.958	10.17	9.87E-105
ENV/1	13	112.8	8.678	17.79	4.99E-34
ENV/2	13	63.3	4.869	9.98	8.91E-19
ENV/3	13	86.7	6.67	13.68	3.36E-26
ENV/4	13	98.1	7.543	15.47	1.16E-29
ENV/5	13	66.3	5.098	10.45	9.71E-20
ENV/6	13	66.5	5.118	10.5	7.97E-20
ENV/7	13	50.5	3.883	7.96	1.44E-14
ENV/8	13	51	3.924	8.05	9.55E-15
ENV/9	13	64	4.923	10.1	5.26E-19
ENV/10	13	46.2	3.557	7.29	3.65E-13
ENV/11	13	59.7	4.596	9.42	1.28E-17
ENV/12	13	39.5	3.042	6.24	6.22E-11
ENV/13	13	54.7	4.207	8.63	5.85E-16
ENV/14	13	59.9	4.607	9.45	1.15E-17
ENV/15	13	75.8	5.829	11.95	8.83E-23
ENV/16	13	72.2	5.551	11.38	1.25E-21
ENV/17	13	60.5	4.654	9.54	7.23E-18
ENV/18	13	38.6	2.972	6.09	1.24E-10
ENV/19	13	58.2	4.477	9.18	4.09E-17
Residuals	504	245.8	0.488		
CV(%)	17.12				

Table 15.11 Estimated descriptive statistics for grain yield

Mean	4.08
SE	0.05
SD	1.4
CV	34.26
Min	1.11 (5 in 3)
Max	9.29 (16 in 3)
Min ENV	1 (2.03)
Max ENV	7 (6.27)
Min GEN	11 (3.59)
Max GEN	3 (4.61)

In addition, the following scripts provide AMMI-based biplots for visual interpretation of the interaction structure:

AMMI-I biplot

```
> a <- plot_scores(model)
```

AMMI-II biplot

```
> b <- plot_scores(model, type = 2)
> arrange_ggplot(a, b)
```

Nominal yield plot

```
> plot_scores(model, type = 4)
```

AMMI-based stability parameters

```
> model <- performs_ammi (data, ENV, GEN, REP, resp = c(GY))
> model_indexes <- ammi_indexes(model)
> model_indexes
```

or

```
> res_ind <- performs_ammi(data, ENV, GEN, REP, resp = c(GY), verbose =
FALSE) %>% ammi_indexes
> res_ind
```

Cross-Validation Procedure

The cross-validation procedure is performed using the ***cv_ammif*** function, which performs comprehensive cross-validation on replicate-based data within the framework of AMMI-family models. By default, the procedure initiates with the AMMIF model, which incorporates all possible IPC axes. Under this model, the original dataset is divided into two subsets: a training set and a validation set. The training set includes all genotype-by-environment combinations, excluding one replication (*N*-1), while the validation set contains of the omitted replication. The specific method of partitioning the dataset into modeling and validation subsets depends on the experimental design. For the default completely randomized block design or for alpha-lattice designs (specified via block arguments), complete replicates are selected within each environment, with the remaining replicate used for validation. If the design is specified as 'RCD', samples are randomly selected for each genotype-by-environment combination. Estimated values from each AMMI-family model are compared to the corresponding validation dataset, and the root mean square prediction difference (RMSPD) is calculated as an indicator of model performance. Upon completion of all bootstrapping steps, a list of results is returned. If is important to note that if dataset is unbalanced (e.g., a genotype is missing in any environment), the function will return an error. Similarly, an error will occur if the number of replications for any genotype-by-environment combination does not match the structure observed in the experimental design.

```
> model <- cv_ammif(data,
env = ENV,
gen = GEN,
rep = REP,
resp = c(GY),
nboot = [Number of boots: for example, 200])
> plot(model)
```

Estimating the Weighted Average Absolute Scores (WAAS) Index

The WAAS index was developed to facilitate a more comprehensive assessment of genotype performance in METs. Traditional models often emphasize individual PCAs without accounting for their collective contribution to genotype stability and adaptability. The WAAS index addresses this limitation by (i) integrating multiple significant PCAs to better capture the complexity of genotype–environment interactions, (ii) balancing stability and performance through the weighting of absolute scores according to the proportion of variance explained, and (iii) supporting more information decision-making in plant breeding programs by providing a holistic perspective on genotype behavior across divers environments (Olivoto et al. 2019). The ***wass()*** function calculates the WAAS index based on the methodology proposed by Olivoto and Lucio (2020).

```
> waas_index <- waas(data, ENV, GEN, REP, GY, verbose = FALSE)
> wass_index
```

The ***waas()*** function may optionally be applied to the output of AMMI-family models to compute the WAAS index, incorporating the significant IPCs identified in the analysis.

```
> waas_index2 <- waas(data, ENV, GEN, REP, GY, naxis = [Numbers of significant IPCAs in AMMI model], verbose = FALSE)
> waas_index2
```

In this implementation, only the significant IPCA axes were considered in the computation of the WAAS index. As a result, only the WAAS and WAASY values are expected to exhibit substantial variation.

A key feature of the WAAS index lies in its ability to classify genotypes and test environments into four distinct quadrants, based on the joint interpretation of mean performance and stability. Genotypes or environments located in quadrant I are characterized by below-average productivity and high instability, yet they possess a strong discriminatory capacity. Quadrant II contains unstable genotypes or environments that, despite their variability, demonstrate above-average productivity. Environments in this quadrant are particularly valuable, as they combine high response variable values with substantial discrimination power. Genotypes in quadrant III exhibit below-average productivity but are considered stable, as indicated by low WAASB values—where lower WAASB values denote higher stability. Environments in this quadrant are typically unproductive and lack discriminatory capacity. In contrast, quadrant IV comprises genotypes that combine high productivity with wide adaptability, characterized by high response values and stable performance, as reflected by low WAASB values (Olivoto et al. 2019). The following script creates the WAAS-GY biplot:

```
> e <- plot_scores(waas_index2, type = 3)
> f <- plot_scores(waas_index2, type = 3, size.shape.gen = 4, col.gen =
"gray50", size.tex.gen = 4)
> arrange_ggplot(e, f, tag_levels = list(c("e", "f")))
```

Note: functions 'col.gen', 'size.shape.gen',and 'size.text.gen' are optional, and users can modify each of them as needed.

Estimating the WAASY Index

For the joint evaluation of genotype mean performance and stability, the WAASY index offers a powerful and integrative selection tool (Olivoto et al. 2019).

```
> i <- plot_waasby(waas_index2)
> j <- plot_waasby(waas_index2, col.shape = c("red", "green"),
plot_theme = theme_metan(grid = "x"))
> arrange_ggplot(i, j, tag_levels = list (c("i", "j")))
```

Note: functions 'col.shape.gen' is optional, and users can modify each of them as needed.

Additionally, two types of heatmaps are available to visualize genotype ranking based on the WAAS and WAASB indices. The first heatmap presents genotype ranking according to the number of IPCs used in the estimating of the WAASB index. A hierarchical clustering dendrogram, constructed using Euclidean distance, is applied to group genotypes as well as principal component axes, facilitating the identification of patterns in genotype performance and stability. The second heatmap displays genotype rankings based on their WAASB/GY ratio. A ratio of 100/0 reflects a selection focused exclusively on stability, whereas a ratio of 0/100 prioritizes productivity.

```
> plot(WAASratio, type = 1)
> plot(WAASratio, type = 2)
```

Practical Example

To assess the G × E interaction, the AMMI model was applied to the experimental dataset described in Sect. 15.3.1.4. This approach involves a detailed evaluation of genotype performance in response to environmental variability (Table 15.12).

```
> model <- performs_ammi(data, ENV, GEN, REP, resp = c(GY))
```

The application of the AMMI model revealed that E, G, and G × E interaction significantly influenced grain yield (GY). The G × E interaction was further divided into six significant interaction principal component axes (IPCAs), explaining

Table 15.12 The results of AMMI model for grain yield data in the 19 investigated barley genotypes across 14 test environments

S.O.V.	df	SS	MS	*F*-value	Pr (> *F*)	Proportion	Accumulated
ENV	13	933.5	71.807	51.92	4.44E-16		
REP (ENV)	28	38.73	1.383	2.84	2.95E-06		
GEN	18	46.3	2.572	5.27	2.87E-11		
GEN: ENV	234	291.06	1.244	2.55	1.32E-18		
PC1	30	100.83	3.361	6.89	0.00E+00	34.6	34.6
PC2	28	52.12	1.862	3.82	0.00E+00	17.9	52.5
PC3	26	41.54	1.598	3.28	0.00E+00	14.3	66.8
PC4	24	27.04	1.126	2.31	5.00E-04	9.3	76.1
PC5	22	20.46	0.93	1.91	7.90E-03	7	83.1
PC6	20	16.82	0.841	1.72	2.70E-02	5.8	88.9
PC7	18	12.61	0.7	1.44	1.07E-01	4.3	93.2
PC8	16	7.75	0.484	0.99	4.66E-01	2.7	95.9
PC9	14	4.83	0.345	0.71	7.65E-01	1.7	97.6
PC10	12	3.18	0.265	0.54	8.89E-01	1.1	98.7
PC11	10	2.08	0.208	0.43	9.32E-01	0.7	99.4
PC12	8	1	0.125	0.26	9.78E-01	0.3	99.7
PC13	6	0.81	0.135	0.28	9.46E-01	0.3	100
Residuals	504	245.79	0.488				
Total	1031	1846.44	1.791				

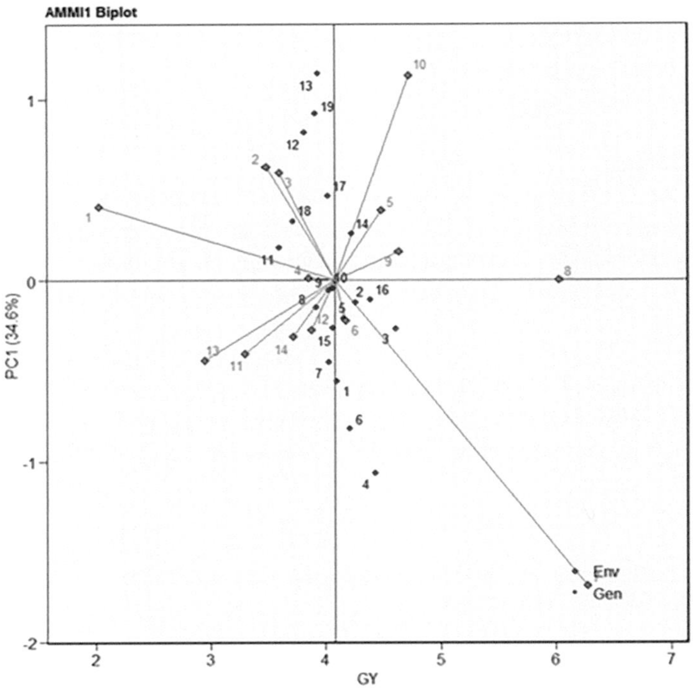

Fig. 15.4 The AMMI-I shows the additive main effects and the first principal component axis (PC1) to visually represent the performance and stability of genotypes across different environments

IPCA1-IPCA6 were 34.6, 17.9, 14.3, 9.3, 7, and 5.8% of the interaction variance, respectively.

Moreover, AMMI-I (Fig. 15.4) and AMMI-2 (Fig. 15.5) biplots are drawn using the following scripts:

```
> a <- plot_scores(model)
> b <- plot_scores(model, type = 2)
> arrange_ggplot(a, b)
```

Based on the results, we receive information, among other things, that genotype 11 positively adapted to environment 3 and did not adapt to environment 14. Genotypes 6, 7, 15, and 16 adapted to environments 6, 11, and 12 and did not adapt to environment 10 (Fig. 15.5). In the AMMI model, the nominal yield plot serves as a graphical tool for analyzing the G × E interactions. This plot displays the nominal yields of genotypes against environmental scores, offering insights into the adaptability and performance of genotypes across various environments. It facilitates the identification of genotypes that excel in specific environments conditions.

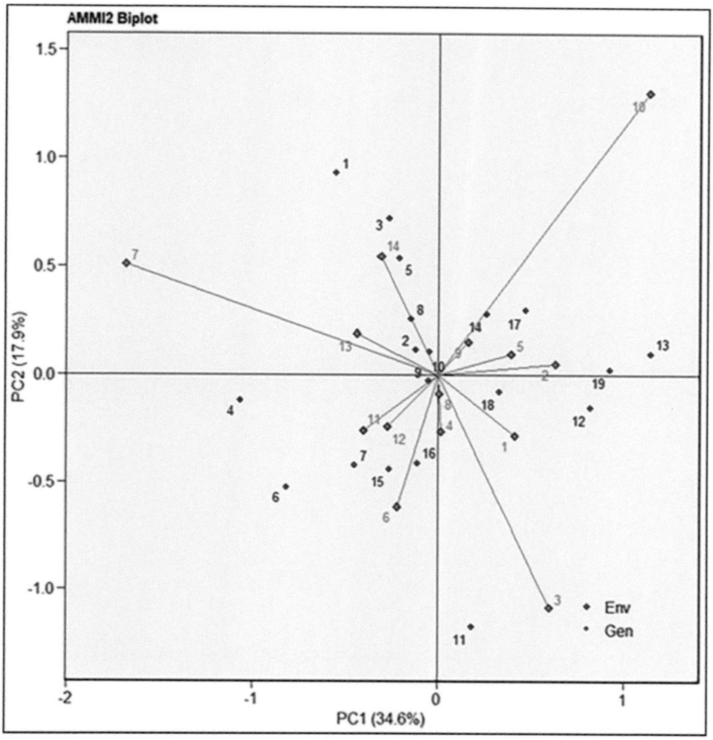

Fig. 15.5 The AMMI-II provides a more detailed representation of the data, offering deeper insights into the relationships between genotypes and environments

Figure 15.6 presents the nominal yield plot for the investigated 19 barley genotypes evaluated across 14 test environments. Based on the results, it can be observed, among other things, that the highest yield in environment 7 was obtained for genotype 4, and the lowest in this environment for genotype 13. The opposite relationship was observed in environment 10. The greatest differentiation of the average nominal yield can be observed in environment 7, and the lowest in environment 1 (Fig. 15.6).

```
> plot_scores(model, type = 4)
```

The following script results in estimate several AMMI-based stability parameters (Table 15.13):

```
> model_indexes <- ammi_indexes(model)
> model_indexes
```

Based on the results, it can be observed that genotype 10 was characterized by the best stability, for which the highest ranking was noted for nine of the 13 parameters

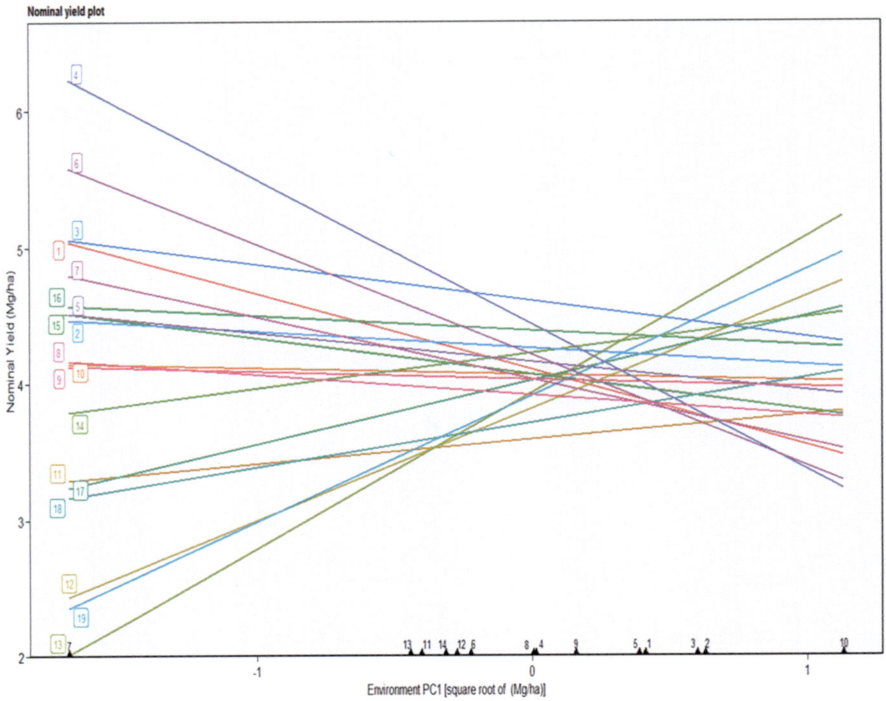

Fig. 15.6 The nominal yield plot for investigated 19 barley genotypes

considered. This genotype is characterized by average yield. On the other hand, genotype 3, which yielded the best, is characterized by poor stability parameters.

Furthermore, the cross-validation procedure (Fig. 15.7) is performed using the following script. In this example, the number of boots is determined as 10.

```
> model <- cv_ammif(dat, env = ENV, gen = GEN, rep = REP, resp = c(GY),
n = 10)
> plot (model)
```

Based on the obtained results, it was observed that the highest mean root mean square prediction difference was obtained for the AMMI9 model, and the lowest for AMMI6. The smallest root mean square prediction difference dispersion was also observed for the AMMI6 model. However, the largest dispersion of these values was obtained for the AMMI12 model (Fig. 15.7).

Two WAAS and WAASY indices (Tables 15.14) and related plots (Figs. 15.8, 15.9, 15.10, and 15.11) are calculated based on the following scripts:

```
> waas_index <- waas(data, ENV, GEN, REP, GY, naxis = 6, verbose = FALSE)
> wass_index
```

Table 15.13 Estimated AMMI-based stability parameters for 19 investigated barley genotypes across 14 test environments

Genotype code	GY	ASTAB	ASI	ASV	AVAMGE	DA	DZ	EV	FA	MASI	MASV	SIPC	ZA	WAAS
1	4.10	1.90	0.26	1.42	7.76	2.76	0.71	0.08	7.61	0.27	1.95	3.12	0.25	0.58
2	4.26	0.98	0.05	0.27	5.11	1.77	0.56	0.05	3.13	0.12	1.35	2.13	0.13	0.28
3	4.61	1.01	0.16	0.89	5.46	1.96	0.53	0.05	3.84	0.17	1.36	2.14	0.16	0.37
4	4.43	1.65	0.37	2.06	7.70	2.83	0.62	0.06	8.01	0.37	2.24	2.40	0.22	0.54
5	4.16	1.30	0.12	0.67	5.83	2.06	0.65	0.07	4.24	0.15	1.45	2.54	0.17	0.38
6	4.21	1.31	0.30	1.66	6.65	2.46	0.56	0.05	6.06	0.30	1.88	2.37	0.22	0.53
7	4.03	0.50	0.17	0.97	4.10	1.50	0.35	0.02	2.25	0.18	1.10	1.43	0.13	0.32
8	3.92	0.28	0.07	0.38	2.55	0.98	0.29	0.01	0.96	0.08	0.72	1.18	0.09	0.19
9	4.03	0.28	0.02	0.11	2.69	0.86	0.33	0.02	0.73	0.04	0.60	0.93	0.05	0.10
10	4.07	0.16	0.02	0.14	2.21	0.72	0.23	0.01	0.52	0.05	0.54	0.85	0.06	0.12
11	3.59	2.23	0.22	1.23	8.39	2.91	0.78	0.10	8.47	0.24	1.94	3.13	0.23	0.50
12	3.82	0.94	0.28	1.59	5.47	2.16	0.46	0.03	4.67	0.29	1.71	1.90	0.18	0.44
13	3.93	1.33	0.40	2.21	7.85	2.77	0.49	0.04	7.65	0.40	2.22	1.48	0.19	0.49
14	4.22	1.21	0.10	0.57	5.18	1.97	0.62	0.06	3.89	0.14	1.49	2.02	0.14	0.31
15	4.07	0.94	0.12	0.67	5.06	1.74	0.56	0.05	3.02	0.14	1.26	1.89	0.14	0.30
16	4.39	2.39	0.08	0.46	7.94	2.97	0.81	0.11	8.79	0.22	2.32	2.65	0.19	0.42
17	4.02	0.94	0.17	0.95	5.08	1.78	0.57	0.05	3.17	0.18	1.28	1.85	0.14	0.33
18	3.71	0.57	0.11	0.64	3.49	1.36	0.44	0.03	1.85	0.12	1.03	1.60	0.11	0.25
19	3.91	1.75	0.32	1.78	8.29	2.72	0.70	0.08	7.41	0.33	2.12	2.73	0.22	0.53

GY grain yield; *STAB* AMMI-based stability parameter; *ASU* AMMI stability index; *ASV* AMMI stability value; *AVAMGE* sum across environments of absolute value of G × E interaction by AMMI; *DA* Annicchiarico's D parameter value; *DZ* Zhang's D parameter; *EV* sums of the averages of the squared eigenvector values; *FA* stability measures based on fitted AMMI model; *MASI* modified AMMI stability index; *SIPC* sums of the absolute value of the IPC scores; *ZA* absolute value of the relative contribution of IPCs to the interaction; *WAAS* weighted average of absolute scores

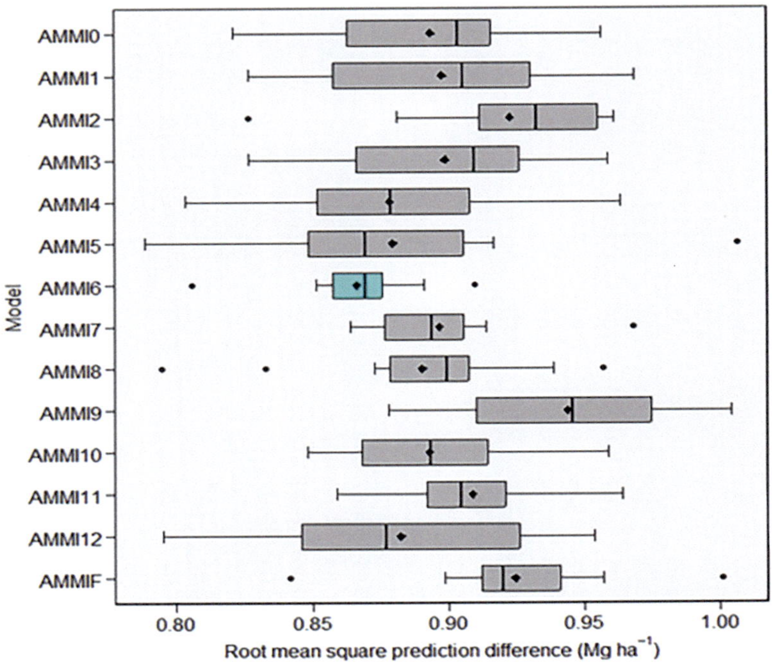

Fig. 15.7 Results of cross-validation procedure based on the AMMI model. The mean (black rhombus), the median (black line), the lower and upper hinges that correspond to the first and third quartiles (the 25th and 75th percentiles, respectively)

```
> e <- plot_scores(waas_index, type = 3)
> f <- plot_scores(waas_index, type = 3, size.shape.gen = 4, col.gen =
"gray50", size.tex.gen = 4)
> arrange_ggplot(e, f, tag_levels = list(c("e", "f")))
> i <- plot_waasby(waas_index)
> j <- plot_waasby(waas_index, col.shape = c("red", "green"),
plot_theme = theme_metan(grid = "x"))
> arrange_ggplot(i, j, tag_levels = list (c("i", "j")))
```

```
> plot(WAASratio, type = 1)
```

Based on the obtained results, four genotype groups were distinguished. One consisted of genotypes 9 and 8. The second: 7, 6, 5, 4, 3, and 19; the third consisted of six genotypes: 2, 18, 17, 16, 15, 14, and 10. The fourth group consisted of the remaining four genotypes: 13, 12, 11, and 1.

```
> plot(WAASratio, type = 2)
```

Table 15.14 The estimated WAAS and WAASY indices for 19 barley genotypes

Genotype code	WAAS	OrWAAS	WAASY	OrWAASY
1	0.578	19	25.06	17
2	0.2799	5	63.68	4
3	0.3734	10	71.22	2
4	0.5421	18	45.02	12
5	0.3757	11	48.93	9
6	0.5277	16	35.46	14
7	0.3164	8	48.76	10
8	0.1905	3	56.2	6
9	0.096	1	71.56	1
10	0.1184	2	70.96	3
11	0.4996	15	8.124	19
12	0.4413	13	25.19	16
13	0.4868	14	26.2	15
14	0.3121	7	58.53	5
15	0.3023	6	51.93	8
16	0.416	12	55.94	7
17	0.3251	9	47.32	11
18	0.2459	4	40.32	13
19	0.5324	17	20.31	18

OrWAAS and OrWAASY: Ranking patters based on WAAS and WAAASY indices, respectively

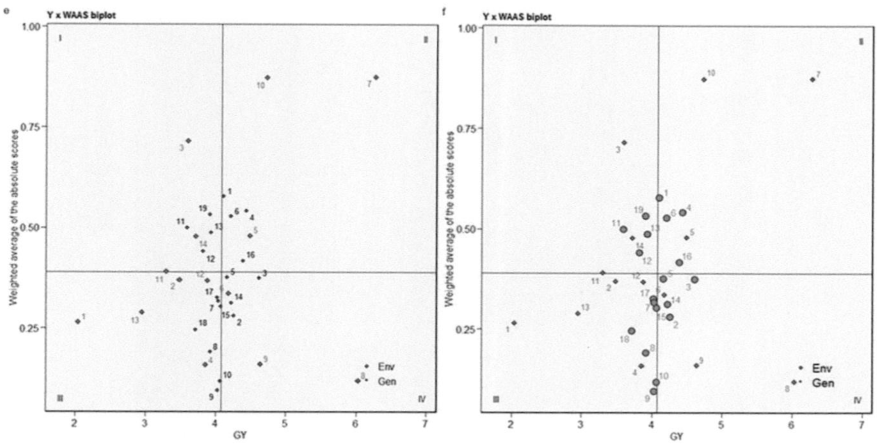

Fig. 15.8 The WAAS-based biplot for identification of high-yielding and stable genotypes of barley

Based on the obtained results, four genotype groups were distinguished. One consisted of genotype 18. The second: 7, 5, 4, 17, 16, and 15; the third consisted of

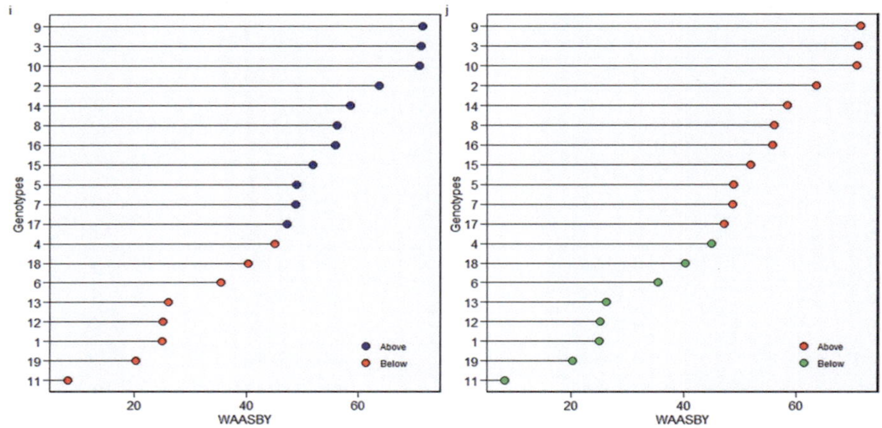

Fig. 15.9 Simultaneous selection for mean performance and stability. The values of WAASY in the plot above were computed considering equal weights for mean performance and stability

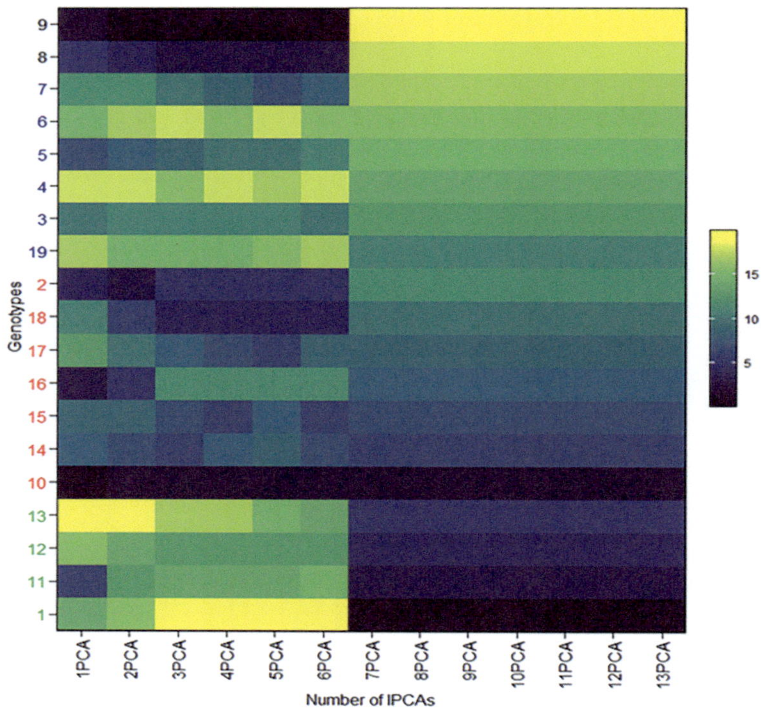

Fig. 15.10 Ranks of genotypes depending on the number of PCA ($n = 6$) used to estimate the WAAS. Genotypes highlighted with green, red, blue, and black colors indicate four distinguished clusters

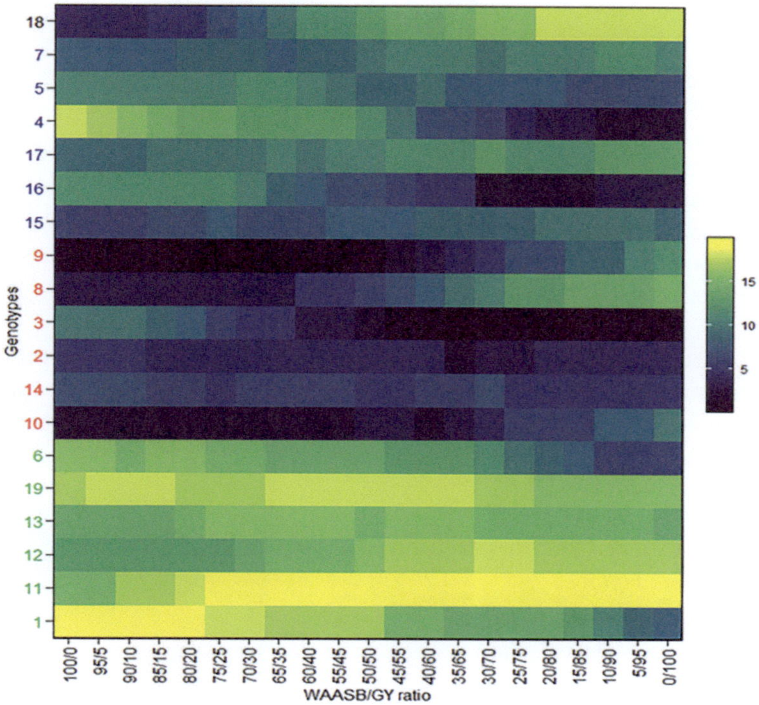

Fig. 15.11 Ranks of genotypes depending on the WAASB/GY ratio. Genotypes highlighted with green, red, blue, and black colors indicate four distinguished clusters

six genotypes: 9, 8, 3, 2, 14, and 10. The fourth group consisted of the remaining six genotypes: 6, 19, 13, 12, 11, and 1.

GGE Biplot Model

The genotype and genotype-by-environment interaction (GGE) model is extensively employed in MET to evaluate genotypic performance and identify mega-environments. This model utilizes a GGE biplot, which combines G and G × E effects. The biplot is constructed using the first two symmetrically scaled principal components (PC1 and PC2) obtained from the singular value decomposition of environment-centered MET data. This graphical representation effectively illustrates the G and G × E components of MET data, thereby enhancing genotype evaluation and facilitating the identification of mega-environments (Yan and Kang 2003).

In the ***'metan'*** package, the ***gge()*** function produces a GGE model from a dataset containing at least the columns for genotypes, environments, and the response variable(s). To visualize the resulting biplots, the ***plot()*** function is utilized. According to Yan and Kang (2003), biplots are categorized into ten distinct types, each serving specific analytical purposes:

Type 1: A basic biplot
Type 2: Mean performance versus stability
Type 3: Which-won-where
Type 4: Discriminativeness versus representativeness
Type 5: Examine an environment
Type 6: Ranking environments
Type 7: Examine a genotype
Type 8: Ranking genotypes
Type 9: Compare two genotypes
Type 10: Relationship among environments

First, execute the GGE model using the specified script. The next step is to create the appropriate biplots once the model has been successfully run. The following scripts can be executed to access each of these biplots.

General GGE model

```
> gge_model <- gge(data, ENV, GEN, GY)
```

Type 1: A basic biplot

```
a <- plot(gge_model)
b <- plot(gge_model,
 col.gen = "orange",
 size.text.env = 2,
 plot_theme = theme_metan(grid = "both"))
arrange_ggplot(a, b, tag_levels = "a")
```

Note: functions 'col.gen' and 'size.text.env' are optional, and users can modify each of them as needed.

Type 2: Mean performance versus stability

This biplot employs the average environment coordinate (AEC) method to evaluate both the mean performance and stability of genotypes. Initially, an average environment—depicted by a small circle—is established using the mean PC1 and PC2 scores of the environments. A line extending from the biplot origin through the AEC serves as the average reference. The projections of genotype markers onto this axis should approximately reflect the mean yield of the genotypes. The AEC ordinate is a line passing through the biplot origin and positioned perpendicular to the AEC abscissa. If the AEC abscissa represents G, then the AEC ordinate approximates the G × E interaction corresponding to each genotype, which reflects its variability or instability (Yan et al. 2007). A greater projection onto the AEC ordinate, irrespective of direction, signifies greater instability.

```
> gge_model <- gge(data, ENV, GEN, GY, svp = "genotype")
c <- plot(gge_model, type = 2)
d <- plot(gge_model,
```

```
 type = 2,
 col.gen = "black",
 col.env = "red",
 axis_expand = 1.5,
 plot_theme = theme_metan_minimal())
arrange_ggplot(c, d, tag_levels = list(c("c", "d")))
```

Note: functions 'col.gen', 'col.env', and 'size_expand' are optional, and users can modify each of them as needed.

Type 3: Which-won-where

In this biplot, a polygon is constructed by connecting the genotypes that are farthest from the biplot origin, effectively enclosing all other genotypes within its perimeter. These vertex genotypes exhibit the longest vectors in their respective directions, indicating their level of responsiveness to environmental conditions. Consequently, these vertex genotypes are considered the most responsive, whereas genotypes located within the polygon display comparatively lower responsiveness in specific directions. A genotype located at the origin would maintain consistent performance across all environments, signifying a complete lack of responsiveness to environmental changes. Perpendicular lines drawn from the sides of the polygon divide the biplot into distinct sectors, each containing a vertex genotype. According to established guidelines, the vertex genotype within any given sector is identified as the highest-yielding genotype for all environments that fall within that sector (Yan et al. 2007).

```
   > gge_model <- gge(data, ENV, GEN, GY, svp = "symmetrical")
e <- plot(gge_model, type = 3)
f <- plot(gge_model,
 type = 3,
 size.shape.win = 5,
 large_label = 6,
 col.gen = "black",
 col.env = "gray",
 title = FALSE)
arrange_ggplot(e, f, tag_levels = list(c("e", "f")))
```

Note: functions 'col.gen', 'col.env', 'large_label', and 'size_shape.win' are optional, and users can modify each of them as needed.

Type 4: Discriminativeness versus representativeness

In the evaluation of test environments through GGE biplot analysis, environments are categorized into three distinct types based on their discrimination ability and representativeness. Type I environments are characterized by short vectors, indicating limited differentiation among genotypes. As a result, these environments are not conducive to the effective evaluation of genetic performance. Type II environments, on the other hand, are defined by long vectors that form small angles with the AEC axis. This configuration makes them particularly valuable for identifying genotypes that are not only high-yielding but also exhibit stability across varying conditions. Finally, Type III environments also feature long vectors, but they are distinguished by larger angles relative to the AEC axis. This arrangement renders them inefficient for genotype assessment, much like the Type I environments. While Type II environments are ideal for discovering superior genotypes, the other two types do not effectively facilitate meaningful evaluations (Yan et al. 2007).

```
g <- plot(gge_model, type = 4)
h <- plot(gge_model,
 type = 4,
 plot_theme = theme_metan_minimal(), title = FALSE)
arrange_ggplot(g, h, tag_levels = list(c("g", "h")))
```

Type 5: Examine an environment

This biplot provides a tool to visualize the performance of different genotypes in a given environment. The genotypes can be sorted based on their performance projections onto the target environment axis.

```
> gge_model <- gge(data, ENV, GEN, GY, svp = "symmetrical")
i <- plot(gge_model, type = 5, sel_env = "TARGET ENVIRONMENT: For
example E2")
j <- plot(gge_model,
 type = 5,
 sel_env = "E2",
 col.gen = "black",
 col.env = "black",
 size.text.env = 10,
 axis_expand = 1.5, title = FALSE)
arrange_ggplot(i, j, tag_levels = list(c("i", "j")))
```

Type 6: Ranking environments

In this biplot, the "ideal" environment is depicted as the central reference point, encircled by concentric rings that serve as a scale to evaluate the proximity of each the environment to this ideal. Given that the primary focus of this biplot is on environments, the SVP applied is the "environment" partition, which is the default selection.

```
> gge_model <- gge(data, ENV, GEN, GY)
k <- plot(gge_model, type = 6)
l <- plot(gge_model,
 type = 6,
 col.gen = "black",
 col.env = "black",
 col.circle = "red",
```

```
col.alpha.circle = 0.5,
title = FALSE,
size.text.env = 5,
axis_expand = 1.5,
plot_theme = theme_metan(color.background = "white"))
arrange_ggplot(k, l, tag_levels = list(c("k", "l")))
```

Note: functions 'col.gen', 'col.env', 'col.circle', 'col.alpha.circle', 'size.text.env', 'size_expand' and 'color.background' are optional, and users can modify each of them as needed.

Type 7: Examine a genotype

Just as it is possible to illustrate the performances of different genotypes within a certain environment through a biplot (Type 5: Examine an environment), we can also enhance this analysis by integrating an average environment coordinate (AEC) into the genotype-centered biplot. This approach allows for a comprehensive visualization that captures both the average performance and the stability of the genotypes across varying conditions. By doing so, researchers can better assess how each genotype responds to environmental changes, thereby gaining deeper insights into their adaptability and overall effectiveness (Yan et al. 2007).

```
> gge_model <- gge(data, ENV, GEN, GY, svp = "genotype")
m <- plot(gge_model, type = 7, sel_gen = " TARGET Genotype: For example
G8")
n <- plot(gge_model,
type = 7,
sel_gen = "G8",
col.gen = "black",
col.env = "black",
title = FALSE,
size.text.env = 10,
axis_expand = 1.5,
plot_theme = theme_metan(grid = "both"))
arrange_ggplot(m, n, tag_levels = list(c("m", "n")))
```

Note: functions 'col.gen', 'col.env', 'size.text.env', and 'size_expand' are optional, and users can modify each of them as needed.

Type 8: Ranking genotypes

This biplot provides a visual comparison of various genotypes with a defined "ideal" genotype. The ideal genotype is represented by a small circle with an arrow pointing toward it, signifying its role as the benchmark for optimal performance. This benchmark is characterized by exceptional yield across all tested environments, achieving the highest average yield while demonstrating complete stability. In this diagram, the genotypes are arranged according to their distance from the ideal genotype, enabling a clear evaluation of each genotype's performance based on the criteria established by Yan et al. (2007).

```
o <- plot(gge_model, type = 8)
p <- plot(gge_model,
 type = 8,
 col.gen = "black",
 col.env = "gray",
 size.text.gen = 6,
 plot_theme = theme_metan_minimal())
arrange_ggplot(o, p, tag_levels = list(c("o", "p")))
```

Note: functions 'col.gen', 'col.env', and 'size.text.gen' are optional, and users can modify each of them as needed.

Type 9: Compare two genotypes

To compare two genotypes, a connector line is drawn between them, along with a perpendicular line that passes through the origin of the biplot and intersects the connector line at a right angle. Environments positioned on the same side of the perpendicular line as one genotype indicate that this genotype yields better results in those environments. Conversely, environments located on the opposite side of the perpendicular line indicate that the alternative genotype exhibits better performance under those situations (Yan et al. 2007).

```
> gge_model <- gge(data, ENV, GEN, GY, svp = "symmetrical")
q <- plot(gge_model, type = 9, sel_gen1 = "[Number or code of target
genotype]", sel_gen2 = ""[Number or code of target genotype]")
r <- plot(gge_model,
 type = 9,
 sel_gen1 = ""[Number or code of target genotype]",
 sel_gen2 = ""[Number or code of target genotype]",
 col.gen = "black",
 title = FALSE)
arrange_ggplot(q, r, tag_levels = list(c("q", "r")))
```

Type 10: Relationship among environments

This biplot provides a comprehensive visualization of the relationships among various test environments by illustrating the angles formed between their corresponding vectors. The orientation and proximity of these vectors reveal how closely related the environments are to one another. In this way, the cosine of the angle between two vectors represents a measure of the relationship between them. When vectors form a 90-degree angle, this indicates that two environments are independent or uncorrelated. Besides, when vectors are 180 degrees apart, this implies a strong negative correlation between the environments. Furthermore, when the angle between vectors is less than 90 degrees, the cosine of the angle is positive. This indicates a positive correlation between the environments represented by the vectors.

```
> gge_model <- gge(data, ENV, GEN, GY)
s <- plot(gge_model, type = 10)
t <- plot(gge_model,
 type = 10,
 col.gen = "black",
 title = FALSE)
arrange_ggplot(s, t, tag_levels = list(c("s", "t"))
```

Practical Example

Similar to Sect. 15.3.1.4, this section presents a practical example of analyzing G × E interaction using the '***metan***' package. To facilitate visualization of genotypes and environments, the data columns corresponding to environments and genotypes are labeled as 'E' and 'G', respectively.

General model

```
> gge_model <- gge(data, ENV, GEN, GY)
```

Type 1: A basic biplot

```
a <- plot(gge_model)
b <- plot(gge_model,
 col.gen = "orange",
 size.text.env = 2,
 plot_theme = theme_metan(grid = "both"))
arrange_ggplot(a, b, tag_levels = "a")
```

Based on the results, it can be concluded that the first two principal components explain 55.60% of the total variability (Fig. 15.12). Genotype G1 positively adapted to environments E7, E12, and E13 and did not adapt to environment E10. Genotypes G4, G6, G7, and G15 adapted to environments E6 and E11 and did not adapt to environments E1, E2, and E10 (Fig. 15.12).

Type 2: Mean performance versus stability

```
gge_model <- gge(data, ENV, GEN, GY, svp = "genotype")
c <- plot(gge_model, type = 2)
d <- plot(gge_model,
 type = 2,
 col.gen = "black",
 col.env = "red",
 axis_expand = 1.5,
 plot_theme = theme_metan_minimal())
arrange_ggplot(c, d, tag_levels = list(c("c", "d")))
```

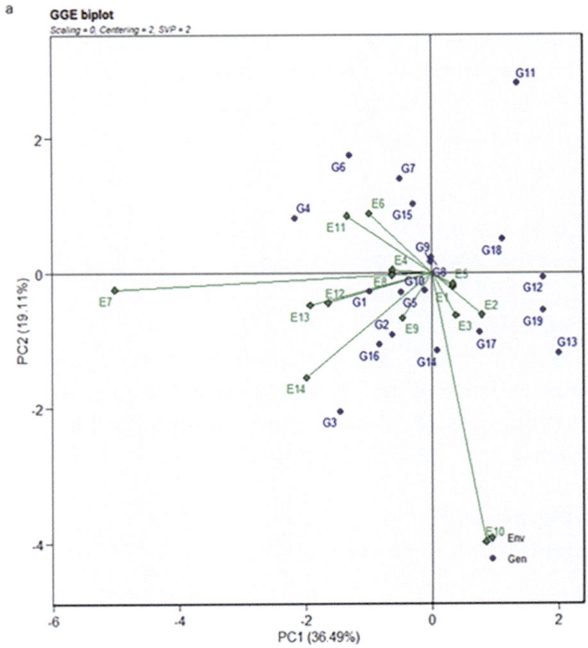

Fig. 15.12 The basic GGE biplot rendered based on the 19 investigated barley genotypes and 14 test environments

Based on the results obtained from the GGE biplot analysis, we were able to identify some of the superior genotypes such as G8 and G9 in terms of their grain performance and stability (Fig. 15.13).

Type 3: Which-won-where

```
gge_model <- gge(data, ENV, GEN, GY, svp = "symmetrical")
e <- plot(gge_model, type = 3)
f <- plot(gge_model,
 type = 3,
 size.shape.win = 5,
 large_label = 6,
 col.gen = "black",
 col.env = "gray",
 title = FALSE)
arrange_ggplot(e, f, tag_levels = list(c("e", "f")))
```

Based on the obtained results, a graphical representation of the GGE "which-won-where" of the winning genotypes for grain yield in each sector was obtained (Fig. 15.14). In this example, the vertex genotypes were G3, G4, G6, G11, and G13. Moreover, three mega-environments (M) were found. The first mega-environment (M-I) included environments E1, E2, E3, and E5. The M-II comprised environments

Fig. 15.13 Biplot for simultaneous selection of grain yield and stability

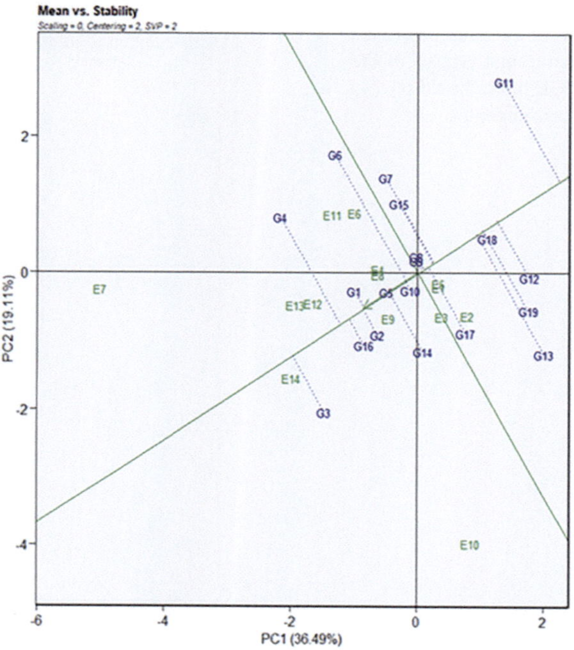

Fig. 15.14 View of the 'which-won-where' biplot of winning genotypes for grain yield and identified the mega-environments

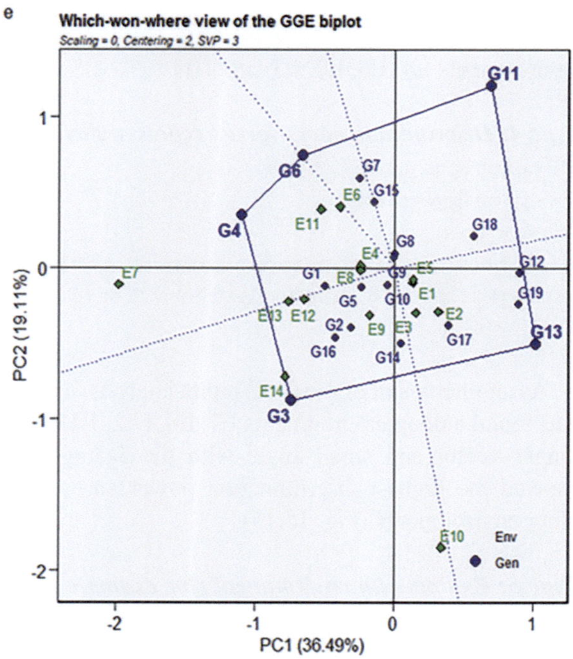

g

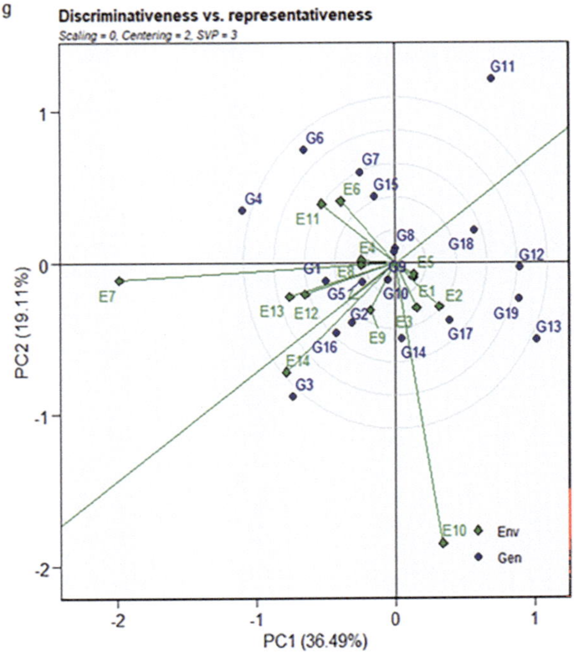

Fig. 15.15 View of the 'discriminating power and representativeness' of the GGE biplot based on 14 test environments

E9, E10, E12, E13, and E14. The third mega-environment (M-III) included environments E4, E6, E7, E8, and E11.

Type 4: Discriminativeness versus representativeness

```
g <- plot(gge_model, type = 4)
h <- plot(gge_model,
 type = 4,
 plot_theme = theme_metan_minimal(), title = FALSE)
arrange_ggplot(g, h, tag_levels = list(c("g", "h")))
```

As another result of the GGE biplot analysis, a positive and significant correlation was found among environments E7, E8, E12, E13, and E14. Furthermore, E7 with a longer vector and small angel with the average environment coordinate (AEC) showed the highest discrimination power and representative ability compared to other environments (Fig. 15.15).

Type 5: Examine an environment (for example E2)

```
gge_model <- gge(data, ENV, GEN, GY, svp = "symmetrical")
i <- plot(gge_model, type = 5, sel_env = "E2")
j <- plot(gge_model,
 type = 5,
```

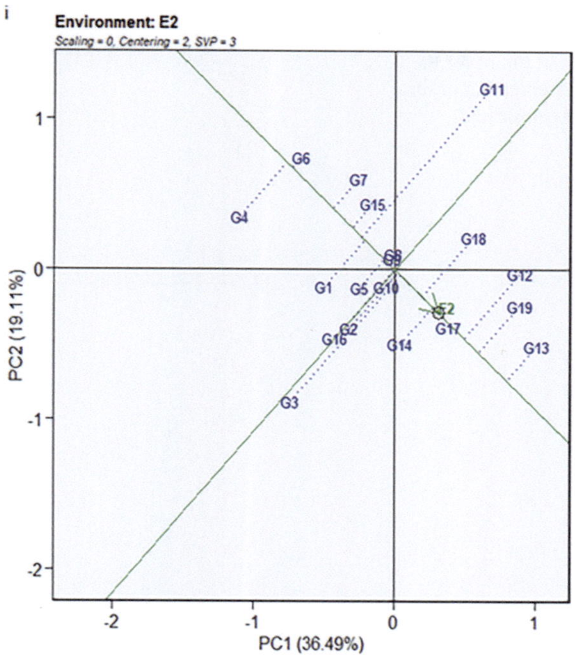

Fig. 15.16 View of the 'examine an environment' of the GGE biplot based on 14 test environments

```
 sel_env = "E2",
 col.gen = "black",
 col.env = "black",
 size.text.env = 10,
 axis_expand = 1.5, title = FALSE)
arrange_ggplot(i, j, tag_levels = list(c("i", "j")))
```

Based on the obtained results, it can be determined which genotypes have positively adapted to a specific environment (in this case E2). In addition, it can be indicated which genotypes have not adapted to the conditions prevailing in this specific environment (Fig. 15.16).

Type 6: Ranking environments

```
gge_model <- gge(data, ENV, GEN, GY)
k <- plot(gge_model, type = 6)
l <- plot(gge_model,
 type = 6,
 col.gen = "black",
 col.env = "black",
 col.circle = "red",
 col.alpha.circle = 0.5,
```

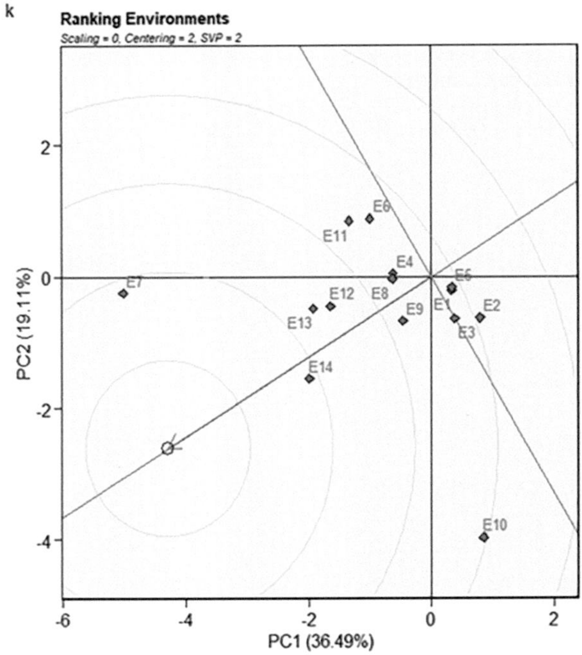

Fig. 15.17 View of the 'ranking environments' of the GGE biplot for the identification of ideal environments

```
 title = FALSE,
 size.text.env = 5,
 axis_expand = 1.5,
 plot_theme = theme_metan(color.background = "white"))
arrange_ggplot(k, l, tag_levels = list(c("k", "l")))
```

Based on the presented results, it is very easy to determine the ranking of environments. Thus, determining which environments are 'universal' and which are 'specific' gives a quick insight into the possibilities of considering (or not considering) given environments in subsequent stages of research and selection (Fig. 15.17).

Type 7: Examine a genotype (for example G3)

```
gge_model <- gge(data, ENV, GEN, GY, svp = "genotype")
m <- plot(gge_model, type = 7, sel_gen = "G3")
n <- plot(gge_model,
 type = 7,
 sel_gen = "G3",
 col.gen = "black",
 col.env = "black",
 title = FALSE,
```

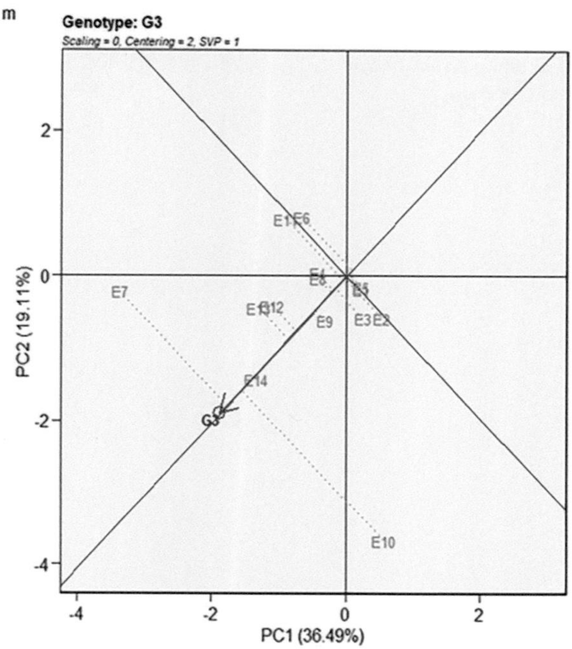

Fig. 15.18 View of the 'examine a genotype' of the GGE biplot across 14 test environments

```
 size.text.env = 10,
 axis_expand = 1.5,
 plot_theme = theme_metan(grid = "both"))
arrange_ggplot(m, n, tag_levels = list(c("m", "n")))
```

Based on the results, it is very easy to determine in which environments a specific genotype (in the presented example—genotype G3) has positively adapted. Similarly, it is easy to indicate in which conditions (corresponding to the conditions of specific environments) genotype G3 has not adapted (Fig. 15.18).

Type 8: Ranking genotypes

```
o <- plot(gge_model, type = 8)
p <- plot(gge_model,
 type = 8,
 col.gen = "black",
 col.env = "gray",
 size.text.gen = 6,
 plot_theme = theme_metan_minimal())
arrange_ggplot(o, p, tag_levels = list(c("o", "p")))
```

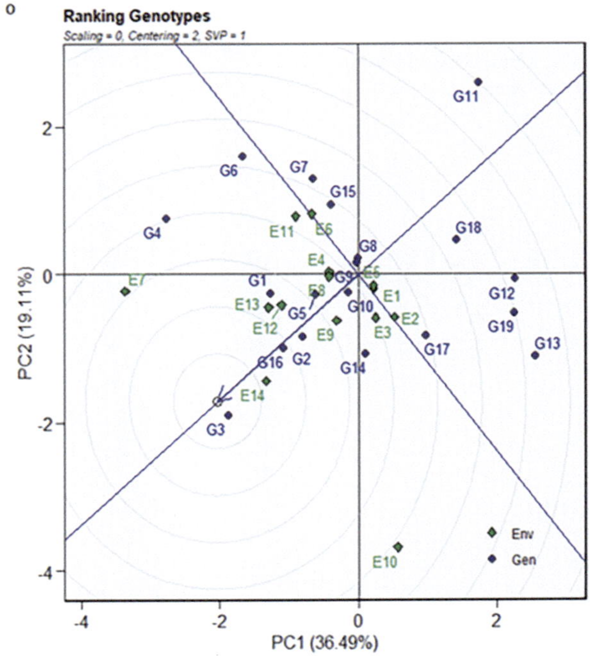

Fig. 15.19 View of the 'ranking genotypes' of the GGE biplot across 14 test environments

The obtained results indicate the ranking of genotypes. The arrangement of genotypes in concentrically growing circles can be observed. The closest to the central point is genotype G3, and the furthest is genotype G11 (Fig. 15.19).

Type 9: Compare two genotypes (for example genotypes G3 and G13)

```
> gge_model <- gge(data, ENV, GEN, GY, svp = "symmetrical")
q <- plot(gge_model, type = 9, sel_gen1 = "G3", sel_gen2 = "G13")
r <- plot(gge_model,
 type = 9,
 sel_gen1 = "G3",
 sel_gen2 = "G13",
 col.gen = "black",
 title = FALSE)
arrange_ggplot(q, r, tag_levels = list(c("q", "r")))
```

In this example, environments E1, E2, E3, and E5 are on the same side of the perpendicular genotype as G13, and the other 10 environments are on the other side of the perpendicular line, together with G3. This indicates that G13 yielded more than G3 in environments E1, E2, E3, and E5 (Fig. 15.20).

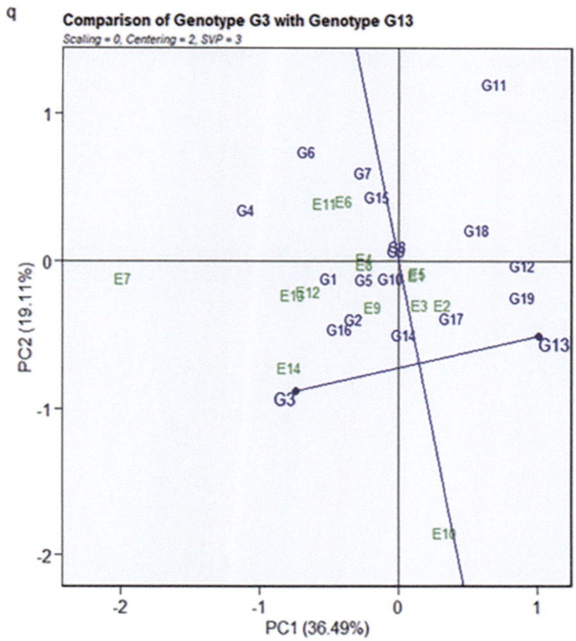

Fig. 15.20 View of the 'compare two genotypes' of the GGE biplot across 14 test environments

Type 10: Relationship among environments

```
> gge_model <- gge(data, ENV, GEN, GY)
s <- plot(gge_model, type = 10)
t <- plot(gge_model,
 type = 10,
 col.gen = "black",
 title = FALSE)
arrange_ggplot(s, t, tag_levels = list(c("s", "t"))
```

Based on the results presented in Fig. 15.21, it is possible to observe similar environments and those that differ in terms of conditions from others. In the presented example, it was observed that the E7 environment differs from the others. Similarly, E10 is a clear outlier (Fig. 15.21).

15.4 Practical Applications

Analysis of G × E interaction is integral to plant breeding, including:

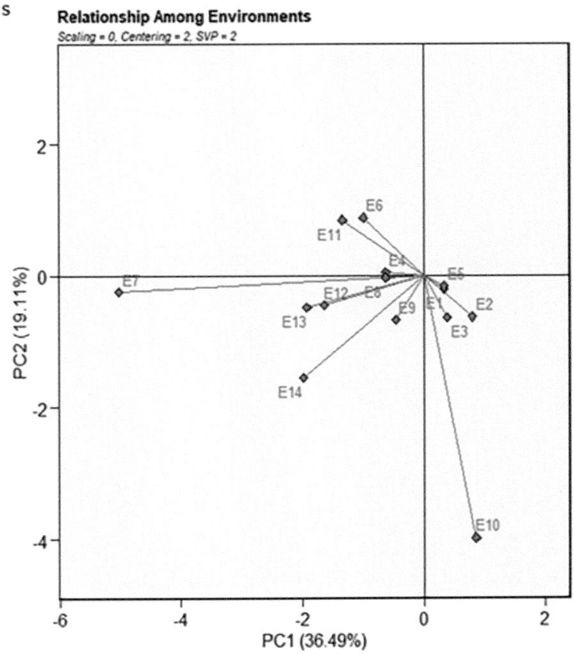

Fig. 15.21 View of the 'relationships among environments' of the GGE biplot based on 14 test environments

15.4.1 Yield Atability Assessment

- AMMI (additive main effects and multiplicative interaction) analysis enables the identification of genotypes that are stable and well adapted to specific environments.
- GGE biplot analysis provides tools to visualize specific adaptation of genotypes.

15.4.2 Optimization of Breeding Programs

- Identification of genotypes with specific adaptation to specific regions.
- Selection of genotypes with high-yield stability in variable conditions.

15.4.3 Predicting the Impact of Environmental Factors

- Factorial regression models allow for the estimation of the influence of weather conditions on yield variability.
- QTL × E analysis supports the identification of loci responsible for responses to environmental factors.

15.5 Challenges and Future Directions

The use of statistical software in the study of G × E interactions is associated with certain challenges, including:

- The need to select appropriate analytical models for specific data.
- Understanding the biological basis of G × E interactions.
- Integration of multidimensional data (genotypic, environmental, and crop management).

Future research should focus on developing analytical tools that incorporate emerging technologies such as big data analysis, machine learning, facilitating the integration of genomic data with phenotypic and environmental data.

15.6 Summary

Statistical software is essential in G × E interaction studies, enabling advanced analysis and interpretation. Tools like R and Genstat facilitate effective assessment of yield stability and specific adaptation of genotypes. The application of modern statistical methods supports breeders in selecting genotypes suited to diverse environmental conditions. Advancements in these technologies enhance the efficiency of breeding programs and deepen understanding of genotype–environment interaction mechanisms.

References

Abbasi Z, Bocianowski J (2021) Genotype by environment interaction for physiological traits in sugar beet (*Beta vulgaris* L.) parents and hybrids using additive main effects and multiplicative interaction model. Eur Food Res Technol 247:3063–3081

Barati A, Pour-Aboughadareh A, Arazmjoo E, Tabatabaei SA, Bocianowski J, Jamshidi B (2025) Identification of high yielding and stable barley genotypes for drought Conditions in the moderate climate of Iran using AMMI model and GGE biplot analysis. J Crop Health 77:8

Bocianowski J, Prażak R (2022) Genotype by year interaction for selected quantitative traits in hybrid lines of *Triticum aestivum* L. with *Aegilops kotschyi* Boiss. and *Ae. variabilis* Eig. using the additive main effects and multiplicative interaction model. Euphytica 218(2):11

Bocianowski J, Jakubowska M, Nowosad K, Ławiński H (2015) Vliv poškození bulev osenicemi Agrotis spp. na technologickou jakost cukrové řepy [Influence of root damage of sugar beet by Agrotis spp. on technological quality of raw material]. Listy Cukrovarnické a Řepařské 131(12): 366–372

Bocianowski J, Szulc P, Nowosad K (2018) Soil tillage methods by years interaction for dry matter of plant yield of maize (*Zea mays* L.) using additive main effects and multiplicative interaction model. J Integr Agric 17(12):2836–2839

Bocianowski J, Księżak J, Nowosad K (2019a) Genotype by environment interaction for seeds yield in pea (*Pisum sativum* L.) using additive main effects and multiplicative interaction model. Euphytica 215:191

Bocianowski J, Niemann J, Nowosad K (2019b) Genotype-by-environment interaction for seed quality traits in interspecific cross-derived Brassica lines using additive main effects and multiplicative interaction model. Euphytica 215:7

Bocianowski J, Nowosad K, Szulc P (2019c) Soil tillage methods by years interaction for harvest index of maize (*Zea mays* L.) using additive main effects and multiplicative interaction model. Acta Agric Scand B Soil Plant Sci 69(1):75–81

Bocianowski J, Nowosad K, Tomkowiak A (2019d) Genotype—environment interaction for seed yield of maize hybrids and lines using the AMMI model. Maydica 64:M13

Bocianowski J, Warzecha T, Nowosad K, Bathelt R (2019e) Genotype by environment interaction using AMMI model and estimation of additive and epistasis gene effects for 1000-kernel weight in spring barley (*Hordeum vulgare* L.). J Appl Genet 60(2):127–135

Bocianowski J, Tratwal A, Nowosad K (2020) Genotype by environment interaction for area under the disease-progress curve (AUDPC) value in spring barley using additive main effects and multiplicative interaction model. Australa Plant Pathol 49:525–529

Bocianowski J, Radkowski A, Nowosad K, Radkowska I, Zieliński A (2021a) The impact of genotype-by-environment interaction on the dry matter yield and chemical composition in timothy (*Phleum pratense* L.) examined by using the additive main effects and multiplicative interaction model. Grass Forage Sci 76(4):463–484

Bocianowski J, Tratwal A, Nowosad K (2021b) Genotype by environment interaction for main winter triticale varieties characteristics at two levels of technology using additive main effects and multiplicative interaction model. Euphytica 217:26

Bocianowski J, Wielkopolan B, Jakubowska M (2022) AMMI Analysis of the effects of different insecticidal treatments against *Agrotis* spp. on the technological yield from sugar beet. Agriculture 12(2):157

Bocianowski J, Nowosad K, Rejek D (2024a) Genotype-environment interaction for grain yield in maize (*Zea mays* L.) using the additive main effects and multiplicative interaction (AMMI) model. J Appl Genet 65(4):653–664

Bocianowski J, Waligóra H, Majchrzak L (2024b) Genotype by year interaction for selected traits in sweet maize (*Zea maize* L.) hybrids using AMMI model. Euphytica 220(6):89

Chełkowski J, Kaptur P, Tomkowiak M, Kostecki M, Goliński P, Ponitka A, Ślusarkiewicz-Jarzina A, Bocianowski J (2000) Moniliformin accumulation in kernels of triticale accessions inoculated with *Fusarium avenaceum*, in Poland. J Phytopathol 148:433–439

Gauch HG (1992) Statistical Analysis of Regional Yield Trials! AMMI analysis of factorial designs. Elsevier, Amsterdam

Ghazvini H, Pour-Aboughadareh A, Jasemi SS, Chaichi M, Tajali H, Bocianowski J (2024) A Framework for selection of high-yielding and drought-tolerant genotypes of barley: applying yield-based indices and multi-index selection models. J Crop Health 76:601–616

Golinski P, Waskiewicz A, Wisniewska H, Kiecana I, Mielniczuk E, Gromadzka K, Kostecki M, Bocianowski J, Rymaniak E (2010) Reaction of winter wheat (*Triticum aestivum* L.) cultivars to infection with *Fusarium* spp.: mycotoxin contamination in grain and chaff. Food Addit Contam: Part A 27(7):1015–1024

James AT, Wilkinson GN (1971) Factorisation of the residual operator and canonical decomposition of non-orthogonal factors in analysis of variance. Biometrika 58:279–294

Jędzura S, Bocianowski J, Matysik P (2023) The AMMI model application to analyze the genotype–environmental interaction of spring wheat grain yield for the breeding program purposes. Cereal Res Commun 51:197–205

Liersch A, Bocianowski J, Nowosad K, Mikołajczyk K, Spasibionek S, Wielebski F, Matuszczak M, Szała L, Cegielska-Taras T, Sosnowska K, Bartkowiak-Broda I (2020) Effect of genotype × environment interaction for seed traits in winter oilseed rape (*Brassica napus* L.). Agriculture 10(12):607

Liersch A, Bocianowski J, Spasibionek S, Wielebski F, Szała L, Cegielska-Taras T, Sosnowska K, Matuszczak M, Nowakowska J, Bartkowiak-Broda I, Mikołajczyk K (2024) Evaluation of the stability of quantitative traits of winter oilseed rape (*Brassica napus* L.) by AMMI analysis. Euphytica 220:130

Niemann J, Bocianowski J, Wojciechowski A (2018) Effects of genotype and environment on seed quality traits variability in interspecific cross-derived Brassica lines. Euphytica 214:193

Nowosad K, Liersch A, Popławska W, Bocianowski J (2016) Genotype by environment interaction for seed yield in rapeseed (*Brassica napus* L.) using additive main effects and multiplicative interaction model. Euphytica 208:187–194

Nowosad K, Liersch A, Poplawska W, Bocianowski J (2017) Genotype by environment interaction for oil content in winter oilseed rape (*Brassica napus* L.) using additive main effects and multiplicative interaction model. Indian J Genet. Plant Breed 77(2):293–297

Nowosad K, Tratwal A, Bocianowski J (2018) Genotype by environment interaction for grain yield in spring barley using additive main effects and multiplicative interaction model. Cereal Res Commun 46(4):729–738

Nowosad K, Bocianowski J, Kianersi F, Pour-Aboughadareh A (2023) Analysis of linkage on interaction of main aspects (genotype by environment interaction, stability and genetic parameters) of 1000 kernels in maize (*Zea mays* L.). Agriculture 13(10):2005

Olivoto T, Lucio AD (2020) Metan: an R package for multi-environment trial analysis. Methods Ecol Evol 11:783–789

Olivoto T, Lúcio ADC, Silva JAG et al (2019) Mean performance and stability in multi-environment trials I: combining features of AMMI and BLUP techniques. Agron J 111:2949–2960

Onofri A (2024) Statforbiology. https://www.statforbiology.com

Payne RW, Tobias RD (1992) General balance, combination of information and the analysis of covariance. Scand J Stat 19:3–23

Payne RW, Wilkinson GN (1977) A general algorithm for analysis of variance. Appl Stat 26:251–260

Pour-Aboughadareh A, Khalili M, Poczai P, Olivoto T (2022) Stability indices to deciphering the genotype by-environment interaction (GEI) effect: an applicable review for use in plant breeding programs. Plants 11:414

Pour-Aboughadareh A, Ghazvini H, Jasemi SS, Mohammadi S, Razavi SA, Chaichi M, Ghasemi Kalkhoran M, Monirifar H, Tajali H, Fathihafshjani A, Bocianowski J (2023) Selection of high-yielding and stable genotypes of barley for the cold climate in Iran. Plants 12:2410

Pour-Aboughadareh A, Marzoghian A, Gholipour A, Zali H, Koohkan S, Keirgoo M, Askari-Kelestani A, Barati A (2024) Genotype-by-environment interaction analysis for grain yield of barley genotypes in the warm climate of Iran. Ecolo Genet Genomics 32:1000264

Purchase JL, Hatting H, van Deventer CS (2000) Genotype × environment interaction of winter wheat (*Triticum aestivum* L.) in South Africa: II Stability analysis of yield performance. S Afr J Plant Soil 17:101–107

Rahmati S, Azizi-Nezhad R, Pour-Aboughadareh A, Etminan A, Shooshtari L (2024) Analysis of genotype-by-environment interaction effect in barley genotypes using AMMI and GGE biplot model. Heliyon 18:e38131

Seidler-Łożykowska K, Bocianowski J, Krol D (2013a) The evaluation of the variability of morphological and chemical traits of the selected lemon balm (*Melissa officinalis* L.) genotypes. Ind Crop Prod 49:515–520

Seidler-Łożykowska K, Kędzia B, Karpińska E, Bocianowski J (2013b) Microbiological activity of caraway (*Carum carvi* L.) essential oil obtained from different origin. Acta Scientiarum. Agronomy 35(4):495–500

Szulc P, Bocianowski J, Nowosad K, Michalski T, Waligóra H, Olejarski P (2018) Assessment of the influence of fertilisation and environmental conditions on maize health. Plant Prot Sci 54(3): 174–182

Wang X, Liu H, Ma S (2022) GEInfo: an R package for gene-environment interaction analysis incorporating prior information. Bioinformatics 38:3139–3140

Weber Z, Kostecki M, Bargen S, Gossmann M, Waskiewicz A, Bocianowski J, Knaflewski M, Büttner C, Golinski P (2006) Fusarium species colonizing spears and forming mycotoxins in field samples of asparagus from Germany and Poland. J Phytopathol 154(4):209–216

Wilkinson GN (1970) A general recursive algorithm for analysis of variance. Biometrika 57:19–46

Yan W, Kang MS (2003) GGE biplot analysis: a graphical tool for breeders, geneticists, and agronomists. CRC Press

Yan W, Kang MS, Ma B, Woods S, Cornelius PL (2007) GGE biplot vs. AMMI analysis of genotype-by-environment data. Crop Sci 47(2):641–653

Correction to: An Introductory Guide to the Usage of Statistical Softwares for Investigating G × E Interaction: From Theory to Application

Alireza Pour-Aboughadareh, Jan Bocianowski, and Bita Jamshidi

Correction to:
Chapter 15 in: R. Mohammadi et al. (eds.), ***Genotype x Environment Interactions and its Implications for Plant Breeding,***
https://doi.org/10.1007/978-981-95-5664-9_15

The author's last name in Chapter 15 was inadvertently published with incorrect spelling. It has been corrected. The correct name is Jan Bocianowski.

The updated version of this chapter can be found at
https://doi.org/10.1007/978-981-95-5664-9_15

R. Mohammadi et al. (eds.), *Genotype x Environment Interactions and its Implications for Plant Breeding*, https://doi.org/10.1007/978-981-95-5664-9_16

MIX
Papier aus verantwortungsvollen Quellen
Paper from responsible sources
FSC® C105338

If you have any concerns about our products,
you can contact us on
ProductSafety@springernature.com

In case Publisher is established outside the EU,
the EU authorized representative is:
Springer Nature Customer Service Center GmbH
Europaplatz 3, 69115 Heidelberg, Germany

Printed by Libri Plureos GmbH
in Hamburg, Germany